P9-DVF-793

HAZARDOUS MATERIALS

MANAGING THE INCIDENT

Third Edition

GREGORY G. NOLL ▪ MICHAEL S. HILDEBRAND ▪ JAMES YVORRA

Chester, Maryland 21619
1-800-603-7700, 410-604-2540 • FAX 410-604-2541
www.redhatpub.com

PRODUCED BY:

LIGHTWORKS PHOTOGRAPHY AND DESIGN
Chester, Maryland 21619

Executive Producer: George Dodson

Production Coordinator/Designer: Kathleen Lawyer, Lightworks Design

Special Graphics: George Dodson, Lightworks Photography

Cartoon Illustrations: Mat Brown

Editor: Patricia Daly

Indexer: Mary Hogan Hearle

Copyright © 2005

All rights reserved. No part of this publication may be reproduced or transmitted in any form or by means, electronic or mechanical, including photocopying and recording, or by any information storage and retrieval system, without permission in writing from the author.

The following photos were supplied by Detrick Lawrence:
Chapter 2 opener, Figures 2.12, 2.15, 3.3, 3.4, 5.4, 8.6, Scan 10-A, 10.11, 10.12, 10.16, 10.20, 10.22, Chapter 11 opener, Scan 11-A, 11.9, 11.11, 11.12, 11.3, Scan 11-C, Chapter 12 opener. Other photo credits supplied by individuals are credited next to the photos in the text. All other photos are credited to Hildebrand and Noll Technical Resources, LLC.

Third Edition

LIBRARY OF CONGRESS NUMBER: 2005901954

ALSO AVAILABLE: Instructor Program with PowerPoint® presentations and exercises, Student Workbook, Field Operations Guide and dedicated Web site. See page xiv for details.

For more information visit us on the World Wide Web at **www.redhatpub.com.**

ISBN: 1-932235-04-3

Printed in the United States

TABLE OF CONTENTS

CHAPTER 3

THE INCIDENT COMMAND SYSTEM / 87

CHAPTER 4

THE EIGHT STEP PROCESS: AN OVERVIEW / 129

CHAPTER 5

SITE MANAGEMENT / 149

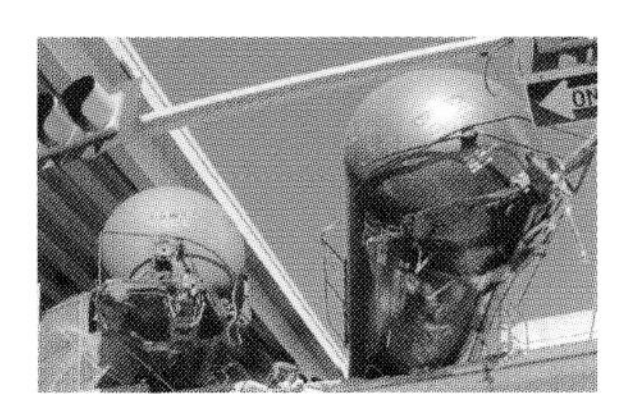

CHAPTER 6

IDENTIFY THE PROBLEM / 193

CHAPTER 7

HAZARD ASSESSMENT AND RISK EVALUATION / 265

CHAPTER 8

SELECT PERSONAL PROTECTIVE CLOTHING AND EQUIPMENT / 349

CHAPTER 9

INFORMATION MANAGEMENT AND RESOURCE COORDINATION / 393

CHAPTER 10

IMPLEMENTING RESPONSE OBJECTIVES / 417

CHAPTER 11

DECONTAMINATION / 509

CHAPTER 12

TERMINATING THE INCIDENT / 555

NOTICE

Hazardous materials emergency response work is extremely dangerous and many emergency responders have died or sustained serious injury and illness while attempting to mitigate an incident. There is no possible way that this text can cover the full spectrum of problems and contingencies for dealing with every type of hazardous materials incident. The user is warned to exercise all necessary cautions when dealing with hazardous materials. Always assume a worst-case scenario and place personal safety first.

It is the intent of the authors that this text be a part of the user's formal training in the management of hazardous materials emergencies. Even though this book is based on commonly used practices, references, laws, regulations, and consensus standards, it is not meant to set a standard of operations for any emergency response organization. The users are directed to develop their own written Standard Operating Procedures which follow all system, agency, or employer guidelines for handling hazardous materials. It is the user's sole responsibility to stay up to date with procedures, regulations, and product developments which may improve personal health and safety.

This document contains copyrighted materials which are protected under Chapter 4 of the United States Copyright Law and the Universal Copyright Convention. The text, art, and photography may not be reproduced, distributed, or sold without written permission. Certain text, art, and photography which has been copyrighted by other organizations and authors has been reproduced in this text with permission from the owners.

The Eight Step Incident Management Process—used throughout this text is copyrighted by Gregory G. Noll, Michael S. Hildebrand, and James G. Yvorra and may not be reproduced for profit without written permission. For more information concerning copyright release contact: Hildebrand and Noll Technical Resources, LLC., P.O. Box 408, Port Republic, Maryland 20676.

ABOUT THE AUTHORS

Greg Noll and Mike Hildebrand each have more than 30 years experience in industry and government and have served as firefighters, hazardous materials technicians, incident commanders, and instructors. Greg and Mike are the co-founders of Hildebrand and Noll Associates, Inc., a consulting firm specializing in emergency planning and response issues. Their experience and expertise includes hazardous materials, weapons of mass destruction, and operations security. They are both Certified Safety Professionals and serve on the NFPA 472 Technical Committee on Hazardous Materials Response.

DEDICATION

This book is dedicated to our wives and children. To Debbie Noll and JoAnne Hildebrand—"For better or worse" and "In sickness and health" pretty much says it all. To Kendra, Sean and Ian, You are what life is all about!

Love—Greg and Mike

ACKNOWLEDGMENTS

IFSTA Validation Committee

The third edition of Hazardous Materials: Managing The Incident has been reviewed and validated by the International Fire Service Training Association (IFSTA) Hazardous Materials Technician Committee. The authors and staff extend a special thanks to the members of the validating committee who contributed their time, wisdom, and knowledge to a thorough review of the final manuscript for this book. All IFSTA validation committees are composed of technical experts who review manual drafts and verify that the contents are valid. Committee members are volunteers who participate because of a commitment to the fire service and its future through training.

Committee Chair

Scott D. Kerwood
Orange Co. Emergency Services, Dist. #1
Vidor, TX

Secretary

Gary M. Courtney
New Hampshire Community Technical College
Laconia, NH

Steven Morikawa
Berkeley Fire Department
Berkeley, CA

Gary E. Allen
Tampa Fire and Rescue
Tampa, FL

Richard D. Armstrong
Maryland Fire and Rescue Institute
College Park, MD

John A. Conte
Stamford Fire & Rescue
Stamford, CT

Steve George
Oklahoma State University Fire Service Training
Stillwater, OK

Steve Hendrix
Arlington Fire Department
Arlington, TX

Barry N. Lindley
DuPont
Belle, WV

Rich Mahaney
Fire Control Services / Slide By Slide Inc.
Las Cruces, NM

W. Wayne Mullannix
Kentucky Fire Commission
Lexington, KY

Jennifer K. Smith
Utah Fire & Rescue Academy (UFRA)
Provo, UT

David Horton
Claremore Fire Department
Claremore, OK

Kirk Johnson
Valdosta Fire Department
Valdosta, GA

Glenn Jirka
Miami Township Fire Division
Miamisburg, OH

IFSTA Staff Liaison
Leslie Miller
IFSTA / Fire Protection Publications
Stillwater, OK

Gene P. Carlson
MIFireE VFIS
York County Haz Mat Team
York, PA

Third Party Reviewers

In the early days of hazmat response our successes at the incident scene were based largely on trial and error. Things usually worked out well, but sometimes they did not. Over the last 30 years emergency responders have paid a heavy price to keep the public and our environment safe from dangerous materials, criminals, and terrorists. Some responders were seriously injured and received debilitating injuries. Others paid with their lives.

This textbook is our attempt to give back to the emergency response community what it has given to us. As authors we take the responsibility for passing on lessons learned as serious business. We view our job as being the "recording secretaries", organizers, and librarians for the "Body of Knowledge" which you have shared with us. If you take the time to look at the Suggested Readings and References at the end of each chapter, you will note that several hundred authors have contributed to our profession. We have attempted to build on their good work.

In our travels around the country we have had an opportunity to learn from thousands of firefighters, law enforcement officers, paramedics, and industrial response personnel. Many people took the time to share their personal successes and failures with us, which helped us improve ourselves as well as this book.

Our thinking and methodology to solving problems has been influenced by some remarkable professionals including Ludwig Benner, Jr., Charlie Wright, Gene Carlson, John Eversole, Bill Hand, Jan Dunbar, Mike Callan, Toby Bevelacqua and Chris Hawley. In addition, to those others who shared their experiences but who cannot be listed—we recognize you and thank you for your service to our country. All of you have contributed so much to the emergency response community, shared your materials with us and encouraged us to think about better and safer ways to get the job done.

We are strong advocates of a third-party review when safety issues are concerned. We have actively applied that philosophy to the third edition The final draft of this book is very different from the first draft and we owe all of the reviewers a big THANK YOU.

Don Abbott
Project Manager, Command Training Center
Phoenix, AZ Fire Department

Chief Robert Andrews, Jr.
Industrial Emergency Services, LLC
Corpus Christi, TX

Eric G. Bachman
Lancaster County Emergency Management Agency
Lancaster, PA

Philip Baker
Hazardous Materials Shift Officer
Prince George's County, MD Fire Department

Armando Bevelacqua
Division Chief Special Operations
Orlando, FL Fire Department

Sean Bradley
Mississauga Fire and Emergency Services
Mississauga, ONT, Canada

Assistant Chief Tim Butters
Fairfax City, VA Fire Department

Michael Callan
Callan Company
Middlefield, CT

Theodore K. Cashel, CFEI
Fire Marshal/OEM Coordinator
Princeton Township, NJ

Gene Carlson
Director—Emergency Community Relations
Volunteer Firemen's Insurance Services (VFIS)
York, PA

Jim Chang, CIH
Emergency Management Coordinator
Duke University Hospital
Durham, NC

Deputy Chief Leonard Deonarine
Industrial Emergency Services, LLC
Corpus Christi, TX

Assist Chief Jan Dunbar (retired)
Special Operations
Sacramento, CA Fire Department

Richard Emery
Emery and Associates
Vernon Hills, IL

Michael Eversole
College Park, MD Fire Department

Kevin Fogarty
Beacon, NY

Rem Gaade
Gaade and Associates
Toronto Fire Department (retired)
Oakville, ONT, Canada

Assistant Chief William Giannini (retired)
Providence, RI Fire Department

Bruce Grabbe
Applied Research Associates
U.S. Air Force Fire Protection (retired)
Panama City, FL

William Hand
Houston, TX Fire Department
Hazardous Materials Response Team

Chris Hawley
FBN Training
Havre de Grace, MD

Dr. David Jaslow, MD, MPH, EMT-P, FAAEM
Clinical Associate Professor of Emergency Medicine, Jefferson Medical College, Philadelphia PA

Mike Kernan, Director
Delaware State Fire School
Dover, DE

Tom McCloskey
The McCloskey Group
Bainbridge Island, WA

Dr. Mary Jo McMullen, MD, FACEP
Medical Director
Summit County, OH
Hazardous Materials Response Team

Fire Chief Paul Mauger
Fire and Rescue Services
Chesterfield County (VA)

Captain J. Arthur Miller
Harrisonburg, VA Fire Department

Mike Moore
Safe Transportation Training Specialists, LLC
Carmel, IN

Henry Morse
Fire Service Testing Company, Inc.
Kathleen, FL

Marty Nevil
PA Task Force 1, Urban Search and Rescue Team
Harrisburg, PA

Vinnie Palmer
Queenstown Volunteer Fire Department
Queenstown, MD

Michael Pirello
Cree, Inc.
Durham, NC
Team Leader—NC RRT 4

Billy Poe
Explosive Service International, Ltd.,
Baton Rouge, LA

Greg Rhoads
Greg Rhoads and Associates, Inc.
Jacksonville, FL

Glen Rudner
Hazardous Materials Officer
VA Department of Emergency Management

Captain David Saenz
Corpus Christi, TX Fire Department

Craig Shelley, CFO, EFO, MFireE
Fire Protection Advisor
Saudi Aramco

Captain Greg Socks
Montgomery County, MD
Fire and Rescue Services
Hazardous Materials Unit

First Officer Chuck Thomas
United Airlines
Santa Rosa, CA

Bill Scholl
Cape Canaveral Volunteer Fire Department
Cape Canaveral, FL

Tamara Sheville
Las Vegas, NV

Danny Simpson
Association of American Railroads
HazMat Services Division
Pueblo, CO

David Wolfe
Safe Transportation Training Specialists, LLC
Carmel, IN

Charles Wright
Manager, HazMat Training
Union Pacific Railroad
Omaha, NE

Jim Ziegler, Ph.D.
DuPont Nonwovens
Richmond, VA

We would like to thank several people who really went the extra mile to help us out with a thorough review of the manuscript. That would include Tamara Sheville, Mike Pirello, Toby Bevalacqua, Charlie Wright, Phil Baker, Craig Shelley, and JoAnne Hildebrand.

As we wrote this stuff up along the way, a lot of people have worked hard to make this project a reality. The many different talents that go into a project like this always impress us. George Dodson and Katie Lawyer from Red Hat Publishing produced the textbook and handled all of the hundreds of little details that go into pulling a book together including layout and design, art, photography, printing and for helping us take the material to a new level of detail. Special thanks also to Patricia Daley, who handles all of our copyediting and Mary Hogan Hearle for preparing the index and final proofing.

JoAnne Hildebrand developed the companion Student Workbook and collaborated with Mike Callan on the Instructors Guide. Mike Callan developed the companion PowerPoint® slide presentation. Gordon Massingham with Emergency Film Group produced the companion video series on the Eight Step Process© and allowed us to tap into his extensive photo collection. Mat Brown was the cartoonist for the O.T. and The Kid series. And lastly we want to thank Toby Bevelacqua who invested hundreds of hours developing the new Field Operations Guide as a companion to the textbook. Toby continues to teach us things about our own material that we did not even know.

A Friend Remembered

Jim Yvorra was one of the most low key people we have ever met. He simply never acted as important as he really was.

As we write this in the Summer of 2004 it seems hard to believe that the first edition of *Hazardous Materials: Managing the Incident* was originally published in 1988 and the second edition was released in 1995. One thing for certain, without Jim Yvorra there would be no Hildebrand and Noll partnership and there would be no *Hazardous Materials: Managing The Incident* textbook.

Jimmy introduced us to each other back in 1979 at the International Association of Fire Chiefs Conference in Kansas City, MO and that was the beginning of a great partnership. We collaborated on many projects including the Eight Step Process©, which was an idea that started out on a cocktail napkin at the National Fire Academy (actually the Ott House) over a couple of beers.

Jim Yvorra was the complete package and he made a big impression on us from the very beginning. He was book smart, street smart, and best off all, the guy could write! It was Jimmy that convinced us that we could "write a book together."

Jimmy had this incredible ability to take complex topics and explain them in simple terms so that as he liked to say, "A guy driving a gravel truck could understand it!" We learned this and other important stuff about the publishing business from him without realizing it then that we were the students and he was really the teacher.

Jimmy was usually happy when he was giving of himself to other people. We saw him in action hundreds of times when he cooked chicken dinners for a Berwyn Heights Fire Department fundraiser, shared a T-Shirt or patch with a visiting firefighter, or when he was risking his life to save someone else's home. But he always seemed the happiest when he created a new textbook, so it is with that same satisfaction and pride that we say that Jim Yvorra still has his fingerprints on the third edition of our textbook. He still teaches us lessons after all these years. We still miss you buddy.

Greg & Mike.

James G. Yvorra was a Deputy Fire Chief with the Berwyn Heights (Maryland) Volunteer Fire Department and a nationally known author and editor in the fields of firefighting, hazardous materials, and emergency medical services. He gave the Supreme Sacrifice and died in the line of duty in January, 1988.

In honor of Jim, his family and friends established a scholarship to support leadership development in the emergency services community. For more information and a scholarship application contact Yvorra Leadership Development on the Internet at www.yld.org.

THE HAZMAT LEARNING SYSTEM™

It is our personal philosophy that training and education should provide quality information, which will allow the student and the instructor to perform their job safely, efficiently and effectively. Likewise, we believe that training and education, particularly when dealing with adults and emergency responders, must also be fun!

LEVELS OF LEARNING

Our survey of users over the years has revealed that *Hazardous Materials: Managing The Incident* is used in three different learning formats. THE HAZMAT LEARNING SYSTEM™ has been designed to provide support at the following levels of training and education:

Self-Guided Study and Promotion Examinations—The Learning System can be used for self-guided study as required reading when studying for a promotion examination. When the textbook is used in conjunction with the Student Workbook, the opportunity to practice examination questions and increase test scores can be enhanced.

Training Academy Programs—The Learning System can be used as part of a formal fire, law enforcement, or military academy training program to support hazardous materials technician and hazardous materials incident commander certification, especially when the delivery is supported by the use of tactical scenarios in the Instructors Guide and the Field Operations Guide.

College Level Curriculums—The Learning System can be used to support either 200 or 400 college level hazardous materials programs, especially when the professor uses the Learning Through Inquiry System included in the Instructors Guide.

THE HAZMAT LEARNING SYSTEM ™ is supported by seven different learning tools that are all fully coordinated with one another. These include: 1) Textbook, 2) Instructors Guide, 3) Power Point Slides, 4) Student Workbook 5) Field Operations Guide, 6) Companion Video Series, and 7) Web Site Support.

TEXTBOOK

Developed by Greg Noll and Mike Hildebrand, the textbook is designed to meet the core competencies for Hazardous Materials Technician and HazMat On-Scene Incident Com-mander as outlined in *OSHA 1910.120 (q)—Hazardous Waste Operations and Emergency Response*, and the educational competencies referenced in *NFPA 472—Professional Competence of Responders to Hazardous Materials*.

The textbook consists of 12 chapters with chapters 1-3 addressing preparing for the incident, and chapters 4–12 addressing how to respond safely to a hazardous materials incident. Chapter 4 is written as a "bridge chapter" and provides an overview of the Eight Step Process© which is a systematic way of approaching a hazmat incident. Chapters 5–12 expands on Chapter 4 by dedicating one chapter to each of the Eight Steps.

INSTRUCTORS GUIDE

Developed by Professor JoAnne Hildebrand, the Instructors Guide is designed for use at the training academy or academic levels. It includes complete learning objectives, an outline of each chapter, learning strategies and The Learning Through Inquiry Method of presenting case studies and scenarios. The Instructors Guide includes 24 new tactical scenarios with questions that can be used by training academy instructors or college professors to explore issues concerning hazard and risk assessment, incident command, and the many political, legal, financial, and ethical issues that confront incident commanders and managers in today's world.

POWERPOINT PRESENTATION

The Power Point presentation, developed by Mike Callan, is fully coordinated with the textbook and the instructor's guide. The package includes learning objectives, key speaking points, and all art, charts, photography, and cartoons included in the textbook. All 24 scenarios described in the Instructors Guide are fully illustrated.

STUDENT WORKBOOK

The Student Workbook, also designed by JoAnne Hildebrand includes several hundred multiple-choice questions, with answer key, to guide the student through all 12 chapters of the textbook. Questions are coded to help the student meet the requirements of NFPA 472. The workbook also outlines a study strategy to prepare for promotion and college level examinations.

FIELD OPERATIONS GUIDE

The Field Operations Guide (FOG), is designed to be used at the incident scene as a reference guide to strategic and tactical decision-making. Developed by Toby Bevelacqua, the FOG includes detailed tactical checklists that follows the Eight Step Process©, a section on identification and recognition of containers, data cards on the top 50 hazardous materials and CBRNE's, as well as a matrix of WMD and drug lab precursor chemicals. The FOG is also designed for use in the classroom to support Hazardous Materials Technician and Incident Commander training.

EIGHT STEP PROCESS© COMPANION VIDEO SERIES

Updated in 2004 for the third edition of the textbook, the companion video series addresses each step in the Eight Step Process© and is fully coordinated with chapters 5–12 in the textbook. Produced by award winning Director, Gordon Massingham with Emergency Film Group and narrated by Noll and Hildebrand, the series was filmed on location with HazMat Teams from around the country.

WEB SITE

The learning system is supported by a dedicated Web site [8stepprocess.org and redhatpub.com]. From the Web site you can send questions to the authors and learn more about the **THE HAZMAT LEARNING SYSTEM™**.

MEET O.T. AND THE KID©

As long as there are emergency response teams there will always be an "Old Timer" and "The Kid".

How do you know if you are O.T. or The Kid? We have a scorecard!

THE KID	O.T.
Is Clear and Articulate	Usually Mumbles and Grumbles
College Graduate	Graduated From The School of Hard Knocks
His Waistline Is Smaller Than His Chest	Would Be Happy To See His Feet
Earring	Ball Cap (Worn The Correct Way)
Likes Books	Likes Magazines, especially "Popular Mechanics"
Likes Anything That Is Digital	Prefers "Radios With Knobs"
Uses a Palm Pilot (PDA)	Sticks To Pad and Pencil
Lives With His Lap Top Computer	Thinks Reference Books Are Just Fine
Prefers Non Polluting Electric Power	Would Buy A Diesel Powered Dish Washer If Caterpillar™ Made One.
Grande Latte	Hot Black Coffee
Argues The Question On The Test Even If He Got It Right ...	Likes To Argue About Anything Just for The "Sport" of it.
Loves America!	Loves America!

NOTE: O.T. and The Kid were drawn by artist Mat Brown.
You can reach Mat at: matbrown19@hotmail.com.

DEDICATION

The O.T. and The Kid cartoon series is dedicated to our friends (current and retired) with the Houston Fire Department Hazardous Materials Response Team (HM-22). To Max, Danny, Bill, Bob, Tommy, Jessie, and all the rest, thank you very much for everything that you have taught us and shared with the emergency response community. You guys are the best of the best.

– Greg & Mike

CHAPTER 1

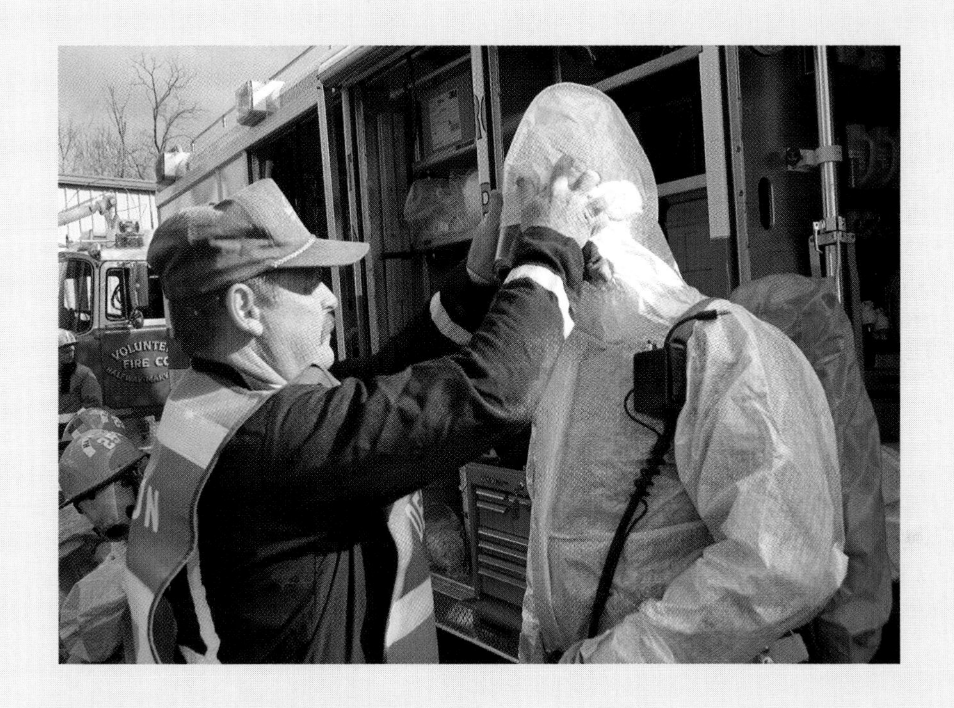

THE HAZARDOUS MATERIALS MANAGEMENT SYSTEM

OBJECTIVES

1. Describe the scope and target audience of this manual.
2. Define and explain the source of, and circumstances for using, the following terms:
   - Hazardous materials (hazmats)
   - Hazardous substances
   - Extremely hazardous substances (EHSs)
   - Hazardous chemicals
   - Hazardous wastes
   - Dangerous goods
3. List the key legislative, regulatory, and voluntary consensus standards that impact hazmat emergency planning, and response operations.
4. Describe the concept of "standard of care" as applied to hazardous materials training, planning and response.
5. List and describe the components of the Hazardous Materials Management System for managing the hazardous materials problem within the facility or community.

ABBREVIATIONS AND ACRONYMS

ACP	Area Contingency Plan
ALS	Advanced Life Support
ANSI	American National Standards Institute
API	American Petroleum Institute
ASTM	American Society for Testing and Materials
BLS	Basic Life Support
CAA	Clean Air Act
CEPPO	Chemical Emergency Preparedness and Prevention Office
CERCLA	Comprehensive Environmental Response, Compensation and Liability Act
CFR	Code of Federal Regulations
CGA	Compressed Gas Association
DOT	Department of Transportation
EHS	Extremely Hazardous Substance
EMS	Emergency Medical Services
EMT-B	Emergency Medical Technician—Basic
EMT-I	Emergency Medical Technician —Intermediate
EMT-P	Emergency Medical Technician —Paramedic
EOP	Emergency Operations Plan
EPA	Environmental Protection Agency
EPCRA	Emergency Planning and Community Right-to-Know Act
ERG	Emergency Response Guidebook
ERT	Emergency Response Team
FAA	Federal Aviation Administration

FBI	Federal Bureau of Investigation
FEMA	Federal Emergency Management Agency
FRA	Federal Railroad Administration
FMECA	Failure Modes, Effects, and Criticality Analysis
HAZCOM	Hazard Communication Regulations
HAZMAT	Hazardous Materials
HAZOP	Hazard and Operability Study
HAZWOPER	Hazardous Waste Operations and Emergency Response
HMRT	Hazardous Materials Response Team
HMS	Office of Hazardous Materials Safety (DOT/RSPA)
HMT	Hazardous Materials Technician
IAEM	International Association of Emergency Managers
ICS	Incident Command System
ICP	Integrated Contingency Plan
IMS	Incident Management System
LEPC	Local Emergency Planning Committee
MSDS	Material Safety Data Sheet
NCP	National Contingency Plan
NFPA	National Fire Protection Association
NRC	National Response Center
NRT	National Response Team
NTSB	National Transportation Safety Board
OHME	Office of Hazardous Materials Enforcement (DOT/RSPA)
OPA	Oil Pollution Act of 1990
OPS	Office of Pipeline Safety
OSC	On-Scene Coordinator
OSHA	Occupational Safety and Health Administration
PSM	Process Safety Management
RCRA	Resource Conservation and Recovery Act
RMP	Risk Management Program
RPM	Remedial Project Manager
RRT	Regional Response Team
RSPA	Research and Special Programs Administration (DOT)
SARA	Superfund Amendments and Reauthorization Act of 1986
SEI	Safety Equipment Institute
SERC	State Emergency Response Commission
USCG	United States Coast Guard
WMD	Weapons of Mass Destruction

INTRODUCTION

This is a text about hazardous materials incident response. It is designed to provide public and private sector emergency response personnel with a logical, building block system for managing hazardous materials emergencies. It is not a chemistry-oriented text. In fact, it assumes that most of the first-arriving emergency responders will have little or no formal chemistry training.

It is designed to begin at the point where responders recognize that they are, in fact, dealing with a hazardous materials emergency, even when the specific hazardous materials have not been identified. Otherwise, normal fire, rescue, and emergency medical services (EMS) guidelines will be followed.

Our primary target audience includes Hazardous Materials Technicians, the Hazmat Group Supervisor or Branch Director, the On-Scene Incident Commander, and members of organized hazardous materials response teams (HMRTs). Other special operations teams, such as Bomb Squads and Confined Space Rescue Teams, will also find specific chapters of interest (e.g., Chapter 7—Hazard and Risk Assessment).

This third edition has been expanded to include additional information to assist the reader in meeting the cognitive skill requirements of Occupational Safety and Health Administration (OSHA) 1910.120(q) and the National Fire Protection Association (NFPA) 472 competencies for the Hazardous Materials Technician and the On-Scene Incident Commander.

WHAT IS A HAZARDOUS MATERIAL?

You might assume that everyone knows what a hazardous material is—or at least knows one when they see one. However, if we were to review the various state and federal regulations that govern the manufacture, transportation, storage, use, and clean-up of chemicals in the United States, the number of terms, definitions, and lists would be overwhelming. Key definitions from various governmental agencies are shown in Figure 1.1.

HAZARDOUS MATERIALS DEFINITIONS

Hazardous materials—Any substance or material in any form or quantity that poses an unreasonable risk to safety and health and property when transported in commerce (Source: U.S. Department of Transportation [DOT], 49 Code of Federal Regulations (CFR) 171).

Hazardous substances—Any substance designated under the Clean Water Act and the Comprehensive Environmental Response, Compensation and Liability Act (CERCLA) as posing a threat to waterways and the environment when released (Source: U.S. Environmental Protection Agency [EPA], 40 CFR 302). Note: Hazardous substances as used within OSHA 1910.120 refers to every chemical regulated by EPA as a hazardous substance and by DOT as a hazardous material.

Extremely hazardous substances (EHS)—Chemicals determined by the EPA to be extremely hazardous to a community during an emergency spill or release as a result of their toxicities and physical/chemical properties (Source: EPA 40 CFR 355).

Hazardous chemicals—Any chemical that would be a risk to employees if exposed in the workplace (Source: OSHA, 29 CFR 1910).

Hazardous wastes—Discarded materials regulated by the EPA because of public health and safety concerns. Regulatory authority is granted under the Resource Conservation and Recovery Act (RCRA). (Source: EPA, 40 CFR 260–281).

Dangerous goods—In international transportation, hazardous materials are commonly referred to as "dangerous goods."

Figure 1.1 Hazardous materials definitions.

Each term in Figure 1.1 has its applications and limitations. In reality, we must recognize that hazmats can be found virtually anywhere—in industry, in transportation, in the workplace, and even in the home—so a broad definition is necessary to cover all the bases. In addition, hazardous materials can also be used as a weapon for criminal or terrorist purposes.

Hazmat emergency response primarily focuses on the interaction of the hazmat and its container. Therefore, for the purposes of this text, we will use the definition of a hazardous material developed by Ludwig Benner, Jr., a former hazardous materials specialist with the National Transportation Safety Board (NTSB) in Washington, D.C.:

> **Hazardous materials—Any substance that jumps out of its container when something goes wrong and hurts or harms the things it touches.**

Benner's definition was developed in the 1970s, and it can still be applied today to all hazmats in all circumstances. The definition recognizes that emergency response is as much a container behavior problem as it is a chemical problem.

> **A hazardous materials incident can then be defined as the release, or potential release, of a hazardous material from its container into the environment.**

Figure 1.2 Hazardous material—any substance that jumps out of its container when something goes wrong and hurts or harms the things it touches.

HAZMAT LAWS, REGULATIONS, AND STANDARDS

Operations involving the manufacture, transport, and use of hazardous materials, as well as the response to hazardous materials incidents, are impacted by a large body of laws, regulations, and voluntary consensus standards. For most of us, discussing rules and regulations is like watching paint dry...it's boring! But as a professional, it's important to understand what the law requires and be able to back up what you think is the right thing to do on the street. These rules are important to responders because they influence virtually every facet of the hazardous materials business.

Because of their importance to emergency planning and response operations, senior hazmat management personnel must have a working knowledge of how the regulatory system works. First, what is the difference between a law, regulation, and standard? These three terms are sometimes used interchangeably, but they do have distinctly different meanings.

Laws are primarily created through an act of Congress, by individual state legislatures, or by local government bodies. Laws typically provide broad goals and objectives, mandatory dates for compliance, and established penalties for noncompliance. Federal and state laws enacted by legislative bodies usually delegate the details for implementation to a specific federal or state agency. For example, the U.S. Occupational Safety and Health Act enacted by Congress delegates rule-making and enforcement authority on worker health and safety issues to OSHA.

Regulations, sometimes called rules, are created by federal or state agencies as a method of providing guidelines for complying with a law that was enacted through legislative action. A regulation permits individual governmental agencies to enforce the law through audits and inspections, which may be conducted by federal and state officials.

Voluntary consensus standards are normally developed through professional organizations or trade associations as a method of improving the individual quality of a product or system. Within the emergency response community, the National Fire Protection Association is recognized for its role in developing consensus standards and recommended practices that impact fire safety and hazmat operations. In the United States, standards are developed primarily through a democratic process whereby a committee of subject matter specialists representing varied interests writes the first draft of the standard. The document is then submitted to either a larger body of specialists or the general public, who then may amend, vote on, and approve the standard for publication. Collectively, this procedure is known as the Consensus Standards Process.

When a consensus standard is completed, it may be voluntarily adopted by government agencies, individual corporations, or organizations. Many hazmat consensus standards are also adopted by reference in a regulation. In effect, when a federal, state, or municipal government adopts a consensus standard by reference, the document becomes a regulation. An example of this process is the adoption of NFPA 30—*The Flammable and Combustible Liquids Code,* and NFPA 58—*The Liquefied Petroleum Gas Code.*

FEDERAL HAZMAT LAWS

Hazmat laws have been enacted by Congress to regulate everything from finished products to hazardous waste. Because of their lengthy official titles, many simply use abbreviations or acronyms when referring to these laws. The following summaries outline some of the more important laws impacting hazmat emergency planning and response.

- **RCRA—The Resource Conservation and Recovery Act (1976).** This law establishes a framework for the proper management and disposal of all waste materials (i.e., solid, medical, hazardous), including treatment, storage, and disposal facilities. It also establishes installation, leak prevention, and notification requirements for underground storage tanks.

- **CERCLA—The Comprehensive Environmental Response, Compensation and Liability Act (1980)**. Known as Superfund, this law addresses hazardous substance releases into the environment and clean-up of inactive hazardous waste disposal sites. It also requires those individuals responsible for the release of the hazardous materials (commonly referred to as the responsible party) above a specified "reportable quantity" to notify the National Response Center (NRC), which is the single point of contact for spill reporting to the federal government.

- **SARA—Superfund Amendments and Reauthorization Act of 1986.** SARA has had the greatest impact upon hazmat emergency planning and response operations. As the name implies, SARA amended and reauthorized the Comprehensive Environmental Response, Compensation, and Liability Act of 1980 (CERCLA, or Superfund). While many of the amendments pertained to hazardous waste site clean-up, SARA's requirements also established a national baseline with regard to hazmat planning, preparedness, training, and response.

  Title I of this law required OSHA to develop health and safety standards covering numerous worker groups who handle or respond to chemical emergencies and led to the development of 29 CFR OSHA 1910.120, *Hazardous Waste Operations and Emergency Response (HAZWOPER).*

  Most familiar to the emergency response community is SARA, Title III. Also known as the Emergency Planning and Community Right-to-Know Act (EPCRA), SARA Title III led to the establishment of the State Emergency Response Commissions (SERCs) and the Local Emergency Planning Committees (LEPCs).

- **CAA—The Clean Air Act**. This law establishes requirements for airborne emissions and the protection of the environment. The Clean Air Act Amendments of 1990 addressed emergency response and planning issues at certain facilities with processes using highly hazardous chemicals. This included the establishment of a National Chemical Safety and Hazard Investigation Board, EPA's promulgation of 40 CFR Part 68—*Risk Management Programs for Chemical Accidental Release Prevention,* and OSHA's promulgation of 29 CFR 1910.119—*Process Safety Management of Highly Hazardous Chemicals, Explosives and Blasting Agents.* In addition, certain facilities are required to make information available to the general public regarding the manner in which chemical risks are handled within a facility.

- **OPA—Oil Pollution Act of 1990**. Commonly referred to as OPA, this law amended the Federal Water Pollution Control Act. Its scope covers both facilities and carriers of oil and related liquid products, including deepwater marine terminals, marine vessels, pipelines, and railcars. Requirements include the development of emergency response plans, regular training and exercise sessions, and verification of spill resources and contractor capabilities. The law also requires the establishment of Area Committees and the development of Area Contingency Plans (ACPs) to address oil and hazardous substance spill response in coastal zone areas.

HAZMAT REGULATIONS

Laws delegate certain details of implementation and enforcement to federal, state or local agencies, who are then responsible for writing the actual regulations that enforce the legislative intent of the law. Regulations will either (1) define the broad performance required to meet the letter of the law (i.e., performance-oriented standards); or (2) provide very specific and detailed guidance on satisfying the regulation (i.e., specification standards).

FEDERAL REGULATIONS

The following summary includes several of the more significant federal regulations that affect hazmat emergency planning and response.

Hazardous Waste Operations and Emergency Response (29 CFR 1910.120)

Also known as HAZWOPER, this federal regulation was issued under the authority of SARA, Title I. The regulation was written and is enforced by OSHA in those 23 states and 2 territories with their own OSHA-approved occupational safety and health plans. In the remaining 27 "non-OSHA" states, public sector personnel will be covered by a similar regulation enacted by EPA (40 CFR Part 311).

The regulation establishes important requirements for both industry and public safety organizations that respond to hazmat or hazardous waste emergencies. This includes firefighters, law enforcement and EMS personnel, hazmat responders, and industrial Emergency Response Team (ERT) members. Requirements cover the following areas:

- Hazmat Emergency Response Plan
- Emergency Response Procedures, including the establishment of an Incident Management System (IMS), the use of a buddy system with back-up personnel, and the establishment of a Safety Officer
- Specific training requirements covering instructors and both initial and refresher training
- Medical Surveillance Programs
- Postemergency termination procedures

Of particular interest to hazmat managers and responders are the specific levels of competency and associated training requirements identified within OSHA 1910.120(q)(6). See Figure 1.3.

OSHA 1910.120 LEVELS OF EMERGENCY RESPONDERS

First Responder at the Awareness Level. These are individuals who are likely to witness or discover a hazardous substance release and who have been trained to initiate an emergency response notification process. The primary focus of their hazmat responsibilities is to secure the incident site, recognize and identify the materials involved, and make the appropriate notifications. These individuals would take no further action to control or mitigate the release.

First Responder—Awareness personnel shall have sufficient training or experience to demonstrate objectively the following competencies:

a. An understanding of what hazardous materials are and the risks associated with them in an incident.

b. An understanding of the potential outcomes associated with a hazardous materials emergency.

c. The ability to recognize the presence of hazardous materials in an emergency and, if possible, identify the materials involved.

d. An understanding of the role of the First Responder–Awareness individual within the local Emergency Operations Plan. This would include site safety, security and control, and the use of the Emergency Response Guidebook (ERG).

e. The ability to realize the need for additional resources and to make the appropriate notifications to the communication center.

The most common examples of First Responder–Awareness personnel include law enforcement and plant security personnel, as well as some public works employees. There is no minimum hourly training requirement for this level; the employee would have to have sufficient training to demonstrate objectively the required competencies.

First Responder at the Operations Level. Most fire department suppression personnel fall into this category. These are individuals who respond to releases or potential releases of hazardous substances as part of the initial response for the purpose of protecting nearby persons, property, or the environment from the effects of the release. They are trained to respond in a defensive fashion without actually trying to stop the release. Their primary function is to contain the release from a safe distance, keep it from spreading, and protect exposures.

First Responder–Operations personnel shall have sufficient training or experience to demonstrate objectively the following competencies:

a. Knowledge of basic hazard and risk assessment techniques.

b. Knowledge of how to select and use proper personal protective clothing and equipment available to the operations-level responder.

c. An understanding of basic hazardous materials terms.

d. Knowledge of how to perform basic control, containment, and/or confinement operations within the capabilities of the resources and personal protective equipment available.

e. Knowledge of how to implement basic decontamination measures.

f. An understanding of the relevant standard operating procedures and termination procedures.

First responders at the operations level shall have received at least 8 hours of training or have had sufficient experience to demonstrate objectively competency in the previously mentioned areas, as well as the established skill and knowledge levels for the First Responder–Awareness level.

Hazardous Materials Technician. These are individuals who respond to releases or potential releases for the purposes of stopping the release. Unlike the operations level, they generally assume a more aggressive role in that they are often able to approach the point of a release in order to plug, patch, or otherwise stop the release of a hazardous substance.

Hazardous Materials Technicians are required to have received at least 24 hours of training equal to the First Responder–Operations level and have competency in the following established skill and knowledge levels:

a. Capable of implementing the local Emergency Operations Plan
b. Able to classify, identify, and verify known and unknown materials by using field survey instruments and equipment (direct reading instruments)
c. Able to function within an assigned role in the Incident Management System
d. Able to select and use the proper specialized chemical personal protective clothing and equipment provided to the Hazardous Materials Technician
e. Able to understand hazard and risk assessment techniques
f. Able to perform advanced control, containment, and/or confinement operations within the capabilities of the resources and equipment available to the Hazardous Materials Technician
g. Able to understand and implement decontamination procedures
h. Able to understand basic chemical and toxicological terminology and behavior

Many communities and facilities have personnel trained as Emergency Medical Technicians (EMTs), yet do not have the primary responsibility for providing basic or advanced life support medical care. Similarly, Hazardous Materials Technicians may not necessarily be part of a hazardous materials response team. However, if they are part of a designated team as defined by OSHA, they must also meet the medical surveillance requirements within OSHA 1910.120.

Hazardous Materials Specialists. These are individuals who respond with and provide support to Hazardous Materials Technicians. While their duties parallel those of the Technician, they require a more detailed or specific knowledge of the various substances they may be called upon to contain. This individual would also act as the site liaison with federal, state, local, and other governmental authorities in regard to site activities.

Similar to the technician level, Hazardous Materials Specialists shall have received at least 24 hours of training equal to the technician level and have competency in the following established skill and knowledge levels:

a. Capable of implementing the local Emergency Operations Plan
b. Able to classify, identify, and verify known and unknown materials by using advanced field survey instruments and equipment (direct reading instruments)

c. Knowledge of the state emergency response plan
d. Able to select and use the proper specialized chemical personal protective clothing and equipment provided to the Hazardous Materials Specialist
e. Able to understand in-depth hazard and risk assessment techniques
f. Able to perform advanced control, containment, and/or confinement operations within the capabilities of the resources and equipment available to the Hazardous Materials Specialist
g. Able to determine and implement decontamination procedures
h. Able to develop a site safety and control plan
i. Able to understand basic chemical, radiological, and toxicological terminology and behavior

Whereas the Hazardous Materials Technician possesses an intermediate level of expertise and is often viewed as a "utility person" within the hazmat response community, the Hazardous Materials Specialist possesses an ad-vanced level of expertise. Within the fire service, the Specialist will often assume the role of the HazMat Group Supervisor or the HazMat Group Assistant Safety Officer, while an industrial Hazardous Materials Specialist may be "product specific." Finally, the Specialist must meet the medical surveillance requirements outlined within OSHA 1910.120.

On-Scene Incident Commander. Incident Commanders, who will assume control of the incident scene beyond the First Responder–Awareness level, shall receive at least 24 hours of training equal to the First Responder–Operations level. In addition, the employer must certify that the Incident Commander has competency in the following areas:

a. Know and be able to implement the local Incident Management System
b. Know how to implement the local Emergency Operations Plan (EOP)
c. Understand the hazards and risks associated with working in chemical protective clothing
d. Knowledge of the state emergency response plan and of the Federal Regional Response Team
e. Know and understand the importance of decontamination procedures

Skilled Support Personnel. These are personnel who are skilled in the operation of certain equipment, such as cranes and hoisting equipment, and who are needed temporarily to perform immediate emergency support work that cannot reasonably be performed in a timely fashion by emergency response personnel. It is assumed that these individuals will be exposed to the hazards of the emergency response scene.

Although these individuals are not subject to the HAZWOPER training requirements, they shall be given an initial briefing at the site prior to their participation in any emergency response effort. This briefing shall include elements such as instructions in using personal protective clothing and equipment, the chemical hazards involved, and the tasks to be performed. All other health and safety precautions provided to emergency responders and on-scene workers shall be used to assure the health and safety of these support personnel.

Specialist Employees. These are employees who, in the course of their regular job duties, work with and are trained in the hazards of specific hazardous substances, and who will be called upon to provide technical advice or assistance to the Incident Commander at a hazmat incident. This would include industry responders, chemists, and related professional or operations employees. These individuals shall receive training or demonstrate competency in the area of their specialization annually.

Figure 1.3 OSHA 1910.120 Levels of Emergency Responders.

Individuals seeking further information on the application of OSHA standards to hazardous materials emergency response situations should consult the OSHA Web site at http://www.osha.gov. Specific attention should be paid to (1) the HAZWOPER Preamble, (2) OSHA interpretations of the HAZWOPER Standard, and (3) OSHA Directive Number CPL 2-2/59A—*Inspection Procedures for the Hazardous Waste Operations and Emergency Response Standard, 29 CFR 1910.120 and 1926.65, Paragraph (q): Emergency Response to Hazardous Substance Releases* (April 4, 1998).

Community Emergency Planning Regulations (40 CFR parts 300 through 399)

This requirement is the result of SARA, Title III and mandates the establishment of both state and local planning groups to review or develop hazardous materials response plans. The state planning groups are referred to as the State Emergency Response Commission (SERC). The SERC is responsible for developing and maintaining the state's emergency response plan. This includes ensuring that planning and training are taking place throughout the state, as well as providing assistance to local governments, as appropriate. States generally provide an important source of technical specialists, information, and coordination. However, they typically provide only limited operational support to local government in the form of equipment, materials, and personnel during an emergency.

The coordinating point for both planning and training activities at the local level is the Local Emergency Planning Committee. Among the LEPC membership are representatives from the following groups:

Elected state and local officials
Fire Department
Law Enforcement
Emergency management
Public health officials
Hospital
Industry personnel, including facilities and carriers
Media
Community organizations

The LEPC is specifically responsible for developing and/or coordinating the local emergency response system and capabilities. A primary concern is the identification, coordination, and effective management of local resources. Among the primary responsibilities of the LEPC are the following:

- Develop, regularly test, and exercise the Hazmat Emergency Operations Plan.
- Conduct a hazards analysis of hazmat facilities and transportation corridors within the community.
- Receive and manage hazmat facility reporting information. This includes chemical inventories, Tier II reporting forms required under SARA, Title III, material safety data sheets (MSDSs) or chemical lists, and points of contact.
- Coordinate the Community Right-to-Know aspects of SARA, Title III.

In a number of communities, the LEPC has expanded it scope and responsibilities to adopt an all-hazards approach to emergency planning and management. Individuals desiring more information on both hazmat and all-hazards planning should consult the following Web sites:

- EPA Chemical Emergency Preparedness and Prevention Office (CEPPO)—http://www.epa.gov/swercepp/
- Federal Emergency Management Agency (FEMA)—http://www.fema.gov/
- International Association of Emergency Managers (IAEM) http://www.iaem.com/
- U.S. National Response Team—http://www.nrt.org/
- U.S. Chemical Safety and Hazard Investigation Board—http://www.chemsafety.gov/

Federal installations and military bases are also required to follow EPA's right-to-know and pollution prevention regulations under Executive Order 12856 (August 3, 1993).

Risk Management Programs for Chemical Accidental Release Prevention (40 CFR Part 68)

Promulgated under amendments to the Clean Air Act, this regulation requires that facilities that manufacture, process, use, store, or otherwise handle certain regulated substances above established threshold values develop and implement risk management programs (RMPs). The regulation is similar in scope to the OSHA Process Safety Management (PSM) standard, with the primary focus being community safety as compared to employee safety.

Risk management programs consist of three elements:

- *Hazard assessment* of the facility, including the worst-case accidental release and an analysis of potential off-site consequences.
- *Prevention program*, which addresses safety precautions, maintenance, monitoring, and employee training. EPA believes that the prevention program should adopt and build upon the OSHA Process Safety Management standard.
- *Emergency response considerations*, including facility emergency response plans, informing public and local agencies, emergency medical care, and employee training.

Hazard Communication (HAZCOM) Regulation (29 CFR 1910.1200)

HAZCOM is a federal regulation that requires hazardous materials manufacturers and handlers to develop written Material Safety Data Sheets (MSDS) on specific types of hazardous chemicals. These MSDSs must be made available to employees who request information about a chemical in the workplace. Examples of information on MSDSs include known health hazards, the physical and chemical properties of the material, first aid, firefighting and spill control recommendations, protective clothing and equipment requirements, and emergency telephone contact numbers. Under the HAZCOM requirements, hazmat health exposure information should be provided to emergency responders during the termination phase, and all exposures should be documented.

Hazardous Materials Transportation Regulations (49 CFR 100–199)

This series of regulations is issued and enforced by the U.S. DOT. The regulations govern container design, chemical compatibility, packaging and labeling requirements, shipping papers, transportation routes and restrictions, and so forth. The regulations are comprehensive and strictly govern how all hazardous materials are transported by highway, railroad, pipeline, aircraft, and by water.

National Contingency Plan or NCP (40 CFR 300, Subchapters A through J)

This plan outlines the policies and procedures of the federal agency members of the National Oil and Hazardous Materials Response Team (also known as the National Response Team, or the NRT). The regulation provides guidance for emergency responses, remedial actions, enforcement, and funding mechanisms for federal government response to hazmat incidents. The NRT is chaired by EPA, while the vice-chairperson represents the U.S. Coast Guard (USCG).

Each of the ten federal regions also has a Regional Response Team (RRT) that mirrors the make-up of the NRT. RRTs may also include representatives from state and local government and Indian tribal governments.

When the NRT or RRT is activated for a federal response to an oil spill, hazmat, or terrorism event, a federal On-Scene Coordinator (OSC) will be designated to coordinate the overall response. For hazmat incidents, the On-Scene Coordinator will represent either EPA or the USCG based upon the location of the incident. If the release or threatened release occurs in coastal areas or near major navigable waterways, the USCG will usually assume primary OSC responsibility. If the situation occurs inland and away from navigable or major waterways, the EPA will serve as the OSC. Local emergency responders should contact EPA or USCG personnel within their region to determine which agency has primary responsibility and will act as the federal OSC for their respective area.

If the incident is a terrorism-related event, the Federal Bureau of Investigation (FBI) will assume the role as federal OSC during the emergency response phase, while FEMA would assume the role for the postemergency response phase.

STATE REGULATIONS

Each of the 50 states and the U.S. territories maintains an enforcement agency that has responsibility for hazardous materials. The three key players in each state usually consist of the State Fire Marshal, the State Occupational Safety and Health Administration, and the State Department of the Environment (sometimes known as

Natural Resources or Environmental Quality). While there are many variations, the fire marshal is typically responsible for the regulation of flammable liquids and gases due to the close relationship between the flammability hazard and the fire prevention code, while the state environmental agency would be responsible for the development and enforcement of environmental safety regulations.

While known by various titles, most states have a government equivalent of the federal OSHA. Approximately 23 states have adopted the federal OSHA regulations as state law. This method of adoption has increased the level of enforcement of hazardous materials regulations, such as the Hazardous Waste and Emergency Response regulation described previously.

State governments also maintain an environmental enforcement agency and environmental crimes unit that usually enforces the federal RCRA, CERCLA, and CAA laws at the local level. Increased state involvement in hazardous waste regulatory enforcement has significantly increased the number of hazardous materials incidents reported. This increase is expected to continue in the future and will continue to generate more fire service activity at the local level.

VOLUNTARY CONSENSUS STANDARDS

Standards developed through the voluntary consensus process play an important role in increasing both workplace and public safety. Historically, a voluntary standard improves over time as each revision reflects recent field experience and adds more detailed requirements. As users of the standard adopt it as a way of doing business, the level of safety gradually improves over time.

Consensus standards are also updated more regularly than governmental regulations and can usually be developed more quickly to meet issues of the day. For example, in response to the need for emergency response personnel operating at terrorism incidents involving dual-use industrial chemicals or chemical or biological agents, NFPA approved *NFPA 1994—Protective Ensembles for Chemical/Biological Terrorism Incidents* in 2001 to provide guidance in the selection and use of protective clothing and equipment.

In many respects, a voluntary consensus standard provides a way for individual organizations and corporations to self-regulate their business or profession. Interestingly, all of the national fire codes in the United States are developed through the voluntary consensus standards process. Historically, the two key players have been the NFPA and Western Fire Chiefs Association. Many of the standards developed by these two organizations address hazardous materials storage and handling, personal protective clothing and equipment, and hazardous materials professional competencies.

Among the most important consensus standards used within the hazmat response are the following:

NFPA 471—Recommended Practice for Responding to Hazardous Material Incidents.

The document covers planning procedures, policies, and application of procedures for incident levels, personal protective clothing and equipment, decontamination, safety, and communications. The purpose of NFPA 471 is to outline the minimum requirements that should be considered when dealing with responses to hazardous materials incidents and to specify operating guidelines.

It should be noted that NFPA 471 is a recommended practice and not a technical standard. A recommended practice is a document that is similar in content and structure to a code or standard but that contains only recommended or non-mandatory provisions. This difference is usually illustrated by using the term should to indicate recommendations in the document (as compared to shall in a standard).

NFPA 472—Standard for Professional Competence of Responders to Hazardous Material Incidents.

The purpose of NFPA 472 is to specify minimum competencies for those who will respond to hazardous material incidents. The overall objective is to reduce the number of accidents, injuries, and illnesses during response to hazmat incidents and to prevent exposure to hazmats to reduce the possibility of fatalities, illnesses, and disabilities affecting emergency responders.

It is important to recognize that NFPA 472 is not limited to fire service personnel but covers all hazardous materials emergency responders from both the public and private sector. This is a common misconception within the emergency response community.

NFPA 472 provides competencies for the following levels of hazmat responders. These levels parallel those listed within OSHA 1910.120, with the exception that the Hazardous Materials Specialist has been deleted and the Private Sector Specialist Employee has been expanded upon:

- *First Responder at the Awareness Level.* These are individuals who, in the course of their normal duties, may be the first on scene of an emergency involving hazardous materials. They are expected to recognize hazardous materials presence, protect themselves, call for trained personnel, and secure the area.
- *First Responder at the Operational Level.* These are individuals who respond to releases or potential releases of hazardous materials as part of the initial response to the incident for the purpose of protecting nearby persons, the environment, or property from the effects of the release. They shall be trained to respond in a defensive fashion to control the release from a safe distance and keep it from spreading.
- *Hazardous Materials Technician.* These are individuals who respond to releases or potential releases of hazardous materials for the purpose of controlling the release. Hazardous materials technicians are expected to use specialized chemical protective clothing and specialized control equipment.
- *Incident Commander.* The person who is responsible for directing and coordinating all aspects of a hazardous materials incident.
- *Private Sector Specialist Employee.* These are individuals who, in the course of their regular job duties, work with or are trained in the hazards of specific materials and/or containers. In response to incidents involving chemicals, they may be called upon to provide technical advice or assistance to the incident commander relative to their area of specialization. There are three levels of private sector specialist employee:

  Level C are those persons who may respond to incidents involving chemicals and/or containers within their organization's area of specialization.

They may be called upon to gather and record information, provide technical advice, and/or arrange for technical assistance consistent with their organization's emergency response plan and standard operating procedures. Level C individuals are not expected to enter the hot/warm zone at an incident.

Level B are those persons who, in the course of their regular job duties, work with or are trained in the hazards of specific chemicals or containers within their organization's area of specialization. Because of their education, training or work experience, they may be called upon to respond to incidents involving these chemicals or containers. Level B employees may be used to gather and record information, provide technical advice, and provide technical assistance (including working within the hot zone) at the incident consistent with their organization's emergency response plan and standard operating procedures and the local emergency response plan.

Level A are those persons who are specifically trained to handle incidents involving specific chemicals and/or containers for chemicals used in their organization's area of specialization. Consistent with their organization's emergency response plan and standard operating procedures, the Level A employees shall be able to analyze an incident involving chemicals within their organization's area of specialization, plan a response to that incident, implement the planned response within the capabilities of the resources available, and evaluate the progress of the planned response.

- *Hazardous Materials Branch Officer.* The person who is responsible for directing and coordinating all operations assigned to the hazardous materials branch (or group) by the Incident Commander. This individual is trained to the Technician level, with additional competencies in the command and control area.
- *Hazardous Materials Branch Safety Officer.* The person who works within an incident management system to ensure that recognized safe practices are followed within the hazardous materials branch (or group). This individual is trained to the Technician level, with additional competencies in the safety area.
- *Hazardous Materials Technician with a Specialty (Cargo Tank, Tank Car, Intermodal Tank).* These are individuals who provide support to the Hazardous Materials Technician, provide oversight for product removal and movement of damaged hazardous materials containers, and act as a liaison between Technicians and other outside resources.

NFPA 473—Standard for Professional Competence of EMS Personnel Responding to Hazardous Material Incidents.

The purpose of NFPA 473 is to specify minimum requirements of competence and to enhance the safety and protection of response personnel and all components of the emergency medical services system. The overall objective is to reduce the number of EMS personnel accidents, exposures, and injuries and illnesses resulting from hazmat incidents. There are two levels of EMS/HM responders:

- *EMS/HM Level I.* Persons who, in the course of their normal duties, may be called on to perform patient care activities in the cold zone at a hazmat incident. EMS/HM Level I responders provide care to only those individuals who no longer pose a significant risk of secondary contamination. Level I includes different competency requirements for Basic Life Support (BLS) and Advanced Life Support (ALS) personnel.

- *EMS/HM Level II.* Persons who, in the course of their normal duties, may be called on to perform patient care activities in the warm zone at a hazmat incident. EMS/HM Level II responders may provide care to those individuals who still pose a significant risk of secondary contamination. In addition, personnel at this level shall be able to coordinate EMS activities at a hazmat incident and provide medical support for hazmat response personnel. Level II includes different competency requirements for Basic (BLS) and Advanced Life Support (ALS) personnel.

NFPA Technical Committee on Hazardous Materials Protective Clothing and Equipment (NFPA 1991, 1992, 1994).

This Technical Committee is responsible for the development of standards and documents pertaining to the use of personal protective clothing and equipment (excluding respiratory protection) by emergency responders at hazardous materials incidents. The committee scope includes personal protective equipment (PPE) selection, care, and maintenance.

Three hazmat protective clothing standards have been developed:

- NFPA 1991—*Standard on Vapor-Protective Ensembles for Hazardous Materials Emergencies*
- NFPA 1992—*Standard on Liquid Splash-Protective Ensembles for Hazardous Materials Emergencies*
- NFPA 1994—*Standard on Protective Ensembles for Chemical/Biological Terrorism Incidents*

Other Standards Organizations.

There are many other important standards-writing bodies, including the American National Standards Institute (ANSI), the American Society for Testing and Materials (ASTM), the Compressed Gas Association (CGA), the Safety Equipment Institute (SEI), and the American Petroleum Institute (API). Each of these organizations approves or creates standards ranging from hazardous materials container design to personal protective clothing and equipment.

Individuals desiring more information on the standards writing organizations discussed here should consult the following Web sites:

- American National Standards Institute (ANSI)—http://www.ansi.org/
- American Petroleum Institute (API)—

  http://api-ec.api.org/intro/index_noflash.htm
- American Society for Testing and Materials (ASTM)—http://www.astm.org
- The Chlorine Institute —http://www.cl2.com/
- Compressed Gas Association (CGA)—http://www.cganet.com
- National Fire Protection Association (NFPA)—http://www.nfpa.org/
- Safety Equipment Institute (SEI)—http://www.seinet.org/

STANDARD OF CARE

"Standard of Care" is a widely accepted practice or standard that is followed by the majority of U.S. emergency response organizations. It represents the *minimum* accepted level of hazardous materials emergency service that should be provided regardless of location or situation.

Standard of Care is established by existing laws and regulations, as well as voluntary consensus standards and recommended practices (e.g., NFPA 471, 472, and 473). Many emergency responders are surprised to learn that the Standard of Care is also determined by local protocols and practices and what has been accepted in the past (i.e., precedent). So, when you consider the broad definition of *standard of care*, you can see how important it is for an Incident Commander or HMRT Team Leader to be well versed in regulatory requirements, as well as what other response organizations are doing to manage hazardous materials responses.

Standard of care is also influenced by legal findings and case law precedents established through the judicial system. This, in turn, allows your actions to be judged based upon what is expected of someone with your level of training and experience acting in the same or similar situation.

Standard of care is a dynamic element and historically has improved over time. Looking back in history, consider the hazardous materials "washdowns" of the 1970s that were standard practices. We just flushed the spill into the storm sewer system and "made it go away." Today, that same procedure is viewed as a poor operating practice and a potential environmental crime.

Emergency responders must recognize that (1) a standard of care exists; and (2) that the "highbar" is constantly moving upward. Training and continuing education are among the best ways to ensure that you will be able to provide the standard of care mandated by both society and the hazmat profession over time.

HOW DO I KNOW IF I'M MEETING THE STANDARD OF CARE? HERE'S A SIMPLE SCORE CARD:

1. *Our operations must be legal and within the requirements of the law.* Responders must be familiar with existing laws and regulations, including 29 CFR 1910.120 or HAZWOPER. Anyone who works in the hazmat response business should know this regulation well.
2. *Our actions and decisions must be consistent with voluntary consensus standards and recommended practices.* Most of what we do in hazmat planning and response is guided by national consensus standards developed by organizations such as NFPA, ANSI, and others. Command personnel working in the emergency response profession should be intimately familiar with not only what the standard requires, but the basis for that requirement. Don't just read the standard—you should also be familiar with the rationale and application of the requirement. Consulting publications as the NFPA Hazardous Materials Handbook can be helpful.
3. *Our actions and decisions to control a problem should have a technical foundation.* This is especially important when conducting hazard and risk evaluation. Responders must have a solid background in basic chemistry and physical science.
4. *Our actions and decisions must be ethical.* Most of us learned what is right and wrong and fair or unfair from individuals we respect (e.g., parents, guardians, or

mentors). If the decision you are making doesn't feel right and your internal "ethics meter" pegs over in the red zone, you need to reevaluate your actions and decisions.

THE HAZARDOUS MATERIALS MANAGEMENT SYSTEM

The fire problem in the United States has traditionally been managed by fire suppression operations at the expense of prevention activities. In 1973, Congress issued *America Burning*, a historical report on the nation's fire problem. The report significantly influenced the way we manage the fire problem today. Community master planning, public education, residential sprinklers, improved fire code enforcement, and fire protection engineering are some examples of the changes influenced by this landmark report.

The lessons learned in analyzing the U.S. fire problem can also be applied to managing the hazardous materials problem. If we want to be effective in managing and controlling the hazmat problem, we must be approach the problem from a larger, coordinated perspective that views the problem from a systems perspective.

There are four key elements in a hazardous materials management systems approach: (1): planning and preparedness; (2) prevention; (3) response; and (4) clean-up and recovery.

If the community or a facility is performing its responsibilities within the planning and prevention functions, you should see a reduction in the number and severity of response and clean-up activities.

PLANNING AND PREPAREDNESS

Planning is the first and most critical element of the system. The ability to develop and implement an effective hazmat management plan depends upon two elements: hazards analysis and the development of a hazmat emergency operations plan.

Hazards analysis—Analysis of the hazardous materials present in the community, including their location, quantity, specific physical (i.e., how they behave) and chemical (i.e., how they harm) properties, previous incident history, surrounding exposures and risk of release.

Contingency (emergency) planning—A comprehensive and coordinated response to the hazmat problem. This response builds upon the hazards analysis and recognizes that no single public or private sector agency is capable of managing the hazmat problem by itself.

The data and information generated by these activities will allow emergency managers to assess the potential risk to the community or facility and to develop and allocate resources as necessary. Many communities and facilities have adopted an "all-hazards" approach to emergency planning that develops a basic framework that is then applied to any emergency scenario, including natural hazards (e.g., floods, storms), technological hazards (e.g., hazmat, utility outages), and terrorist events.

HAZARDS ANALYSIS

Hazards analysis is the foundation of the planning process. It should be conducted for every location designated as having a moderate or high probability for a hazmat incident. In addition to risk evaluation, vulnerability—what is susceptible to damage should a release occur—must also be examined.

A hazards analysis provides the following benefits:

- It lets emergency responders know what to expect.
- It provides planning for less frequent incidents.
- It creates an awareness of new hazards.
- It may indicate a need for preventive actions, such as monitoring systems, remote isolation and process modifications.
- It offers an opportunity to evaluate using reduced chemical inventories or alternative chemicals to lower the consequences of an event.
- It increases the chance of successful emergency operations.

An evaluation team familiar with the facility or response area can facilitate the hazards analysis process. For example, within a facility this team may include safety, environmental, and industrial engineering professionals. Similarly, fire officers and members from each battalion or district, as well as representatives from prevention and the hazmat section, would be appropriate for a fire department. The primary concern here is geographic, as most responders are very familiar with their "first due" area.

There are four components of a hazards analysis program:

1. *Hazards identification*—Provides specific information on situations that have the potential for causing injury to life or damage to property and the environment due to a hazardous materials spill or release. Hazards identification will initially be based upon a review of the history of incidents within the facility or industry, or evaluating those facilities that have submitted chemical lists and reporting forms under SARA, Title III and related state and local right-to-know legislation. Information should include the following:
   - Chemical identification
   - Location of facilities that manufacture, store, use, or move hazardous materials
   - The type(s) and design of chemical container or vessel
   - The quantity of material that could be involved in a release
   - The nature of the hazard associated with the hazmat release (e.g., fire, explosion, toxicity)
   - The presence of any fixed suppression or detection systems
   - The level of physical security to prevent theft by criminals or terrorists
2. *Vulnerability analysis*—Identifies areas that may be affected or exposed and what facilities, property, or environment may be susceptible to damage should a hazmat release occur. A comprehensive vulnerability analysis provides information on:
   - The size/extent of vulnerable zones—specifically, what size area may be significantly affected as a result of a spill or release of a known quantity of a specific hazmat under defined conditions
   - The population, in terms of numbers, density, and types—for example, facility employees, residents, special occupancies (e.g., schools, hospitals, nursing homes)

- Private and public property that may be damaged, including essential support systems (e.g., water supply, power, communications) and transportation corridors and facilities
- The environment that may be affected, and the impact of a release on sensitive environmental areas and wildlife

3. *Risk analysis*—Assesses (1) the probability or likelihood of an accidental release, and (2) the actual consequences that might occur. The risk analysis is a judgment of incident probability and severity based upon the previous incident history, local experience, and the best available hazard and technological information.

   In today's environment, we must also consider the threat of hazardous materials and weapons of mass destruction (WMD) being used for criminal or terrorist activity. Emergency planners should coordinate and share risk analysis findings with law enforcement officials in their local jurisdiction.

4. *Emergency response resources evaluation*—Based upon the potential risks, considers emergency response resource requirements. These would include personnel, equipment, and supplies necessary for hazmat control and mitigation, EMS, protective actions, traffic control, and inventories of available equipment and supplies and their ability to function. For example, are firefighting foam supplies adequate to control and suppress vapors from a gasoline tank truck rollover? What are our capabilities to initiate mass decon operations? What level of personal protective clothing is provided to law enforcement personnel?

   Time and resources will dictate the depth and extent to which the hazards analysis can be completed. The focus is on the hazards created by the most common and most hazardous substances. Even the simplest plan will be better than no plan at all.

   A completed hazards analysis should allow emergency managers and planners to determine what level of response to emphasize, what resources will be required to achieve that response, and what type and quantity of mutual aid and other support services will be required.

PROCESS SAFETY MANAGEMENT HAZARDS ANALYSIS TECHNIQUES

Hazards analysis is also an integral element of the Process Safety Management (PSM) process required by OSHA 1910.119—*Process Safety Management of Highly Hazardous Chemicals, Explosives and Blasting Agents and EPA Part 68—Risk Management Programs for Chemical Accidental Release Prevention.* Both regulations impact industries that manufacture, store, and use highly hazardous chemicals and explosives, including refineries and chemical and petrochemical manufacturing plants.

Hazards analysis methods commonly used by safety professionals within industry include the following:

- *What if Analysis.* This method asks a series of questions, such as, "What if Pump X stops running?" or "What if an operator opens the wrong valve?" to explore possible hazard scenarios and consequences. This method is often used to examine proposed changes to a facility.

- *HAZOP Study*. This is the most popular method of hazard analysis used within the petroleum and chemical industries. The hazard and operability (HAZOP) study brings together a multidisciplinary team, usually of five to seven people, to brainstorm and identify the consequences of deviations from design intent for various operations. Specific guide words ("No," "More," "Less," "Reverse," etc.) are applied to parameters such as product flows and pressures in a systematic manner. The study requires the involvement of a number of people working with an experienced team leader.
- *Failure Modes, Effects, and Criticality Analysis (FMECA)*. This method tabulates each system or unit of equipment, along with its failure modes, the effect of each failure on the system or unit, and how critical each failure is to the integrity of the system. Then the failure modes can be ranked according to criticality to determine which are the most likely to cause a serious accident.
- *Fault Tree Analysis*. A formalized deductive technique that works backward from a defined accident to identify and graphically display the combination of equipment failures and operational errors that led up to the accident. It can be used to estimate the likelihood of events.
- *Event Tree Analysis*. A formalized deductive technique that works forward from specific events or sequences of events that could lead to an incident. It graphically displays the events that could result in hazards and can be used to calculate the likelihood of an accident sequence occurring. It is the reverse of fault tree analysis.

Under the original rule making, RMP results were to be made available to the public to ensure that the community was aware of the potential risks posed by specific facilities and chemicals. However, given the potential for this same information to be used as background and intelligence for a terrorist attack, its distribution and release to the general public is now being more closely controlled. Individuals desiring the latest information should consult the EPA Chemical Emergency Preparedness and Prevention Office (CEPPO) Web site at http://www.epa.gov/swercepp/.

FIGURE 1.4 Process Safety Management Hazards Analysis Techniques.

CONTINGENCY AND EMERGENCY PLANNING

Hazardous materials management is a multidisciplined issue that goes beyond the resources and capabilities of any single agency or organization. There will be a variety of "players" responding to a major hazmat emergency, and the emergency operations plan and related procedures will establish the framework for how the emergency response effort will operate. To manage effectively the overall hazmat problem within the community, a comprehensive planning process must be initiated. This effort is usually referred to as contingency planning or emergency planning.

There are many federal, state, and local requirements that apply to emergency planning. The one that most directly affects ERP is Title III of the Superfund Amendments and Reauthorization Act of 1986. SARA Title III requires the establishment of state emergency response commissions (SERC) and local emergency planning committees (LEPC). Title III also outlines specific requirements covering factors such as extremely hazardous substances, threshold planning quantities, make-up of LEPCs, dissemination of the planning, chemical lists and MSDS information to the community and the general public, facility inventories, and toxic chemical release reporting.

Figure 1.5 provides an overview of the emergency planning process, including the following:

1. **Organizing the planning team**—Planning requires community involvement throughout the process. Experience has shown that plans prepared by only one person or agency are doomed to failure. Emergency response requires trust, coordination, and cooperation. Remember, there is no single agency (public or private) that can effectively manage a major hazmat emergency alone.
2. **Defining and implementing the major tasks of the planning team**—These include reviewing any existing plans, identifying hazards, and analyzing and assessing current prevention and response capabilities.
3. **Writing the plan**—There are two approaches to this step: (a) Develop or revise a hazmat appendix or a section of a multihazard emergency operations plan, or (2) develop or revise a single-hazard plan specifically for hazardous materials. Once the plan is written, the respective planning groups involved must approve it.

   In recent years there has been a push toward the development of single integrated contingency plans (ICPs) or the "one-plan" concept. Readers desiring further information on the ICP concept should consult the EPA Chemical Emergency Preparedness and Prevention Office (CEPPO) Web site at http://www.epa.gov/swercepp/.
4. **Revising, testing, and maintaining the plan**—Every emergency plan must be evaluated and kept up to date through the review of actual responses, simulation exercises, and the regular collection of new data and information.

While emergency planning is essential, the completion of a plan does not guarantee that the facility or community is actually prepared for a hazmat incident. Planning is only one element of the total hazmat management system.

PREVENTION

The responsibility for the prevention of hazmat releases is shared between the public and private sectors. Because of their regulatory and enforcement capabilities, however, public sector agencies generally receive the greatest attention and often "carry the biggest stick."

Prevention activities often include the following:

HAZMAT PROCESS, CONTAINER DESIGN, AND CONSTRUCTION STANDARDS

Almost all hazardous materials facilities, containers, and processes are designed and constructed to some standard. This standard of care may be based upon voluntary consensus standards, such as those developed by the NFPA and ASTM, or on government regulations. Many major petrochemical companies, hazardous materials companies, and industry trade associations (e.g., Chlorine Institute, Compressed Gas

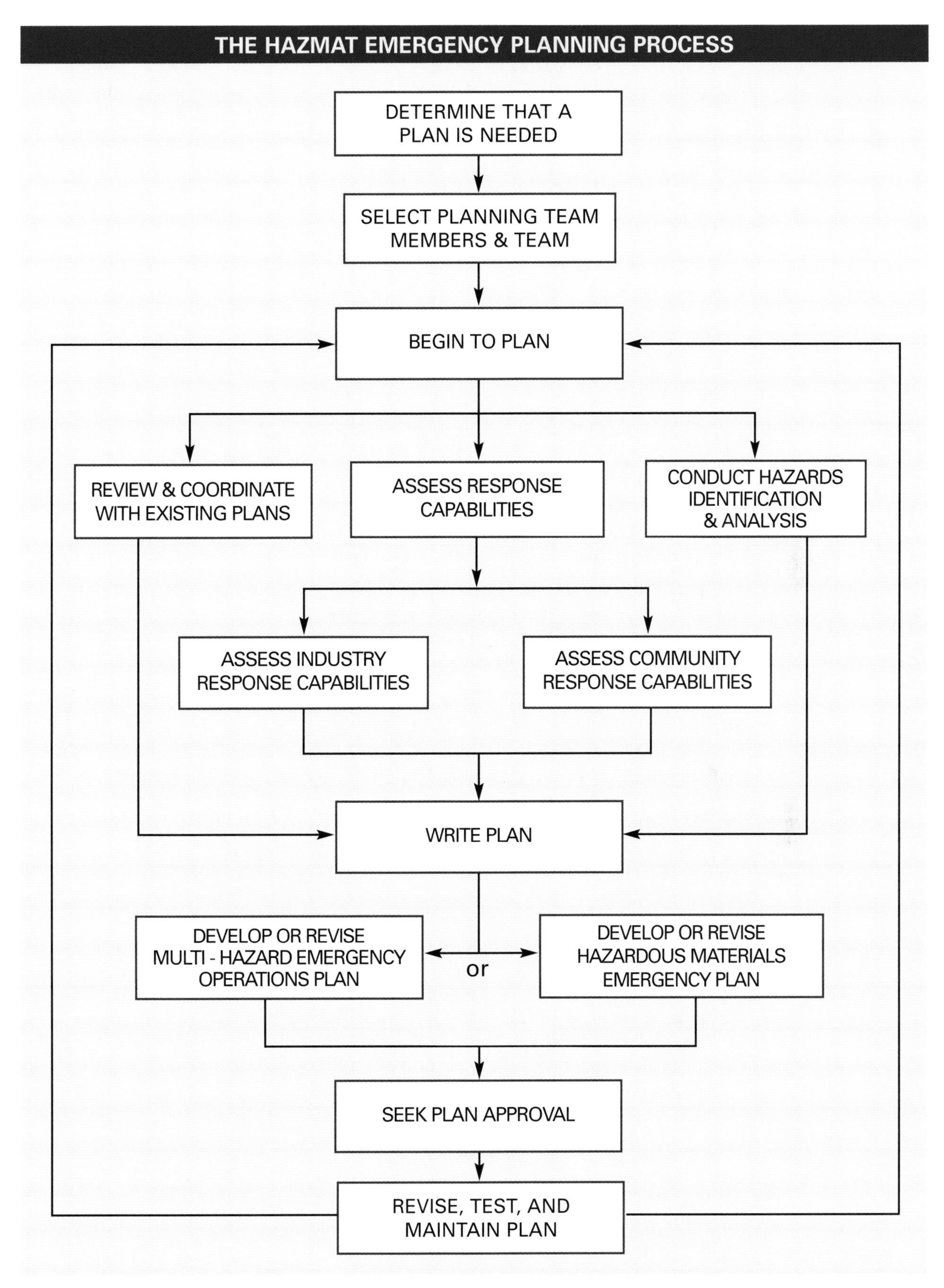

FIGURE 1.5 The Hazmat Planning Process.

Association) have also developed their own respective engineering standards and guidelines.

All containers used for the transportation of hazardous materials are designed and constructed to both specification and performance regulations established by U.S. DOT. These regulations can be referenced in CFR Title 49. In certain situations, hazardous materials may be shipped in non-DOT specification containers that have received a DOT exemption.

INSPECTION AND ENFORCEMENT

Fixed facilities, transportation vehicles, and transportation containers are normally subject to some form of inspection process. These inspections can range from comprehensive and detailed inspections at regular intervals to visual inspections each time a container is loaded. Fixed facilities will commonly be inspected by state and federal OSHA and EPA inspectors, in addition to state fire marshals and local fire departments. It should be recognized that many of these inspections will focus upon fire safety and life safety issues and may not adequately address either the environmental or process safety issues.

Transportation vehicle inspection is generally based upon criteria established within Title 49 CFR. The enforcing agency is often the state police, but this will vary according to the individual state, the hazardous materials being transported, and the mode of transportation. Some local fire departments, such as that in Aurora, Colorado, routinely perform inspections of hazardous materials cargo tank trucks.

Among the U.S. DOT agencies with hazardous materials regulatory responsibilities are the following:

- *Office of Hazardous Materials Safety (HMS)* of the Research and Special Programs Administration (RSPA). Responsible for all hazardous materials transportation regulations except bulk shipment by ship or barge. Includes designating and classifying hazardous materials, container safety standards, label and placard requirements, and handling, stowing and other in-transit requirements. Serves as the DOT representative to the National Response Team, supports the National Response Center operation in coordination with the U.S. Coast Guard, and serves as the DOT liaison with the Federal Emergency Management Agency on hazmat transportation issues.

- *Office of Hazardous Materials Enforcement (OHME)* of the Office of Hazardous Materials Safety (HMS). Inspection and enforcement staff determine compliance with safety standards by inspecting entities that offer hazardous materials for transportation; manufacture, rebuild, repair, recondition, or retest packaging used to transport hazardous materials; and handle intermodal transfers of hazardous materials. Responsible for the management and coordination of RSPA's hazardous materials inspection and enforcement program.

- *Office of Pipeline Safety (OPS)* of the Research and Special Programs Administration (RSPA). Administers DOT's national regulatory program to assure the safe transportation of natural gas, petroleum, and other hazardous materials by pipeline. OPS develops regulations and other approaches to risk management to assure safety in design, construction, testing, operation, maintenance, and emergency response of pipeline facilities.

- *Federal Railroad Administration (FRA).* Responsible for enforcement of regulations relating to hazardous materials carried by rail or held in depots and freight yards.
- *Federal Aviation Administration (FAA).* Responsible for the enforcement of regulations relating to hazardous materials shipments on domestic and foreign carriers operating at U.S. airports and in cargo handling areas.
- *U.S. Coast Guard (USCG).* Responsible for the inspection and enforcement of regulations relating to hazardous materials in port areas and on domestic and foreign ships and barges operating in the navigable waters of the United States. Supports the operation of the National Response Center in Washington, D.C.

PUBLIC EDUCATION

Hazmat is a concern not only for industry but also for the community. The average homeowner contributes to this problem by improperly disposing of substances such as used motor oil, paints, solvents, batteries, and other chemicals used in and around the home. As a result, many communities have initiated full-time household chemical waste awareness, education, and disposal programs. In other instances, communities have established used motor oil collection stations and chemical clean-up days in an effort to reclaim and recycle these materials.

An example of a highly successful public education program is the Wally Wise Guy program developed by the Deer Park, Texas LEPC. An example of industry and government partnership, Wally Wise Guy is a friendly turtle who teaches children and their parents how to shelter-in-place in case of a chemical emergency. The program has been produced and delivered in both English and Spanish and has been a highly successful educational tool. Further information on the Wally Wise Guy Program can be obtained by consulting the Deer Park LEPC Web site at http:/ / www.wally.org/.

HANDLING, NOTIFICATION, AND REPORTING REQUIREMENTS

These guidelines actually act as a bridge between planning and prevention functions. There are many federal, state, and local regulations that require those who manufacture, store, or transport hazardous materials and hazardous wastes to comply with certain handling, notification, and reporting rules. Key federal regulations include CERCLA (Superfund), RCRA, and SARA, Title III. There are also many state regulations that are similar in scope and which often exceed the federal standard requirements.

RESPONSE

When the prevention and enforcement functions fail, response activities begin. Since it is impossible to eliminate all risks associated with the manufacture, storage, and use of hazmats, the need for a well-trained, effective emergency response capability will always exist.

Response activities should be based upon the information and probabilities identified during the planning process. Response activities must be based upon an evaluation of the facility or local hazmat problem. While every community should have access to a hazmat response capability, that capability does not always have to be provided by either local government or the fire service. Numerous states and regions have established both statewide and regional hazmat response team systems that ensure the delivery of a competent and effective capability in a timely manner.

RESPONSE GROUPS

The emergency response community consists of various agencies and individuals who respond to hazmat incidents. They can be categorized based upon their knowledge, expertise, and resources. These responders can be compared to the levels of capability found within a typical EMS system. In that system, an injury such as a fractured arm can be effectively managed by a First Responder or Emergency Medical Technician–Basic (EMT-B), while a cardiac emergency will require the services of an EMT-I (Intermediate) or an EMT-P (Paramedic).

In the same way, a diesel fuel spill can usually be contained by First Responder-Operations level personnel, such as a fire department engine company using dispersants or absorbent. An accident involving a poison gas or reactive chemical will, however, require the on-scene expertise of a hazardous materials technician or hazardous materials response team. In simple terms, we try to match the nature of the problem with the capabilities of the responders. Figure 1.6 illustrates this comparison.

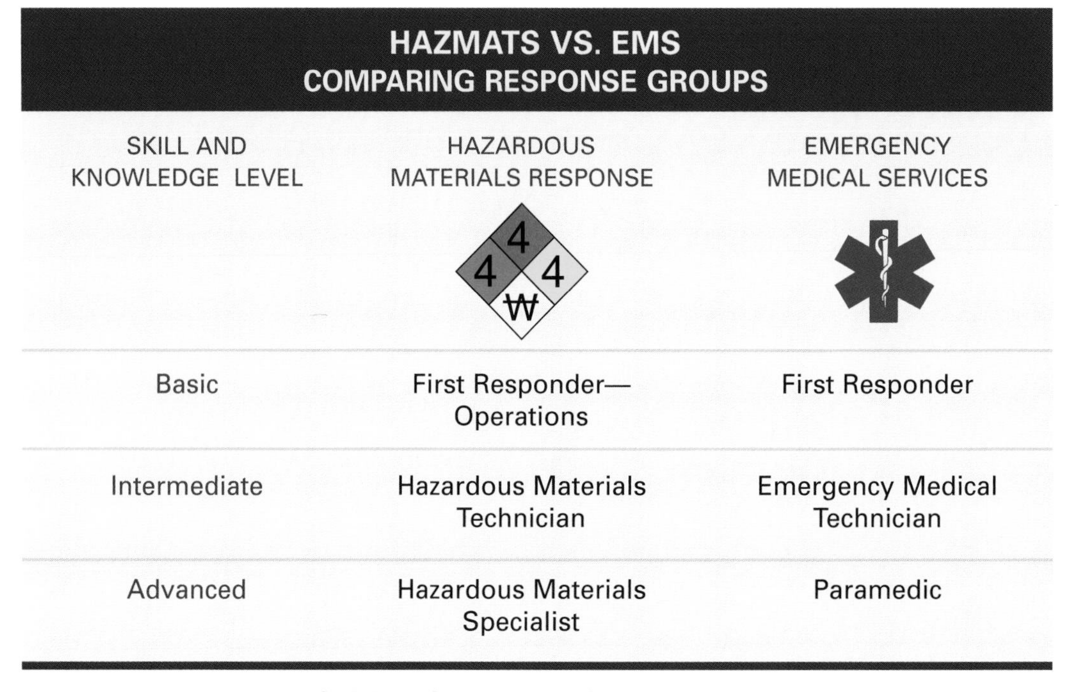

HAZMATS VS. EMS
COMPARING RESPONSE GROUPS

SKILL AND KNOWLEDGE LEVEL	HAZARDOUS MATERIALS RESPONSE	EMERGENCY MEDICAL SERVICES
Basic	First Responder—Operations	First Responder
Intermediate	Hazardous Materials Technician	Emergency Medical Technician
Advanced	Hazardous Materials Specialist	Paramedic

FIGURE 1-6 Comparing EMS and Hazmat Emergency Response Groups.

LEVELS OF INCIDENT

Fortunately, not every incident is a major emergency. Response to a hazmat release may range from a single-engine company responding to a natural gas leak in the street, to a railroad derailment involving dozens of government and private agencies and their associated personnel.

These incidents can be categorized based upon their severity and the resources they require. Figure 1-7 (a and b) outlines these response levels.

LEVELS OF HAZARDOUS MATERIALS INCIDENTS—COMMUNITY

RESPONSE LEVEL	DESCRIPTION	RESOURCES	EXAMPLES
 I Potential Emergency Conditions	An incident or threat of a release which can be controlled by the first responder. It does not require evacuation, beyond the involved structure or immediate outside area. The incident is confined to a small area and poses no immediate threat to life and property.	Essentially a local level response with notification of the appropriate local, state, and federal agencies. Required resources may include: • Fire Department • Emergency Medical Services (EMS) • Law Enforcement • Public Information Officer (PIO) • Chemtrec • National Response Center	• 500-gallon fuel oil spill • Inadvertent mixture of chemicals • Natural gas leak in a building
 II Limited Emergency Conditions	An incident involving a greater hazard or larger area than Level I which poses a potential threat to life and property. It may require a limited protective action of the surrounding area.	Requires resources beyond the capabilities of the initial local response personnel. May require a mutual aid response and resources from other local and state organizations. May include: • All Level I Agencies • Hazmat Response Teams • Public Works Department • Red Cross • Regional Emergency Management Staff • State Police • Public Utilities	• Minor chemical release in an industrial facility • A gasoline tank truck rollover • A chlorine leak at a water treatment facility
  Full Emergency Conditions	An incident involving a severe hazard or a large area which poses an extreme threat to life and property and which may require a large-scale protective action.	Requires resources beyond those available in the community. May require the resources and expertise of regional, state, federal, and private organizations. May include: • All Level I and II Agencies • Mutual Aid Fire, Law Enforcement, and EMS • State Emergency Management Staff • State Department of Environmental Resources • State Department of Health • Environmental Protection Agency (EPA) • U.S. Coast Guard • Federal Emergency Management Agency (FEMA)	• Major train derailment with fire • Explosion or toxicity hazard • A migrating vapor cloud release from a petrochemical processing facility

FIGURE 1-7 (A) Levels of Hazardous Materials Incidents—Community.

LEVELS OF HAZARDOUS MATERIALS INCIDENTS—PETROCHEMICAL INDUSTRY

RESPONSE LEVEL	DESCRIPTION	RESOURCES	EXAMPLES
 I Incident	Minimal danger to life, property, and the environment. Problem is limited to immediate work area and public health, safety, and environment are not affected.	Handled by On-Shift Emergency Response Team (ERT) with no off-shift or mutual aid response.	• Minor spills and releases less than 55 gallons • Small pump seal fire • Minor vapor release during product transfer operations
 II Serious Incident	Moderate danger to life, property, and the environment on the plant site. Problem is currently limited to plant property, but has the potential for involving additional exposures or migrating off-site and affecting public health, safety, and environment for a short period of time.	Requires On-Shift ERT response. Additional assistance required from off-duty ERT personnel and/or mutual aid units. Plant EOC may be activated. Corporate Crisis Emergency Plan may be activated.	• Large release of flammable, corrosive, or toxic vapors • Releases of over 100 gallons of hazardous material • Large spill fire or a seal fire on a floating roof tank
 III Crisis Situation	Extreme danger to life, property, and the environment. Problem goes beyond the refinery property and can impact public health, safety and the environment or a large geographic area for an indefinite period of time.	Requires multi-organizational response from plant, local fire service, industrial mutual aid units, and public safety resources. Plant EOC is activated. Corporate Crisis Emergency Plan activated.	• Process unit fire or explosion • Major release of flammable, corrosive or toxic vapors • Shipboard fire, major oil spill, or HM release which can impact major waterways

FIGURE 1-7 (B) Levels of Hazardous Materials Incidents—Petrochemical Industry.

HAZMAT RESPONSE TEAM (HMRT)

In order to respond effectively and efficiently to hazmat emergencies, many facilities and communities have established hazmat response teams. NFPA 472 defines an HMRT as an organized group of trained response personnel operating under an Emergency Operations Plan and appropriate standard operating procedures, who are expected to perform work to handle and control actual or potential leaks or spills of hazardous materials requiring close approach to the material. The HMRT members respond to releases for the purpose of control or stabilization of the incident. Among the specialized equipment carried by an HMRT are reference libraries, computers and communications equipment, personal protective clothing and equipment, direct-reading monitoring and detection equipment, control and mitigation supplies and equipment, and decontamination supplies and equipment.

In evaluating the need for an HMRT, consider the following points:

- There is no single department or agency that can effectively manage the hazmat issue by itself. Regardless of what agency operates the HMRT, it must work closely with other local, state, and federal governmental agencies. A unified command organization is a necessity.
- Every community does not require a HMRT. However, every community should have access to a HMRT capability through local, regional, state or contractor resources.
- An HMRT will not necessarily solve the hazmat problem. Remember the hazardous materials management system—planning, prevention, response and recovery.
- There are numerous constraints and requirements associated with developing an effective HMRT capability. These include legal, insurance, and political issues; both initial and continuing funding sources; resource determination and acquisition; personnel and staffing, and initial and continuing training requirements.
- Successful HMRT response programs are those that truly understand what services an HMRT can provide at all emergencies, not just those involving hazardous materials. For example, no organization within the emergency response community better understands and routinely practices (1) the process of risk evaluation; (2) the use of air monitoring and detection equipment and the interpretation of its results; and (3) the fundamentals of safe operating practices in a field environment better than HMRT personnel. **The HMRT is *Not* a chemical resource; it is a health and safety resource with capabilities that can be used in a variety of response scenarios, including hazardous materials, confined space, structural collapse, aircraft accidents, and other significant fires and emergencies.**

Regional and statewide hazardous materials response systems have been developed in some areas of the country, including North Carolina, Massachusetts, and California. A number of these response systems have different levels of HMRTs based upon the HMRTs staffing and equipment inventory. The tiered response concept is also used in a number of metropolitan fire departments, where the HMRT is supported by a number of ladder or rescue companies designated as Hazardous Materials Support Companies, which are trained to handle the lower-risk incidents as well as provide support to the HMRT for a "working" hazardous materials response.

HMRTs typically function as a group or sector within the Incident Command System (ICS) under the direct control of a Hazardous Materials Group Supervisor. Based upon their assessment of the hazardous materials problem, the HMRT, through the Hazardous Materials Group Supervisor, provides the Incident Commander or Operations Section Chief with a list of options and a recommendation for mitigation of the hazardous materials problem. However, the final decision always remains with the Incident Commander. This topic will be discussed further in Chapter 3.

HMRT members must be properly trained and must participate in a medical surveillance program based upon the requirements of 29 CFR 1910.120. The two primary information sources for establishing an HMRT training program are OSHA 1910.120(q) and NFPA 472. While OSHA 1910.120 outlines the regulatory requirements (what you have to do), NFPA 472 spells out the specific training and educational competencies for the training program (how you can do it).

Both OSHA 1910.120 and NFPA 472 recommend that HMRT personnel be trained to the Hazardous Materials Technician level. According to OSHA 1910.120(q), the Hazardous Materials Technician requires a minimum of 24 hours of initial training at the First Responder Operations level. A survey of Hazardous Materials Technician training courses would find that they range from 24 to 200+ hours. Industry emergency response team members are regularly trained to the Hazardous Materials Technician-level in a 24- to 40-hour course because they respond to a limited number of chemicals and response scenarios within the confines of their facility. In contrast, given the wide range of potential scenarios that may occur in the community, public safety Technician-level training requirements will most likely be in the 40 to 200+ hour area.

CLEAN-UP AND RECOVERY

Clean-up and recovery operations are designed to (1) clean up or remove the hazmat spill or release, and (2) restore the facility and/or community back to normal as soon as possible. In many instances, chemicals involved in a hazmat release will be eventually classified as hazardous wastes.

Clean-up operations fall under the guidelines of HAZWOPER, CERCLA (Superfund), and RCRA. Clean-up activities can be classified as follows:

- **Short term**—Those actions immediately following a hazmat release that are primarily directed toward the removal of any immediate hazards and restoring vital support services (i.e., reopening transportation systems, drinking water systems, etc.) to minimum operating standards. Short-term activities may last up to several weeks.

  These activities are normally the responsibility of the "responsible party"—usually the facility operator or transportation carrier (e.g., railroad, truck company). In situations where the responsible party has not been identified or does not have sufficient financial resources, they may be assumed by state or federal environmental agencies.
- **Long term**—Those remedial actions that return vital support systems back to normal or improved operating levels. Examples would include groundwater treatment operations, the mitigation of both aboveground and underground spills, and the monitoring of flammable and toxic contaminants. These activities may not be directly related to a specific hazmat incident but are often the result of abandoned industrial or hazardous waste sites. These operations may extend over months or years.

Recovery operations focus upon restoring the facility, the community, and/or emergency response organization to normal operating conditions. Tasks would include restocking all supplies and equipment, compilation and documentation of resources purchased and/or used during the emergency response, and financial restitution, where appropriate.

ROLE OF EMERGENCY RESPONDERS DURING CLEAN-UP OPERATIONS

Many plant-level industrial responders are also responsible for the clean-up of minor spills and releases so that facility operations may continue. In contrast, public safety response personnel are usually not directly responsible for the final clean-up and recovery of a hazardous materials release. Depending upon the nature of the incident, however, they may continue to be responsible for site safety until all risks are stabilized and the emergency phase is terminated. Once the emergency phase is terminated, there must be a formal transfer of command from the lead response agency (e.g., fire department) to the lead agency responsible for post-emergency response operations (e.g., responsible party, state or federal EPA, etc.).

At short-term operations immediately following an incident, the Incident Commander should ensure that the work area is closely controlled, that the general public is denied entry, and that the safety of emergency responders and the public is maintained during clean-up and recovery operations. When interfacing with both industry responders and contractors, the Incident Commander should ensure that they are trained to meet the requirements of OSHA 1910.120 and/or NFPA 472.

Long-term clean-up and recovery operations do not normally require the continuous presence of the fire service. Depending upon the size and scope of the clean-up, a contractor or government official (Remedial Project Manager, or RPM) will be the central contact point. Emergency responders should be familiar with the clean-up operation, including its organizational structure, the OSC/RPM, work plan, time schedule, and site safety plan.

Clean-up operations should conform to the general health and safety requirements of both state and federal EPA and OSHA standards. Although ERP generally do not have the authority to conduct inspections or issue citations at clean-up operations, they can bring specific concerns to the attention of the state or federal regulatory agency having jurisdiction.

SUMMARY

Hazardous materials are a multidisciplined problem requiring an organized facility-level and community-level approach. Although this textbook is primarily oriented towards managing and implementing emergency response operations, one should recognize that response accounts for only a small portion of an effective, comprehensive hazmat management program.

REFERENCES AND SUGGESTED READINGS

Callan, Michael, STREET SMART HAZMAT RESPONSE, Chester, MD: Red Hat Publishing (2002).

Fire, Frank L., Nancy K. Grant, and David H. Hoover, SARA TITLE III—INTENT AND IMPLEMENTATION OF HAZARDOUS MATERIALS REGULATIONS, Tulsa, OK: Fire Engineering Books and Videos (1990).

Lesak, David M., HAZARDOUS MATERIALS STRATEGIES AND TACTICS, Upper Saddle River, NJ: Brady/Prentice Hall (1999).

National Fire Protection Association, HAZARDOUS MATERIALS RESPONSE HANDBOOK (4th edition), Quincy, MA: National Fire Protection Association (2002).

National Institute for Occupational Safety and Health (NIOSH), OCCUPATIONAL SAFETY AND HEALTH GUIDANCE MANUAL FOR HAZARDOUS WASTE SITE ACTIVITIES, Washington, DC: NIOSH, OSHA, USCG, EPA (1985).

National Response Team, HAZARDOUS MATERIALS EMERGENCY PLANNING GUIDE (NRT-1), Washington, DC: National Response Team (1987).

Stringfield, William H., A FIRE DEPARTMENT'S GUIDE TO IMPLEMENTING SARA, TITLE III AND THE OSHA HAZARDOUS MATERIALS STANDARD, Ashland, MA: International Society of Fire Service Instructors (1987).

Stringfield, William H., EMERGENCY PLANNING AND MANAGEMENT—ENSURING YOUR COMPANY'S SURVIVAL IN THE EVENT OF A DISASTER, (2nd edition), Rockville, MD: Government Institutes (2000).

U.S. Environmental Protection Agency, HAZMAT TEAM PLANNING GUIDANCE, Washington, DC: EPA (1990).

U.S. Environmental Protection Agency, HAZARDOUS MATERIALS PLANNING GUIDE, Washington, DC: EPA (2001).

U. S. Environmental Protection Agency et al., TECHNICAL GUIDANCE FOR HAZARDS ANALYSIS—EMERGENCY PLANNING FOR EXTREMELY HAZARDOUS SUBSTANCES, Washington, DC: EPA, FEMA, DOT (1987).

U.S. Occupational Safety and Health Administration, OSHA DIRECTIVE NUMBER CPL 2-2/59A—INSPECTION PROCEDURES FOR THE HAZARDOUS WASTE OPERATIONS AND EMERGENCY RESPONSE STANDARD, 29 CFR 1910.120 AND 1926.65, PARAGRAPH (Q): EMERGENCY RESPONSE TO HAZARDOUS SUBSTANCE RELEASES, Washington, DC: OSHA (April 4, 1998).

MAGNESIUM
DANGER
USE NO WATER
SEE THE EXOTIC METALS $5.00
THIS KID IS
Sooo.o.oo
EASY!

logical agents. Anyone with large open cuts, rashes, or abrasions should be prohibited from working in areas where they may be exposed. Smaller cuts or abrasions should be covered with a nonporous dressing.

The rate of skin absorption can vary depending on the body part that's exposed. For example, assuming the area of skin exposed and the duration of exposure are equal, the rate of skin absorption through the scalp or genitals area will be considerably faster than through the forearm. Absorption through the eyes is one of the fastest means of exposure since the eye has a high absorbency rate. This type of exposure may occur when a chemical is splashed directly into the eye, when a chemical is "carried" on toxic smoke particles into the eyes from a fire, or when gases or vapors are absorbed through the eyes. Likewise, this may be seen as an early warning signal for either PPE or respiratory protection failures.

- **Ingestion**—The introduction of a chemical into the body through the mouth or inhaled chemicals trapped in saliva and swallowed. Exposed personnel should be prohibited from smoking, eating, or drinking except in designated rest and rehab areas after being decontaminated.
- **Direct contact**—Direct skin contact with some chemicals, such as corrosives, will immediately damage skin or body tissue on contact. Acids have a strong affinity for moisture and can create significant skin and respiratory tract burns. However, the injury process also creates a clotlike barrier that blocks deep skin penetration. In contrast, caustic or alkaline materials dissolve the fats and lipids that make up skin tissue and change the solid tissue into a soapy liquid (think of how drain cleaners using caustic chemicals dissolve grease and other materials in sinks and drains!). As a result, caustic burns are often much deeper and more destructive than acid burns.
- **Injection**—Other chemicals may be injected directly through the skin and into the bloodstream. Mechanisms of injury include needle stick cuts at medical emergencies and the injection of high pressure gases and liquids into the body similar to the manner in which flu shots are injected with pneumatic guns.

DOSE/RESPONSE RELATIONSHIP

A fundamental relationship exists between the chemical dose (i.e., dose = concentration x time) and the response produced by the human body. Given the broad range of toxicities any substance might cause, the wisdom of Paracelsus (1493–1541) becomes clear: "All substances are poisons; there is none which is not a poison. The right dose differentiates a poison and a remedy." Translation: Dose makes the poison.

The dose/response concept is based upon the following assumptions:

- The magnitude of the response is dependent upon the concentration of the chemical at the biological site of action (i.e., target organ).
- The concentration of the chemical at the biological site of action is a function of the dose administered.
- Dose and response are essentially a cause/effect relationship.

Remember that dose is the concentration or amount of material to which the body is exposed over a specific time period. In simple terms, dose = concentration x time. The human body's response to a dose is either toxic or nontoxic. Human response may also be influenced by the age, state of health, and nutrition state of the individual. Typically, as the size of the dose increases, the potential for a toxic response also increases. For example, one gram of aspirin is an accepted dose for medicinal purposes, yet quantities in excess of 10 grams (approximately 25 tablets) have caused death.

With many chemicals, increasing moderate doses will cause no apparent biological damage. However, increasing the dose rate will eventually reach a point defined as the threshold level. At this concentration and above, chemical exposure can result in biological harm. These effects may range from slight irritation to death, depending on the dose and properties of the chemical. Figure 2.2 illustrates a typical dose/response curve, and Figure 2.3 illustrates the dose/response relationship using alcohol as an example.

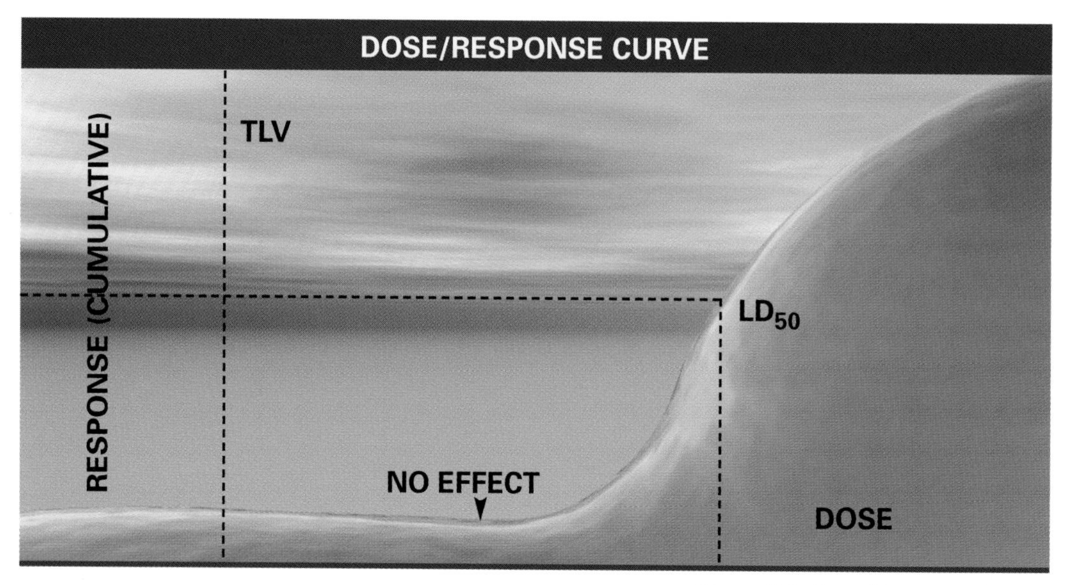

Figure 2.2 Dose/Response Curve.

DOSE/RESPONSE RELATIONSHIP

DOSE	ACUTE EFFECT	CHRONIC EFFECT
1 oz. of Bourbon Consumed in 60 mins.	Minimal	None
1 qt. of Bourbon Consumed in 60 mins.	Illness or Death	Minimal
1 oz. of Bourbon Consumed every 60 mins. for 12 hrs. Each day, 365 days a year	Minimal	Brain / Liver Damage
1 qt. of Bourbon Consumed over a year	None	None

Figure 2.3 Dose/Response Relationship.

EFFECTS OF HAZARDOUS MATERIALS EXPOSURES

Health effects of a hazardous materials exposure can be described in terms of how a hazmat attacks the body. A local effect implies an effect at the point of contact—for example, a corrosive burn. A systemic effect occurs when a chemical enters the bloodstream and attacks target organs and internal areas of the human body. Multiple systemic effects are also a distinct possibility.

Systemic effects often show up at target organs. Target organs are organs/tissues where a toxin exerts its effects; it is not necessarily the organ/tissue where the toxin is most highly concentrated. For example, over 90% of the lead in the adult human body is in the skeleton, but lead exerts its effects on the kidneys, the central nervous system, and the blood system.

Examples of target organs that may be affected by a systemic poison include the following:

TERM	TARGET ORGAN	EXAMPLES
Hepatotoxins	Liver	Carbon tetrachloride, Vinyl chloride monomer, and nitroamines
Nephrotoxins	Kidneys	Halogenated hydrocarbons and mercury
Neurotoxins	Central Nervous System (CNS)	Lead, toluenenerve agents, and organophosphate pesticides
Respiratory Toxins	Lungs and Pulmonary System	Asbestos, chlorine, and hydrogen sulfide
Hematotoxins	Blood System	Benzene, chlordane, and cyanides
Skeletal System	Bones	Hydrofluoric acid and selenium
Dermatotoxins	Skin (Note: Act as irritants, ulcers, chloracne, and/or cause skin pigmentation disturbances)	Halogenated hydrocarbons, coal tar compounds, and high levels of ultraviolet light
Teratogens	Fetus	Lead and ethylene oxide
Mutagens	Genetic damage to cells or organisms	Radiation, lead, and ethylene dibromide

The human body can be subject to seven types of harm events:

- **Thermal**—Those events related to temperature extremes. High temperatures are common at fire-related emergencies involving flammable liquids, gases, and solids, as well as explosions. Thermal harm resulting in frostbite can be found as a result of exposures to the extremely low temperatures associated with liquefied gases and cryogenic materials.

- **Mechanical**—Those events resulting from direct contact with fragments scattered because of a container failure, explosion, bombing or shock wave. Remember that small fragments and shrapnel can cause significant damage to the body.
- **Poisonous**—Those events related to exposure to toxins. Some chemicals affect the body by causing damage to specific internal organs or body systems. Examples include hepatotoxins and nephrotoxins. Other chemicals, such as benzene and phenol, are considered blood toxins because of their effects on the circulatory system. Similarly, neurotoxins—such as organophosphate and carbamate pesticides—affect the central nervous system. A number of flammable liquids have some form of toxicity well before they become fire hazards (e.g., toluene, methyl alcohol).
- **Corrosive**—Those events related to chemical burns and/or tissue damage from exposure to corrosive chemicals. Corrosives are divided into two chemical groups: acids and bases. Acids—such as strong inorganic acids like nitric, sulfuric, hydrochloric, and hydrofluoric—can cause severe tissue burns to the skin and permanent eye damage. Bases, or caustic or alkaline materials, break down fatty skin tissue and penetrate deeply. Sodium and potassium hydroxide are examples. Acids generally cause greater surface tissue damage while bases produce deeper, slower healing burns.

  Inhaled corrosive gases and vapors can also cause acute swelling of the upper respiratory tract and chemically induced pulmonary edema. High water soluble materials, such as anhydrous ammonia, will affect the upper respiratory tract, while low water soluble materials, such as nitrogen dioxide and phosgene, will affect the lower respiratory tract. Lower respiratory tract injuries can lead to chemically induced pulmonary edema and may be delayed for 24 to 72 hours after the exposure.
- **Asphyxiation**—Those events related to oxygen deprivation within the body. Asphyxiants can be categorized as simple or chemical. Simple asphyxiants act on the body by displacing or reducing the oxygen in the air for normal breathing. Examples include carbon dioxide, nitrogen, and natural gas. Chemical asphyxiants disturb the normal body chemistry processes that control respiration. They can range from chemicals that inhibit oxygen transfer from the lungs to the cells (carbon monoxide) or prevent respiration at the cellular level (hydrogen cyanide) to those that incapacitate the respiratory system (hydrogen sulfide in large concentrations).
- **Radiation**—Those events related to the emission of radiation energy. Radiation energy is defined as the waves or particles of energy that are emitted from radioactive sources. It may be in the form of alpha or beta particles or gamma waves, depending on the intensity of the source material. Examples include plutonium and uranium hexafluoride. Nonionizing radiation, such as microwaves and lasers, may also create potential harm in certain emergency situations.
- **Etiological**—Those events created by uncontrolled exposures to living microorganisms. Etiologic/biological harm is normally associated with diseases such as typhoid fever and tuberculosis, with bloodborne pathogens such as the hepatitis (A, B or C) virus, and with weapons-grade biological agents (e.g., anthrax). It is often difficult to detect when and where the physical exposure to the biological agent occurred and the route(s) of exposure.

TOXICITY AND EXPOSURE TERMINOLOGY

The terms used to describe chemical toxicity and exposures can seem complicated, and some have similar meanings, further complicating the issue. All relate to how long an individual can safely work in a chemical or hazardous atmosphere. There is a safety factor that is incorporated into each of these values, as each person can react differently based upon their age, sensitivity, and pre-existing medical conditions. All of these terms and values are based upon a person wearing NO skin or respiratory protection.

Remember, HEALTH HAZARD = EXPOSURE + TOXICITY

Poison lines are an excellent tool for visually illustrating the significance and inter-relationship between health exposures and other hazards. Examine the following poison lines for chlorine (poison gas), methanol (flammable liquid with health hazards), and anhydrous ammonia (corrosive gas that is toxic by inhalation and flammable in certain atmospheres). Note that (1) as concentrations rise above the TLV, harm occurs; and (2) as the concentration increases, harm increases. At some point, the IDLH is reached.

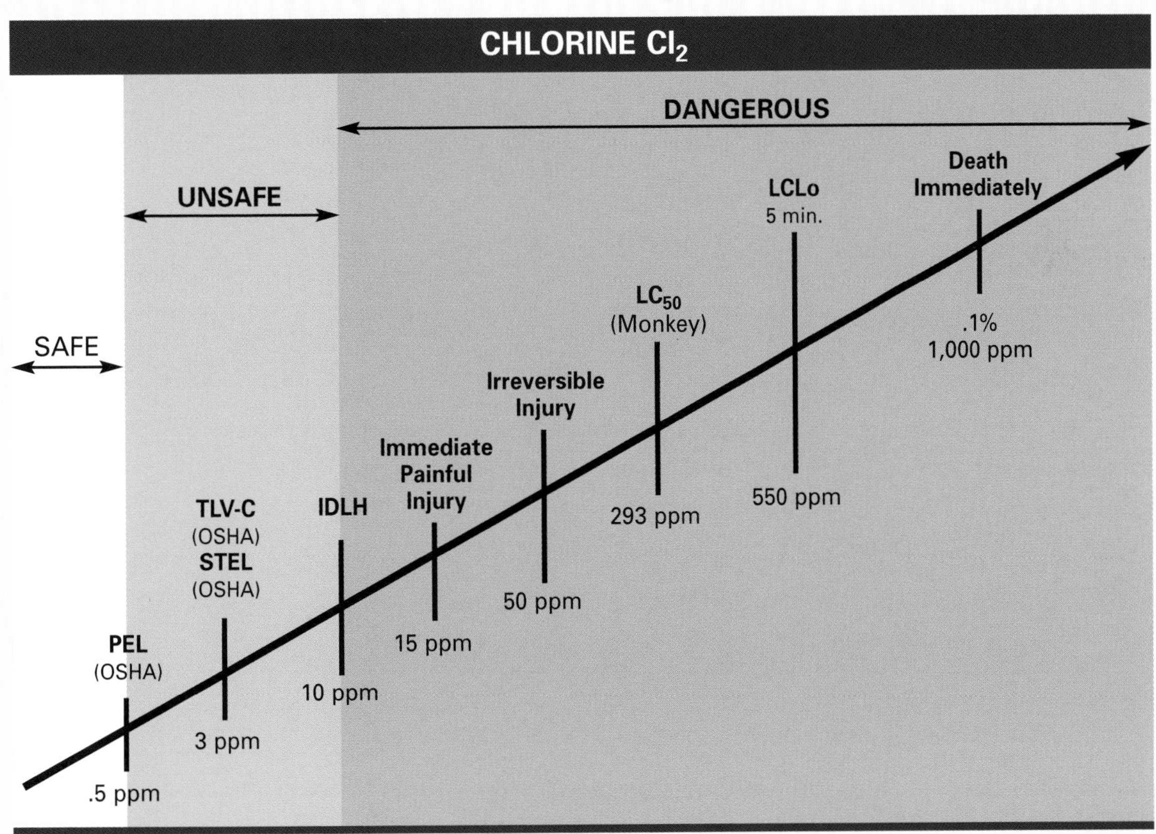

MICHAEL CALLAN

Poison Line—Concentration and physiological effects of chlorine.

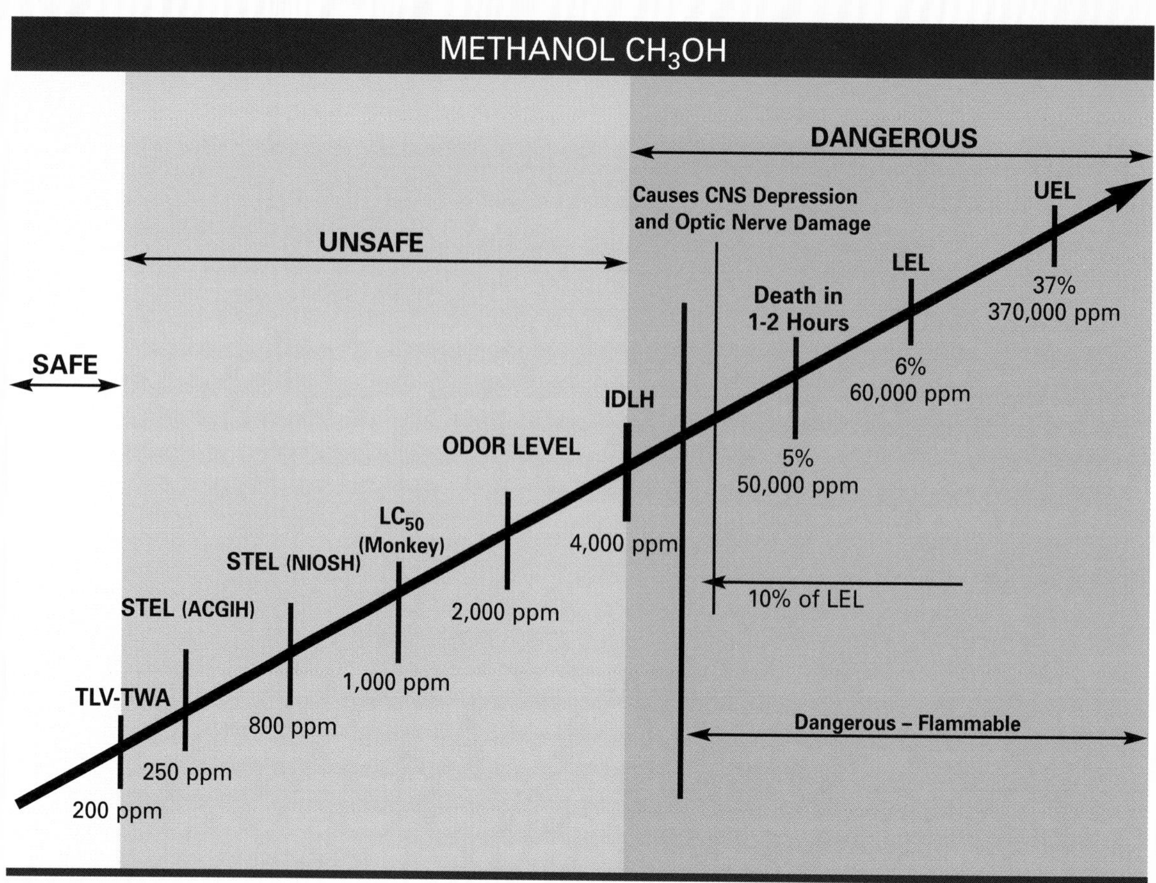

MICHAEL CALLAN

Poison Line—Concentration and physiological effects of Methanol.

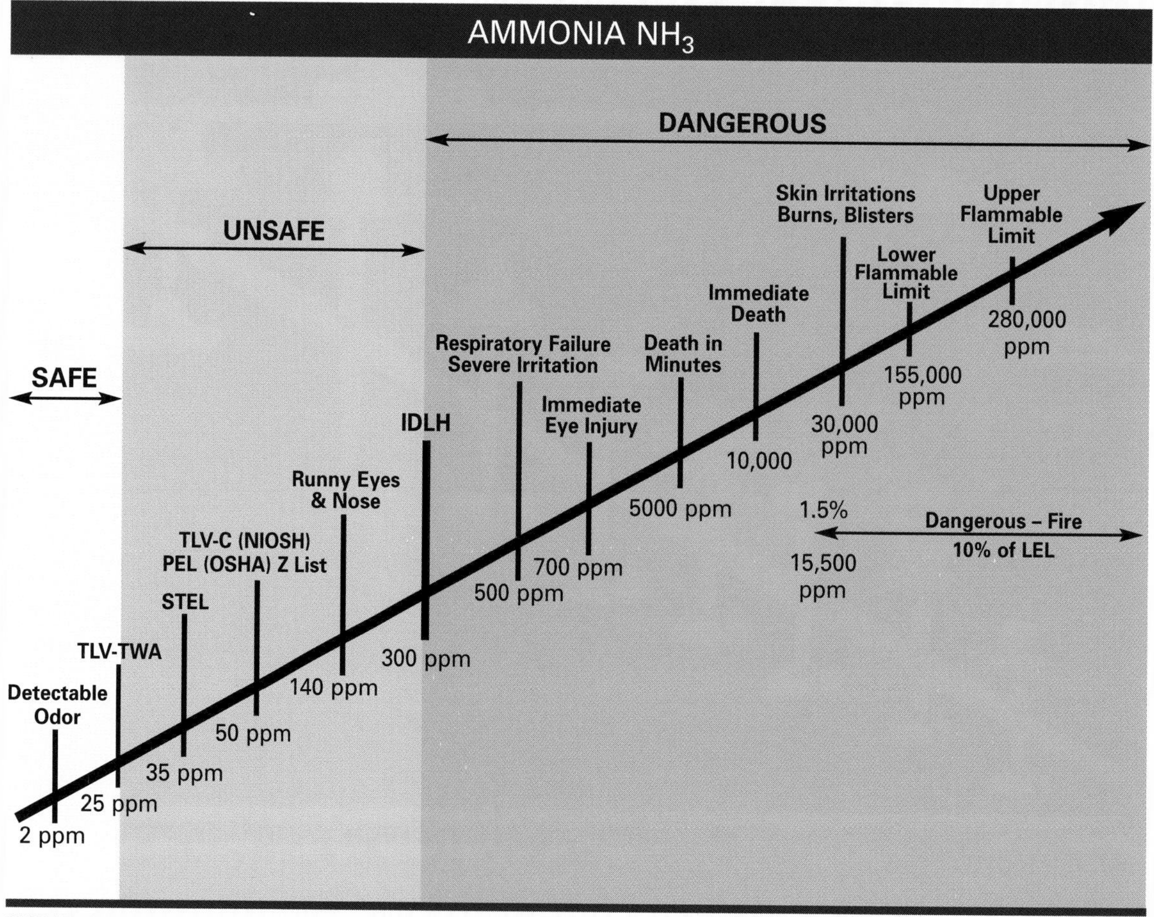

MICHAEL CALLAN

Poison Line—Concentration and physiological effects of Ammonia.

EMERGENCY CARE OF HEAT STRESS EMERGENCIES

HEAT CRAMPS

- Move patient to a nearby cool place, such as in shade or an air-conditioned ambulance.
- Administer high flow oxygen.

HEAT EXHAUSTION

- Move the patient to a nearby cool place, such as an air-conditioned ambulance.
- Remove enough clothing to cool the patient without chilling him or her (watch for shivering).
- Cool the individual through fanning. An electric-powered fan may be used to assist in cooling. Some response organizations also use precooled towels to assist with cooling.
- Place the individual in a supine position with the legs elevated. Keep at rest.
- If responsive and not nauseated, have the individual drink water or other fluids as specified by local medical protocols. If nauseous or vomiting, give nothing by mouth.
- If patient is unconscious, fails to recover rapidly, has other injuries, or has a history of medical problems, transport as soon as possible.
- Responders exhibiting the signs and symptoms of heat exhaustion should not return to service at the incident.

HEAT STROKE

- Cool the patient—in any manner—rapidly, move the patient out of the sun or away from the heat source. Use the maximum setting on the air conditioner in the patient compartment of the ambulance.
- Remove the patient's clothing, as appropriate. If cold packs or ice bags are available, apply them to the neck, groin, and armpits.
- Keep the skin wet by applying water with sponges or wet towels, or by wrapping the person in sheets soaked in cool water.
- Fan aggressively. Give the individual nothing by mouth.
- Administer high flow oxygen.
- Monitor vital signs and transport as soon as possible. Should transport be delayed, find a tub or container and immerse the responder up to the neck in cooled water.
- If not managed appropriately, heat stroke will likely result in the responder's death. Early recognition of the condition and rapid cooling are essential to survival.

SOURCES: Edward T. Dickinson and Michael A. Weider, Emergency Incident Rehabilitation, Upper Saddle River, NJ: Brady/Prentice Hall (2000) and Michael F. O'Keefe et al., Emergency Care (8th edition), Upper Saddle River, NJ: Brady/Prentice Hall (1998).

Figure 2.11 Emergency care of heat stress emergenices.

To minimize the effects of heat stress, EMS personnel should be on site to monitor and screen response personnel. Heat stress should be managed through a series of both administrative controls and through the use of PPE. Administrative controls, including the need for acclimatization or conditioning the body to working in hot environments, work/rest scheduling, rehab, and fluid replacement are outlined in Figure 2.13. PPE options are outlined below:

Figure 2.12 Heat stress should be managed through a series of controls.

- **Air-cooled jackets and suits**—Consist of small airlines attached to either vests, jackets, or CPC to provide convective cooling of the user by blowing cool air over the body inside of the suit. Cooling may be enhanced by the use of a vortex cooler or by refrigeration coils and a heat exchanger. Although sometimes found at remediation operations, they are typically not used for emergency response applications.

  These units require an airline and large quantities of breathing air (10 to 25 cubic feet per minute) and are not as effective as the active and passive cooling units in controlling body core temperatures.

- **Ice-cooled vests**—These consist of frozen ice or synthetic coolant packs that are part of a vest or jacket. This passive cooling system operates on the principle of conductive heat cooling. Although not as effective as the full-body cooling suit in controlling the body core temperature, studies have shown that ice vests are better than both the air cooled units and water cooled jackets. In addition to the physiological advantages, there may also be psychological benefits; however, people may say they "feel better" even though their body core temperature is actually increasing.

  Ice cooled units are relatively inexpensive and lightweight, improve worker comfort, decrease lens fogging, and are "user friendly" (that means responder-proof!). On the negative side, some vests require frozen coolant packs or an ice source at the scene of the emergency (unfortunately, the frozen coolant packs often leave the hazmat unit for someone's lunch box or cooler) and may add bulk underneath the CPC.

  These vests can also be used with heat packs for operations in extremely cold working environments.

MEDICAL SURVEILLANCE PROGRAM	
COMPONENT	**RECOMMENDATION**
PRE-EMPLOYMENT SCREENING	• Medical History • Occupational History • Physical Examination • Determination of fitness to wear protective equipment (e.g., respirator fit testing per OSHA 1910.134) • Baseline monitoring for specific exposures
PERIODIC MEDICAL EXAMINATION	• Annual update of medical and occupational history; annual physical examination; testing based on (1) examination results, (2) exposures, and (3) job class and task • More frequent testing may be required based upon specific exposures • Exams may be biannual based on a physician's recommendation.
EMERGENCY TREATMENT	• Provide medical care on site. • Develop liaison with local hospital and medical specialists. • Arrange for decontamination of victims. • Arrange in advance for the transport of victims. • Provide for the transfer of medical records; provide details of the incident and medical history to the next-care provider. • Provide for postincident surveillance of potentially exposed emergency responders and civilians.
NON-EMERGENCY TREATMENT	• Develop a mechanism for nonemergency health care.
RECORDKEEPING AND REVIEW	• Maintain and provide access to medical records in accordance with OSHA, state, and provincial records. • Report and record occupational injuries and illnesses. • Review medical surveillance program periodically. Focus on current site hazards, exposures, and industrial hygiene standards.

Figure 2.14 Medical surveillance is an essential element of a comprehensive employee health and safety program.

The success of any medical program depends on management support and employee involvement. Occupational health physicians and specialists, emergency medicine physicians, safety professionals, local or regional poison control center specialists, as well as advanced EMS personnel should be consulted for their expertise. Many hazmat units have a physician with a background or interest in hazardous materials who serves as the medical director for their unit.

Confidentiality of all medical information is paramount. Prospective employees and new hazmat response team members must be provided a complete, detailed occupational and medical history so a baseline profile can be established. Responders should be encouraged to document any suspected exposures, regardless of the degree, along with any unusual physical or physiological conditions. Training programs must emphasize that even minor complaints (e.g., headaches, skin irritations) may be important.

Additional information on medical surveillance programs can be referenced from the OSHA Web site at http://www.osha.gov.

PRE-EMPLOYMENT SCREENING

The objectives of pre-employment screening are to determine an individual's fitness for duty, including respirator and protective clothing use, and to provide baseline data for future medical comparisons. The screening should focus on the following areas:

- **Occupational and medical history**—This questionnaire should be completed with attention toward prior exposures to chemical and physical hazards. Also note previous illnesses, chronic diseases, hypersensitivity to specific substances, ability to use PPE, family history, and general lifestyle habits such as smoking and drug use.
- **Physical examination**—Complete a comprehensive physical examination focusing on the pulmonary, cardiovascular, and musculoskeletal systems. Additional tests that can help gauge the capacity to perform emergency response duties while wearing protective clothing include pulmonary function, electrocardiograms (EKG), hearing, and cardiac "stress tests."
- **Baseline laboratory profile**—This verifies the effectiveness of protective measures and determines whether the responder is adversely affected by previous exposures. The profile may include medical screening and biological monitoring tests based on potential exposures, such as liver, renal, and blood forming functions. Pre-employment blood and serum specimens may be frozen for later testing and comparison.

PERIODIC MEDICAL EXAMINATIONS

Periodic exams must be used in conjunction with pre-employment screening. Their comparison with the baseline physical may detect trends and early warning signs of adverse health effects. Under the OSHA 1910.120 requirements, such exams shall be administered annually, and no longer than every 2 years if the attending physician believes a longer interval is appropriate. In addition, more frequent intervals may be required depending on the nature of potential or actual exposures, type of chemicals involved, and the individual's medical and physical profile.

If an individual develops signs or symptoms indicating possible overexposure to hazardous substances or health hazards or has been injured or exposed to substances above accepted exposure values in an emergency, medical examinations and consultations shall be provided as necessary. Periodic screening exams can include medical history reviews that focus on health changes, illness and exposure-related symptoms, physical examinations, and specific tests such as pulmonary function, audiometric, blood, and urine.

To ensure the completion of a comprehensive medical profile, a medical exam is required to be given to all personnel when they are removed from active duty as a hazmat responder and at the termination of their employment or membership.

EMERGENCY TREATMENT

EMS personnel and units must be available at each hazmat incident. OSHA 1910.120 (q)(3)(vi) requires that "advanced first aid support personnel, as a minimum, shall stand-by with medical equipment and a transportation capability at hazmat emergencies." The level of emergency medical support may be influenced by the nature of the incident, risks involved, tasks to be performed, and the intensity and/or duration of the tasks. Informal interpretations by OSHA have indicated that the following factors will be considered in determining whether responders or a facility are in compliance with this requirement:

- Advanced first aid personnel are considered as individuals who have been trained to the Red Cross Advanced First Aid level or higher (e.g., First Responder, EMT-Basic, etc.) and are capable of providing the basic "ABCs" of medical care.
- Medical equipment is not required to be on scene but must be available for immediate response. As a general rule, medical treatment should be provided within 3 to 4 minutes of the incident, while a transportation capability should be on-site in approximately 15 to 20 minutes.

Although most public safety response agencies will have an EMS unit on scene as part of the response, that is not always the case with facility Emergency Response Teams (ERT) who typically deal with small spills of hazardous materials commonly found or used within the facility.

Figure 2.15 EMS personnel must be on scene at hazmat emergencies.

When possible, an EMS responder with a background in hazmat operations should be in charge of the EMS operations and coordinate closely with the Hazmat Group Supervisor. At "working" incidents the Incident Commander may establish a Medical Group to coordinate all EMS activities. Specific responsibilities of EMS or Medical Group personnel include the following:

- Provide technical assistance to responders in the development and analysis of EMS-related data and information. This shall include signs and symptoms of exposure, medical treatment procedures, antidote information, patient handling guidelines, transportation recommendations, and medical resource requirements.

- Designate a treatment and triage area in proximity to the decontamination area.
- Perform pre-entry and post-entry medical monitoring of all entry and backup team personnel, as appropriate.
- Coordinate and supervise all patient handling activities, including decontamination, treatment, handling and transportation of contaminated victims. This should include recommendations for the protection of all EMS personnel.
- Communicate and coordinate with local hospitals and specialized treatment facilities, including the poison control center, as necessary.

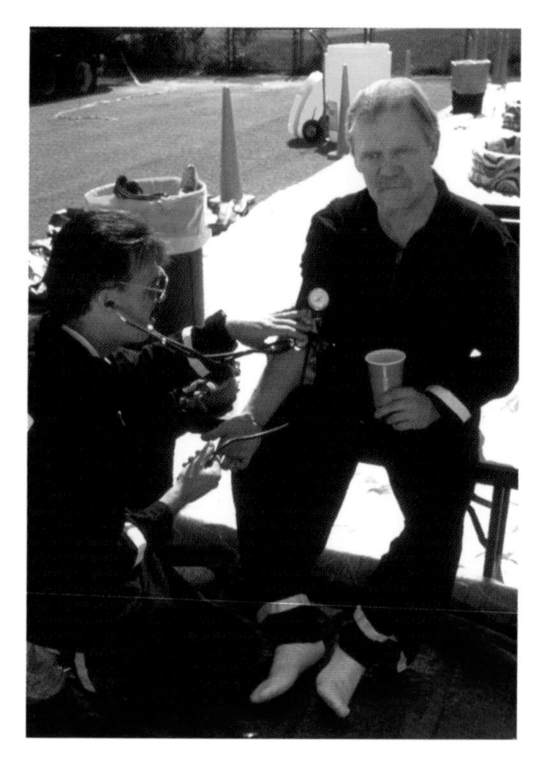

Figure 2.16 Pre-entry and post-entry medical monitoring should be performed on personnel.

Some regions and industrial facilities provide either advanced life support (ALS) units or individuals specially trained and equipped for hazardous materials emergencies. They often carry drugs and antidotes for chemicals commonly found within the plant or community, along with medical information and baseline profiles of all hazmat response team personnel. Hazmat training competencies for EMS personnel can be found in NFPA 473, *Competencies for EMS Personnel Responding to Hazardous Materials Incidents.* Another good source of training and information is Advanced Hazmat Life Support Course offered by the University of Arizona Health Services Center (see www.ahls.org for additional information).

Standard Operating Procedures (SOPs) for the clinical management and transportation of chemically contaminated patients must be developed as part of the planning process. In addition, the specific roles, responsibilities, and capabilities of hazmat personnel, EMS personnel, and local medical facilities need to be determined.

The handling of chemically contaminated patients will be addressed in Chapter 11—Decontamination. However, remember these basic principles:

- Always ensure that EMS personnel are properly protected—both skin and respiratory.
- When dealing with victims in a contaminated environment, determine whether it is a rescue operation or a body recovery operation.
- Although certain situations may exist where decontamination may aggravate patient care or further delay priority treatment, as a rule of thumb all patients should receive gross decontamination.
- The ABCs can be administered to a contaminated victim if rescuers and EMS personnel are protected. It's a much better option than having a fully decontaminated but dead patient.
- Always coordinate with your local medical facilities.

Under OSHA 1910.120, paragraph b, components of a site safety plan should include site map or sketch, hazard and risk analysis of the identified hazardous materials, site monitoring, establishment of control zones, site safety practices and procedures, communications, implementation of an incident management organization and the location of the incident command post, decontamination practices, EMS support, and other relevant topics. Standard site safety practices and procedures are outlined in Figure 2-17.

STANDARD SITE SAFETY PRACTICES

- Minimize the number of personnel operating in the contaminated area.
- Avoid contact with all contaminants, contaminated surfaces, or suspected contaminated surfaces. Avoid walking through any suspected releases or placing equipment on contaminated surfaces.
- Advise all entry personnel of all site control policies including entry and egress points, decon layout and procedures, and working times.
- Always have an escape route. Ensure that everyone knows the emergency evacuation signals.
- Ensure that all tasks and responsibilities are identified before attempting entry. If necessary, practice unfamiliar operations prior to entry.
- Use the buddy system for all entry operations. Always ensure that properly staffed and equipped back-up crews are in place.
- Maintain radio communications between entry, backup crews, and the Safety Officer (whenever possible).
- Prohibit drinking, smoking, and any other practices that increase the possibility of hand-to-mouth transfer in all contaminated areas.
- Follow decontamination and personal cleanliness practices before eating, drinking, or smoking after leaving the contaminated area.

Figure 2.17 Standard site safety practices for operations in a contaminated environment.

SAFETY OFFICER AND SAFETY RESPONSIBILITIES

Under the OSHA HAZWOPER regulation, the safety function must be addressed at every incident in which hazardous materials are involved. At small, Level 1 incident scenarios, the safety function can be easily managed by the Incident Commander. However, as the scope and complexity of the hazmat problem increase, it will be necessary to designate an Incident Safety Officer.

At incidents where an HMRT is operating, safety responsibilities will often be divided into two areas—first, the safety of all units operating on the incident scene and under the control of the Incident Safety Officer; and second, the safety of those operating within the ICS Hazmat Group and under the control of the Hazmat Group Safety Officer.

Although the Hazmat Group Safety Officer is subordinate to the Incident Safety Officer, he or she has certain responsibilities within the Hazmat Group that may circumvent the normal chain of command. In either case, both Safety Officers must have the authority to stop any operations that are deemed unsafe.

Among the primary responsibilities of the Safety Officer are the following:

OVERALL SITE SAFETY

- Ensure that the Safety Officer is identified to all personnel. The IC should advise all operating personnel, as appropriate. The use of command vests for identification is recommended.
- Ensure that all personnel and equipment are positioned in a safe location. Remember the basics—upwind, uphill—and always have an escape route and predesignated withdrawal signal. Consider having vehicles and apparatus back into the incident scene for a quick exit.
- Ensure that hazard control zones are established, identified, and constantly monitored and that their locations are communicated to all personnel. Consider the location of the Incident Command Post, Hazmat Group, and Staging Area in relation to the hazard control zones and the potential worsening of the emergency.
- When necessary, designate a security officer to maintain overall site security. Delegate to plant security or law enforcement whenever possible.
- Ensure that all personnel in controlled areas are in the proper level of personal protective clothing.
- Use a Personal Accountability System to track all responders throughout the incident. Conduct regular personal accountability reviews (PAR).

ENTRY OPERATIONS

- Coordinate with the Medical Officer to ensure that pre-entry medical monitoring has been conducted.
- Hold a pre-entry safety briefing prior to recon or entry operations. This may be provided by either the Hazmat Group Safety Officer or other Hazmat Group personnel (e.g., Entry). All entry and backup personnel must be familiar with the objectives, tasks, and procedures to be followed. Topics should include objectives of the entry operation, a review of all assignments, verification of radio procedures (designated channels) and emergency signals (both hand signals and audible), emergency escape plans and procedures, protective clothing requirements, immediate signs and symptoms of exposure, and the location and layout of the decon area.
- Coordinate entry operations with backup crews and the Decon Unit. The Entry Team should be permitted to enter the hot zone only when backup crews are in place and the decon area is prepared.
- Monitor entry operations and advise entry personnel and the IC of any unsafe practices or conditions.
- During the termination phase, advise all personnel of the possible signs and symptoms of exposure and ensure that health exposure forms are documented.

EMERGENCY RESCUE CAPABILITIES AT INTERIOR FIREFIGHTING OPERATIONS "TWO-IN / TWO-OUT"

An emergency rescue capability must be provided when operations are conducted in an IDLH atmosphere. In permit-required confined space entry operations where the IDLH atmosphere has been characterized and controlled, this requirement is usually satisfied by having one or more stand-by persons who are trained and equipped to provide an effective emergency rescue (see *OSHA 1910.146—Permit-Required Confined Space Entry Operations*). Similarly, emergency rescue requirements for hazmat incidents are outlined in OSHA 1910.120(q).

The release of the *OSHA Respiratory Protection Standard* (1910.134) brought the question of emergency rescue capabilities to interior firefighting operations. Commonly referred to as the "two-in / two-out rule," it requires that a backup and rescue capability be provided for interior structural fires. *OSHA Compliance Directive 2-0.120—Inspection Procedures for the Respiratory Protection Standard*, makes the following notes:

- It is the Incident Commander's responsibility, based on training and experience, to judge whether a fire is an interior structural fire, and how it will be attacked.
- There must always be at least two firefighters stationed outside during interior structural firefighting, and they must be trained, equipped, and prepared to enter, if necessary, to rescue the firefighters inside. However, the IC has the responsibility and flexibility to determine when more than two outside firefighters are necessary given the circumstances of the fire. The two-in / two-out rule does not require an arithmetic progression for every firefighter inside (i.e., the Standard should not be interpreted as 4-in / 4-out, 8-in / 8-out, etc.).
- Life-saving activities in interior structural fire fighting are not precluded by the OSHA standard. There is an explicit exemption in the standard that if life is in jeopardy, firefighters have the discretion to perform the rescue, and the "two-in / two-out" requirement is waived. There is no violation of the standard under such life-saving rescue circumstances.
- The two-in / two-out provision is not intended as a staffing requirement. The two-in / two-out rule is a worker safety practice requirement, not a staffing requirement.
- The standard allows one of the standby firefighters to have other duties such as serving as the IC, Safety Officer, or fire apparatus operator. However, one of the outside firefighters must actively monitor the status of the inside firefighters and may not be assigned additional duties. The second outside firefighter may be involved in a wide variety of activities. Both of the outside personnel must be able to provide support and assistance to the two interior firefighters; any assignment of additional duties for one of the outside firefighters must be weighed against the potential for interference with this requirement.
- The two firefighters (buddies) entering an IDLH atmosphere to perform interior structural firefighting must maintain visual or voice communication at all times. Electronic

methods of communication such as the use of radios shall not be substituted for direct visual contact between the team members in the danger area. However, reliable electronic communication devices are not prohibited and certainly have value in augmenting communication and may be used to communicate between inside team members and outside standby personnel.

For further explanation, refer to the preamble of the *OSHA Respiratory Protection Standard and the Respirator Question and Answer* Document (August 3, 1998). Both documents can be found at the OSHA Web site at http://www.osha.gov.

THE PHOENIX, ARIZONA EXPERIENCE

As a result of a firefighter fatality during a 2001 supermarket fire, the Phoenix, Arizona Fire Department conducted 200 rapid intervention drills that evaluated the department's ability to remove two firefighters in trouble in a similar occupancy. The results show that rapid intervention may not be rapid and revealed three consistent ratios: (1) It takes 12 firefighters to rescue one; (2) 1 in 5 rescuers will get into trouble themselves; and (3) a 3000-psi air cylinder has approximately 18.7 minutes of air (+ or –30%).

How do these experiences relate to hazmat emergency response? As a result of the rapid intervention drills, the Phoenix Fire Department HMRT conducted informal drills on removing a downed hazmat responder from a contaminated environment. The results showed the following:

- For hazmat incidents outdoors in an open-air environment, the two-in/two-out back-up procedure is still effective.
- For hazmat incidents inside a structure, a two-in/four-out back-up team procedure is required.
- Incidents that require the backup team to remove downed entry personnel via a stairwell will significantly complicate the timing and effectiveness of any rescue operation. Operations may be further complicated if entry personnel are wearing chemical vapor (EPA Level A) protective clothing.

PRE- AND POST-ENTRY MEDICAL MONITORING

Medical monitoring may be defined as an ongoing, systematic evaluation of individuals at risk of suffering adverse effects of exposure to heat, stress, or hazardous materials as a result of working at a hazardous materials emergency. The objectives of medical monitoring are (1) to obtain baseline vital signs; (2) to identify and preclude from participation individuals who are at increased risk to sustain either injury or illness; and (3) to facilitate the early recognition and treatment of personnel with adverse physiological and/or emotional responses.

Pre- and post-entry medical monitoring is normally the responsibility of the Hazmat Medical Group. Medical monitoring provides baseline vital signs of all entry personnel, and identifies, evaluates, and eliminates those individuals who are at increased risk from the effects of heat stress or hazmat exposure. However, be practical—scenarios requiring rapid entry to facilitate rescue should not be delayed to conduct a pre-entry examination! Some career fire department HMRTs conduct pre-entry medical monitoring at the start of each shift to facilitate the rapid response concept.

Components of the pre-entry exam should include the following:

- Vital signs, including blood pressure, pulse, respiratory rate, temperature, and body weight. If available, a 10-second EKG rhythm strip or a 12-lead EKG may also be taken. Preferred by many physicians, the 12-lead EKG can be interpreted by a paramedic and compared to baseline traces to indicate possible cardiac abnormalities.
- Skin evaluation, with an emphasis on rashes, lesions, and open sores or wounds.
- Lung sounds, including wheezing, unequal breath sounds, and so on.
- Mental status (alert and oriented to time, location, and person).
- Recent medical history, including medications, alcohol consumption, any new medical treatment or diagnosis within the last 2 weeks, and symptoms of fever, nausea, diarrhea, vomiting, or coughing within the past 72 hours.
- Prehydration with 8 to 16 ounces of water. Electrolyte fluids may also be used.

Criteria should be established for evaluating responders prior to entry operations. These criteria should be reviewed by the HMRT Medical Director or an occupational health physician/specialist who is familiar with the duties and tasks of hazmat responders. **The following exclusion criteria are used by a number of Hazmat Medical Groups; however, they should not supersede any existing criteria established by the local medical control.**

Entry shall be denied if the following criteria are not satisfied:

- Blood pressure—BP exceeds 100 mm Hg diastolic.
- Pulse—Greater than 70% maximum heart rate (>115) or irregular rhythm not previously known.
- Respirations—Respiratory rate is greater than 24 per minute.
- Temperature—Oral temperature less than 97°F or exceeds 99.5°F. Core temperature less than 98°F or greater than 100.5°F.
- Body Weight—No pre-entry exclusion.
- EKG—Dysrhythmias not previously detected must be cleared by medical control.

- Mental Status—Altered mental status (e.g., slurred speech, clumsiness, weakness).
- Other criteria, including:

  Skin—Open sores, large skin rashes, or significant sunburn.

  Lungs—Wheezing or congested lung sounds.

  Medical history—Recent onset of heart or lung problems, hypertension, diabetes, etc. Experienced nausea and vomiting, diarrhea, fever, or heat exhaustion within the last 72 hours. Use of prescription medication and over-the-counter medicines (e.g., decongestants, antihistamines, etc.) must be cleared through local medical control. Heavy alcohol consumption within the previous 24 hours or any alcohol within the past 2 hours.

Post-entry medical monitoring is performed following decontamination to determine if the responder has suffered any immediate effects from heat stress or a chemical exposure, and to determine the individual's health status for future assignment during or after the incident. Components of the post-entry exam should include the following:

- Any signs or symptoms of chemical exposures, heat stress, or cardiovascular collapse.
- Vital signs, including blood pressure, pulse, respiratory rate, temperature, and body weight. If available, a 10-second EKG rhythm strip may also be taken.
- Skin evaluation, with an emphasis on rashes, lesions, and open sores or wounds.
- Lung sounds, including wheezing, unequal breath sounds, and so on.
- Mental status (alert and oriented to time, location, and person). One cognitive test often used in the field is to have the individual spell any five-letter word backward.
- Body weight.
- Hydration—Provide plenty of liquids. Replace body fluids (water and electrolytes); use a 0.1% salt solution or commercial electrolyte mixes.

Vital signs should be monitored every 5 to 10 minutes, with the person resting, until they return to approximately 10% of the baseline. If vital signs do not return to normal, it may be necessary to transport the individual to a medical facility. Medical control should be consulted for direction and recommendations, as necessary.

EMERGENCY INCIDENT REHABILITATION

The IC should consider the circumstances of the incident and make adequate provisions early in the incident for the rest and rehabilitation of all personnel operating at the scene. This is particularly critical for "campaign" emergencies that extend over a period of hours. The Incident Commander should establish a Rehabilitation Sector or group to coordinate rest and rehab activities. At most hazmat incidents, rehabilitation will be the responsibility of the Hazmat Medical Group.

In addition to coordinating for EMS support, treatment, and monitoring, the Rehab Sector is responsible for providing food and fluid replenishment, mental rest, and relief from the extreme environmental conditions associated with the incident.

CHAPTER 3

MANAGING THE INCIDENT: INCIDENT COMMAND SYSTEM

OBJECTIVES

1. List the categories of players and participants at a hazardous materials incident.
2. Identify the key organizational elements of the Incident Command System.
3. Describe the concept of unified command and its application and use at a hazardous materials incident.
4. Identify the duties and responsibilities of the following Hazmat Group functions within the Incident Command System: [NFPA 472-6.4.1.2].
   - Backup
   - Decontamination
   - Entry
   - Hazmat group management
   - Hazmat group safety
   - Information/research
   - Reconnaissance
   - Resources
5. Identify the key elements of the Incident Command System necessary to coordinate response activities at a hazardous materials incident.

ABBREVIATIONS AND ACRONYMS

CHEMTREC	Chemical Transportation Emergency Response Center
CRM	Crew Resource Management
DECON	Decontamination
DEQ	Department of Environmental Quality
DHS	Department of Homeland Security
EOC	Emergency Operations Center
EOD	Explosive Ordinance Disposal
ERP	Emergency Response Plan
FAA	Federal Aviation Administration
IAP	Incident Action Plan
ICP	Incident Command Post
ICS	Incident Command System
IMT	Incident Management Team
IST	Incident Support Team
LEPC	Local Emergency Planning Committee
NIMS	National Incident Management System
NOAA	National Oceanic and Aeronautical Administration
NOP	Next Operational Period
NRP	National Response Plan
OIM	Off-Shore Installation Manager
PIO	Public Information Officer
RIT	Rapid Intervention Team
SWAT	Special Weapons and Tactics
ROE	Rules of Engagement
UC	Unified Commanders
USAR	Urban Search and Rescue
WMD	Weapons of Mass Destruction

INTRODUCTION

Direct, effective command and control operations are essential at every type of incident. However, hazardous materials incidents place a special burden on the command system since they often involve communications among separate agencies and the coordination of many different functions and personnel assignments—from public protective actions to the use of specialized personal protective equipment (PPE).

This chapter reviews the fundamental concepts of incident management and its application at a hazmat incident. Primary topics include the various players who characteristically appear at a hazmat incident, the elements of the Incident Command System (ICS), the functions and responsibilities of the Hazmat Group, and "street smart" tips.

MANAGING THE INCIDENT: THE PLAYERS

A hazmat incident often attracts an interesting collection of participants. Some can be found at the scene of any major emergency, while others are unique to hazmat events. Each of these participants also brings his or her own agendas, organizational structures, and priorities to the incident scene. The key to success is to have a coordinated incident command structure in which all of the players integrate their resources to make the problem go away in a safe and effective manner. In simple terms, you need an ICS structure.

The basic ICS organization that must be created to bridge these potential gaps and problems includes the following:

The Incident Commander (Command or IC)—The individual responsible for establishing and managing the overall incident action plan (IAP). This process includes developing an effective organizational structure, developing an incident strategy and tactical action plan, allocating resources, making appropriate assignments, managing information, and continually attempting to achieve the basic command goals. Everyone working at the event reports through the chain of command to this individual. It doesn't matter *who* is in charge; the most important concept is that regardless of one's normal position within the organization, the person serving as IC is the highest authority on the scene.

Unified Commanders (UCs)—Command-level representatives from each of the primary responding agencies who present their agency's interests as a member of a unified command organization. Depending on the scenario and incident timeline, they may be the lead IC or play a supporting role within the command function. The unified commanders manage their own agency's actions and make sure all efforts are coordinated through the unified command process.

ICS General Staff—ICS provides a mechanism to divide and delegate tasks and develop a management structure to handle the overall control of the incident. Section Chiefs are members of the IC's general staff and are responsible for the broad response functions of Operations, Planning, Logistics, and Finance/ Administration. Individuals below the section level are the front-line supervisors who implement tactical objectives to meet the strategies established by the IC within a branch, group or division (e.g., Hazmat Group Supervisor).

ICS Command Staff—Those individuals appointed by and directly reporting to the IC. These include the Safety Officer, the Liaison Officer, and the Public Information Officer (PIO).

THE PLAYERS

Regardless of who they are and how they materialize on the scene, the IC must also be able to quickly identify and categorize the various players and participants who will interact within the ICS organization. These include the following:

Fire/Rescue/EMS Companies—Provide resources for fire suppression, rescue, and medical triage, treatment and transport. They implement assigned tasks, provide support to specialized assets, and help to coordinate overall response efforts. Examples include firefighters and fire officers, EMS personnel, and other knowledgeable responders on scene.

Police Officers and Law Enforcement Personnel—Resources for ensuring scene safety (i.e., scene and traffic control), perpetrator arrest or control, evidence preservation, etc. They implement assigned tasks, provide support to specialized assets, and help to coordinate overall response efforts. Examples include police and security personnel in both government and industry who provide fundamental law enforcement services.

Figure 3.1 Hazmat incidents bring many different players to the scene.

Emergency Response Team (ERT)—Crews of specially trained personnel used within business and industrial facilities for the control and mitigation of emergency situations. They may consist of full-time personnel, shift personnel with ERT responsibilities as part of their job assignment (e.g., plant operators), or volunteer members. ERTs may be responsible for any combination of fire, hazmat, medical, and technical rescue emergencies, depending on the nature, size, and operation of their facility.

Hazardous Materials Response Teams (HMRTs)—Crews of specially trained and medically evaluated individuals responsible for directly managing and controlling hazmat problems. They may include people from the emergency services, private industry, governmental agencies, environmental contractors, or any combination. They generally perform more complex and technical hazmat response functions than fire, rescue, and EMS companies.

Special Operations Teams—Highly trained and equipped response teams who deliver a highly specialized response service and capability. Examples include urban search and rescue (USAR) teams, bomb squads and explosive ordinance disposal (EOD) units, water rescue, and law enforcement tactical units (e.g., SWAT—Special Weapons and Tactics).

Communications Personnel—The central communications function for emergency services. They receive "911" calls for assistance and dispatch appropriate units to the incident locations. They provide a crucial link for those working on scene by dispatching additional needed resources, including product and container specialists, and providing response data and information to field units. Some large industrial complexes may have a central communications center that monitors plants operations and alarms and would forward any facility emergency alarms to the local public safety communications center.

Responsible Party—Organization legally responsible under government law for the clean-up of a hazmat release. Depending on the scenario, this may be a transportation carrier or a facility owner.

Facility Managers—Individuals who normally do not have an on-scene emergency response function, but who are key players within the plant environment. In the event of a plant emergency, they report to the plant Emergency Operations Center (EOC) and are responsible for providing overall plant command and logistical support to field emergency response units and for coordinating external issues, including community liaison, media relations, and agency notifications.

Support Personnel—Individuals who provide important support services at the incident. Water and utility company employees, heavy equipment operators, and food service/rehab personnel are some examples.

Technical Information Specialists—Individuals who provide specific expertise to the IC either in person, by telephone, or through other means. They usually are product and/or container specialists representing the manufacturer or shipper or are familiar with the chemicals, containers or problems involved. CHEMTREC™ and the Local Emergency Planning Committee (LEPC) often provide an excellent way to reach information resources in any given area of expertise.

Environmental Clean-up Contractors—Individuals who may provide both mitigation and support services at the incident. Capabilities may include spill control, product transfer operations, site clean-up and recovery, and remediation operations. They are usually retained by the responsible party, the IC, or government environmental agencies (e.g., EPA). Personnel should be trained to meet the training requirements of OSHA 1910.120, paragraphs (b) through (o).

Government Officials—Individuals who normally do not have an emergency response function but who bring a lot of political clout to the incident. Examples include mayors, city/county managers, or other elected officials who may be involved. For large-scale events they may play a command role, or they may delegate this responsibility to an emergency manager. Failure to professionally address their questions and concerns within the ICS organization can have significant political and other impacts both during and after the response.

News Media—Individuals representing various elements of the media who work to inform the public of major happenings within their community or region. Because of the unusual and often frightening nature of hazardous materials incidents, it is very important that the public be accurately informed quickly and regularly of the incident. Television and radio are excellent methods with which to coordinate and manage large-scale public protective actions activities. On-scene media may also be willing to provide quality videotape equipment and helicopter shots for emergency service use in exchange for coverage.

Investigators—Individuals who are responsible for determining the origin and cause of the hazmat release, including any related evidence collection and preservation. A hazmat incident is not really concluded until the investigation is complete. Future legal proceedings, possible regulatory citations or criminal charges, and financial reimbursement for the time, equipment, and supplies of emergency services may well depend on investigators' efforts. Certain types of incidents require interaction between investigators on the federal, state, and local levels, as well as in the private sector.

Victims—Individuals who may be exposed, contaminated, injured, killed, or displaced as a result of the hazardous materials incident. Once emergency responders begin to provide treatment and care, they themselves may become patients. Special care should be given to their welfare and to informing them of potential short- and long-term signs and symptoms of exposure.

Spectators—Curious, usually well-meaning members of the facility and/or general public who arrive at the scene to assist or watch the adventure. Since they are often difficult to control—especially during campaign incidents—spectators need to be monitored and managed constantly to ensure their safety.

The Bad Guy—Also referred to as suspect, perpetrator, criminal, terrorist or whatever else the local lingo is. A potentially harmful person (or group of persons) who is intent on bringing harm, destruction, injury, or death to both the general public and/or public safety professionals. There may be one or many Bad Guys at the scene, and they may conceal their identity and blend in with victims and spectators after an attack.

The Hazardous Material—A potentially harmful substance or material that has escaped or threatens to escape from its container. It should be considered an active, mobile opponent that must be carefully monitored at all times. Whenever its container is stressed or it has already escaped, the hazardous material should be considered a threat to the other players.

MANAGING THE INCIDENT: THE INCIDENT COMMAND SYSTEM

Why so much interest in incident command? One primary reason is the OSHA 1910.120(q) requirement that both public safety and industrial emergency response organizations use a "nationally recognized Incident Command System for emergencies involving hazardous materials." But beyond regulatory requirements, experience has shown that the normal, day-to-day business organization is not well suited to meeting the broad demands created by "working" hazmat incidents.

The information presented in this chapter is based on the National Incident Management System (NIMS). NIMS is a baseline incident management organization that is utilized by federal, state, and local governments, as well as many private sector organizations throughout North America.

The Incident Command System is an organized system of roles, responsibilities, and procedures for the command and control of emergency operations. It is a procedure-driven system based upon the same business and organizational management principles that govern organizations on a daily basis. As is the case with the day-today management of any organization, ICS has both technical and political aspects that must be understood by the key players.

In the past, these players were concerned primarily with the technical or operational aspects of an emergency. Today, however, the playing field has changed significantly. "Working" hazmat incidents have very real political, legal, and financial effects on how both the public and the corporate shareholders view emergency response professionals. Recent incidents have taught us that an emergency can have a favorable technical or operational outcome and still be a political disaster. These political issues will be reviewed later in this chapter.

Homeland Security Presidential Directive / HSPD-5 Management of Domestic Incidents and a National Incident Management System

In February 2003, President George W. Bush signed Homeland Security Presidential Directive 5 pertaining to the management of domestic incidents. The purpose of HSPD-5 is to enhance the ability of the United States to manage domestic incidents by establishing a single, comprehensive national incident management system.

Policy

The following is a summary of the key points of HSPD-5 as it pertains to the development of a National Incident Management System (NIMS).

- To prevent, prepare for, respond to, and recover from terrorist attacks, major disasters, and other emergencies, the United States government shall establish a single, comprehensive approach to domestic incident management. The objective of the United States government is to ensure that all levels of government across the nation have the capability to work efficiently and effectively together, using a national approach to domestic incident management. In these efforts, with regard to domestic incidents, the United States government treats crisis management and consequence management as a single, integrated function rather than as two separate functions.
- The Secretary of Homeland Security is the principal federal official for domestic incident management. Pursuant to the Homeland Security Act of 2002, the Secretary is responsible for coordinating federal operations within the United States to prepare for, respond to, and recover from terrorist attacks, major disasters, and other emergencies. The Secretary shall coordinate the federal government's resources utilized in response to or recovery from terrorist attacks, major disasters, or other emergencies if and when any one of the following four conditions applies:
  1. A federal department or agency acting under its own authority has requested the assistance of the Secretary;
  2. The resources of state and local authorities are overwhelmed and federal assistance has been requested by the appropriate State and local authorities;

3. More than one federal department or agency has become substantially involved in responding to the incident; or
4. The Secretary has been directed to assume responsibility for managing the domestic incident by the President.

- Nothing in this directive alters, or impedes the ability to carry out, the authorities of federal departments and agencies to perform their responsibilities under law. All federal departments and agencies shall cooperate with the Secretary in the Secretary's domestic incident management role.
- The federal government recognizes the roles and responsibilities of State and local authorities in domestic incident management. Initial responsibility for managing domestic incidents generally falls on state and local authorities. The federal government will assist state and local authorities when their resources are overwhelmed, or when federal interests are involved. The Secretary will coordinate with State and local governments to ensure adequate planning, equipment, training, and exercise activities. The Secretary will also provide assistance to state and local governments to develop all-hazards plans and capabilities, including those of greatest importance to the security of the United States, and will ensure that state, local, and federal plans are compatible.
- The federal government recognizes the role that the private and nongovernmental sectors play in preventing, preparing for, responding to, and recovering from terrorist attacks, major disasters, and other emergencies. The Secretary will coordinate with the private and nongovernmental sectors to ensure adequate planning, equipment, training, and exercise activities and to promote partnerships to address incident management capabilities.

The National Response Plan (NRP)

- The Secretary shall develop, submit for review to the Homeland Security Council, and administer a National Incident Management System (NIMS). This system will provide a consistent nationwide approach for federal, state, and local governments to work effectively and efficiently together to prepare for, respond to, and recover from domestic incidents, regardless of cause, size, or complexity. To provide for interoperability and compatibility among Federal, State, and local capabilities, the NIMS will include a core set of concepts, principles, terminology, and technologies covering the incident command system; multiagency coordination systems; unified command; training; identification and management of resources (including systems for classifying types of resources); qualifications and certification; and the collection, tracking, and reporting of incident information and incident resources.
- The heads of federal departments and agencies shall adopt the NIMS within their departments and agencies and shall provide support and assistance to the Secretary in the development and maintenance of the NIMS. All federal departments and agencies will use the NIMS in their domestic incident management and emergency prevention, preparedness, response, recovery, and mitigation activities, as well as those actions taken in support of State or local entities.

- The NRP is based on the following planning assumptions and considerations:
  1. *A single "all hazards/all disciplines" plan.* The Secretary shall develop, submit for review to the Homeland Security Council, and administer a National Response Plan. This plan shall integrate federal government domestic prevention, preparedness, response, and recovery plans into one all-discipline, all-hazards plan for conducting activities throughout the "life cycle" of an incident. The NRP shall be unclassified. If certain operational aspects require classification, they shall be included in classified annexes to the NRP.
  2. *A plan that emphasizes unity of effort among all levels of government.* The NRP is a national plan that emphasizes unity of effort among all levels of government. Under the plan, federal, state and local governments, along with private organizations and the American public, work as partners to manage domestic contingencies efficiently and effectively.
  3. *A plan that integrates crisis and consequence management.* In contrast to previous approaches, the NRP treats crisis management and consequence management as a single, integrated function rather than two separate functions.
  4. *A plan that places the same emphasis on awareness, prevention, and preparedness as traditionally has been placed on response and recovery.* The incident life cycle contains five basic domains within which domestic incident management activities occur—awareness, prevention, preparedness, response, and recovery.
- Additional information and resources on NIMS can be referenced at http://www.fema.gov/nims/.

Figure 3.2 Homeland Security Presidential Directive 5 (HSPD-5) pertains to the development of a national response plan and a national incident management system to provide a consistent approach to incident management at all levels of government.

INCIDENT MANAGEMENT VERSUS CRISIS MANAGEMENT

Crisis management is an integral element of most corporate and industrial organizations. Experience has shown that there is a direct relationship between incident management and crisis management concerns as well as the organizational structure for managing each event.

What is the difference between an incident and a crisis? In the broadest sense, *an incident can be defined as an occurrence or event, either natural or human made, which requires action by emergency response personnel to prevent or minimize loss of life or damage to property and/or natural resources.* Essentially, an incident interrupts normal procedures, has limited and definable characteristics, and has the potential to precipitate a crisis. Examples of incidents may include fires, hazmat, medical, and rescue emergencies. In short, an incident does not necessarily mean that an organization or a community has a crisis.

The definition of a crisis will vary significantly depending on the type or organization (e.g., public vs. private sector) and situations anticipated. A good starting point is the following definition: *A crisis is an unplanned event that can exceed the level of available resources and has the potential to impact significantly an organization's operability, credibility, and reputation, or pose a significant environmental, economic, or legal liability.*

The transition from incident management to crisis management may not be easy. The scope or severity of an incident is not the sole factor that determines its potential to develop into a crisis. Other factors may come into play, including the occurrence of recent incidents in the area, the occurrence of similar incidents nationally or within the region, and the political environment in which the incident occurs.

Crisis management builds on the philosophy of incident management. If ICS is not utilized on a regular basis for the "more routine" incidents, it will be very difficult to implement it successfully during a major incident or crisis situation.

ICS LESSONS LEARNED

ICS must be an inherent element in any successful hazmat response program. Response experience has provided us with the following lessons learned:

1. **A variety of different players will respond to a working hazmat incident. Therefore, what occurs during the planning and preparedness phase will establish the framework for how the emergency response effort will operate**.

   Title III of the Superfund Amendments and Reauthorization Act of 1986 (SARA, Title III) and OSHA 1910.120 focused much attention on the need for hazmat planning and the implementation of an ICS organization. Communities and facilities are now required to develop planning documents to meet pre-established regulatory criteria. Unfortunately, the result has sometimes been emergency response plans (ERP) that look good on paper but don't really work on the street. In the rush to develop ERPs that meet the letter of the law, some have lost sight of the importance of the ERP's operational utility.

   In some cases, more emphasis has been placed on "Do we meet the requirements of the law?" as compared to "Do we meet the requirements of the law, *and* can our personnel perform the duties that we expect?". In fact, a subjective assessment of emergency planning programs could be summarized as follows: The majority of industrial emergency response programs are compliance oriented, where the letter of the law is satisfied, but the performance of facility emergency responders is often untested and not verified through either exercises or actual experience. In contrast, many public safety response programs are "operationally oriented," where personnel are able to perform the expected emergency response tasks, but sometimes lack the required regulatory documentation.

   Planning and preparedness establish the framework for how the emergency response effort will function. While ERPs must satisfy minimum regulatory requirements, the response plan must be both operationally oriented and representative of actual personnel and resource capabilities. Response plans that are not user friendly do not get used.

2. **There is no single agency that can effectively manage a major emergency alone**.

   A major community or facility emergency is going to require the resources and expertise of various organizations and agencies. All emergency response organizations bring their own agendas to the emergency scene. Each of these agendas represents real, valid, and significant concerns. Problems are often created, however, when there is no communication prior to the emergency (i.e., you don't know the key players and their emergency response mission) and everyone feels that his or her specific agenda or interest is the most important. Remember, the ability to mount a safe and effective response builds on what is accomplished during planning and preparedness activities. The real issue is not just command…it is also coordination.

In the absence of planning, organizational relationships are often based on perceptions, and perceptions are often based on our experiences with one individual or one incident. If that experience was positive, we tend to view the respective organization in a positive light until proven otherwise. Similarly, if that experience was negative…well, you know the rest!

There is no single organization that can effectively manage a major hazmat incident alone. Organizations that attempt to maintain the normal organization or bureaucracy in managing a major event will have inherent problems in implementing a timely and effective emergency response. There is no excuse for not knowing who the key players are within your area, and with whom you are going to interface with on scene.

3. **Many special operations teams, including HMRTs, tend to be people-dependent programs.**

Emergency response programs can be categorized as being either "people-dependent" or system-dependent. Special operations teams, such as hazmat, bomb squads, and technical rescue teams, are often very people-dependent when they are initially formed. These organizations rely almost totally on the experience of a few key individuals and can result in failed emergency response efforts if these key individuals are not present at an incident.

Figure 3.3 Many HMRTs are people-dependent programs.

In contrast, a system-dependent organization has clearly defined objectives, specific duties and responsibilities that are spelled out in standard operating procedures (SOP's) and operational checklists, and resources based upon probable response scenarios. A system-dependent response allows individuals to assume different roles in an emergency regardless of their daily activities. Written procedures, operational checklists, and an effective training and critique program ensure that less experienced personnel can get the job done with an acceptable level of safety and efficiency.

In essence, a system-dependent organization delivers a consistent level of quality and service, regardless of personnel or location. For example, when you order a "Big Mac" from any McDonald's throughout the world, there will be little difference in either quality or taste. How do they do it several billion times in a row? By emphasizing procedures and personnel training. In short, Mickey D's is the epitome of a system-dependent organization.

An organizational philosophy and management goal of emergency response programs should be the development of operational procedures that will bring consistency to emergency operations. The components of this system are the following:

- The development of SOPs;
- Training all personnel in the scope, application, and implementation of the SOPs;
- The execution of the SOPs on the emergency scene;
- Post-incident review and critique of their operational effectiveness; and
- Revision and updating of SOPs on a regular basis.

This standard management cycle helps build an organization with the ability to self-improve over time. This is critical, as the accepted standard of care keeps rising over time. What was considered an adequate emergency response program five years ago may be viewed as inadequate by today's standards.

Of course, having procedures alone is not sufficient, as the procedures must reflect the ability to handle not only the major emergency but also the day-to-day operations. Many organizations prepare for the "big one" but still can't handle the everyday occurrence. For example, you probably won't fare well managing a mass casualty and mass decon involving hundreds of potential victims if you cannot handle a multi-car or school bus accident involving 20 or 30 kids.

Remember, if you don't have your act together on the day-to-day operations, you are not going to pull it out of your hat for a major emergency.

4. **In those cases where ICS has not resulted in the operational improvements expected, the problems are typically associated with planning, training, and the organization buying into the ICS program, as compared to the ICS system itself.**

   ICS is not a panacea but is an organizational process and a resource management tool. As with any new effort, it will be necessary to establish and communicate a policy regarding the application and use of ICS for incident management and crisis management purposes. This policy must be established and supported by the highest levels of management and communicated throughout the organization. If the boss doesn't buy in to the program, neither will the troops!

5. **The management and control of routine, day-to-day incidents establishes the framework for how the larger, more significant events will be managed.**

   ICS should be the basic operating system for all emergencies. The incident management structure should logically expand as the nature and complexity of the emergency expands, resulting in a smooth transition with minimum organizational changes or adjustments.

   The corollary of this activity is quite simple—if ICS is not utilized for all routine emergencies, don't expect the organizational structure to function and adapt effectively, efficiently, and safely when a major emergency occurs. The routine establishes the foundation on which the nonroutine must build. The more routine decisions that are made prior to the emergency, the more time the IC will have to make critical decisions during the emergency.

ICS ELEMENTS

ICS has common characteristics that permit different organizations to work together safely and effectively in order to bring about a favorable outcome to the emergency. ICS is predicated on basic management concepts, including the following:

- Division of labor—Work is assigned based on the functions to be performed, the equipment available, and the training and capabilities of those performing the tasks.
- Lines of authority are clearly defined, including the delegation of authority and responsibility, as appropriate. However, ultimate responsibility always rests with the Incident Commander.
- Unity of command—Every person reports to only one supervisor (and only one IC). This provides for a proper chain of command, helps to eliminate confusion and freelancing, and facilitates personnel and organizational accountability.
- Optimum span of control of five individuals (range of 3 to 7), depending on the tasks to be performed, the associated danger or difficulty, and the level of delegated authority. When span-of-control problems arise, they can be addressed by expanding the organization in a modular fashion.
- Establishment of both line and staff functions within the organization. Line functions are directly associated with the implementation of incident operations to "make the problem go away," while staff functions support incident operations.

The following are the key organizational elements of ICS:

COMMON TERMINOLOGY

A hazmat response program must be built around an ICS organization that uses standardized terminology for organizational functions, resource elements, and incident facilities. This is particularly critical at multiagency and multijurisdiction incidents. Basic ICS organizational terms include the following:

1. *Incident Commander (IC)*: The individual responsible for the management of on-scene emergency response operations. The IC must be thoroughly trained to assume these responsibilities and is not automatically authorized to perform these activities by virtue of his/her position within the organization.

   At most routine incidents, the IC will be located at the emergency scene and will operate from a designated incident command post (ICP) location. However, as the scope of the incident escalates and senior officers and managers are activated, overall Command may be transferred from the emergency scene to the EOC. In this instance, on-scene command will now be under the direction of the Operations Section Chief or the On-Scene Commander, depending on local terminology. In simple terms, life in the field doesn't change much; what does change is the establishment of a broader response management organization that can (1) rapidly support the field response and (2) effectively deal with the external world impacts caused by the incident.

   A single command structure is used when one response agency has total responsibility for the overall incident. However, some hazmat emergencies will require that command be unified or shared between several organizations. Unified command is described in greater detail later in this chapter.

2. *Sections*—The organizational level that has functional responsibility for primary functions of emergency incident operations. Sections and their respective unit-level positions are only activated when their respective functions are required by the incident.

   Within ICS, sections are part of the IC's general staff and represent broad functional areas, such as Operations, Planning, Logistics, or Administration/Finance. Section Chiefs report directly to Command.

   Primary responsibilities of each respective section include the following:

   - Operations Section—The Operations Section delivers the required tactical-level "services" in the field to make the problem go away, including fire, hazmat, oil spill, technical rescue, and emergency medical operations. Until the Operations Section is established, the IC has direct control of all tactical re-sources. (*Note*: This ICS terminology should not be confused with an industrial facility's Operations Department, and does not specifically refer to process/operations personnel or activities.)
   - Planning Section—Typically regarded as the research and development (R&D) arm of the ICS process. The Planning Section conducts assessments and identifies the future needs and then develops the plans required to support the response. In the early phase of an incident, planning will focus on what the requirements are for one or two hours into the future. Once the incident stabilizes, planning begins to develop plans for the next operational period (NOP) in the future. At major incidents, this may be 12+ hours into the future.

     Another important role of the Planning Section is to develop plans that maximize the resources needed. For example, an event involving a large geographic area or multiple locations must have someone thinking about how scarce resources will be allocated and operations coordinated. Failure to establish a Planning Section for a moderate to large incident is similar to a military General Officer attempting to manage the battleground without any reconnaissance or military intelligence support. Planning units include situation status, resource status, documentation and demobilization, and technical specialists.

     The Planning Section plays a critical role in providing specialized expertise based on the needs of the incident, such as health and safety, industrial hygiene, environmental, weapons of mass destruction, and process engineering.

   - Logistics Section—Once staged resources are exhausted, the response effort becomes extremely dependent on the availability of the resupply and logistics effort. Logistics is responsible for providing all incident support needs, including facilities, services, and materials. This would encompass communications, rehabilitation and food unit concerns within the Service Branch, and supply, facilities, and ground support unit concerns within the Support Branch.

     The Logistics Section should not be viewed as an "order taker" to support the response effort. To be effective, the Logistics Section must be proactive and anticipate future needs, be knowledgeable of both the emergency and project procurement process, and be able to facilitate the transition from the emergency response to project phase of a long-term emergency.

- Administration/Finance Section—Responsible for all costs and financial actions of the incident. Their primary responsibility is to get funds where they are needed; ensure that adequate, yet simple, financial controls are in place; and keep track of funds. Primary finance units would include time, procurement, cost, and compensation/claims. In some corporate organizations, the legal unit may also be assigned to the Finance Section.

  Some incidents may be large enough that a federal disaster is declared, thereby allowing the local jurisdiction to be reimbursed for some of its costs. To receive reimbursement, expenses must be tracked and accounted for. Unfortunately, many costs are not tracked or documented during the initial stages of a campaign event.

3. *Branch*: The organizational level having functional or geographic responsibility for major segments of incident operations. The branch level is organizationally between the section and division/group/sector levels and is supervised by a Branch Director. Branches are often established when the number of divisions/groups or sectors exceeds the recommended span of control. Hazmat operations may be combined into a Hazmat Branch at a large Level III incident.

4. *Division/Groups/Sectors*: Divisions are the organizational level having responsibility for operations within a defined geographic area. A building floor, plant location, or process area may be designated as a division, such as Division 4 (i.e., 4th floor area) or the Alky Division (i.e., Alkylation Process Unit). Divisions are under the direction of a Supervisor.

   Groups are the organizational level responsible for a specified functional assignment at an incident. Hazmat units may operate as a Hazmat Group at Level I and smaller Level II hazmat incidents. Groups are under the direction of a Supervisor and may move between divisions at an incident.

   Sectors are the organizational level having responsibility for operations within a defined geographic area *or* with a specific functional assignment. At major incidents, sectors are essentially subunits within a section or branch and are under the supervision of a Sector Officer. Examples of functional sectors within the Operations Section would be the Safety, Rescue, or Medical Sectors. A location may also be designated as a geographic sector, such as the North Sector or Marine Dock Sector.

5. *Command Staff Officers*: Based on response requirements, the IC may assign personnel to serve on the Command Staff to provide Information, Safety, and Liaison services for the entire organization. Although there is only one Command Staff position for each of these functions, the Command Staff may have one or more assistants, as necessary. For example, hazmat incidents will often have an Assistant Safety Officer, who functions as a member of the tactical unit (e.g., Hazmat Group Safety Officer).

Figure 3.4 The Incident Safety Officer has the authority to stop unsafe actions.

- Safety Officer—Responsible for the safety of all personnel, including monitoring and assessing safety hazards, unsafe situations, and developing measures for ensuring personnel safety. The Incident Safety Officer (ISO) has the authority to terminate any unsafe actions or operations and is a required function based upon the requirements of OSHA 1910.120 (q).
- Liaison Officer—On large incidents or events, representatives from other agencies (usually called Agency Representatives) may be assigned to the incident to coordinate their agency's involvement. The Liaison Officer serves as a coordination point between the IC and any assisting or coordinating agencies who have responded to the incident, but who are not part of unified command or are represented at the ICP. In many respects, Liaison can often become the IC's "political officer," who may need to run interference for the IC.
- Information Officer—The Information Officer will be the point of contact for the media or other organizations seeking information directly from the incident or event. In those incidents where multiple Incident Officers from different agencies are present, a Joint Information Center (JIC) should be established and all of the individual agency Information Officers work jointly and cooperatively from one location.

MODULAR ORGANIZATION

The ICS organizational structure develops in a modular fashion based on the size and nature of the incident. The system builds from the top down, with initial responsibility and performance placed upon the IC. At the very least, an IC must be identified on all incidents, regardless of their size. As the need exists, separate sections can be developed, each with several divisions/groups or sectors that may be established.

The specific ICS organizational structure will be based on the management needs of the incident. For example, at a Level I incident, such as a 30-gallon carboy spill of sulfuric acid on a loading dock, personnel are not required to staff each major functional ICS area. The operational demands and the number of resources do not require delegation of management functions. However a Level III incident, such as a major train derailment or a large toxic vapor cloud release, may require staffing all sections to manage each major functional area and delegate management functions.

PREDESIGNATED INCIDENT FACILITIES

Emergencies require a central point for communications and coordination. Depending on the incident size, several types of predesignated facilities may be established to meet ICS requirements. These are as follows:

1. *Incident Command Post (ICP)*: The on-scene location where the IC develops goals and objectives, communicates with subordinates, and coordinates activities between various agencies and organizations. The ICP is the field office for on-scene response operations and requires access to communications, information, and both technical and administrative support. It may range from the front seat of a Suburban to a mobile command vehicle that could pass as a bus for a rock star.

   An ICP should provide Command with the following:

   - A place safe from the hazardous material(s) or problem
   - A quiet (relatively) place in which to think, discuss, and decide

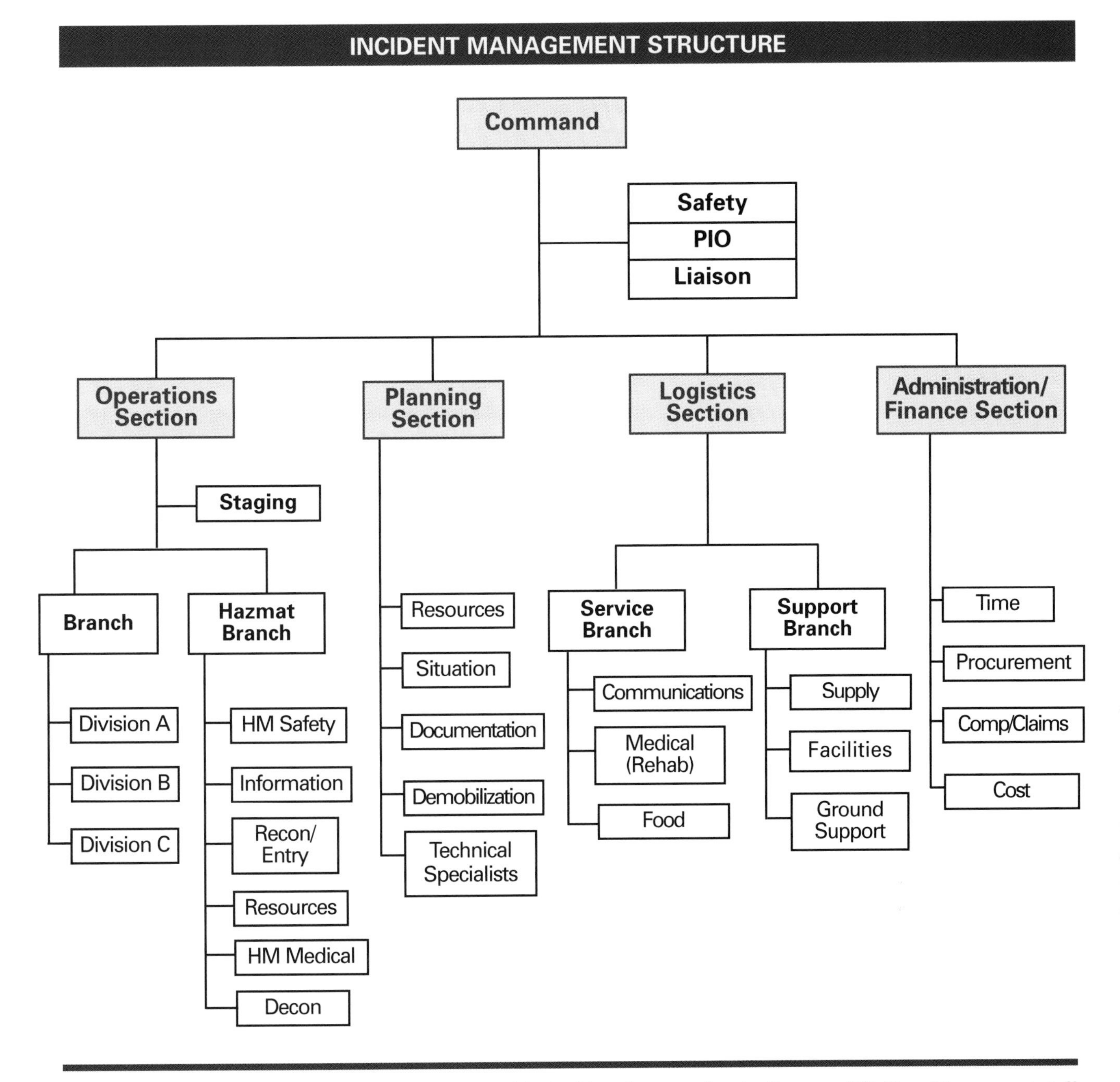

Figure 3.5 The Incident Command System consists of the Incident Commander, the Command Staff, and the general staff consisting of the Operations, Planning, Logistics and Finance/Administration Sections.

- Advantage point from which to see (when possible)
- Inside lighting
- A place to write and record
- Protection from the weather
- Protection from the media
- Staff space

The IC should remain at the ICP so that he or she is readily accessible to all personnel.

As noted, a number of emergency response organizations have specialized command vehicles that function as a mobile ICP. However, the necessary equipment for an ICP can be prepackaged into an incident command post kit. Minimum equipment should include the following:

- Radio capability to communicate with responders, mutual aid units, and facility maintenance/operations personnel. This should include mutual aid radios, programmable scanner radios for monitoring emergency radio frequencies, and access to the NOAA weather frequencies, where appropriate. Agency interoperability is critical!
- Cellular telephone capability. Remember, unless you have an encryption capability, all cell phone communications can be monitored.
- Copies of appropriate emergency response guidebooks and other reference sources. These should include the ERP and a response folder/booklet containing copies of all ICS checklists and worksheets.
- Technical and administrative support, including the possible use of laptop computers, personal data assistants (PDAs), and related electronic equipment.
- ICS command vests
- Tactical command chart
- Pair of binoculars

2. *Emergency Operations Center (EOC)*: The ICP is the nerve center of on-scene operations and is usually located near the scene of the emergency. However, if the scope of the incident increases, the plant or community EOC would then be activated. In this situation, overall command would be transferred to the EOC, while on-scene response operations would continue to be managed from the ICP. All communications with the media and outside agencies would now be coordinated through the EOC.

   Based on physical needs and safety requirements, the EOC is normally remote from the emergency scene. The EOC should provide phone, radio and fax communications, information resources, and the ability for a large number of personnel to work in a comfortable and secure area. These elements become essential as the number of players increases and the incident stretches into days as opposed to hours.

   It is important to understand the differences and relationship between the ICP and the EOC when both are operating simultaneously at a major emergency. The ICP is primarily oriented towards tactical control issues pertaining to the on-scene response, while the EOC deals with both strategic and external world issues and coordinates all logistical and resource support for on-scene operations.

   Emergencies have occurred where the EOC was impacted by the incident and could not be used. Perhaps the most vivid example was the loss of the New York City EOC in Building 7 of the World Trade Center Complex on September 11, 2001. As a result, many facilities and some communities have identified both a primary and alternate EOC location. Both EOCs should be similarly equipped and provided with comparable information and resource capabilities. In evaluating potential EOC locations, consideration should be given to the impact of potential fire, hazardous materials spills, vapor releases or terrorist attacks on the site. It should be noted that control rooms are typically poor options for an EOC within high-risk process industries.

An EOC should be equipped with the following:

- Radio, phone, and fax communications. This should include mutual aid radios, programmable scanner radios for monitoring emergency radio frequencies, and access to the NOAA weather frequencies, where appropriate. Sufficient telephone lines should be available for both incoming and outgoing calls.
- Detailed copies of area and facility maps, site plot plans, emergency preplans, hazard analysis documentation, and other related information.
- Copies of appropriate emergency response guidebooks and other reference sources. These should include the Emergency Response Plan, the LEPC Plan, material safety data sheets (MSDS), and other pertinent plans and procedures.
- General administrative support, including writing boards, incident status and documentation boards, telefax and copying machines.
- Electronic communication capabilities, including the use of computers and e-mail, the Internet and intranet, and the development of incident or agency-specific Web sites. Always consider the security of your electronic communication system.
- Television sets and AM/FM radios to monitor local and national news coverage.
- Backup emergency power capability to support EOC lighting, telephone, and radio base stations.

3. *Staging Area*—The designated location where emergency response equipment and personnel are assigned on an immediately available basis until they are needed. Staging is effective when the IC anticipates that additional resources may be required and orders them to respond to a predesignated area approximately three minutes from the scene. In simple terms, staging is the IC's tool box.

   The Staging Area should be clearly identified through the use of signs, color-coded flags or lights, or other suitable means. The exact location of the Staging Area will be based upon prevailing wind conditions and the nature of the emergency and are assigned to the emergency scene from the Staging Area as needed. Staging ensures that resources are close by but not in the way.

   Staging becomes an element within the Operations Section. The Staging Officer is responsible to account for all incoming emergency response units, to dispatch resources to the emergency scene at the request of the IC, and to request additional emergency resources, as necessary.

INTEGRATED COMMUNICATIONS

Communications are critical to the safe and efficient incident management. Ideally, the IC should be able to communicate directly with all on-scene units and support personnel. However, the more players at the incident, the less likely that they will share the same radio frequency. The more people using the same frequency, the more crowded it becomes.

Communications are managed most effectively through the use of a common communications center and network. Radio and communications interoperability are the key. Where common or mutual aid radio frequencies are unavailable, the IC should request that a designated individual report to the ICP with a radio from his or her organization. Using these individual radios, the IC can ensure that communications flow horizontally and vertically within the command structure. When common radio channels do not exist between all on-scene units, it may be more effective to have all companies operating on a common radio frequency to work together as a group or division, rather than dividing their resources between functions or areas.

Whenever a situation is encountered that could immediately cause or has caused injuries to emergency response personnel, the term *emergency traffic* should precede the radio transmission. This will be given priority over all other radio traffic. If an "emergency traffic" message is issued by the Communications Center, it may be preceded by a radio alert tone.

While difficult, it is critical that the IC be provided with regular and timely progress reports throughout the course of the emergency. Information is power; when the Operations Section becomes a "black hole" for all incident information, it breaks down the ability of the overall ICS structure to support the operational response effort and address the myriad of external issues in a timely and effective manner.

Communications of a sensitive nature should not be given over nonsecure cellular telephones or radios which can be monitored.

UNIFIED COMMAND STRUCTURE

Hazmat incidents often involve situations where more than one organization shares management responsibility or where the incident is multijurisdictional in nature. A unified command structure simply means that the key agencies that have statutory or jurisdictional responsibility jointly contribute to the process of

Figure 3.6 Unified command operations are a critical benchmark in the successful management of a hazmat incident.

- Determining overall incident priorities and strategic goals
- Selection of tactics for achieving those incident priorities and strategic goals
- Ensuring joint planning for tactical activities
- Ensuring that integrated tactical operations are conducted
- Maximizing use of all assigned resources
- Resolving conflicts between the players

The sooner a unified command structure is established, the better. Unified command is *not* management by committee; there will always be a lead agency or one agency that has 51% of the vote as compared to the other players. A representative of the lead agency should serve as the IC and should be supported by the senior officers from the other agencies involved. For example, during a chemical agent terrorism scenario where rescue and medical operations are underway, the lead IC will most likely be a fire department officer. However, once these life safety issues are completed and the focus switches to crime scene management and incident investigation, the lead IC should transfer to a law enforcement representative.

As in a single-agency command structure, the Operations Section Chief will have responsibility for implementation of the Incident Action Plan. When multiple agencies are involved in the response, the selection of the Operations Section Chief must be made by the mutual agreement of the unified command team. This may be done on the basis of greatest jurisdictional involvement, number of resources involved, existing statutory authority, or by mutual knowledge of the individual's qualifications.

If the players aren't coordinated and don't play as a team, the result is often ugly. While the members of unified command should only perform command level tasks and avoid getting involved in direct activities at the incident, they can and often are supported by aides in the field who are observing on-scene operations and serve as their "eyes and ears."

CONSOLIDATED PLAN OF ACTION

Every emergency incident needs some form of an IAP. The IAP consists of incident priorities, strategic goals, tactical objectives, and resource requirements. For a Level I hazmat incident, the plan is usually simple and can be communicated directly to the individual(s) carrying out the IC's orders. However, on large-scale incidents or campaign events this may not be practical, and a formal, written IAP may be required.

Emergencies involving multiple organizations or jurisdictions working within a unified command structure require consolidated action planning. As more organizations arrive at the emergency scene, they bring with them individual agendas and objectives. These may be driven by

- Facility responsibilities
- Legally mandated requirements
- Financial interests
- Contractual responsibilities
- Specific mission goals and charters

For example, local law enforcement agencies arriving at an incident may be primarily interested in traffic management and safety, while U.S. EPA and the State Department of Environmental Quality (DEQ) are both interested in the environmental impacts of the release. All three agencies have a legal right to be involved in the emergency, but neither group has the same objective.

A consolidated action plan is used to ensure that

- Everyone works together toward a common emergency response goal; that is, protecting life safety, the environment, and property.
- Individual response agendas are coordinated so that personnel and equipment are used effectively and in a spirit of cooperation and mutual respect.
- Everyone works safely at the scene of the emergency.

The most effective way to ensure that a consolidated plan of action is implemented is to have the senior representative of each major player at the incident present at the ICP and/or EOC at all times. Command runs the incident like a business meeting, making sure that every organization has its say and that the entire group works toward a resolution of the emergency as quickly and as safely as possible. The IC should remain focused on realistic objectives and ensure that each entity or special interest has input into the plan.

COMPREHENSIVE RESOURCE MANAGEMENT

The Incident Commander must analyze overall incident resource requirements and deploy available resources in a well-coordinated manner. In the case of a major hazmat incident or weapons of mass destruction (WMD) event, resources will include personnel, equipment, and supplies. The bottom line is simple—the proper type and level of resources must be available to support emergency operations in a timely manner.

Logistics and resource management have been the Achilles heel of many a response. Once the initial supply of resources are exhausted, responders are dependent on the ability of the ICS organization, specifically the Logistics Section, to provide a stead, consistent and reliable flow of resources. Purchasing procedures that work well for day-to-day operations may be totally inadequate in an emergency or crisis environment.

Among the resource management lessons learned as a result of previous incidents and exercises are the following:

- Get it done rather than argue about whose problem it is.
- It is easier to "gear down" operations than it is to play "catch-up." If you think you will need it, call for it!
- Overreact until the emergency situation is fully assessed and completely understood. React to the incident potential, not the existing situation. If you have to play "catch-up," you'll usually come in second place.
- Accept help from others. Mrs. Smith doesn't really care from where an asset or resource is obtained...neither should you. Remember, our customer just wants the problem to go away!
- Don't pay for stuff you don't need—downsize once the emergency has been stabilized and it is safe to do so. Demobilization is as important as the initial ramp-up and mobilization.

MANAGING THE INCIDENT: HAZMAT GROUP OPERATIONS

The IMS Model Procedures Guide for Hazardous Materials Incidents is available through Oklahoma State University—Fire Protection Publications at (800) 654-4055.

Depending on the scope and complexity of an incident, special operations (e.g., hazmat, bomb squad, technical rescue) may be managed as either a Branch or Group within the ICS organization. This section provides a brief overview on the application and use of a Hazardous Materials Group at a hazmat incident. In some organizations, this may also be referred to as the Hazardous Materials Sector. Readers desiring additional information on Hazardous Materials Group operations should consult the IMS Model Procedures Guide for Hazardous Materials Incidents.

The Hazardous Materials Group is normally under the command of a senior

Hazmat Officer (known as the Hazardous Materials Group Supervisor), who, in turn, reports to the Operations Section Chief or the Incident Commander. The Hazardous Materials Group is directly responsible for all tactical hazmat operations that occur in the hot and warm zones of an incident. Tactical operations outside of these hazard control zones (e.g., public protective actions) are not the responsibility of the Hazardous Materials Group.

Although a number of resources may be assigned to the Hazardous Materials Group, they are typically HMRT personnel and resources. The scope and nature of the problem will determine which roles are staffed.

Primary functions and tasks assigned to the Hazardous Materials Group include the following:

- **Safety function**—Primarily the responsibility of the Hazardous Materials Group Safety Officer (may also be referred to as the Assistant Safety Officer—Hazmat). Responsible for ensuring that safe and accepted practices and procedures are followed throughout the course of the incident. Possesses the authority and responsibility to stop any unsafe actions and correct unsafe practices.
- **Entry/back-up function**—Responsible for all entry and backup operations within the hot zone, including reconnaissance, monitoring, sampling, and mitigation.
- **Decontamination function**—Responsible for the research and development of the decon plan, set-up, and operation of an effective decontamination area capable of handling all potential exposures, including entry personnel, contaminated patients, and equipment. If necessary, will include the coordination of a Safe Refuge Area.
- **Site access control function**—Establish hazard control zones, establish and monitor egress routes at the incident site, and ensure that contaminants are not being spread. Monitor the movement of all personnel and equipment between the hazard control zones. Manage the Safe Refuge Area, if established.
- **Information/research function**—responsible for gathering, compiling, coordinating and disseminating all data and information relative to the incident. This data and information will be used within the Hazardous Materials Group for assessing hazards and evaluating risks, evaluating public protective options, selecting PPE, and developing the incident action plan.

Secondary support functions and tasks that may be assigned to the Hazardous Materials Group will include the following:

- **Medical function**—Responsible for pre- and post-entry medical monitoring and evaluation of all entry personnel, and provides technical medical guidance to the Hazardous Materials Group, as requested.
- **Resource function**—Responsible for control and tracking of all supplies and equipment used by the Hazardous Materials Group during the course of an emergency, including documenting the use of expendable supplies and materials. Coordinates, as necessary, with the Logistics Section Chief (if activated).

Figure 3.7 illustrates the standard ICS positions found at a typical hazardous materials incident. Remember, ICS is a modular organization. Only those functions necessary for the control and mitigation of the incident would be activated, as appropriate.

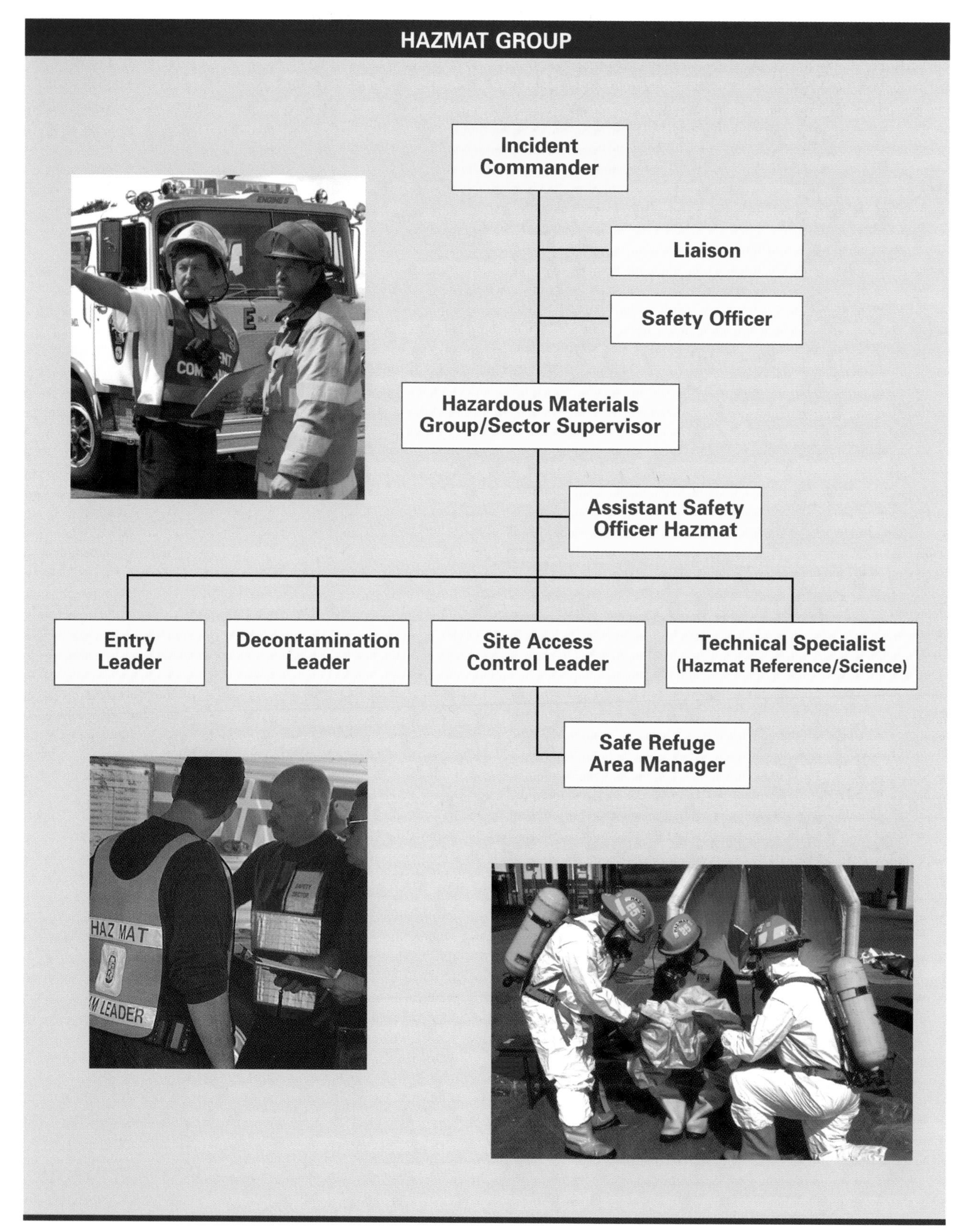

Figure 3.7 The Hazmat Group is responsible for all tactical level hazmat response operations.

HAZARDOUS MATERIALS GROUP STAFFING

❒ **Hazardous Materials Group Supervisor**. Responsible for the management and coordination of all functional responsibilities assigned to the Hazardous Materials Group, including safety, site control, research, entry, and decontamination. The Hazardous Materials Group Supervisor must have a high-level of technical knowledge and be knowledgeable of both the strategic and tactical aspects of hazardous materials response.

The Hazardous Materials Group Supervisor will be trained to the Hazardous Materials Technician level and will normally be filled by either the HMRT Team Leader or HMRT Officer. Depending on the scope and nature of the incident, the Hazardous Materials Group Supervisor will usually report to either the Operations Section Chief or the Incident Commander.

Based on the IC's strategic goals, the Hazardous Materials Group Supervisor develops the tactical options to fulfill the hazmat portion of the IAP and is responsible for ensuring that the following tasks are completed:

- Hazard control zones are established and monitored.
- Site monitoring is conducted to determine the presence and concentration of contaminants.
- Site Safety Plan is developed and implemented.
- Establish tactical objectives for the Hazardous Materials Entry Team within the limits of the team's training and equipment limitations.
- Ensure that all hot zone operations are coordinated with the Operations Section Chief or Incident Commander to ensure tactical goals are being met.

❒ **Hazardous Materials Group Safety Officer (i.e., Assistant Safety Officer—Hazmat)**. The Hazardous Materials Group Safety Officer reports to the Hazardous Materials Group Supervisor and is subordinate to the Incident Safety Officer. This individual is responsible for coordinating safety activities within the Hazardous Materials Group but also has certain responsibilities that may require action without initially contacting the normal chain of command. The Incident Safety Officer is responsible for the safety of all personnel operating at the incident, while the Hazardous Materials Group Safety Officer is responsible for all operations within the Hazardous Materials Group and within the hot and warm zones. This includes having the authority to stop or prevent unsafe actions and procedures during the course of the incident.

The Hazardous Materials Group Safety Officer must have a high level of technical knowledge to anticipate a wide range of safety hazards. This should include being hazmat trained, preferably to the Hazardous Materials Technician level, and being knowledgeable of both the strategic and tactical aspects of hazmat response. This position is typically filled by a senior HMRT Officer or member. While it is not the Hazardous Materials Group Safety Officer's job to make tactical decisions or to set goals and objectives, it is his or her responsibility to ensure that operations are implemented in a safe manner.

Specific functions and responsibilities of the Hazardous Materials Group Safety Officer include the following:

- Advise the Hazardous Materials Group Supervisor of all aspects of health and safety, including work/rest cycles for the Entry Team.

- Coordinate site safety activities with the Incident Safety Officer, as appropriate.
- Possess the authority to alter, suspend, or terminate any activity that may be judged to be unsafe.
- Participate in the development and implementation of the Site Safety Plan.
- Ensure the protection of all Hazardous Materials Group personnel from physical, chemical, and / or environmental hazards and exposures.
- Identify and monitor personnel operating within the hot zone, including documenting and confirming both "stay times" (i.e., time using air supply) and "work times" (i.e., time within the hot or warm zone performing work) for all entry and decon personnel.
- Ensure that EMS personnel and / or units are provided, and coordinate with the Hazardous Materials Medical Leader.
- Ensure that health exposure logs and records are maintained for all Hazardous Materials Group personnel, as necessary.

❒ **Entry Team**. The Entry Team is managed by the Entry Leader (aka Entry Officer). This individual is responsible for all entry operations within the hot zone and should be in constant communication with the Entry Team. The Entry Team and the Entry Leader are responsible for the following:

- Recommend actions to the Hazardous Materials Group Supervisor to control the emergency situation within the Hot Zone.
- Implement all offensive and defensive actions, as directed by the Hazardous Materials Group Supervisor, to control and mitigate the actual or potential hazmat release.
- Direct rescue operations within the hot zone, as necessary.
- Coordinate all entry operations with the Decon, Hazmat Information, Site Access, and Hazardous Materials Medical Units.

Personnel assigned to the Entry Team will include the entry and back-up teams, and personnel assigned for entry support. The Entry Team consists of all personnel who will enter and operate within the hot zone to accomplish the tactical objectives specified within the Incident Action Plan. Entry Teams will always operate using a buddy system.

The Back-Up Team is the safety team that will extract the entry team in the event of an emergency. Depending upon the scenario, the back-up capability may be provided by a Rapid Intervention Team or "RIT Team" as commonly used within the fire service. The Back-Up Team must be in-place and ready whenever entry personnel are operating within the hot zone. Entry support personnel (may also be known as the Dressing Team) are responsible for the proper donning and outfitting of both the Entry and Back-Up Teams.

❒ **Decontamination Team**. The Decon Team is managed by the Decon Leader (aka Decon Officer), and are responsible for the following:

- Determine the appropriate level of decontamination to be provided.
- Ensure that proper decon procedures are used by the Decon Team, including decon area set-up, decon methods and procedures, staffing, and protective clothing requirements.

- Coordinate decon operations with the Entry Leader, Site Access Control and other personnel within the Hazardous Materials Group.
- Coordinate the transfer of decontaminated patients requiring medical treatment and transportation with the Hazardous Materials Medical Group.
- Ensure that the decon area is established before any entry personnel are allowed to enter the hot zone. If rapid rescue operations are required, establish an emergency decon capability until a formal decon area can be set up.
- Monitor the effectiveness of decon operations.
- Control all personnel entering and operating within the decon area.

❐ **Site Access Control**. The Site Access Control Unit is managed by the Site Access Control Leader and is responsible for the following:

- Monitor the control and movement of all people and equipment through appropriate access routes at the incident scene to ensure that the spread of contaminants is controlled.
- Based upon recommendations from the Entry, Decon, and Info/Research Units, oversee the placement of the hazard control zone lines.
- As necessary, establish a safe refuge area and appoint a Safe Refuge Area Manager. This would include coordinating with Decon on decon and medical priorities for contaminated persons.
- Ensure that injured or exposed individuals are decontaminated prior to departure from the incident scene.

❐ **Hazardous Materials Information/Research Team**. The Hazardous Materials Information/Research Team is managed by the Information Leader (may also be known as Research or Science). Depending upon the level of the incident and the number of hazardous materials involved, the Information Team may consist of several persons or teams. The Hazardous Materials Information Team and the Information Leader are responsible for the following:

- Provide technical support to the Hazardous Materials Group.
- Research, gather and compile technical information and assistance from both public and private agencies.
- Provide and interpret environmental monitoring information, including the analysis of hazardous materials samples and the classification and/or identification of unknown substances.
- Provide recommendations for the selection and use of protective clothing and equipment.
- Project the potential environmental impacts of the hazardous materials release.

❐ **Hazardous Materials Medical Unit**. Medical support services may be provided by either a Hazardous Materials Medical Unit or Medical Group within the ICS organization. Hazardous Materials Medical Unit personnel will be located in the Entry Team dressing area and in the Rehabilitation Area. The Hazardous Materials Medical Unit and Hazardous Materials Medical Leader are responsible for the following:

- Provide pre-entry and post-entry medical monitoring of all entry and backup personnel.

- Provide technical assistance for all EMS-related activities during the course of the incident.
- Provide emergency medical treatment and recommendations for ill, injured, or chemically contaminated civilians or emergency response personnel.
- Provide EMS support for the Rehab Area.

The Hazardous Materials Medical Unit will conduct post-entry medical monitoring, cooling and rehydration of entry and back-up personnel in the Rehab Area. All operating personnel should not be given anything to eat or drink unless approved by Hazardous Materials Medical personnel. Medical findings and personal exposure forms should be forwarded to the Hazardous Materials Group Safety Officer and/or the Entry Team Leader.

❒ **Hazardous Materials Resource Unit**. At some "working" incidents, a hazardous materials resource function may be established to support Hazardous Materials Group activities and is directed by the Hazardous Materials Resource Leader. This unit will be located in the cold zone and will be responsible for acquiring all supplies and equipment required for Hazardous Materials Group operations, including protective clothing, monitoring instruments, leak control kits, and so on. In addition, the Hazardous Materials Resources Leader will also be responsible for documenting all supplies and equipment expended as part of the emergency response effort. The Hazardous Materials Resources Unit and Leader must work closely with the Logistics Section Chief.

MANAGING THE INCIDENT: STREET SMARTS

Ever watch Olympic figure ice skating? You would find that the contestants are judged in two broad areas: (1) technical merit, which accounts for the technical precision and quality of their program; and (2) artistic impression, which accounts for how the program is choreographed and the relationship between the skater's performance and music. In simple terms, a poor score in one area can offset a great performance in the other.

Does hazmat response resemble figure ice skating? Hopefully not! However, emergency response operations are also increasingly being judged in how responders perform in two similar areas. Consider the following comparison:

- *Technical merit*—In simple terms did responders make the "problem" go away. While we can often argue about safety and response efficiencies, most would agree that the emergency response community is very effective! At the end of the day, the problem usually goes away....sometimes because of a lack of fuel, but the problem still goes away!
- *Artistic impression*—Our version of artistic impression is how well we manage the external world impacts of the problem. This could cover everything from media communications to community relations and intergovernmental relationships.

To restate this concept in another manner, the overall performance of a hazmat response program will be based on two interrelated factors: (1) the implementation of a timely, well-trained and equipped emergency response effort; and (2) the effective management of the interpersonal and organizational dynamics created by the event, particularly those dealing with external groups and audiences (e.g., the media, government agencies, and the public at large). An effective field response effort can be compromised or completely negated by poor management of the political and external issues. Or to put it another way, an incident can become a crisis when the political and external issues are not effectively addressed.

Remember, perceptions are reality. Emergency response efforts that do not address external and political concerns will often be perceived by the public, elected officials, and management as a failed response, regardless of how effective the on-scene emergency response effort may actually perform. This section will examine some of the factors that may influence the political and external impacts of an emergency response effort.

COMMAND AND CONTROL

Experienced officers often regard the problem in enemy-oriented, pessimistic terms. Simply, the hazardous materials involved are your enemy and will behave in both known and unknown ways. Command and control efforts must be applied toward achieving results—that is, defeating the "enemy." Confident ICs and hazmat officers refuse to be overwhelmed as they assume command. As soon as possible, they delegate certain responsibilities and empower their subordinates to make decisions and do the job they have been assigned. Effective ICs recognize that a few minutes spent establishing effective command and control at the beginning may save hours in the course of a long-term incident.

A fire service colleague once mentioned that the best officers he had worked with during his career had one thing in common—their presence was so strong that it inspired people to perform on the emergency scene. If you look like you know what you are doing, it sets a tone for the management of the emergency. Many refer to this as command presence. If command presence is not strong, both individual and organizational freelancing, and on-scene "rubber-necking" can result.

Constant reassessment and possible revision of tactical operations are needed to maximize response effectiveness. Both Command and the Hazmat Group Leader must be able to integrate evaluation and revision into the overall management approach. However, review and evaluation are useful only when the tactical plan is kept open-ended. Both Command and the Hazmat Group Leader need to plan ahead and operate with a backup plan, constantly asking questions such as these:

- Where will the incident be in 30 minutes? 60 minutes?
- Given the problem, what is our worst-case scenario?
- What is Plan B and Plan C if Plan A doesn't work?

Figure 3.8 Where will this incident be in 30 minutes? What will plan B and C be?

Don't accept a bad situation; on the emergency scene, things go right and things go wrong. As an officer, you must be willing to admit that a mistake has been made or that conditions have been changed and then modify the placement and deployment of available personnel and resources. Failure to constantly evaluate can eventually place the IAP and the safety of responders at a great disadvantage.

INCIDENT POTENTIAL

The development of strategies and tactics to defeat the enemy must be based on the assessment of incident potential. Elements of incident potential can include incident severity, magnitude and duration of the event, as well as the nature and degree of incident impacts. In addition, don't forget about the "artistic" side of the equation, including community impacts, external world and media affairs impacts, and legal concerns.

What are some of the causes of a delayed assessment of incident potential? They include the following:

- Responders get so focused on the tactical problem that they fail to consider the big picture or strategic aspects of the incident.
- Response agencies that think they can do it all.
- Response agencies believe that a request for mutual aid or assistance might involve acknowledging a mistake.
- Responders define a problem down to a manageable level.
- Responders are inexperienced.
- There is a lack of information.

You cannot defeat the enemy if you only focus on where the problem is now... always consider where the enemy is going and incident potential.

DECISION MAKING

The decision making process begins with both Command and the Hazmat Group Leader recognizing the need to avoid dead-end decisions. Whenever possible, decisions must be open ended, allowing for expansion or reversal. Having to make quick decisions may worry the inexperienced officer, but they become easier to make once you recognize the following:

- *Distinguish between assumptions and facts.* Response operations must sometimes be based upon incomplete or assumed information. Factual information is often not available or incomplete, particularly in the initial stages of an emergency. Realize that both information and decision-making will have the opportunity to improve as the incident grows older.
- *Maintain a flexible approach to decision making.* The overall IAP must be constantly updated as more and better information is received from operations units and outside sources. Feedback allows for revisions to the general strategy, specific tactics, and all major decisions.
- *Shift to a management role after initiating action.* An IC cannot make *all* ongoing response decisions. The efficiency of command decisions will improve once the IC delegates tactical responsibility. Otherwise Command will quickly become overwhelmed with both people and information. The same concept holds true within the Hazmat Group.

The IC must quickly prioritize problems and develop solutions. This requires the effective gathering, recording, and organizing of information.

INFORMATION IS POWER

Emergency scene "intelligence" can rapidly supply Command and the Hazmat Group with random data and information. Both functions must develop and use an information gathering and processing routine that is within their own mental limits. Tactical worksheets can facilitate this process. However, without effective information management, mental overload will quickly occur and decision making will suffer.

Unfortunately, information quality and quantity during the initial stages of an incident are sometimes incomplete. In other instances, time is required to gather the information required to develop an initial IAP. The term *fog of war* is not limited to the military battlefield but is present at most working incidents.

Equally critical is the ability to manage and disseminate information in a timely manner. Most information has a "half-life," in that it is valuable for a limited period of time. If critical information is received after a tactical window of opportunity has closed, it will not have a positive impact on defeating the enemy. How many times have you heard the statement, "It would have been nice to have this information 15 minutes ago?"

There is often a reluctance for on-scene personnel to provide regular and timely updates to the ICP or EOC. This is a critical problem, as both Command and the EOC cannot support tactical response operations and address the external issues (e.g., media relations, community notifications, etc.) if they do not have a regular and reliable flow of information. If they don't know the "true" picture, they may begin to "make it up" based on what they think is happening!

In the heat of battle, the Operations Section may start to reduce the flow of information to the remainder of the ICS organization. In essence, it becomes an information "black hole"—questions flow to the Operations Section but nothing ever comes out! Ways to avoid this problem include assigning a communications aide to the Operations Section Officer, and establishing procedures that require that updates and status reports be provided at timely intervals (e.g., every 10 minutes).

Information is power. Whoever controls the flow of information controls the incident.

THE RULES OF ENGAGEMENT

HMRTs are often requested to respond on a mutual aid basis into other areas or jurisdictions to provide technician-level response services. At one time in our response history, these requests were "far and few"—local IC's often stated that "...there is no way that any hazmat team is going to come into my town/plant/sandbox and run my incident." Today, some ICs are willing to give majority command and control of the incident to the HMRT, often making comments like "Please make my problem go away" or "If you need me I'll be over there."

Figure 3.9 Regular situation status reports must flow from the Operations Section to the Incident Commander.

Law enforcement, special ops and military operations typically lay out rules of engagement (ROE) for all of the players prior to initiating actions. The ROE provide the structure for engaging the enemy, including the chain of command, decision-making authority, accountability and responsibility. In a similar manner, ROE may need to be established for both hazmat and special operations events, particularly those where the special ops unit is playing in someone else's sandbox.

The best scenario is one where the ROE are outlined and agreed on as part of the planning process. If the incident is a "home game," departmental policies and procedures will usually address this issue. However, if the incident is an "away game," the ROE should be reviewed and established between Command and the HMRT before hazmat response operations are initiated.

What should ROE look like? At a minimum, they should include the following:

- Relationship between the IC and the HMRT. In simple terms, hazmat should always work for the IC. Don't allow a well-meaning and well-intentioned IC to saddle you with command and control of the entire operation.
- Clearly identify what objectives, tasks, or areas are the responsibility of hazmat.
- HMRTs are often reliant on other response units for various personnel and resources, such as people, water and supplies. In other cases, unusual or specialized resources may be required, including heavy equipment. Command must understand that if you ask for it, you need it.
- Finally, always have a "get out of jail" card when playing at an away game. Be prepared for those occasions where the local IC may reject the recommendations of the HMRT and pursue tactical options with unacceptable risks to hazmat responders. Although not common, this scenario has tremendous potential career, political, and legal liabilities.

LIAISON OFFICER

Hazmat incidents pose a multitude of technical, managerial, and political issues. As previously noted, emergencies that have been effectively managed from a technical perspective have been perceived to be poorly managed by the public and governmental agencies because the political issues and external impacts of the incident were not adequately addressed.

ICS is an effective and necessary tool for ensuring that the internal, technical, and external/political aspects of an emergency are addressed. A key member of the ICS command staff is the Liaison Officer. The Liaison Officer is sometimes viewed as the "Political Officer" who serves as the point of contact for all assisting and cooperating external and governmental representatives who are not represented within the unified command structure. This ICS position allows the IC and general staff to focus on problem resolution while ensuring that political sensitivities are still addressed. The Liaison Officer's ability to effectively coordinate, handle, and, if necessary, "stroke" these individual agencies and representatives will have a significant impact on how the incident will be perceived from both a political and external perspective.

"WHAT YOU SEE IS NOT NECESSARILY WHAT YOU GET"

Underestimating the significance of a hazmat emergency can increase the level of risk to both responders and the public. Human tendency is to downplay the potential of "minor" emergencies; however, there have been instances where some individuals have not acted on early warnings for fear of "screwing up" or because they were not sure if they had the authority to take action. Of course, consistently overreacting is equally a problem and can damage the credibility of emergency responders.

The absence of physical indicators of a hazard (i.e., smoke, vapors, odors, etc.) can influence both public and political perception of an emergency. If visible physical indicators are present, fewer individuals will initially question responder decisions. However, in the absence of such indicators, all bets are off.

Of course, the opposite can be true when there is not an emergency but physical indicators are present that may be perceived as a problem. For example, an unusually large or smokey flare stack at a refinery may be perceived by the public as an on-site problem. If the facility does not notify local public safety agencies, the public perception is "they're hiding something." Facilities with good community relationships have procedures for notifying the community when there are both actual emergencies and nonevents that may be perceived as an emergency by the community.

The duration of the incident can also be an influencing factor, even in the presence of a physical indicator. Hazmat releases often create significant operational and community disruptions until the incident is controlled and terminated. Public intolerance is directly proportional to the length of time citizens are inconvenienced. See Figure 3.10.

Figure 3.10 Public tolerance of response operations is directly proportional to the amount of time people are inconvenienced.

WORKING WITH TECHNICAL SPECIALISTS

When gathering information, responders are often the true nonbelievers. They have been lied to so many times that they automatically question most information when it is initially presented to them. This is not necessarily a bad trait, as long you use a structured procedure to guide you through the information gathering process. In many respects, the role of responders during the information gathering process is similar to that of a detective.

A likely source of hazard information will be product or container specialists, or other individuals who have specific knowledge needed by responders. When evaluating these sources and the information they provide, consider the following observations:

- Individuals who are specialists in a narrow, specific technical area may not have an understanding of the broad, multi-disciplined nature of emergency response. For example, information sources may provide extensive data on process engineering and design, yet may be unfamiliar with basic site safety and personal protective clothing practices.
- Some technical specialists have knowledge that is totally based upon dealing with a chemical or process in a structured and controlled environment. When faced with the same chemical or process in an uncontrolled or emergency response situation, they may provide only limited information.
- There are no experts, but only information sources! Each individual source will have its own advantages and limitations. A colleague once provided this response after being referred to as an expert: "We aren't experts, but we do have good judgment. Recognize, however, that good judgment is based upon experience, and experience is often the product of bad judgment." Translation—we screwed up enough to know what will probably work and what won't.
- Sometimes responders interact with individuals with whom they have had no previous contact. Before relying upon their recommendations, ascertain their level of expertise and job classification by asking specific questions. Remember these two points: (1) Technical smarts is not equivalent to street smarts! Having an alphabet behind one's name (e.g., PhD., M.S., CSP, etc.) does not automatically mean that an individual necessarily understands the world of emergency response and operations in a field setting; and (2) Twenty years of experience may actually be one year of experience repeated twenty times.
- Questioning information sources is an art and requires the skills of both a detective and a diplomat. While this is certainly not an interrogation process, you must be confident of the source's authority and expertise. Always conduct the interview with respect to the person's rank or position within the field or organization. One method is to ask questions for which you already know the answer in order to evaluate that person's competency and knowledge level. Remember, final accountability rests with Command.
- A time-tested rule for minimizing political vulnerability is the "rule of threes." Simply, when faced with significant or politically sensitive decisions, consult at least three independent reference sources. The more politically sensitive the incident, the greater the need for the reference sources to be respected and reputable individual(s). For example, using a single emergency response guidebook as the sole technical justification for evacuating 5,000 people is only asking for well-justified technical and political criticism of the emergency

response effort, even if it was the correct decision. To minimize your political vulnerability, also seek input from CHEMTREC™, the shipper, or local technical information specialists.

Finally, remember that everybody brings their own agenda and scorecard to a hazmat incident. Don't assume that your concerns (1) are the most important, and (2) are always going to be the same as "their" concerns.

"EVERYBODY HAS THE ANSWER FOR YOUR PROBLEM"

Hazmat emergencies attract a great deal of attention. They also bring a number of entrepreneurs, salespeople, managers, and "do gooders" to the attention of the IC, many of them professing to have the answer to your problem.

The reality of incident command is that while these people can be helpful, they are usually a major distraction to the IC, especially at long-term emergencies or if they have the ear of a local political or governmental official.

The IC may have to designate the Liaison Officer to address these external contacts, as they occasionally do provide worthwhile information or resources, and can influence both public and political opinions. At the least, failure to seriously address these people may generate bad publicity and political backlash.

EVERYONE HAS A BOSS

Regardless of one's position within an organization, everyone has a boss. If the bosses don't have a clearly defined role, organization and structure, you may have as much trouble with your own management and its dynamic as well as the incident.

THE ETERNAL OPTIMISTS

Be aware of the eternal optimists within your own ranks. Don't allow your own people to "walk you down the yellow brick road" without exploring all viable options and alternatives. Remember, initial observations often underestimate the significance of a problem. Command should always be prepared to implement an alternative action plan if the current plan fails.

Experience has shown that there can be an inadequate flow of timely and accurate information from the hot zone or the incident scene. When asking subordinates for progress reports, don't continuously ask questions which can be answered with a simple "yes" or "no." If you only ask if everything is okay, don't be surprised when your people consistently say it is.

Both the IC and the Safety Officer have a unique view of the incident that many of the other players do not. Remember the difference between "street smarts" and "technical smarts"—you don't need a degree in chemistry to know when something just doesn't look right.

LONG-TERM INCIDENTS AND PLANNING

The majority of hazmat incidents are "high intensity—short duration" events that are terminated in 8 hours or less. Activities are typically oriented towards planning ahead of current events and ensuring that the IAP is adequately staffed and resourced. Future events and needs are often assessed using a "what if" approach.

Campaign incidents extending over a period of days or weeks create different challenges for emergency responders. Issues such as developing a shift schedule, determining short-term and long-term logistical requirements, and establishing a

CREW RESOURCE MANAGEMENT: AN ICS TOOL

The term "Crew Resource Management (CRM)" was originally defined in 1977 by aviation psychologist Dr. John Lauber as "...using all available resources—information, equipment and people—to achieve safe and efficient flight operations." Dr. Lauber noted that "...CRM is the stuff that takes two competent, proficient pilots and turns them into one competent, proficient crew."

CRM was an outgrowth of aviation accident investigations that showed that a significant percentage of losses (70%) were due to flight crew failures rather than technical problems. These failures included a lack of situational assessment and awareness, failure to follow or complete SOPs, failure to adapt to unusual or emergency situations, and interpersonal communications and coordination problems. CRM focuses upon the cognitive and interpersonal skills needed to manage the flight within an organized aviation system. These skills include the mental processes for gaining and maintaining situational awareness, for solving problems, and for making decisions.

Experience shows that there are many parallels between in-flight decision-making and emergency services decision-making. A number of emergency response professionals are now looking at opportunities for integrating the concepts of CRM into the emergency response community. For example, a CRM training approach has been used since the early 1990's for training Offshore Installation Managers (OIMs) and their respective teams who operate on offshore oil production platforms in the North Sea. The U.S. Department of Interior has recently incorporated the basic concepts of CRM into its ICS and leadership training curriculums as a result of several wildland fire incidents that resulted in firefighter injuries and fatalities.

The underlying tenets of CRM are leadership and resource management.

LEADERSHIP

Leadership can directly influence the quality of the ICS organization. While the exercise of leadership is typically associated with the IC, every ICS supervisor has a certain amount of authority, responsibility, and opportunity to exercise leadership. Leadership styles will vary depending upon the situation.

CRM elements that can be used to improve leadership effectiveness include problem definition, inquiry, and advocacy.

- **Problem Definition**. The first and most critical step in problem solving is defining the problem ("A problem well-defined is half-solved."). Once the problem is understood, alternative solutions can be evaluated. Involving the ICS organization in the "diagnosis" increases both decision-making effectiveness and organizational "buy in."
- **Inquiry**. Human error is inevitable. Unfortunately, the cost of human error at a hazmat incident is often measured in lives. Inquiry is a mental process that involves constantly evaluating and re-evaluating everything that can be anticipated during an incident. Commonly known as "playing the devil's advocate" or "what if's", this process can help the IC sense a difference between what actually is happening or about to happen, and what should be occurring. Inquiry is the responsibility of every individual within the ICS organization.
- **Advocacy**. Advocacy is the responsibility of personnel to speak out in support of a course of action different than that currently being planned or implemented (e.g., " I'm not comfortable with this" or "Let me push back...") Reasons may include technical concerns, safety

issues, risk versus gain, etc. Likewise, advocacy also includes listening to viewpoints that might be contradictory to one's position.

Inquiry and advocacy are essential to each other. Inquiry which results in the detection of potential safety problems is of little use unless the information is advocated so others can react. Advocacy is constructive questioning and is not a resentment of authority.

Effective ICs and ICS supervisors should work to create an environment where subordinates are encouraged, and feel free to raise concerns or ask questions about a particular course of action. The input received may be used to enhance the problem definition and decision-making process. Subordinates must remember, however, that while advocacy is important, the final decision still remains with the IC or the ICS supervisor.

RESOURCE MANAGEMENT

Resource management includes the effective use of emergency responder skills, knowledge, and expertise. Communications, coordination, conflict resolution, and critique are critical CRM elements that can enhance the effectiveness of both individuals and the ICS organization.

- *Communications* within the ICS organization is an essential prerequisite for good CRM. In addition to transferring information, it is the vehicle for both situational awareness and problem solving. Communication involves both talking and listening. To talk when no one is listening is worthless. To have deaf ears to what is being said is equally ineffective. One must always be sensitive to the "atmosphere" of the discussion. When the atmosphere is open, it allows for a free flow of communication and encourages input.
- *Coordination* is the process by which information, plans, and operational activities are considered and shared throughout the ICS organization. Coordination minimizes the likelihood of confusion because the players understand the Incident Action Plan. It also reduces the potential for error because of overlooked or disregarded information.
- *Conflict Resolution*. Conflict is inevitable. Differences in thoughts, opinions, values, or actions—whether actual or perceived—can lead to disagreements and disputes. Differences in personality alone may create conflict.

  Conflict is not necessarily negative or destructive. What makes conflict "bad" is the inability to constructively cope with it, such as when the conflicting positions are passively given up or aggressively suppressed. The concept of "legitimate avenue of dissent" is critical for clearing the air and maintaining lines of communication. When effectively channeled by the IC, conflict can be transformed into an effective comparison of viewpoints, problem definition, options, and sound solutions.
- *Critique*. Many have said that, "experience is the best teacher." This may be true, but only if one takes full advantage of the lessons learned. Critique involves studying a plan of action before, during, and after its implementation. Feedback is at the heart of a critique, but don't confuse critique with criticism.

In summary, CRM is a training tool that can improve our teamwork and performance when operating at an emergency. In 1989, United Airlines Captain Al Haynes, in coordination with his crew and an air traffic controller, was able to land a DC-10 aircraft damaged by an in-flight explosion in Sioux City, IA despite having limited control of the aircraft. He later acknowledged the role of CRM in this accomplishment:

> *"I am firmly convinced that CRM played a very important part in our being able to land at Sioux City with any chance of survival. I also believe that its principles apply no matter how many crew members are in the cockpit."*

formal IAP development flow and process are foreign to many responders. Where can you go for help? Significant expertise in the management of long-term events exists through the U.S. Forest Service and its system of overhead Incident Management Teams (IMT). In addition, since the events of September 11, 2001, a number of regions and states have also established concepts such as Incident Support Teams (IST) and state-based IMTs to assist local IC's with the management of long-term emergencies.

RANDOM THOUGHTS

The following are some random thoughts and comments pertaining to the management of a hazmat incident: These include:

- Command must be aware of all major decisions and operations made under his/her jurisdiction. If the IC is unsure or uncomfortable with any part of the IAP or any of the information received, the strategy and tactics should be put on hold until the IC is satisfied. The IC is a risk evaluator and a resource manager.
- From the time of arrival on scene, the IC must prioritize problems and develop solutions by collecting information. The effective IC will:
  1. Seek out data that is current, accurate, and specific.
  2. Delegate information retrieval.
  3. Know how to find reference data and how to use it.
  4. Collect the right information in the right order.
  5. Use a wide variety of sources.
- Managing a major hazmat incident is no different than taking an army to war—the emergency response effort (Operations Section) will be no better than the information, forecasting, and technical expertise available (Planning Section), the physical and personnel resources available for timely response and support (Logistics Section), and the financial and administrative support provided (Administrative/Finance Section).
- Solicit opinions and ideas—they foster both individual and organizational "buy-in" into the decision-making process. Allow everyone (through the ICS organization) to voice opinions, particularly when dealing with situations where the hazards are exceptionally high. Remember, the people being asked to take the risks should have a voice in the decision-making process. This collective input will strengthen the decision-making process, as well as present the IC with more options. Be careful to avoid "group-think," as voiced by those "yes-men" who surround every Incident Commander.
- Emergency responders often seek direction on how to handle a decision-making situation where the IC or a senior officer does not agree with a subordinate's recommendation, or where a senior plant manager has no on-scene responsibility and does not belong on the emergency scene. There are no easy answers here other than these two fundamental points: (1) always be professional in this situation (i.e., don't make your boss look bad!); and (2) always try to have any conversation on a one-to-one basis away from the rest of the troops.

- Never say never, particularly when dealing with a long-term, campaign operation. History is full of incidents where tactical options that appeared totally unrealistic during the first hour eventually looked real good and were actually implemented during the twentieth hour.
- Consider the art of communications. Effective communications is one part talking and ten parts listening. Beware of individuals whose hearing is affected by management position or promotion, as well as the "yes men" who show up at major emergencies and often flock around the IC.
- When an incident goes bad or is particularly politically sensitive, anticipate being the scapegoat. In order to minimize political vulnerability, the IC must continuously (1) consult and build a consensus on the IAP; (2) document; and (3) not assume anything.

SUMMARY

The successful management of a hazmat incident is directly linked to the rapid development of an effective incident management process and organization. From the arrival of the first emergency responder, the IC must match and balance the size and structure of the organization with the number of units and organizations present on the incident scene. A functional hazmat command system must allow the Incident Commander to use the standard elements present at every hazmat incident to establish and maintain control, to be responsive to the special characteristics of each incident, and, finally, to apply the same ICS system and process to every incident, regardless of its nature and size.

This chapter also provided an overview of the common interpersonal, organizational, and external issues associated with the management of a hazmat incident. It is based upon "lessons learned" while managing emergencies, and it supports the fundamental ICS concepts. Many of the issues raised are not necessarily new or unique, but experience shows they can be overlooked by some command and hazmat officers.

Remember, your emergency response performance will be evaluated on two interrelated factors: (1) the implementation of a timely, well-trained and equipped emergency response effort in the field, and (2) the effective management of the interpersonal, organizational, and external impacts created by the incident. An effective response effort can be compromised or completely negated by poor management of the political and external issues.

REFERENCES AND SUGGESTED READINGS

Air Force Civil Engineer Support Agency (AFCESA) and PowerTrain, Inc., HAZARDOUS MATERIALS INCIDENT COMMANDER EMERGENCY RESPONSE TRAINING CD-ROM. Tyndall Air Force Base, FL: Headquarters AFCESA (2002).

American Chemistry Council, CRISIS MANAGEMENT PLANNING FOR THE CHEMICAL INDUSTRY., Washington, DC: Chemical Manufacturers Association (1991).

BP Exploration (Alaska) Inc., BP SEALS—Skills Enhancement and Leadership Seminar Materials. Anchorage, AK: BP Exploration (Alaska)—Crisis Management Center (2003).

Brunacini, Alan V., FIRE COMMAND (2nd edition). Quincy, MA: National Fire Protection Association (2002).

ConocoPhillips Inc., TIGER—Training for Integrated Group Emergency Response Materials. Anchorage, AK: ConocoPhillips (Alaska)—Crisis Management Center (2003).

Daimler, Gottlieb and Karl Benz Foundation et al. WORKSHOP ON GROUP INTERACTION IN HIGH RISK ENVIRONMENTS, Zurich, Switzerland: Gottleib Daimler And Karl Benz Foundation (July 5–6, 2001).

Department of Homeland Security (DHS), THE NATIONAL RESPONSE PLAN. Washington, DC: Department of Homeland Security (2004).

Emergency Film Group, Industrial Incident Management (videotape). Plymouth, MA: Emergency Film Group (1999).

Federal Aviation Administration (FAA), AIR CIRCULAR NO. 120-51B—CREW RESOURCE MANAGEMENT TRAINING. Washington, DC: Federal Aviation Administration (January 3, 1995).

Flin, Rhona, SITTING IN THE HOT SEAT—LEADERS AND TEAMS FOR CRITICAL INCIDENT MANAGEMENT. New York: John Wiley & Sons (1996).

Flin, Rhona, et. al, DECISION MAKING UNDER STRESS—EMERGING THEMES AND APPLICATIONS. Aldershot, England: Ashgate Publishing Ltd (1997).

Hildebrand, Michael S., and Gregory G. Noll, "Incident Management and ICS: A Petrochemical Perspective." INDUSTRIAL FIRE SAFETY (January / February, 1993), pages 38–48.

International Association of Fire Fighters, TRAINING FOR HAZARDOUS MATERIALS RESPONSE: TECHNICIAN. Washington, DC: IAFF (2002).

National Fire Protection Association, HAZARDOUS MATERIALS RESPONSE HANDBOOK (4th edition). Quincy, MA: National Fire Protection Association (2002).

National Fire Protection Association, NFPA 1500—FIRE DEPARTMENT OCCUPATIONAL SAFETY AND HEALTH PROGRAM HANDBOOK. Quincy, MA: National Fire Protection Association (2002).

National Fire Protection Association, NFPA 1561—TECHNICAL STANDARD ON EMERGENCY SERVICES INCIDENT MANAGEMENT SYSTEMS. Quincy, MA: National Fire Protection Association (2000).

National Fire Service Incident Management System Consortium Model Procedures Committee, IMS MODEL PROCEDURES GUIDE FOR HAZARDOUS MATERIALS INCIDENTS. Stillwater, OK: Fire Protection Publications, Oklahoma State University (2000).

National Fire Service Incident Management System Consortium Model Procedures Committee, IMS MODEL PROCEDURES GUIDE FOR STRUCTURAL FIREFIGHTING. Stillwater, OK: Fire Protection Publications, Oklahoma State University (1997).

National Institute for Occupational Safety and Health (NIOSH), OCCUPATIONAL SAFETY AND HEALTH GUIDANCE MANUAL FOR HAZARDOUS WASTE SITE ACTIVITIES. Washington, DC: NIOSH, OSHA, USCG, EPA (1985).

Royal Aeronautical Society—Crew Resource Management Standing Group, "A Paper on Crew Resource Management." London, England: Royal Aeronautical Society (October 1999).

Tippett, John, CREW RESOURCE MANAGEMENT—A Positive Change for the Fire Service. Fairfax, VA: International Association of Fire Chiefs (2002).

CHAPTER 4

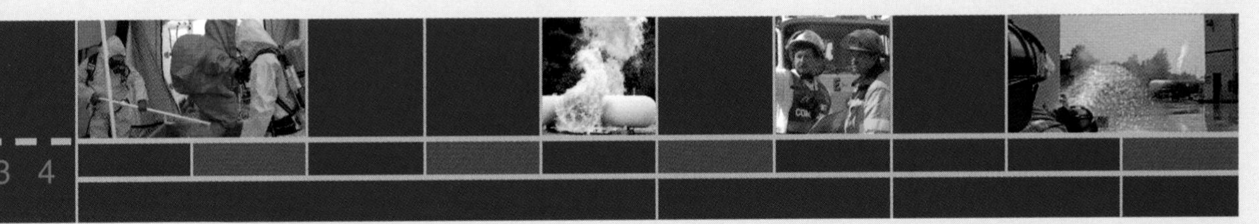

THE EIGHT STEP PROCESS©: AN OVERVIEW

OBJECTIVES

1. Describe the Eight Step Process© and its application as a tactical incident management tool for managing on-scene operations at a hazardous materials incident.
2. Describe the critical success factors in managing the first hour of a hazardous materials incident.

ABBREVIATIONS AND ACRONYMS

APR	Air Purifying Respirator
DECON	Decontamination
EOC	Emergency Operations Center
IAP	Incident Action Plan
IC	Incident Commander
ICP	Incident Command Post
IED	Improvised Explosive Devices
LPG	Liquefied Petroleum Gas
PASS	Personal Alert Safety System
PERO	Postemergency Response Operations
PPA	Public Protective Actions
PPE	Personal Protective Equipment
PPV	Positive Pressure Ventilation
SBCCOM	U.S. Army Soldiers Biological and Chemical Command
SOP	Standard Operating Procedures
SWAT	Special Weapons and Tactics Team

INTRODUCTION

On-scene response operations must always be based on a structured and standardized system of protocols and procedures. Regardless of the nature of the incident and response, a reliance on standardized procedures will bring consistency to the tactical operation. If the situation potentially involves hazardous materials or weapons of mass destruction (WMD), this reliance upon standardized tactical response procedures will help to minimize the risk of exposure to all responders.

This chapter will introduce the reader to a system for the tactical management of incidents or emergencies that may involve hazardous materials classified as "hazardous" in some manner. This includes hazardous materials, hazardous substances, and materials capable of being used as WMD agents.

The Eight Step Process© is widely used throughout the country by public and private sector hazardous materials response teams for the tactical management of hazardous materials emergencies. It also serves as an example of a structured system that can be used by law enforcement and special operations personnel at incidents involving hazardous substances and materials.

MAKING THE TRANSITION

Experience has shown that the critical success factors in hour 1 of a hazmat response will typically be (1) the ability of responders to recognize "clues" that the incident may involve hazardous materials; (2) the ability of responders to quickly gain control of the incident scene and separate responders from the problem; and (3) the ability to establish command and control in a timely manner.

The ability of emergency responders initially to size-up and assess the clues that hazardous materials may be involved starts with the quantity and quality of information provided by Communications (or Dispatch). What do responders need to know while en-route? Basic considerations during the alerting and response phase include the following:

- Where is the location of the incident? Is the incident at a target occupancy or target hazard event? Is there a preincident plan for the location?
- Are hazmats involved? If unsure, are hazmats found at the location?
- Are there reports of any unusual odors? Explosions? Hazardous materials?
- Are any injuries or casualties involved? Are the reasons known or unknown?
- Are initial responders down?

These factors will assist responders and the Incident Commander (IC) in determining whether to follow "normal" or hazmat response protocols.

THE EIGHT STEP PROCESS©

The Eight Step Process© outlines the basic tactical functions to be evaluated and implemented at incidents involving hazardous materials. Like all SOPs, the Eight Step Process should be viewed as a flexible guideline and not as a rigid rule. Individual departments and agencies should decide what works best for them.

The Eight Step Process© offers several benefits. First, it recognizes that the majority of incidents involving hazardous materials are minor in nature and generally involve limited quantities. It also builds on the action of first responding units and facilitates identifying the roles and responsibilities of each level of response. The Eight Step Process provides a flexible management system that expands as the scope and magnitude of the incident grows and, finally, it provides a consistent management structure, regardless of the classes of hazardous materials involved.

Essentially, there are eight basic functions that must be evaluated at emergencies involving, or suspected of involving, hazmats or WMD agents. These eight functions typically follow an implementation timeline at the incident:

1. Site Management and Control
2. Identify the Problem
3. Hazard and Risk Evaluation
4. Select Personal Protective Clothing and Equipment
5. Information Management and Resource Coordination
6. Implement Response Objectives
7. Decon and Clean-Up Operations
8. Terminate the Incident

STEP 1: SITE MANAGEMENT AND CONTROL

FUNCTION: Site management and control involves managing and securing the physical layout of the incident. Reality is that you cannot safely and effectively manage the incident if you do not have control of the scene. As a result, site management and control is a critical benchmark in the overall success of the response and is the foundation on which all subsequent response functions and tactics are built.

GOAL: To establish the playing field so that all subsequent response operations can be implemented both safety and effectively.

CHECKLIST:

- ❑ During the approach to the incident scene, avoid committing or positioning personnel and units in a hazardous position. Assess the situation and consider having an escape route out of the area if conditions should deteriorate suddenly.

> CAUTION: Emergency responders must be aware that certain chemical releases may travel throughout the scene and impact response routes. In addition, some chemicals may produce vapor clouds that may be mistaken for fog or other normal weather and environmental conditions (e.g., anhydrous ammonia, liquefied petroleum gas or LPG). The danger area may extend beyond the visible vapor cloud.

- ❑ Establish command of the incident and establish an Incident Command Post (ICP). If other public safety units are already on scene, ensure that operations are coordinated and unified.
- ❑ Establish a Staging Area (Level I, II) for additional responding equipment and personnel.
- ❑ Establish an isolation perimeter (i.e., outer perimeter) to isolate the area and deny entry. Establish access control—restrict emergency site access to authorized essential personnel; all nonessential personnel should be isolated from the problem. Isolation perimeters should include land, water, and air areas.

- ❑ Establish a hot zone or inner perimeter as the "playing field." The location of the inner hot zone should be identified and communicated to all personnel operating on the site. Methods of identifying the restricted area include visible landmarks, barricade tape, traffic cones, and so on.
- ❑ Do not attempt to enter the area unless having the appropriate level of respiratory and skin protection, based on the hazards present. If people are down, SCBA will be considered the minimum level of respiratory protection for initial emergency response operations.
- ❑ If civilians are injured or are casualties and personal contamination is suspected, isolate all personnel until emergency decon can be established.
- ❑ Initiate public protective actions (PPA).

STREET SMART TIPS

- Site Management establishes the playing field for the players (responders) and the spectators (everyone else).
- The initial 10 minutes of the incident will determine operations for the next 60 minutes, and the first 60 minutes will determine operations for the first 8 hours.
- Don't try to control more real estate than you can effectively isolate and control. Smaller and tighter may be better than bigger and looser.
- Remember the basics. The more time, distance, and shielding between you and the material, the lower the risk will be.
- Designate an emergency evacuation signal and identify rally points if emergency evacuation is necessary.
- Remember the first law of hot zone operations when dealing with hazardous materials: To play in the game you must
  - Be trained to play
  - Be dressed to play
  - Have a buddy system with backup personnel (minimum of 2 in/2 out)
  - Have decon established
  - Coordinate with Command and Safety

STEP 2: IDENTIFY THE PROBLEM

FUNCTION: Identify the scope and nature of the problem. This includes

- Recognition, identification, and verification of the hazardous materials involved in the incident
- Type of container, as appropriate
- Exposures

Methods of identification include analyzing container shapes, markings, labels and placards, and facility documents (e.g., Material Safety Data Sheets or MSDS); using monitoring and detection equipment; and identifying by the senses (i.e., physical observations, smell, reports from victims, etc.). Responders should remember that even when the hazardous substances involved have been identified, the information should always be verified.

GOAL: To identify the scope and nature of the problem, including the type and nature of hazardous materials involved as appropriate.

CHECKLIST:

❑ Survey the incident—identify the nature and severity of the immediate problem, including the recognition, identification, and verification of the material(s) involved, type of container involved, and any potential or existing life hazards. If multiple problems exist, prioritize them and make independent assignments.

❑ Clues for determining the identity of the materials involved include
- Occupancy and location
- Container shapes
- Markings and colors
- Placards and labels
- Shipping papers or facility documents
- Monitoring and detection equipment
- Senses, including physical observations, smell, and so on.

❑ Factors to consider in assessing the type of container involved include:
- Bulk versus nonbulk
- Pressurized versus nonpressurized
- Number of compartments
- Material(s) of construction
- Pressure relief devices

❑ Conduct offensive or defensive reconnaissance, as necessary, to gather intel on the situation.

STREET SMART TIPS

- A problem well defined is half solved.
- Assume that initial information is not correct. Always verify your initial information. Verify, verify, verify.
- Conduct recon operations, as necessary.

  Defensive Recon. Objective is to obtain information on site layout, weapon condition, physical hazards, access, and other related conditions from beyond the inner perimeter. This is normally obtained through threat assessments, interviews, physical observations, and so on.

  Offensive Recon. Objective is to obtain intel and incident information by physically entering the inner perimeter. At incidents that involve both hazmat and explosive risks, this may be a joint entry operation with both bomb squad and hazmat personnel to conduct monitoring, sampling, and video or photo documentation for analysis.

- Be alert for the presence of improvised explosive devices (IED) and secondary events. IED clues can include
  - Abandoned container out of place for the surroundings
  - Obvious explosive device components, such as batteries, timers, blasting caps, charges
  - Partially exploded devices found
  - Unusual or foreign devices attached to hazmat containers, especially liquefied and compressed gas cylinders, flammable liquid containers and bulk storage tanks and vessels
  - Unattended vehicles not appropriate to the immediate environment

STEP 3: HAZARD AND RISK EVALUATION

FUNCTION: This is the most critical function that public safety personnel perform. The primary objective of the risk evaluation process is to determine whether or not responders should intervene, and what strategical objectives and tactical options should be pursued to control the problem at hand. You can't get this wrong. If you lack the expertise to do this function adequately, get help from someone who can provide that assistance, such as local HMRTs and product/container specialists.

GOAL: To assess the hazards present, evaluate the level of risk, and establish an Incident Action Plan (IAP) to make the problem go away.

CHECKLIST:

❑ Assess the hazards posed by the problem (health, physical, chemical, weapons, other).

❑ Collect, prioritize. and manage hazard data and information from all sources, as appropriate, including

- Technical reference manuals
- Technical information sources
- Hazmat databases
- Technical information specialists
- MSDS
- Monitoring and detection equipment

❑ Primary technical information centers available to public safety personnel include

- CHEMTREC—(800) 924-9300
- National Response Center (NRC) which serves as the federal single point of contact for accessing federal assistance
- State single point of contact
- Product or container specialists
- Regional poison control centers

❑ Air monitoring and the *General Hazmat Behavior Model* are critical in implementing a "risk-based response." Understand the relationship between the container and the hazmat(s) involved.

- Stress
- Breach
- Release
- Engulf
- Impingement
- Harm

❑ Based on the risk evaluation process, develop your IAP. Determine whether the incident should be handled offensively, defensively, or by nonintervention. Remember that offensive tactics increase the risks to emergency responders.

- **Offensive tactics**—Require responders to control/mitigate the emergency from within/inside the area of high risk.
- **Defensive tactics**—Permit responders to control/mitigate the emergency remote from the area of highest risk.
- **Non-intervention tactics**—Pursuing a passive attack posture until the arrival of additional personnel or equipment, or allowing the fire to burn itself out.

STRATEGY	OFFENSIVE	DEFENSIVE	NON-INTERVENTION
Rescue	X		
Public protective actions	X	X	X
Spill control	X	X	
Leak control	X		
Fire control	X	X	
Clean-up and recovery	X	X	

STREET SMART TIPS

- Focus on those things that you can change and that will make a positive difference to the outcome.
- Every incident will arrive at some outcome, with or without your help. If you can't change the outcome, why get involved?
- There's nothing wrong with taking a risk. If there is much to be gained, there is much to be risked. If there is little to be gained, then little should be risked.
- Pubic safety personnel should view their roles as that of risk evaluators, rather than risk takers, where hazardous materials are involved. Bad risk takers get buried; effective risk evaluators come home.

- Hour 1 priorities within the IAP are as follows:
  - Establish Site Management and Control.
  - Determine the materials/agents involved.
  - Ensure the safety of all personnel from all hazards.
  - Ensure that PPE is appropriate for the hazards.
  - Initiate tactical objectives to accomplish initial rescue, decon, medical, and public protective action needs.
  - If criminal activities are involved (e.g., terrorism incidents), maintain the integrity of potential evidence.

STEP 4: SELECT PERSONAL PROTECTIVE CLOTHING AND EQUIPMENT

FUNCTION: Based on the results of the hazard and risk assessment process, emergency response personnel will select the proper level of personal protective clothing and equipment. Two primary types of personal protective clothing are commonly used at hazmat incidents: (1) structural firefighting protective clothing, and (2) chemical protective clothing.

GOAL: To ensure that all emergency response personnel have the appropriate level of personal protective clothing and equipment for the expected tasks.

CHECKLIST:

❑ The selection of personal protective clothing will depend on the hazards and properties of the materials involved and the response objectives to be implemented (i.e., offensive, defensive, or nonintervention). In evaluating the use of specialized protective clothing, the following factors must be considered:

- The hazard to be encountered, including the specific tasks to be performed
- The tasks to be performed (e.g., entry, decon, support)
- The level and type of specialized protective clothing to be utilized
- The capabilities of the individual(s) who will use the PPE in a hostile environment. Remember—specialized protective clothing places a great deal of both physiological and psychological stress on an individual.

❑ The following levels of personal protective clothing are typically utilized by emergency responders at hazmat incidents, as appropriate:

- **Structural firefighting clothing**—Includes helmet, fire retardant hood, turnout coat and pants, personal alert safety system (PASS device), and gloves. Positive-pressure, SCBA should be considered the minimum level of respiratory protective clothing.
- **Chemical vapor protective clothing**—Specialized chemical protective clothing, which when used in conjunction with air supplied respiratory protection devices offers a sealed, integral level of full-body protection from a hostile environment. It is primarily designed to offer protection from both gases and vapors, as well as total body splash protection. It may also be referred to as EPA Level A chemical protective clothing.

- **Chemical splash protective clothing**—This is specialized protective clothing that protects the wearer against chemical liquid splashes but not against chemical vapors or gases. It is primarily designed to provide personal protection against liquid splashes, solids, dusts and particles. It can be found in both single- and multipiece garment ensembles and may also be referred to as EPA Level B chemical protective clothing when air supplied respiratory protection is provided or EPA Level C chemical protective clothing when air-purifying respirators (APR) are provided.

❑ Ensure that all emergency response personnel are using the proper protective clothing and equipment equal to the hazards present. Do not place personnel in an unsafe emergency condition.

❑ Order additional personnel and other specialized equipment and expertise early in the incident. If you are unsure of your requirements, always call for the highest level of assistance available. Do not wait to call for emergency assistance.

STREET SMART TIPS

- Remember that structural firefighting protective clothing is not designed to provide protection against chemical hazards.
- There is no one single barrier that will effectively combine both chemical and thermal protection.
- Wearing any type and level of impermeable protective clothing creates the potential for heat stress injuries.
- Personal protective clothing is your last line of defense!

STEP 5: INFORMATION MANAGEMENT AND RESOURCE COORDINATION

FUNCTION: Refers to proper management, coordination, and dissemination of all pertinent data and information within the ICS in effect at the scene. In simple terms, this function cannot be effectively accomplished unless a unified ICS organization is in place. Of particular importance is the ability to determine the incident factors involved, which functions of the Eight Step Process© have been completed, what additional information must be obtained, and what incident factors remain unknown.

GOAL: To provide for the timely and effective management, coordination, and dissemination of all pertinent data, information, and resources between all of the players.

CHECKLIST:

❑ Confirm that the ICP is in a safe area.

NOTE: Personnel not directly involved in the overall command and control of the incident should be removed from the ICP area.

- ❑ Confirm that a unified command organization is in place and all key response and support agencies are represented directly or via a liaison officer.
- ❑ Expand the ICS and create additional branches, divisions, or groups, as necessary.
- ❑ Ensure that all appropriate internal and external notifications have been made. Coordinate information and provide briefings to other agencies, as appropriate.
- ❑ Confirm emergency orders and follow through to ensure that they are fully understood and correctly implemented. Maintain strict control of the situation.
- ❑ Make sure that there is continuing progress toward solving the emergency in a timely manner. Do not delay in calling for additional assistance if conditions appear to be deteriorating.
- ❑ If activated, provide regular updates to the local Emergency Operations Center (EOC).

STREET SMART TIPS

- Consider the security of the ICP and all other incident response areas (e.g., Staging, Rehab) of the incident.
- Don't look stupid because you didn't have a plan.
- Bad news doesn't get better with time. If there's a problem, the earlier you know about it the sooner you can start to fix it.
- Don't allow external resources to "freelance" or do the "end run."
- Don't let your lack of a Planning Section become the Achilles heel of your response. Establish it early, particularly if the incident has the potential to become a "campaign event."
- Play nice together!

STEP 6: IMPLEMENT RESPONSE OBJECTIVES

FUNCTION: The phase where responders implement the best available strategic goals and tactical objectives, which will produce the most favorable outcome. If the incident is in the emergency phase, this is where we "make the problem go away." Common strategies to protect people and stabilize the problem include rescue, public protective actions, spill control, leak control, fire control, and recovery operations. In simple terms, these strategies are typically implemented by fire and rescue units, with law enforcement responsible for all security and criminal-related issues.

If the incident is in the postemergency response phase, the focus of response personnel will likely become scene safety, clean-up, evidence preservation (as appropriate), and incident investigation. Specific tasks will include (1) initial site entry and monitoring to determine the extent of the hazards present; (2) an evaluation of the scene to locate evidence that can be used to reconstruct the events leading up to the incident; (3) identification of the contributing factors that caused the incident;

(4) interviewing of on-scene personnel and witnesses to corroborate the information obtained and opinions formed based on the available data; and (5) documentation of preliminary results.

GOAL: To ensure that the incident priorities (i.e., rescue, incident stabilization, environmental and property protection) are accomplished in a safe, timely, and effective manner.

CHECKLIST:

- ❑ Implement response objectives. Remember that offensive tactics increase the risks to emergency responders; evaluate the risks of offensive control tactics before sending emergency response crews into the hazard area.
  - **Offensive tactics**—Require responders to control/mitigate the emergency from within/inside the area of high risk.
  - **Defensive tactics**—Permits responders to control/mitigate the emergency remote from the area of highest risk.
  - **Nonintervention tactics**—Pursuing a passive attack posture until the arrival of additional personnel or equipment, or allowing the fire to burn itself out.

STRATEGY	OFFENSIVE	DEFENSIVE	NON-INTERVENTION
Rescue	X		
Public protective actions	X	X	X
Spill control	X	X	
Leak control	X		
Fire control	X	X	
Clean-up and recovery	X	X	

> **NOTE:** Rapidly changing incident conditions may require using multiple tactics simultaneously or switching from one tactic to another. Defensive tactics are always desirable over offensive tactics if they can accomplish the same objectives.

- ❑ Ensure that properly equipped backup personnel wearing the appropriate level of personal protective clothing are in-place before initiating operations.
- ❑ Ensure that Entry Teams have been briefed prior to being allowed to enter the hot zone. For hazardous materials emergencies, this should include the following:
  - All watches, jewelry, and personal valuables have been removed
  - Objectives of the entry operation
  - Safety procedures, including radio communications, SCBA, and PPE checks
  - Emergency procedures, including placement of back-up teams, escape signals and escape corridors
  - Decontamination area location, set-up, and procedures

CAUTION: Decon should be set up prior to initiating entry operations. For situations where civilians are down and chemical exposures are suspected, emergency decon must be established as soon as possible.

- ❑ Conduct regular monitoring of the hazard area to determine if conditions are changing.

STREET SMART TIPS

- Positive pressure ventilation (PPV) can be used to significantly reduce chemical vapor levels within a building and increase the safety of emergency responders who must enter a structure to effect rescue operations.
- Never touch or handle anything in a clandestine lab operation until the area has been evaluated and cleared by bomb squad personnel who have proper training in identifying booby traps.
- What will happen if I do nothing? Remember—this is the baseline for hazmat decisionmaking and should be the element against which all strategies and tactics are compared.
- There is a very fine line between explosives, oxidizers, and organic peroxides. All are capable of releasing tremendous amounts of energy.
- Surprises are nice on your birthday but not on the emergency scene. Always have a Plan B in case Plan A doesn't work!

STEP 7: DECON AND CLEAN-UP OPERATIONS

FUNCTION: Decontamination (decon) is the process of making personnel, equipment, and supplies "safe" by reducing or eliminating harmful substances (i.e., contaminants) that are present when entering and working in contaminated areas (i.e., hot zone or inner perimeter). Although decon is commonly addressed in terms of "cleaning" personnel and equipment after entry operations, response personnel should remember that in some instances, due to the nature of the materials involved, decontamination of clothing and equipment may not be possible and these items may require disposal.

All personnel trained to the First Responder Operations level should be capable of delivering an emergency decon capability. At most "working" hazmat incidents, decon services will be provided by HMRTs or fire and rescue units working under the direction of a Hazmat Technician. Questions pertaining to disposal methods and procedures should be directed to environmental officials and technical specialists, based on applicable federal, state, and local regulations.

GOAL: To ensure the safety of both emergency responders and the public by reducing the level of contamination on scene and minimizing the potential for secondary contamination beyond the incident scene.

CHECKLIST:

- ❑ Ensure that the decon operations are coordinated with tactical operations. This should include the following tasks:
  - The decontamination area is properly located within the warm zone, preferably upslope and upwind of the incident location.
  - The decontamination area is wellmarked and identified.
  - The proper decontamination method and the type of personal protective clothing to be used by the Decon Team have been determined and communicated, as appropriate.
  - All decon operations are integrated within the ICS organization.
- ❑ Incidents involving large numbers of contaminated or potentially contaminated individuals will require the establishment of a mass decon operation. Basic principles of mass casualty decon include the following:
  - Anticipate a 5 : 1 ratio of unaffected to affected casualties (in simple terms, for every person who is physically contaminated there will be five others who may be exposed or who think they are contaminated). Bottom line—lots of folks may be seeking attention.
  - Decontaminate ASAP. It may be necessary to "corral" those contaminated until a mass decon corridor is established.
  - Disrobing is decon. Top to bottom, more is better.
  - Outer clothing may remove up to 80% of the contaminant when chemical agents are suspected (Note: U.S. Army Soldiers Biological and Chemical Command [SBCCOM] findings for chemical agents).
  - Water flushing for a period of 3 minutes is generally the best mass decon method (Note: SBCCOM findings for chemical agents).
  - After a known exposure to a hazardous material, responders must decon ASAP to avoid serious effects.
- ❑ Ensure proper decon of all personnel before they leave the scene. For example, flammable gases and some toxic and corrosive gases can saturate protective clothing and be carried into "safe" areas.
- ❑ Establish a plan to clean up or dispose of contaminated supplies and equipment before cleaning up the site of a release. Federal and state laws require proper disposal of hazardous waste.

STREET SMART TIPS

- Establishing an emergency decon capability should be part of the Incident Action Plan for any incident where hazardous materials are involved (e.g., clan labs, SWAT tactical operations).
- Decon involving large numbers of people will be a challenge and will make for a long day. Remember the basics—separate people from the problem and keep them corralled until emergency decon is established.
- Permeation can occur with any porous material, not just PPE. That includes shoes, belts, fire hose, holsters, gun stocks, and so on.

- Sodium hypochlorite (i.e., bleach) may be used as a decon agent for equipment when dealing with chemical and biological materials. Be aware that bleach will degrade Kevlar™ and will shorten the life of the material.
- Degradation chemicals, such as bleach, should never be applied directly to the skin.
- Never transport contaminated victims from the scene to any medical facility without conducting field decon. At best, the hospital folks will be pretty mad at you; at worst, you will completely shutdown the emergency room and the hospital.
- Law enforcement operations at incidents where hazmats are involved create some unique challenges:
- Be aware of "Bad Guys" possibly being mixed in with civilians when conducting mass decon operations.
- Establish a weapons safety officer as part of decon to ensure that all weapons are properly handled and rendered safe.
- Ensure that procedures are in place when conducting decon of contaminated prisoners.

STEP 8: TERMINATE THE INCIDENT

FUNCTION: This is the termination of emergency response activities and the initiation of postemergency response operations (PERO), including investigation, restoration, and recovery activities. This would include the transfer of command to the agency that will be responsible for coordinating all post-emergency activities.

GOAL: To ensure that overall command is transferred to the proper agency when the emergency is terminated and that all post-incident administrative activities are completed per local policies and procedures.

CHECKLIST:

- ❑ Account for all personnel before securing emergency operations.
- ❑ Conduct incident debriefing session for on-scene response personnel. Provide background information necessary to ensure the documentation of any exposures.
- ❑ Command is formally transferred from the lead response agency to the lead agency for all post-emergency response operations.
- ❑ Ensure that the following elements are documented:
    - All operational, regulatory, and medical phases of the emergency, as appropriate.
    - All equipment or supplies used during the incident.
    - Obtain the names and telephone numbers of all key individuals. This should include contractors, public officials, and members of the media.

- ❑ Ensure that all emergency equipment is reserviced, inspected, and returned to proper locations. Provide a point of contact for all post-incident questions and issues.
- ❑ Conduct a critique of all major and significant incidents based on local protocols.

STREET SMART TIPS

- Although every organization has a tendency to develop its own critique style, never use a critique to assign blame (public meetings are the worst time to discipline personnel).
- Organizations must balance the potential negatives against the benefits that are gained through the critique process. Remember—the reason for doing the critique in the first place is to improve your operations.
- Most critiques fall into one of three categories: (1) We lie to each other about what a great job we just did; (2) we beat up each other for screwing up; or (3) we focus on the lessons that were learned and the changes/improvements that must be made to our response system.

SUMMARY

Emergency response operations at incidents involving hazardous materials must always be based upon a structured and standardized system of protocols and procedures. Regardless of the nature of the incident, the nature of the response, or the personnel involved, a reliance on standardized procedures will bring consistency to the tactical operation. If the situation potentially involves hazardous materials or WMD agents, this reliance on standardized tactical response procedures will help to minimize the risk of exposure to all responders.

The Eight Step Process© is a tool used for the tactical management of hazardous materials emergencies. It serves as an example of a structured system that can be used by response personnel at incidents involving hazardous substances and materials. Although the level of equipment, training, and personnel may vary among organizations, there are fundamental functions and tasks that must be evaluated and implemented on a consistent basis. The Eight Step Process© provides a framework necessary to translate planning and preparedness into the delivery of an effective system for responding to and investigating incidents where hazardous materials and WMD materials may be involved.

The eight functions are

1. Site Management and Control
2. Identify the Problem
3. Hazard and Risk Evaluation
4. Select Personal Protective Clothing and Equipment
5. Information Management and Resource Coordination
6. Implement Response Objectives
7. Decon and Clean-Up Operations
8. Terminate the Incident

CORROSIVE
8
YOU GO FIRST, KID:
- BOOK SMARTS
TAKE PRECEDENCE!
OH NO, O.T.!
—I INSIST!
-EXPERIENCE
COUNTS MORE!

HAZARDOUS MATERIALS

LEARNING SYSTEM™

TEXTBOOK—New Third Edition consists of 12 chapters with chapters 1-3 addressing preparing for the incident, and chapters 4–12 addressing how to respond safely to a hazardous materials incident. Chapters 5–12 dedicate a one chapter to each of the Eight Step Process©, a systematic way of responding to a hazmat incident.

HAZARDOUS MATERIALS
MANAGING THE INCIDENT
Third Edition
GREGORY G. NOLL • MICHAEL S. HILDEBRAND • JAMES YVORRA
International Fire Service Training Association

INSTRUCTORS PROGRAM:

INSTRUCTORS GUIDE—includes complete learning objectives, an outline of each chapter, learning strategies and The Learning Through Inquiry Method of presenting case studies and scenarios. The Instructors Guide includes 24 new tactical scenarios.

POWERPOINT PRESENTATION—is fully coordinated with the textbook chapters and the instructor's guide. The package includes learning objectives, key speaking points, and all art, charts, photography, and cartoons included in the textbook. All 24 scenarios described in the Instructors Guide are fully illustrated. Instructor Program also includes printable checklists from the Field Operations Guide. Cross platform CD-ROM packaged in DVD Amaray case.

STUDENT WORKBOOK—includes several hundred multiple-choice questions, with answer key, to guide the student through all 12 chapters of the textbook. Questions are coded to help the student meet the requirements of NFPA 472. The workbook also outlines a study strategy to prepare for promotion and college level examinations.

FIELD OPERATIONS GUIDE—(FOG), is designed to be used at the incident scene as a reference guide to strategic and tactical decision-making. It includes detailed tactical checklists that follows the Eight Step Process, a section on identification and recognition of containers, data cards on the top 50 hazardous materials and CBRNE's, as well as a matrix of WMD and drug lab precursor chemicals. Designed also for use in the classroom.

WEB SITE—The learning system is supported by a dedicated Web site [8STEP-PROCESS.COM]. From the Web site you can send questions to the authors and learn more about the **THE HAZMAT LEARNING SYSTEM™**.

To order call Fire Protection Publications, 1-800-604-4055
or Red Hat Publishing, 1-800-603-7700

HAZARDOUS MATERIALS

MANAGING THE INCIDENT

DVD or VHS

8 programs on VHS or DVD illustrate each step of the 8 Step Process

An integral part of a Complete Hazmat Learning System!

With Greg Noll & Mike Hildebrand

Step 1 - Site Management & Control Demonstrates how to set up a systematic, coordinated approach to a hazmat incident. Covers taking command, establishing the perimeter, safe approach & positioning, hazard control zones, and protective actions. 18 minutes.

Step 2 - Identifying the Problem Portrays how responders recognize the presence of hazmats at the scene & identifying the material or general hazard class involved. Covers design & constructionof containers, DOT hazard classes, placarding, NFPA 704. 18 minutes.

Step 3 - Hazard & Risk Evaluation How to evaluate risks and determine response objectives. Covers sources for hazard information, air monitoring, and risk factors. How hazmats and their containers behave. 25 minutes.

Step 4 - Protective Clothing & Equipment Examines protective clothing and respiratory protection used at hazmat incidents, the limitations of each, and degradation, penetration, & permeation. Also NFPA standards, EPA levels of protection. 22 minutes.

Step 5 - Information Management & Resource Coordination Studies the types of information needed to manage a hazmat incident and setting up and implementing an incident management system. Covers gathering and evaluating information, coordinating multiple resource groups, controlling the flow of information. 22 min.

Step 6 - Implementing Response Objectives How the incident commander evaluates and implements the best available strategic goals and tactical objectives. Explains the difference between strategy and tactics. Covers rescue and protective actions, spill control & confinement, and fire control. 25 minutes.

Step 7 - Decontamination Discusses how contamination occurs, setting up the decon site, the decon officer, protective clothing requirements, technical and emergency decon, evaluating the effectiveness of decon operations and clean-up. 22 minutes.

Step 8 - Terminating the Incident How to carry out a smooth, safe transition from the emergency phase to restoration and recovery. Covers documenting the incident, post-incident analysis, debriefing, and critique. 21 minutes.

**Formerly sold as "The Eight Step Process"*

To Order, Contact

PO Box 1928, Edgartown, MA 02539

Tollfree: 800 842-0999
Fax: 508 627-8863
Email: info@efilmgroup.com
Online: www.efilmgroup.com

Previews & rentals available. Call for details

This series is essential training for fire fighters, hazmat teams, and industrial emergency response teams! Helps meet the training requirements of OSHA's HAZWOPER rule and NFPA 471 and 472! A must for every training library!!

CHAPTER 5

FDNY

SITE MANAGEMENT

OBJECTIVES

1. Define Site Management and Control.
2. List and describe the six major tasks that must be implemented as part of the site management and control process.
3. Describe the procedures for initially establishing command at a hazmat incident.
4. Describe the guidelines for the safe approach and positioning of emergency response personnel at a hazmat incident.
5. Define the following terms and describe their significance in controlling emergency response resources at a hazmat incident:
   - Staging
   - Level I Staging
   - Level II Staging
6. Describe the procedures for establishing scene control through the use of an Isolation Perimeter at a hazmat incident.
7. Describe the procedures for establishing scene control through the use of Hazard Control Zones at a hazmat incident.
8. Define the following terms and describe their significance in establishing Hazard Control Zones:
   - Hazard Control Zones (NFPA 472–3.3.13)
   - Hot Zone (NFPA 472–3.3.13.2)
   - Warm Zone (NFPA 472–3.3.13.3)
   - Cold Zone (NFPA 472–3.3.13.1)
   - Area of Refuge
9. Describe the role of on-scene security and law enforcement personnel in establishing perimeters at a hazmat emergency. [NFPA 472–7.3.4.3 (6)]
10. Define the following terms and describe their significance in protecting the public at a hazmat incident:
    - Public Protective Actions
    - Protection-in-Place
    - Evacuation
11. Describe three criteria for evaluating Protection-in-Place as a Public Protective Action option. [NFPA 472–7.3.4.3 (1 through 12)]
12. Describe the guidelines and procedures for implementing Protection-in-Place at a hazmat incident.
13. Describe three criteria for evaluating evacuation as a Public Protective Action option.
14. List the four criteria for implementing a Limited Scale Evacuation.
15. Define Sick Building Syndrome and list three indicators of a Sick Building.
16. List four situations that may justify a Full-Scale Evacuation.
17. List four critical issues that must be addressed by the Incident Commander to effectively manage a Full-Scale Evacuation.

ABBREVIATIONS AND ACRONYMS

CPC	Chemical Protective Clothing
EAS	Emergency Alerting System
EBA	Escape Breathing Apparatus
ICP	Incident Command Post
IDLH	Immediately Dangerous to Life and Health
MOU	Memorandum of Understanding
PPA	Public Protective Action
SBS	Sick Building Syndrome
VOC	Volatile Organic Compounds

INTRODUCTION

Site Management is the first step in the Eight Step Process©. The major emphasis of Site Management is on establishing control of the incident scene by assuming command of the incident and isolating people from the problem by establishing an Isolation Perimeter and Hazard Control Zones. Site management and control provide the foundation for the response. Responders cannot safely and effectively implement an incident action plan (IAP) unless the playing field is clearly established and identified for both emergency responders and the public.

The actions taken by the Incident Commander (IC) in the first ten minutes of the incident usually dictate how well the next hour will go. Addressing site management tasks early in the incident helps the IC manage the biggest problem first—people. People too close to the hazmat scene can become potential rescues. The IC should clear the most hazardous areas first and use an isolation perimeter and hazard control zones to protect responders and spectators.

When the IC launches into tactical operations without first addressing basic site management tasks, safety issues will continue to arise throughout the course of the incident and become a problem. The Incident Commander should not begin extended operations until the hazard areas have been identified and the isolation perimeter secured.

TERMINOLOGY AND DEFINITIONS

To understand the material in this chapter better, let's quickly review some terminology. The following terms are listed in the order they will be discussed in this chapter.

Staging—The safe area established for temporary location of available resources close to the incident site to reduce response time.

Level I Staging—The initial location for emergency response units at a multiple unit response to a hazmat incident. The first-arriving unit responds to the incident scene, while all other units are ordered to stage at a safe location close to, but away from the scene.

Level II Staging—The location where arriving units are initially sent when an incident escalates past the capability of the initial response. It is a tool usually reserved for large, complex, or lengthy hazmat operations.

Isolation Perimeter—The designated crowd control line surrounding the Hazard Control Zones. The isolation perimeter is always the line between the general public and the Cold Zone. Law enforcement personnel may also refer to this as the outer perimeter.

Hazard Control Zones—Designation of areas at a hazardous materials incident based on safety and the degree of hazard. These zones are defined as the hot, warm, and cold zones.

Hot Zone—The control zone immediately surrounding a hazardous materials incident that extends far enough to prevent adverse effects from hazardous materials releases to personnel outside the zone. Law enforcement personnel may also refer to this as the inner perimeter.

Warm Zone—The control zone at a hazardous materials incident site where personnel and equipment decontamination and hot zone support takes place.

Cold Zone—The areas at a hazardous materials incident that contains the incident command post and other support functions necessary to control the incident.

Area of Refuge—A holding area within the hot zone where personnel are controlled until they can be safely decontaminated or treated.

Public Protective Actions—The strategy used by the Incident Commander to protect the general population from the hazardous material by implementing a strategy of either (1) Protection-in-Place, (2) Evacuation, or (3) a combination of Protection-In-Place and Evacuation

Protection In-Place (shelter-in-place)—Directing people to go inside a building, seal it up as effectively as possible, and remain there until the danger from a hazardous materials release has passed.

Evacuation—The controlled relocation of people from an area of known danger or unacceptable risk to a safer area or one in which the risk is considered to be acceptable.

SITE MANAGEMENT TASKS

From a tactical perspective, Site Management can be divided into six major tasks:

1. Assuming Command and establishing control of the incident scene
2. Assuring safe approach and positioning of emergency response resources at the incident scene
3. Establishing Staging as a method of controlling arriving resources
4. Establishing an Isolation Perimeter around the incident scene
5. Establishing Hazard Control Zones to assure a safe work area for emergency responders and supporting resources
6. Sizing up the need for immediate rescue and implementing initial Public Protective Actions

Life safety is the highest tactical priority of any Incident Commander. There will always be situations where initial size-up warrants that responders move directly into rescue operations (e.g., a driver who is obviously alive and trapped in the cab of a burning gasoline tank truck). However, even under the most extreme situations, implementing initial Site Management tasks will save lives. Don't let a bad situation become worse by getting sucked into letting emergency responders charge into rescue situations without following safe operating procedures. Rescue is discussed in more detail in Chapter 10, "Implementing Response Objectives."

ESTABLISHING COMMAND

Chapter 3 discusses Incident Command in detail. This section provides a brief review of the Incident Command process as it relates to Site Management.

Hazmat incidents require strong, centralized command. Without it, the scene will usually degenerate into an unsafe, disorganized group of freelancers (people running around the incident scene doing their own thing with no clue about what the objectives are).

The success or failure of emergency operations will depend on the manner in which the first-arriving officer or responder establishes command. Regardless of job title or rank, this individual should always initiate the following command functions:

- Correctly assume command. The person assuming command should be the highest ranking or most experienced person present.
- Confirm command. Confirm that all personnel on the scene and enroute have been notified of the command structure. Responders need to know (1) who is in command, and (2) where to find the location of the Incident Command Post (ICP).
- Select a stationary location for the incident command post. An experienced commander only gives up the advantage of a stationary command post when it is absolutely necessary for the IC to personally provide one-on-one direction to responders operating in forward positions. In either case, the IC must maintain a command presence on the radio. The IC should also be readily identifiable using a command vest or other means.
- Establish a Staging Area. Make sure that Staging is in an easily accessible location and has been announced over the radio for incoming personnel and apparatus. Staging is discussed in more detail later in this chapter.
- Request necessary assistance. Does the level of resources match the scope and nature of the hazmat problem? Is the problem a Level I, II, or III hazmat incident? Levels of Incidents are discussed in more detail later in this chapter.

APPROACH AND POSITIONING

Safe approach and positioning by the initial emergency responders is critical to how the overall incident will be managed. Emergencies that start bad because of poor positioning sometimes stay bad. If initial emergency responders become "part of the problem," the IC has to change the Incident Action Plan (IAP) to deal with new circumstances. For example, if firefighters become contaminated, the IAP shifts from protecting the public to rescuing and decontaminating the responders.

Good approach and positioning is just a common-sense application of basic safe operating principles. Some general guidelines include the following:

- **Approach from uphill and upwind**. We recognize that highway engineers don't always build roads upwind from hazmat incidents, but if the weather and topography are on your side, take advantage of it to make the incident scene safer. If you do find yourself approaching from the downwind side, then use distance from the incident to your maximum advantage or switch to self-contained breathing apparatus (SCBA).
- **Look for physical hazmat clues**. Avoid wet areas, vapor clouds, spilled material, and so on. Use some common sense; if birds are flying in one side of the vapor cloud and not coming out the other side, you probably have a problem.

Conditions can change quickly at hazmat incidents. The wind can shift and the vapor cloud can overrun your position. Don't position too close until a proper size-up has been completed.

ASK YOURSELF: Where is the hazmat now? Where is it going? Where am I in relation to where it is going? What will it do to me when it gets to where I am? If the answer is that you will be hurt, you are in the wrong position and need to move! This thought process should be ongoing throughout the incident.

STAGING AREAS

Staging procedures facilitate safety and accountability by allowing for the orderly, systematic, and deliberate deployment of responders. The Staging Area is the designated location where emergency response resources (people, equipment, and supplies) are assigned until they are needed.

Staging becomes an element within the Operations Section command structure (see Chapter 3). The Staging Area Manager accounts for all incoming emergency response units, dispatches resources to the emergency scene at the request of the Incident Commander, and requests additional emergency resources as necessary.

The ideal Staging Area is close enough to the Isolation Perimeter to reduce response time significantly, yet far enough away to allow units to remain highly mobile for assignment. The Staging Area should be easily accessible to responding apparatus and sufficiently large (expandable) for the anticipated demand. Staging is effective when the Incident Commander anticipates that additional resources may be required and orders them to respond to a predesignated area several minutes from the scene.

Large campaign-style incidents can bring extensive resources to the scene that may be needed at different times throughout the emergency. If resources will not be required for some time, the IC should consider establishing primary and secondary staging areas. These are sometimes referred to as Level I and Level II Staging Areas.

Level I Staging should be used for any multiple unit response to a hazmat incident. The first unit arriving at the scene of the emergency assumes command and begins site management operations. All other responding units stage at a safe distance away from the scene until ordered into action by the Incident Commander. Level I staging places a response unit on scene ASAP, and allows the first arriver to establish command, size up the problem, and begin to formulate an IAP. It also provides the IC with the most options in assigning the remainder of the response.

Normally, Level I Staging takes place when units stop short of the scene (approximately one block in their direction of travel) or remain outside of a facility's main gate. Level I Staging should be in a safer upwind location. Obviously, you should not drive through an unsafe location just to take up a position in a safer location.

Level II Staging is used when an incident escalates past the capability of the initial response. Level II Staging is a management tool usually reserved for large, complex, or lengthy hazmat operations. As additional units arrive near the emergency, they are staged together in a specific location under the command of a Staging Officer. The crews can be briefed on the situation by the Staging Area Manager and wait for assignments in more comfortable surroundings where they are protected from the weather. When resources are required on the scene, they can be moved forward to an area located near the incident perimeter and deployed by the IC.

Staging Areas should be clearly identified through the use of signs, color-coded flags or lights, or other suitable means. The exact location of the Staging Area will be based on prevailing wind conditions and the nature of the emergency.

ISOLATION PERIMETER

The Isolation Perimeter is the designated crowd control line surrounding the incident scene that is set up to maintain the safety and security of the spectators and the players. Designating and establishing the Isolation Perimeter is an Incident Command responsibility. Maintaining the perimeter throughout the incident is usually the responsibility of law enforcement and security professionals. In simple terms, the isolation perimeter separates the players from the spectators. Who makes up each group?

> **Spectators** = The general public. This includes Mom and Dad, Uncle Larry, all "helpful" people, the media, cats and dogs, and anyone else that is not a Player. Spectators should always be kept outside the Isolation Perimeter.
>
> **Players** = Emergency responders and other support personnel who are part of the team brought to the incident scene to safely make the problem go away. The IC controls the Players at the incident scene by establishing Hazard Control Zones.

The Isolation Perimeter should be flexible and becomes larger or smaller to reflect the hazards and risks. Big Hazards and Risks = Big Isolation Pe-rimeter.

The first objective in setting up the isolation procedure is to immediately limit the number of civilian and public safety personnel exposed to the hazmat. This begins by identifying and establishing an isolation perimeter. Utilize existing features whenever practical. For example, when confronted with an incident inside a structure, the best place to begin is at the points of entry such as the main entrance doors. Once the doorways are secured and the entry of unauthorized personnel is denied, crews can begin to isolate above and below the hazard. Proper protective clothing and equipment must be worn. See Figure 5.1.

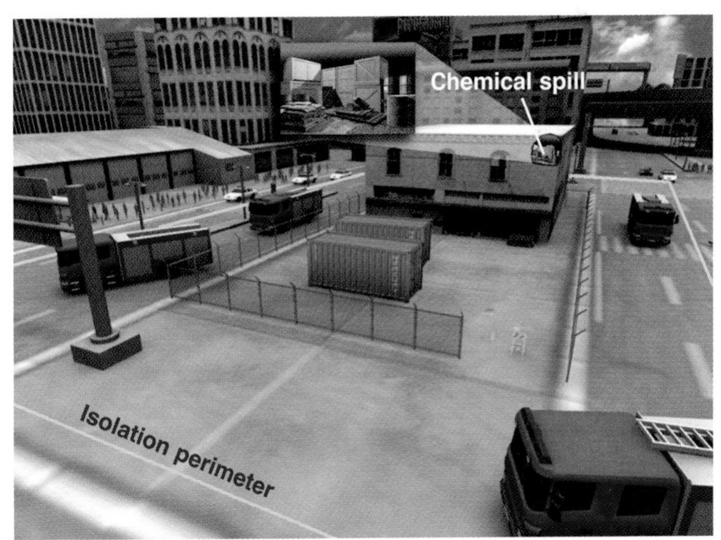

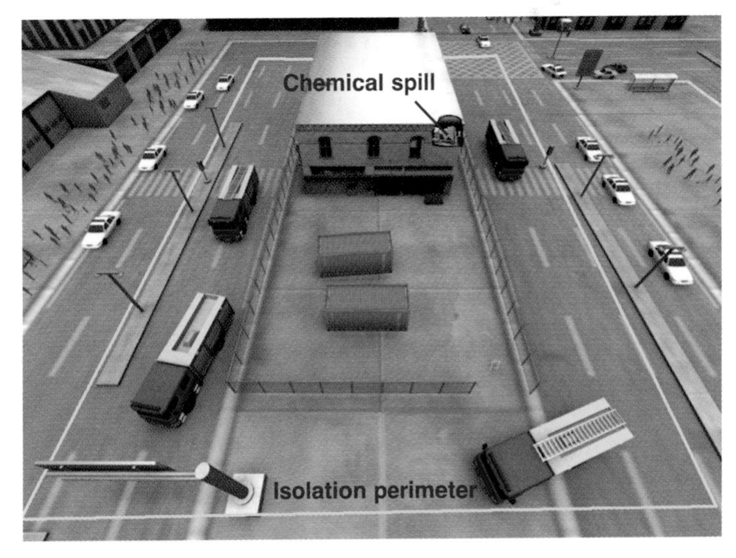

Figure 5.1 Establishing an isolation perimeter outside of a building.

The same concept applies for outdoor scenarios. First, use the existing terrain and roadway features to secure the main entry points into the area and then establish an isolation perimeter around the hazard. Begin by controlling intersections, on/off ramps, service roads, or any other access routes to the scene. Controlling traffic flow early in the incident helps to reduce gridlock. When this task has been completed, a Recon crew can begin a size-up. Injured civilians are still a top priority, but highways and access points can quickly choke up and totally restrict any type of access to the scene. Surrounding the problem with gridlocked, occupied vehicles will quickly compound rescue operations and generally bring the entire operation to a grinding halt. See Figure 5.2.

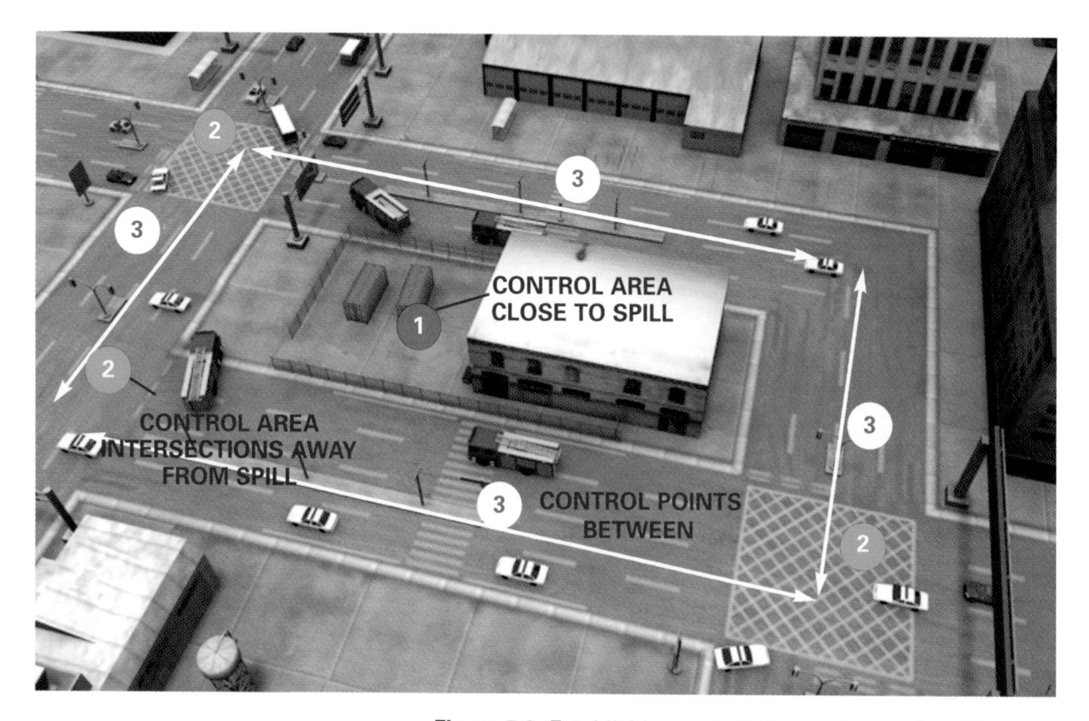

Figure 5.2 Establishing an isolation perimeter for highways.

ESTABLISHING HAZARD CONTROL ZONES

Hazard Control Zones are designated areas at a hazardous materials incident based on the degree of hazard.

DEFINING HAZARD CONTROL ZONES

Hazard Control Zones are three distinctly different zones, beginning at the hazmat problem or release, and working outward toward the perimeter. Hazard Control Zones are designated from the most hazardous to least hazardous as Hot, Warm, and Cold. (Hot = Greatest Risk and Cold = Least Risk.)

The primary purpose of establishing three different Hazard Control Zones within the isolation perimeter is to provide the highest possible level of control and personnel accountability for all responders working at the emergency scene. Defined control zones help ensure that responders do not inadvertently cross into a contaminated area or place themselves in locations that could be quickly endangered by explosions or migrating vapor clouds.

As a general rule, the public should always be located outside of the isolation perimeter, the incident command post and support personnel should be located in the Cold Zone, emergency responders supporting the tactical hazmat response operation should be positioned in the Cold and Warm Zones, and the entry team should be located in the Hot Zone. To work in the hot zone or inner perimeter, remember the First Law of Hot Zone Operations: To play in the game, you must be

- Trained to play
- Dressed to play (i.e., proper level of PPE based on the hazards present)
- Set up in a buddy system with a back up capability
- Decon established
- Coordinated with the IC and Safety Officer

An Area of Refuge (i.e., Safe Refuge Area) should be established within the Hot Zone to control personnel who may have been exposed to the hazmat. They should be held within this limited area until they can be safely handled (e.g., a Decon area has been established). The Hot Zone should be large enough to provide one or more Areas of Refuge as necessary. See Figure 5.3.

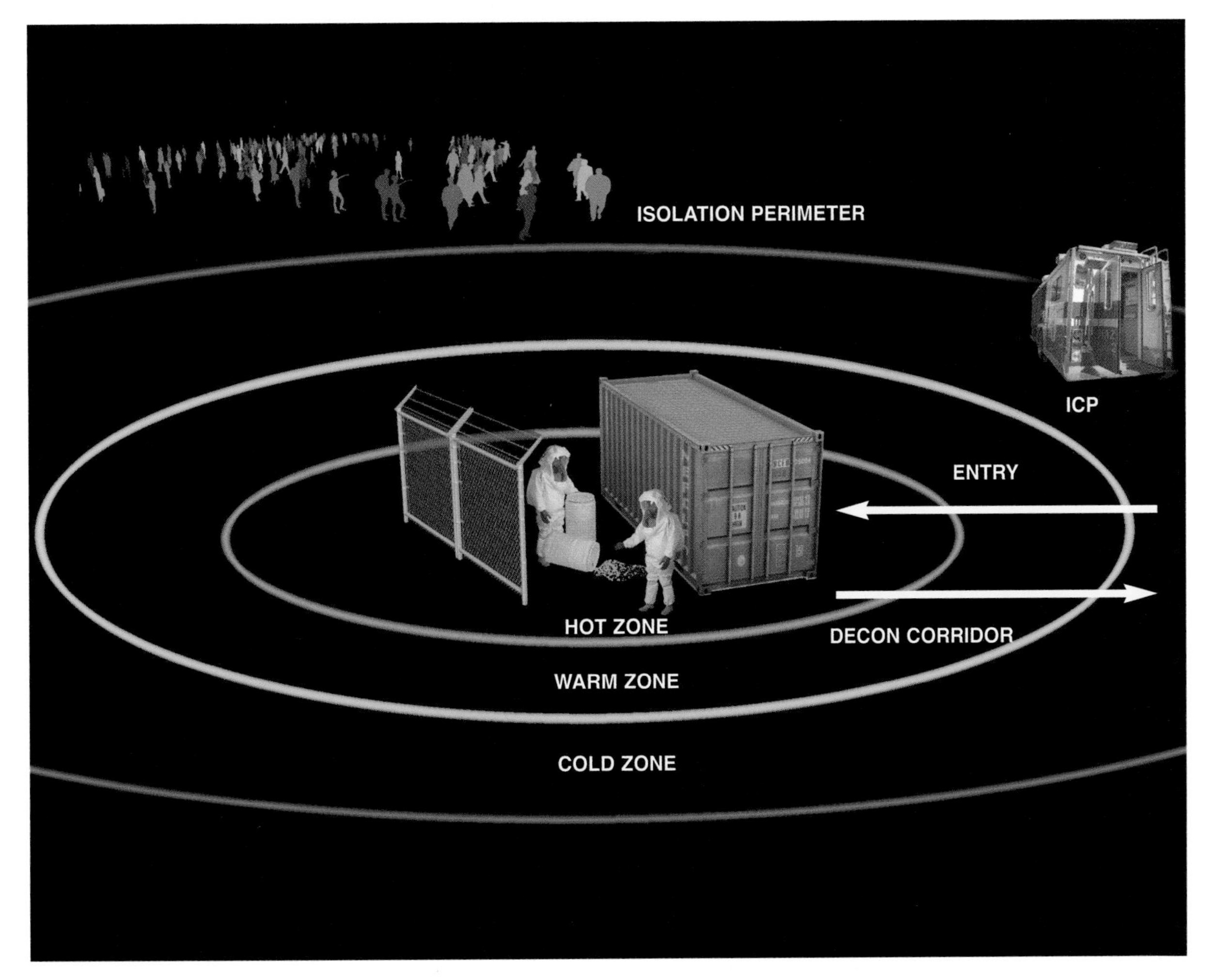

Figure 5.3 Establishing hazard control zones.

IDENTIFYING HAZARD CONTROL ZONES

Hazard Control Zones should be clearly marked and posted on the IC's tactical command worksheet. The Hot Zone can be indicated with colored banner tape, color-coded traffic cones, color-coded light sticks, or temporary fencing for long-term operations.

In outdoor situations, Hazard Control Zones can be designated by using key geographical reference points such as a tank dike wall, fenceline, or street name. Some response agencies also use colored flags or traffic cones to identify control zone locations. Geographic areas should be communicated verbally by radio or in a face-to-face briefing between the IC and other officers.

When the hazard is inside of a building, these zones can be denoted by their location within the structure. For example, a spill in room 321 may dictate that rooms 320 to 322 would be a Hot Zone, the rest of the building would be the Warm Zone, and the area outside of the building itself would be designated the Cold Zone. See Figure 5.1.

While it is acceptable to estimate the size of the Hazard Control Zones early in the incident based on visible clues, the IC should move toward a more definitive assessment using monitoring instruments.

As a guideline, any area with a measurable concentration of a contaminant using available monitoring should initially be considered within the hot zone until additional data are obtained and evaluated. Initial monitoring efforts should concentrate on determining if IDLH concentrations are present. Decisions regarding the size of Hazard Control Zones should be based on the following:

1. **Flammability**—The IDLH action level is 10% of the lower explosive limit (LEL). Any areas above this concentration are clearly inside the Hot Zone.
2. **Oxygen**—An IDLH oxygen-deficient atmosphere is 19.5% oxygen or lower, while an oxygen-enriched atmosphere contains 23.5% oxygen or higher. When evaluating an oxygen-deficient atmosphere, consider that the level of available oxygen may be influenced by contaminants that are present. Areas containing atmospheres that are either oxygen deficient or enriched should be designated as the Hot Zone.
3. **Toxicity**—Unless a published action level or similar guideline is available (e.g., ERPG-2), the STEL or IDLH values should initially be used. If there is no published IDLH value, consider using an estimated IDLH of ten times the TLV/TWA. Initial Hazard Control Zones can be established for toxic materials using the following guideline:
    - **Hot Zone**—Monitoring readings above STEL or IDLH exposure values
    - **Warm Zone**—Monitoring readings equal to or greater than TLV/TWA or PEL exposure values
    - **Cold Zone**—Monitoring readings less than TLV/TWA or PEL exposure values
4. **Radioactivity**—Any positive reading two times above background levels or alpha and/or beta particles that are 200 to 300 counts per minute (cpm) above background would confirm the existence of a radiation hazard and should be used as the basis for establishing a hot zone. Chapter 7 includes a more detailed discussion concerning the application and interpretation of monitoring instruments as it relates to hazard and risk assessment.

Safe operating procedures should strictly control and limit the number of personnel working in the Hot Zone. Most hazmat operations can be accomplished with two to four personnel working for specified time periods using the Buddy System. This is a system of organizing emergency response personnel into work groups so that each person in the work group is designated to be observed by at least one other entry person. The purpose of the Buddy System is to provide rapid assistance in the event of an emergency.

The size of Hazard Control Zones should change over time as the risks increase or decrease. The zones might expand or contract depending on the size of the incident and the nature of the hazards and risks. Retaining large Hazard Control Zones without good technical reasons will create problems with property owners and outside agencies. This is especially true at incidents involving critical infrastructure or industrial facilities. On longer-duration incidents, holding onto large chunks of real estate could generate political problems that can erode the IC's credibility. Don't keep the property owner or law enforcement personnel out of the information loop. Brief them on how and why you have established Hazard Control Zones early in the incident.

While a large perimeter surrounding the incident is desirable, a common mistake is to seal off more real estate than can be effectively controlled. Just as in military operations, it takes a given amount of personnel to patrol a certain perimeter. If the patrols are too sparsely manned or too infrequent, someone is sure to penetrate the line. Given limited personnel, it is better to secure a smaller area completely and expand the perimeter outward as additional resources become available. On the other hand, it is much easier to contract a zone as the hazards decrease than to try to grow one in a hurry when things go bad.

ROLES OF SECURITY AND LAW ENFORCEMENT PERSONNEL

The IC should make perimeter isolation assignments as soon as possible. This usually begins with summoning the senior security or law enforcement supervisor to the Incident Command Post. This individual will become a key player who will help establish communications between agencies and determine what areas will be controlled first, and how they will be managed throughout the incident. Law enforcement officers should be briefed with all available information.

The people who are actually involved in establishing a perimeter or in the hands-on acquisition of buildings need to know exactly what the potential hazards and risks appear to be. If there is a risk that these officers may be exposed to the hazard as the isolation area expands, they must be provided with proper PPE.

Law enforcement personnel are best utilized where traffic and crowd control will involve large groups of people on public property. Another important law enforcement function is patrolling the perimeter for civilians trying to sneak a closer look and for the occasional renegade photographer or camera operator trying to get the closest shot. In today's world, terrorism incidents also pose the potential for secondary devices and the targeting of emergency responders.

In today's threat environment, law enforcement also plays a very important role in providing security for the ICP and emergency responders within the Isolation Perimeter. Police are better trained for perimeter security than firefighters. They look the part and have the credibility to convince people to relocate to safer areas. If the situation gets ugly, they usually have the authority and muscle to relocate people against their wishes. See Figure 5.4.

Figure 5.4 Law enforcement officers are essential for controlling the Isolation Perimeter.

When operating on a private facility such as an industrial plant or government installation, the on-site security force fills the same slots as law enforcement in the system. Most industrial security officers are adequately trained and extremely familiar with the site, its hazards, and the available resources. They often know the employees inside the plant by sight and can provide specifics on evacuation plans, emergency procedures, and availability of special tools. The IC can usually rely on Security to handle perimeter control issues within the plant while local law enforcement officers control the areas outside of the fence. Working together under a Unified Command System, the law enforcement and security team can be a valuable asset to the IC.

RESCUE AND INITIAL PUBLIC PROTECTIVE ACTIONS

Scenarios may occur where immediate response actions are required in order to rescue or remove victims from a hazardous environment. If the hazards are known and the risks can be readily evaluated, initial rescue operations may be implemented at an acceptable level of risk to responders. However, scenarios such as those that may be encountered at a terrorism event involving chemical agents will pose many unknowns and present responders with significantly higher risk levels.

The reality of most tactical hazmat response operations is that they are not well suited for rapid entry and extrication operations. Some hazmat responders have developed in-house programs to enhance their ability to rapidly deploy personnel wearing CPC. However, experience shows that it will most likely be first responders who will be faced with the decision to "go or not go." Critical factors will include the number of responders available and the level of PPE available to responders. Rescue as a strategy will be discussed further in Chapter 10.

INITIATING PUBLIC PROTECTIVE ACTIONS

Public Protective Actions (PPAs) are the strategy used by the Incident Commander to protect the general population from the hazardous material by implementing a strategy of either (1) Protection-in-Place, (2) Evacuation, or (3) Combination of Protection-in-Place and Evacuation. PPA strategies are usually implemented after the IC has established an Isolation Perimeter and defined the Hazard Control Zones for emergency responders.

There are no clear benchmarks for the PPA decision-making process. A combination of factors must be evaluated to select the best strategy based on incident conditions. Some of the more important factors include what has been released, how much, the hazards of the material(s) involved, the population density, time of day, weather conditions, type of facility, and the availability of air- tight structures. As you can see, there is a great deal to consider both before and during an incident.

One common misunderstanding about PPA strategies is that the IC must use one strategy versus the other (e.g., either Protection-in-Place or Evacuation). This is not an either/or choice. Sometimes evacuation is the best strategy, sometimes Protection-In-Place is the best choice, and sometimes the best option is to use both strategies at the same time (e.g., evacuate the area close to the hazards while directing others further away to Protect-In-Place).

INITIAL PPA DECISION MAKING

PPA decisions are very incident specific and require the use of the IC's judgment and experience. For example, if a release occurs over an extended period of time, or if there is a fire that cannot be controlled within a short time, evacuation is typically the preferred option for nonessential personnel. Evacuation may not always be necessary during incidents involving the airborne release of extremely hazardous substances, such as anhydrous ammonia, chlorine, and hydrofluoric acid. Airborne materials can move downwind so rapidly that there may be no time to evacuate a large number of plant employees or the surrounding community. In other situations, evacuating people may actually expose them to greater potential risk. For short-term releases, the most prudent course of action may be to remain inside of a structure. See Figure 5.5.

AP Wide World

Figure 5.5 For short term releases, remaining inside a structure may be the best decision.

The IC's decision to either evacuate or protect-in-place should be based on the following factors:

- Hazardous material(s) involved, including their characteristics and properties, amount, concentrations, physical state, and location of release.

- The population at risk, including facility personnel and the general public. In addition, the IC must consider the resources required to implement the recommended protective action, including notification, movement/transportation, and possible relocation shelters.
- The time factors involved in the release. Consideration must be given to the rate of escalation of the incident, the size and observed or projected duration of the release, the rate of movement of the hazardous material, and the estimated time required to implement protective actions.
- The effects of the present and projected meteorological conditions on the control and movement of the hazardous materials release. These would include atmospheric stability, temperature, precipitation, and wind conditions.
- The capability to communicate with the population at risk and emergency response personnel prior to, during, and after the emergency.
- The availability and capabilities of hazmat responders and other personnel to implement, control, monitor, and terminate the protective action. This should include a size-up of the structural integrity and infiltration rates of structures potentially available for Protection-in-Place throughout the area.

Prior knowledge of the hazmat or the facility through planning information or computer dispersion models acquired through the hazards analysis process can also assist the Incident Commander in this evaluation.

Regardless of the tactic used, achieving PPA objectives translates into gaining control of a specified area beyond the isolation perimeter, securing and clearing that area, and then controlling a second downwind or adjacent area. In this manner, more and more threatened areas can be secured as more resources become available to the IC.

It is imperative that the Incident Commander use a systematic and structured approach to clearing the public away from the hazard area. Establishing priorities and communicating the plan for Public Protective Action tactics are important from the beginning and should be updated on a map at the ICP as new PPA zones are identified and controlled.

In the early stages of an incident, the IC is often preoccupied with size-up and rescue activities and can easily overlook people problems in the immediate area. Be aware that if the situation deteriorates rapidly, exposures within 1,000 feet may be contaminated. In other words, everyone inside the isolation perimeter is a potential rescue.

Areas that should receive immediate attention by the IC include the following:

- Locations within 1,000 feet of the incident that will be rapidly overtaken by the hazmat. This is especially a concern when a flammable or toxic gas or vapors are drifting downwind.
- Locations near the incident where people are already reasonably safe from the hazmat. People near the hazmat should be alerted to keep clear of the hazard and remain indoors until given other instructions.
- Key locations that control the flow of traffic and pedestrians into the hazard area (for example, doorways, on-ramps, and grade crossings).
- Special high-occupancy structures such as schools.
- Structures containing sick, disabled, or incarcerated persons.

The Emergency Response Guidebook (ERG) is a good resource document to guide the Incident Commander in making quick initial judgment calls on which PPA option to implement. The Guidebook also provides some basic guidelines concerning the size of the initial isolation zone based on the type of hazardous material and size of container. The IC should be thoroughly familiar with how to use the Emergency Response Guidebook. The instructions to the ERG provide some useful background information on Public Protective Action decision making.

There is a fine line between isolation objectives and evacuation. For our purposes, Isolation requires quick action to protect the public and first responders from an immediate, life-threatening situation. Isolation is a necessity; failure to act when people are outside, in exposed locations, will result in injuries. In contrast, Evacuation implies a prolonged, precautionary stay away from the affected location.

PROTECTION-IN-PLACE

The Protection-in-Place strategy involves directing people to go inside of a building, seal it up as effectively as possible, and remain there until the danger from a hazardous materials release has passed. Protection-in-place (sheltering in place) is a concept that is a familiar part of people's daily activities. We routinely shelter ourselves indoors when we close windows to keep out noise or dust or to keep the house cool on a hot summer day when the air conditioning is running. The concept of protection-in-place as applied to a hazmat release is identical except the objective is to prevent toxic vapors from entering the structure.

An excellent source of research and technical information on protection-in-place is the National Institute of Chemical Studies (NICS) in Charleston, West Virginia. Information on NICS and its research can be obtained at http://www.nicsinfo.org. In its June 2001 report entitled "Sheltering in Place as a Public Protective Action," NICS reviewed a number of case studies to determine whether it is better to evacuate or protect-in-place during a chemical release. The report concluded as follows:

- Sheltering-in-place is an appropriate public protection tool in the right circumstances. For chemical releases of limited duration, it is faster and usually safer to shelter in place than to evacuate.
- For all case studies examined during this study, there were no fatalities associated with sheltering in place.
- The body of evidence suggests that if there is insufficient time to complete an evacuation, or the chemical leak will be of limited duration, or conditions would make an evacuation more risky than staying in place, then sheltering in place is a good way to protect the public during chemical emergencies.

Research and accident investigations clearly indicate that staying indoors may provide a safe haven during toxic vapor releases. However, it must also be recognized that sustained continuous releases may eventually filter into a structure and endanger the occupants. Protection-in-place may not be the best option if the vapors are flammable, such as with liquefied petroleum gas.

The Incident Commander may have to make critical decisions based on weather conditions and forecasts. High humidity and warm air can force vapors toward the ground. Air ventilation and air conditioning ducts may force toxic vapors into the building before the public is warned and the order to take shelter is issued by the IC.

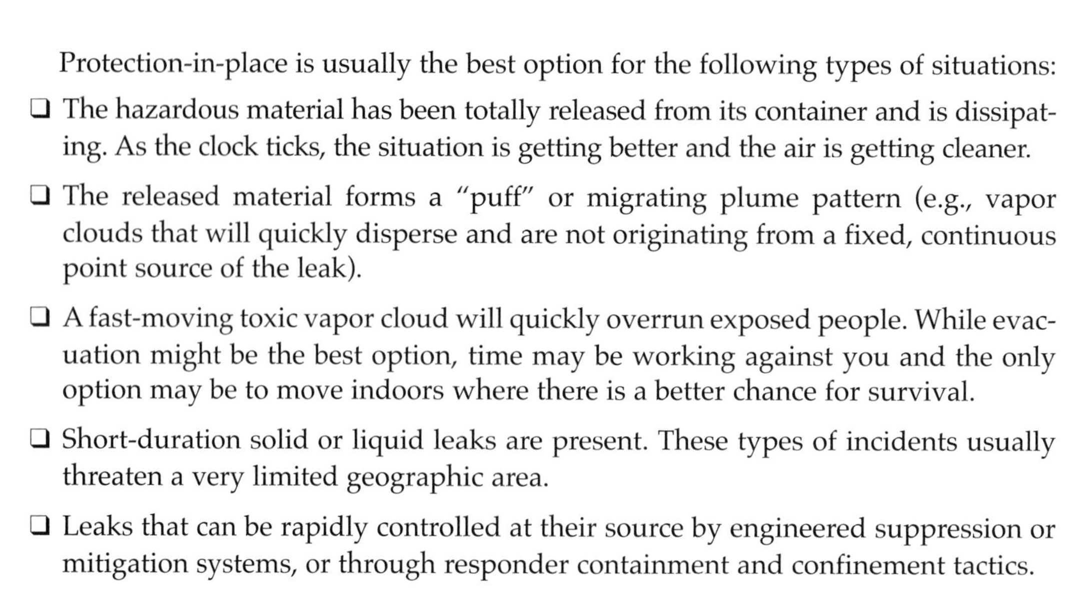

Protection-in-place is usually the best option for the following types of situations:

❑ The hazardous material has been totally released from its container and is dissipating. As the clock ticks, the situation is getting better and the air is getting cleaner.

❑ The released material forms a "puff" or migrating plume pattern (e.g., vapor clouds that will quickly disperse and are not originating from a fixed, continuous point source of the leak).

❑ A fast-moving toxic vapor cloud will quickly overrun exposed people. While evacuation might be the best option, time may be working against you and the only option may be to move indoors where there is a better chance for survival.

❑ Short-duration solid or liquid leaks are present. These types of incidents usually threaten a very limited geographic area.

❑ Leaks that can be rapidly controlled at their source by engineered suppression or mitigation systems, or through responder containment and confinement tactics.

When protection-in-place is the best course of action, the IC must make sure that the public is provided with clear instructions on what they should do. The following are examples:

- Close all doors to the outside and close and lock all windows (windows seal better when locked). Seal any obvious gaps around windows, doors, vents, and so on with tape, plastic wrap, wet towels, or other materials.
- Turn off all heating, ventilating, and air conditioning systems (HVAC) systems and window unit air conditioners. If applicable, place inlet vents in the closed position.
- Close fireplace dampers. If there is a fire in the fireplace, let it burn down without closing the dampers.
- Turn off and cover all exhaust fans.
- Close as many internal doors as possible.
- Pick one room in the house or structure to use as a shelter room. A bathroom can be a good choice because water and a toilet are available, if needed. A master bedroom can also be a good choice if it has a bathroom and a phone. An upstairs room may be a better choice than a basement room, as may chemicals are heavier than air and tend to sink near the ground.
- Monitor the local AM/FM radio or local television stations for further information.

The protect-in-place option sounds good on paper, but experience has shown that the public's compliance with the IC's recommendations and instructions will be dependent on the following factors:

❑ Receipt of a timely and an effective warning message. The public needs to get the word rapidly and the instructions must be clear. For example, a warning message broadcast in English in a Spanish-speaking neighborhood may be misunderstood. A warning to take shelter may be interpreted to mean evacuate.

❑ Clear rationale for the decision to protect-in-place, as compared to an evacuation. If the instructions sound dumb, they won't be implemented. If the public feels threatened (they can see smoke, fire, or can smell the hazmat), they may want to evacuate even though the safest course of action would be to stay where they are.

- ❑ Credibility of emergency response personnel with the general public. If the public's perception of your public safety agency is high, you will have more credibility and trust when you ask them to do something in an emergency.
- ❑ Previous training and education by fixed facility personnel and the public on the application and use of protection-in-place. If the public has never even heard the words "protect-in-place," don't be surprised when they ignore your order. Advance training of the public in what to do when a chemical emergency occurs makes a big difference in establishing your credibility. Citizens are more comfortable following PPA orders if they understand that the community has a well thought out plan and they are familiar with the protection-in-place procedure.

WALLY WISE GUY™

An excellent example of an effective public education effort is the "Wally Wise Guy" program. Originally developed by the Deer Park, Texas, Local Emergency Planning Committee (LEPC), Wally is a turtle that "Knows it's wise to go inside his shell whenever there's danger." This costumed character is a mascot designed to teach children and their parents how to shelter-in-place in case of a chemical emergency. See Figure 5.6.

Wally is targeted to reach children from kindergarten to fourth grade, so Wally makes personal appearances at elementary schools, day care centers, community parades, sporting events, and other civic events. The person wearing the Wally costume could be male or female of any age or nationality, Wally is a non-speaking character. Wally is always accompanied by a spokesperson who presents Wally's messages and answers the audiences questions.

The Wally Wise Guy™ program has been a very effective tool to market the protection-in-place concept. The program has spread across the United States through LEPCs. Wally's message has been projected on everything from coloring books (in Spanish and English) to refrigerator magnets. Wally's image has even been used on fire apparatus compartment doors in industrial areas as a marketing tool. For more information, go to http://www.wally.org.

Wally Wise Guy™

Figure 5.6 Wally Wise Guy program teaches children how to shelter.

Case Study

I-610 AT SOUTHWEST FREEWAY, HOUSTON, TEXAS
MAY 11, 1976

NTSB

Figure 5.7

On a bright sunny day at 11:08 am, on May 11, 1976, a Transport Company of Texas tractor–semitrailer tank truck transporting 7509 gallons of anhydrous ammonia struck and broke through a bridge rail on a ramp connecting I-610 with the Southwest Freeway (U.S. 59) in Houston, Texas. The truck left the ramp, struck a bridge support column of an adjacent overpass, and fell onto the Southwest Freeway, approximately 15 feet below.

The tank truck breached immediately on impact, releasing most of its contents into the atmosphere. At the time of the accident there were about 500 people within 1/4 mile of the release. The released ammonia fumes rapidly penetrated automobiles and buildings. When their occupants left to escape the fumes during the early minutes of the release, many were exposed to fatal doses of ammonia.

The temperature at the time of release was in the low 80s. The released ammonia immediately vaporized and the 7-mph wind gradually decreased the vapor concentration at ground level. Witnesses reported the white ammonia vapor cloud initially reached a height of 100 feet before being carried by the 7-mph wind for approximately 1/2 mile. After five minutes, most of the liquefied ammonia had boiled off and the vapor cloud was completely dispersed.

Seventy-eight of the 178 victims, who were within 1000 feet of the release point, were hospitalized and treated for symptoms of ammonia inhalation. Over 100 persons were treated for less severe injuries. Five of the six fatalities were due to ammonia exposure. Because all fatalities were within 200 feet of the estimated release, it is estimated that the ammonia concentration within this distance was greater than 6500 parts per million for at least two minutes. The IDLH for ammonia is 500 ppm.

A detailed investigation of this incident conducted by the U.S. National Transportation Safety Board in 1979 revealed that there were significant differences in the degree of injury among the exposed victims who evacuated buildings and those who protected-in-place.

The Board's conclusion was that the protection offered survivors by the vehicles and buildings demonstrated that there were alternatives to simply running away from the released hazmat. A detailed investigation conducted by the Board showed that people who sheltered and remained inside buildings received no harm from the ammonia. Also, people who remained inside of their automobiles generally received less severe injuries than those who left their cars and tried to escape the ammonia.

While there have been other investigations conducted with similar conclusions, the Houston case was the first the authors are aware of concerning protection-in-place issues that were documented using forensic science. As a result of NTSB's report, many emergency response and safety professionals began to rethink whether evacuation was always the best tactical option.

The investigation conducted by the Board also documented that the actions taken by the Houston Fire Department (HFD) saved lives. Within 10 minutes the HFD dispatched 14 emergency rescue units and 4 pieces of fire apparatus. The Incident Command System was used to coordinate EMS, fire, and police agencies. As a result, many contaminated and injured victims were located by search and rescue teams and escorted to safety, where they received medical treatment.

UNDERSTANDING WHY PROTECTION IN PLACE WORKS
UNIVERSITY OF ALBERTA RESEARCH

In the late 1980s the Department of Mechanical Engineering at the University of Alberta, Edmonton, Canada, conducted extensive tests on Canadian and American homes to determine whether protection-in-place tactics would actually work. Test findings revealed that for an accidental toxic gas release that occurs over several minutes to half an hour, even a very leaky building contains a sufficient reservoir of fresh air to provide effective sheltering-in-place. However, for longer-duration releases of one to three hours, the average indoor concentration may reach 80% or more of the outdoor average during a steady continuous release.

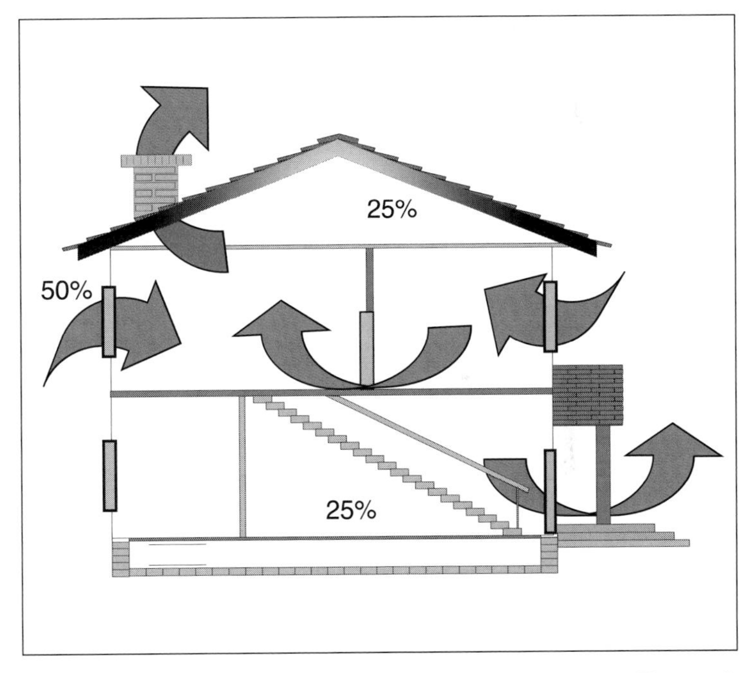

Figure 5.8

For releases that have a long duration of an hour or more, the choice between shelter or evacuation is difficult to make. Typical air exchange rates in a house are about 0.5 air changes per hour (ACH). For a three-hour release, this exchange rate causes the air in the house to be replaced 1.5 times during the event. After this air exchange, the indoor concentration is about 80% of the average outdoor value. Obviously, the more energy efficient or tight the building is, the slower the air exchange rate will be.

Source: "Effectiveness of Indoor Sheltering During Long Duration Toxic Gas Releases," by D. J. Wilson, Department of Mechanical Engineering, University of Alberta, Edmonton, Alberta, T6G 2G8.

EVALUATING STRUCTURES FOR PROTECTION-IN-PLACE

The age and construction of a building greatly influence how successful protection-in-place will be. Responders should become familiar with the types of structures in the community. Preincident plans and hazards analysis surveys should incorporate this type of information. As a general rule, the older a building is, the less likely it will provide a safe place of refuge for periods of longer than one hour.

Hazmat responders should be familiar with the types of structures within the community and consider the following variables:

- Age of the building—Older buildings do not seal well. Structures with leaky windows, doors, and poke-throughs may not be good places to seek refuge during a toxic gas release.
- Prevailing wind direction—Most communities have a predictable prevailing wind direction that is seasonal. You should be familiar with the local Wind Rose, which indicates the direction the wind blows the majority of the time for the time of year in the area you live in. This information is available from the National Oceanic and Atmospheric Administration, National Weather Service (also refer to http://home.pes.com/windroses).
- Building height—The geometry of high-rise buildings and their proximity to one another significantly effects the movement of toxic gas. Canyon and tunnel effects in high-rise areas can make toxic gas plumes go in directions against the prevailing wind that are unpredictable.

LOS ALAMOS NATIONAL LABORATORY RESEARCH

Studies conducted at the Los Alamos National Laboratory by Michael Brown and Gerald Streit indicate that "for an outdoor plume release, closing the windows and doors and staying inside may be initially safer than remaining outdoors. Due to infiltration (e.g., leaks, ventilation) and the eventual passage of the bulk of the plume, there may come a time when it is actually safer to be outdoors rather than indoors." See Figure 5.9.

Emergency response decision makers responsible for public announcements should be aware that safe zones may change from inside to outside buildings during the course of the event, so that strict guidelines for when to go outside may be difficult to derive. [Source: Brown, Michael, J., and Streit, Gerald, E., "Emergency Responders' Rules-of-Thumb for Air Toxics Releases in Urban Environments." Los Alamos National Laboratory, U.S. Department of Energy, Publication LA-UR-98-4539, Los Alamos, NM (1998), page 21]. Note: This report is available at http://www.mipt.org/pdf/la-ur-98-4539.pdf.

In the aforementioned study, the researchers note that when protection-in-place is implemented in urban environments, the IC must recognize that the geometry of the structure, especially in high-rise building areas, can significantly effect the movement of airborne toxic gases. The report includes 12 "Rules of Thumb" on predicting how toxic gas releases in high-rise areas may be affected by the buildings shape and wind direction. The report includes many easy-to-understand graphics.

INDOOR EFFECTS

When the plume is passing over, it is probably safer to remain indoors. After the plume has passed by, it may be safer to move outdoors.

Lesson: For an outdoor release, modeling studies show that concentrations can initially be lower indoors, but then later the concentrations become lower outside. These relationships, however, depend upon the details of the building ventilation.

Figure 5.9 Staying inside may be initially safer than going outdoors.

NEIGHBORHOOD SURVEYS

Neighborhood surveys can reveal a significant amount of information concerning the types of buildings present and they can influence organizational policy concerning whether protection-in-place is a viable option. For example, if an initial survey of the homes surrounding a chemical plant reveals that they are of 1940s construction and have not been retrofitted with energy efficient doors and windows, then protection-in-place may not be the best option to implement in this area. On the other hand, if the same neighborhood is made up primarily of modern, energy efficient homes, the rate of air changes per hour will be substantially lower.

Some jurisdictions and petrochemical companies have adopted a simple survey tool to evaluate the various types of structures. An example is provided in Figure 5.11.

The following Rating System for protection-in-place provides some general guidelines for evaluating structures:

Type I Structure—Modern, energy efficient building constructed since 1970. Or older building with upgraded energy efficiency.

Type II Structure—Older construction, with limited energy efficiency. Constructed between 1950 and 1970.

Type III Structure—Oldest type construction built before 1950.

Type IV Structure—Mobile home, trailer, shed, and so on without energy efficiency.

RATING SYSTEM FOR PROTECTION-IN-PLACE

TYPE I STRUCTURE

Modern, energy efficient building constructed since 1970. Or, older building with upgraded energy efficiency.

TYPE II STRUCTURE

Older construction, with limited energy efficiency. Constructed between 1950 and 1970.

TYPE III STRUCTURE

Oldest type construction built between 1920 and 1950.

TYPE IV STRUCTURE

Mobil home, trailer, shed, etc. without energy efficiency.

Figure 5.10 Rating system for evaluating structures for protection-in-place.

PROTECTION-IN-PLACE FIELD SURVEY

Please briefly describe the location of the area by cross-street, neighborhood or facility.

Complete Family Name ______________________________

Street Address ______________________________

Phone Number ______________________________

Number of Occupants ______________________________

Special Comments ______________________________

____

Are there any additional buildings ☐ YES ☐ NO
How many? ______________________________

1. Check the box which best describes the block, neighborhood or area that you are surveying and list the geographic location.

   - ☐ Residential area consisting primarily of single family structures.
   - ☐ Residential area consisting primarily of multiple family structures.
   - ☐ Commercial area consisting primarily of stores, restaurants, office buildings.
   - ☐ Industrial area consisting primarily of warehouses and manufacturing type structures.
   - ☐ Special structures consisting of high life occupancy such as a hospital, nursing home, school, or prison. For example, structures requiring special assistance.

2. Check the box that best describes the type of building construction for the majority if the structures in the area you are surveying.
   - ☐ TYPE I: Modern construction built since1970.
   - ☐ TYPE II: Older construction built between 1950 and 1970.
   - ☐ TYPE III: Oldest construction or historic built between 1920 and 1950.
   - ☐ TYPE IV: Mobil home, trailer, etc. no energy efficiency.

Figure 5.11

EVACUATION

Evacuation is the controlled relocation of people from an area of known danger or unacceptable risk to a safer area or one in which the risk is considered to be acceptable. Evacuation of both industrial fixed facility personnel and the general public is an attempt to avoid their exposure to any quantity of the released hazardous material. Under ideal conditions, evacuation will remove these individuals from any exposure to the released hazmat for a given length of time. For our purposes, we will describe evacuations in terms of limited-scale and full-scale evacuations.

LIMITED-SCALE EVACUATIONS

Limited-Scale Evacuations are implemented by the IC when the incident affects one or two buildings in the vicinity of the incident. The majority of the evacuations required at hazmat incidents affect a small number of people. See Figure 5.12.

AP Worldwide

Figure 5.12 Limited-scale evacuations usually affect one building.

The U.S. Department of Health and Human Services, Agency for Toxic Substances Disease Registry, conducts a biennial study involving participation from State Health Departments in the United States. The most recent study tracked hazmat incidents reported by 15 State Health Departments in the United States from 1999 to 2000. The study found there were 13,808 hazardous substances emergency events reported that produced 4425 injuries and 74 fatalities.

The study showed that 73% of these events occurred at fixed facilities and the remainder involved transportation incidents. The most frequently released substances from these facilities were ammonia, sulfur dioxide, and sulfuric acid.

In the aforementioned 13,808 incidents, evacuations were ordered in 1,182 events. Of these evacuations, 71% involved a building or an affected part of the building, 14% involved a defined circular radius surrounding the incident scene, 6% were a downwind or downstream area, and the remaining incidents involved had no criteria or were not significant. The median number of people evacuated in these events was just 20 people, which is a small number and easily managed if you have a good plan in place. Interestingly, a contingency plan was followed by the Incident Commander in 95% of the events. This reinforces the importance of having a plan for implementing an evacuation. [Source: U.S. Department of Health and Human Services, Agency for Toxic Substances Disease Registry, "Hazardous Substances Emergency Events Surveillance 1999/2000, Biennial Report," Atlanta, Ga. (2000).]

A limited-scale evacuation may be the best option for the IC under the following conditions:

- ❑ Whenever the building is on fire
- ❑ Whenever the hazmat is leaking inside the building and the material is flammable or toxic
- ❑ Whenever explosives or reactive materials are involved and can detonate or explode, producing flying glass or structural collapse
- ❑ Whenever the building occupants show signs or symptoms of acute illness and there is a known hazmat spill inside the structure

SICK BUILDING SYNDROME

Emergency responders are sometimes called to incidents where several people are exhibiting signs and symptoms of illness but there is no apparent source of the problem (e.g., there is no reported hazmat leak or spill inside or outside the building). This may be indicative of Sick Building Syndrome (SBS).

SBS is a situation in which occupants of a building experience acute health effects that seem to be linked to time spent in a building, but no specific illness or cause can be identified. The complaints may be localized in a particular room or zone or may be widespread throughout the building. Indicators of sick buildings include the following:

- Building occupants complain of symptoms associated with acute discomfort e.g., headache; eye, nose, or throat irritation; dry cough; dry or itchy skin; dizziness and nausea; difficulty in concentrating; fatigue; and sensitivity to odors.
- The cause of the symptoms is not known.
- Most of the complainants report relief soon after leaving the building.

SBS problems are usually a result of the building being operated or maintained in a manner that is inconsistent with its original design or prescribed operating procedures. Sometimes indoor air problems are a result of poor building design or remodeling activities like painting, using glues and maskings for flooring and wall coverings, or the recent installation of new synthetic carpet or plastic materials. Other sources of SBS problems include copy machines, cleaning agents, pesticides, and chemicals that may release volatile organic compounds (VOCs) including formaldehyde.

SBS problems may also be caused by biological contaminants inside the building. Biological contaminants include pollen, bacteria, viruses, and molds. These contaminants can breed in stagnant water that has accumulated in humidifiers, drain pans, and ducts, or where water has collected on ceiling tiles, insulation, or carpet. Biological contaminants can cause fever, chills, cough, chest tightness, muscle aches, and allergic reactions. One indoor air bacterium, Legionella pneumophila, has caused both Pontiac Fever and Legionnaire's disease. [Sources: Environmental Health Center, National Safety Council, Washington, D.C. (http://www.nsc.org). Also see U.S. Environmental Protection Agency, Washington, D.C., Indoor Air Facts No. 4 (revised): Sick Building Syndrome (http://www.epa.gov)].

The IC should consider initiating a Limited-Scale Evacuation of a suspected SBS whenever the signs and symptoms of the affected people become acute. Be aware that many SBS complaints also have a psychological component that can result in a "mass hysteria" reaction with psychogenic symptoms.

EVACUATION OF FIXED INDUSTRIAL FACILITIES

Unlike their public safety counterparts, industrial emergency response teams have the advantage of knowing exactly what types of hazardous materials are in their facilities. This prior knowledge allows evacuation decisions to be more specific.

OSHA requires fixed facilities to have written evacuation procedures. Well-written PPA procedures can provide useful guidelines to supervisors and the facility IC concerning whether a limited or full-scale evacuation is necessary. Many facilities have developed a tiered approach to implementing Public Protective Actions. One method used is to define three levels of Public Protective Action that are associated with the Levels of Incidents described in Chapter 1.

LEVELS OF PROTECTIVE ACTIONS

Fixed industrial facilities should provide written guidelines to employees and contractors concerning when it is appropriate to evacuate or protect-in-place. Industrial facilities are familiar with the hazards and risks of the products they manufacture and can develop specific guidelines. The following example was prepared by a gas plant that processes natural gas with high hydrogen sulfide levels.

Level I Incident Protective Action—A Level 1 Protective Action requires all employees and contractors to evacuate the work site in an upwind direction and report to their pre-designated briefing area as defined in the site-specific emergency plan. Personnel will be accounted for and emergency work assignments will be issued by the on-site supervisor. Individual employees who are working alone at a remote location and must evacuate a work site should contact the Control Room Operator and report that they are leaving the area. Examples of incidents requiring a Level 1 protective action would include a small flammable or toxic leak from a valve or flange that produces concentrations in the 10 ppm range on company property. The leaking gas is not likely to go beyond company property. The leak is easily repairable using safety equipment on-site. There is no risk to the public, but the fire department is notified immediately that there is a Level I Incident in progress.

Level II Incident Protective Action—A Level II incident will require personnel to evacuate the immediate work site and meet at the pre-designated assembly area for accountability and emergency work assignment as specified in the site-specific emergency plan. The public surrounding the work site should be protected-in-place. Examples of incidents requiring a Level II Incident Protective Action would include a moderate release of hydrogen sulfide that is producing atmospheres of up to 300 ppm on company property with atmospheres of 10 ppm immediately adjacent and downwind of company property. The leak can be rapidly repaired, and the hydrogen sulfide gas is rapidly dispersing. The fire department is alerted via 911, and the Fire Chief assumes command of the incident.

Level III Incident Protective Action—A Level III incident is a major emergency which requires the total evacuation of all company personnel and the surrounding public to a pre-designated location(s) outside the immediate area. Depending on the nature of the incident and the potential for the problem to migrate off site, there will also be public protective action instructions given to the surrounding community. Concentrations of hydrogen sulfide are rapidly exceeding 300 ppm immediately adjacent to company property, and the off-site concentrations will reach 50 to 100 ppm. The leak cannot be immediately repaired, atmospheric concentrations are unfavorable for rapid dispersing of hydrogen sulfide gas, and homes within the H_2S release area are not of energy-efficient construction.

FULL-SCALE EVACUATIONS

Full-Scale Evacuations involve the relocation of large populations from a hazardous area to a safe area. Full-Scale Evacuations present two major problems for the Incident Commander:

1. **Life Safety**—In some cases, evacuations may endanger the lives of the people being evacuated. Traffic accidents, stress induced heart attack, and accidental exposure to the hazmat being released are real-world examples. You evacuate them, you own them!
2. **Expense**—Full-scale evacuations are expensive. One study conducted by the Battelle Human Affairs Research Center for the Atomic Industrial Forum indicated that the cost to individual households for evacuation would be expected to be almost seven times the cost of protection-in-place. Costs to the public sector are approximately three times as high and fifteen times as high for the manufacturing sector. An expensive operation is very hard to justify, and there will be no shortage of critics the day after the evacuation (even if you made the right decision).

Fortunately, most Incident Commanders will never be faced with the decision to evacuate thousands of people, but that should not be an excuse for being unprepared. When emergency responders are having their worst day, they usually don't "rise to the occasion", rather they default to their level of planning and training.

Some of the worst days in North American history that involved full scale evacuations include:

- **The World Trade Center**—The terrorist attacks of September 11, 2001 and the subsequent collapse of the New York City's World Trade Center Towers was the worst building disaster in recorded history, killing some 2,800 people. More than 350 emergency response personnel died in the line of duty and are credited with saving approximately 25,000 lives through their extreme heroism when they directly supervised the evacuation of the buildings occupants. It is estimated that between 300,000 and 1 million people evacuated Manhattan Island by powerboat, ferry, barge, or tugboat. One vessel, the Staten Island Ferry Samuel I. Newhouse, evacuated 6,000 people. See Figure 5.13.

- **Mississauga, Ontario**—On November 10, 1979 at 11:53 p.m., a Toronto-bound Canadian Pacific train No. 54 derailed at Mississauga, Ontario (a suburb of Toronto) spilling hazardous materials. Police evacuated 218,384 residents from the area due to the threat presented by chlorine vapors.
- **Three Mile Island**—On March 28, 1979 at 4:00 a.m., a malfunction occurred in the secondary cooling circuit at the Three Mile Island Nuclear Plant near Harrisburg, Pennsylvania and initiated the largest nuclear power facility accident in U.S. history. The accident led to 145,000 people being evacuated from an area that extended about 15 miles from the plant.

AP/World Wide

Figure 5.13 Full-scale evacuations involve potentially thousands of people.

The decision to commit most emergency response resources to a full-scale evacuation should be initially determined by the Incident Commander based upon the specific conditions of the emergency. Some situations that may justify a full-scale evacuation include the following:

- Large leaks involving flammable and/or toxic gases from large-capacity storage containers and process units.
- Large quantities of materials that could detonate or explode, damaging additional process units, structures, and storage containers in the immediate area.
- Leaks and releases that are difficult to control and could increase in size or duration.
- Whenever the Incident Commander determines that the release cannot be controlled and facility personnel and/or the general public are at risk.

When the decision is made to commit to a full-scale public evacuation, four critical issues must be addressed and managed effectively in order for the operation to succeed.

1. **Alerting and Notification**—People must be alerted that an evacuation order has been issued and they must be told where to go and what to do when they get there. Using only one method of alerting and notification is not going to work. You must use a variety of tools to be successful; e.g., radio and television, sirens, door-to-door visits, carrier pigeons, or whatever else works for your community. Alerting methods are discussed later in this chapter.
2. **Transportation**—While it may not seem obvious when you are driving to work in the morning, not everyone owns an automobile and not everyone has more than one vehicle. Transportation must be provided to move lots of people from where they are now to where you want them to go. Sounds simple; it's not.
3. **Relocation Facility**—Once people have been relocated from where they were to where you want them to be, they need to be taken care of. People need to be safe, comfortable, and provided with the basic things everyone needs (e.g., food, water, medicine). Did we mention bathrooms and showers, beds, pets and something to keep the kids out of trouble? The longer people must remain relocated in a temporary facility, the greater their needs will be. See Figure 5.14.
4. **Information**—You must keep displaced people informed about the progress you are making. People need to know what is going on and when will they can go home. They also want to know if their family and friends are safe. Nobody likes not knowing what is going on. Lack of information creates concern, and concern breeds anger, which leads to outrage, which gets the attention of elected officials, who will eventually get your attention. Keep evacuees in the information loop.

Figure 5.14 Relocation facilities need to address the health and welfare of people. Food, refrigeration, sanitation, etc., are all issues that need addressed.

Case Study

THE CSX TUNNEL FIRE, BALTIMORE MARYLAND July 18, 2001

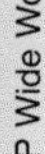

On Wednesday, July 18, 2001, a Northbound 4000-foot-long 60 car CSX Transportation train carrying hazardous materials derailed under the streets of Baltimore, Maryland inside of the hundred-year old 1.7-mile Howard Street railroad tunnel. Although the incident began at 3:10 p.m., the Baltimore City Fire Department was not notified until 4:15 P.M.

Initial arriving companies believed that the derailment ruptured a railroad tank car carrying a flammable liquid. The Fire Department soon learned that nine of the 60 cars in the train contained hazardous materials, including several cars with acids, including hydrochloric acid (5000 gallons were spilled), hydrofluoric acid, glacial acetic acid, and flurosalicic acid. Other materials included ethyl hexyl phthalate (a plasticizer reducing embrittlement in plastics), tripropylene (lubricant similar to paint thinner), and propylene glycol (antifreeze). The fire response quickly elevated to five alarms within the first two hours of the fire. At the height of the incident 150 firefighters were on scene working to extinguish the fire.

Throughout the incident, fire officials were plagued with three problems: (1) fighting the fires inside the tunnel, (2) the presence of hazardous materials in the tunnel, and (3) the weakening structural integrity of the tunnel and the immediate surrounding areas.

Beyond the fire itself, the incident presented many other challenges for public safety agencies. For example, the derailment ruptured a 40-inch water main that ran directly above the tunnel. Flooding from the water main complicated extinguishing efforts and collapsed several city streets. Lombard Street was flooded with 2 feet of water.

The fire disrupted rail traffic along the east coast and Midwest rail corridors as well as interrupting e-mail and phone service as fiber optics cables running through the tunnel were destroyed by the 1,500-degree heat. The failure of the fiber optic lines created problems with Web pages in Manhattan, New York and caused e-mail system failures as far away as South Africa. Thick black smoke pushed out of both ends of the tunnel openings at Camden Yards and Mount Royal Station. Smoke seeped through manhole covers along Howard Street.

Baltimore City firefighters first attempted to fight the fire by entering the tunnel wearing standard turnout gear with SCBA using vehicles equipped with rail wheels. Firefighters entered the south end of the tunnel to recon the situation. They immediately encountered thick black smoke and intense heat. Despite zero visibility in the tunnel, they were able to advance within 300 yards of the burning train when they encountered a burning sensation on their necks from what they suspected was caused by a hazardous material reacting with the perspiration on their necks. Firefighters backed out of the tunnel at either end. At least 22 firefighters were treated at area hospitals for chest pains and heat exhaustion.

Firefighters then tried an alternative plan that involved lowering large diameter hose from the street above into the tunnel where attack lines were set up for suppression operations.

Downtown traffic in Baltimore became a total gridlock. Making the situation worse was the fact that the Baltimore Orioles were scheduled to play the second game of a scheduled double-header at Camden Yards. The game was canceled, downtown stores and the University of Baltimore were forced to close, and the U.S. Coast Guard closed Baltimore Harbor at 5:00 P.M.

At 5:45 P.M., the Baltimore Emergency Management Agency activated the city's Civil Defense sirens to warn citizens of the impending danger (this was the first time the sirens had been activated for an actual emergency since they were installed in 1952). On the night of the derailment, city officials closed down the entrances to the city from all major highways. Many people were trapped downtown, unable to leave the city.

At 6:00 P.M., the main water main broke downtown, causing an electrical power failure, which left 1,200 inner city residents without power.

Firefighters were finally able to reach the burning cars at 10:00 p.m. The final attack significantly lowered the temperature of the burning cars within a few hours. Approximately 60 million gallons of water were eventually used for fire suppression activities

The impact this incident had on the city of Baltimore was significant. The city estimated the clean-up costs to be about $1.3 million. The Baltimore Orioles baseball team was force to postpone four games in three days because of public safety concerns, with an estimated loss of $4.5 million.

KEY LESSONS LEARNED

1. There were no fatalities from the incident. Adherence to Incident Command decisions and cooperation among the various agencies contributed greatly to the low number of injuries—only two significant injuries were reported; both were heat related.
2. Strong partnerships existed between the Baltimore City Fire Department and other agencies and businesses. Having predefined plans in place helped the fire departments succeed in the response and with public protective actions.
3. Strong site management and accountability of personnel on scene played a major role in being able to control firefighters responding to the scene. As a result, there were no injuries to firefighters (other than heat related) and manpower and resources were properly maintained to meet the needs of the incident.

SOURCES

This case study was based on two primary sources. The authors have extracted most of the information from these original sources, with minor editing to organize the information better for presentation.

1. David Michael Ettlin and Del Quentin Wilber, Sun Staff, Published July 19, 2001, Baltimore Sun. The Baltimore Sun provided excellent news coverage of the incident over several days. These articles are available from http://www.sunspot.net.
2. The U.S. Fire Administration's Technical Report Series, "CSX Tunnel Fire, Baltimore, Maryland." Authored by Hilary C. Styron. This 36-page report can be downloaded at http://www.fema.gov. The report number is USFA-TR-140.
3. "CSX Train Derailment and Subsequent Fire in the Howard Street Tunnel, Baltimore, Maryland, July 18, 2001." This report number NTSB/RAB-0408 can be downloaded at www.ntsb.gov.

As seen in the Baltimore Tunnel Fire Case Study, there is no single government or private agency with the charter or expertise to manage and service all four of these areas of responsibility (e.g., alerting, transportation, relocation, and information). Usually a public safety agency takes the lead for overall incident coordination, and other public agencies and private organizations handle specific elements of the evacuation effort. For example, working under a Unified Command structure, the Fire Department may coordinate the tactical aspects of the evacuation, while Emergency Management and Law Enforcement agencies handle the alerting and notification responsibilities. Working under the direction of the Emergency Management Agency, the local Public Transit Authority or School District may provide bus transportation from the affected area to relocation shelters. The School District, in cooperation with the local Red Cross, may provide temporary shelter and food for the displaced, and the Salvation Army or other volunteer groups may provide food and clothing.

Each part of the country has its own variation on the theme of emergency preparedness, and it really doesn't matter who provides what service as long as it is planned in advance and coordinated at a central location using Unified Command. See Chapter 3 for more detail on how Unified Command works.

Good working relationships on a one-on-one basis at the local level can help produce results fast in a crisis, but full-scale evacuations simply have too many moving parts, and there is far too much at stake to simply "wing it." Well-thought-out evacuation plans should be part of every industrial and public safety agency Emergency Preparedness program. The plan should lay out the big picture and clearly spell out areas of responsibility for each agency. These should also be backed up by Memorandums of Understanding (MOU) and written agreements between different organizations who may have overlapping areas of responsibility. For example, if the evacuation plan calls for the School District to provide buses and schools for transportation and relocation of residents, who provides the drivers and opens up the school during summer vacation? Who buys and cooks the food, how will it be requisitioned in the middle of the night, and who will serve it to the people once they arrive? Don't look stupid because you didn't have a plan.

In order to make all these moving parts work, the IC should have direct and frequent communications with representatives of the lead agencies supporting the evacuation. During extended operations, these representatives should relocate from the field command post and operate from the Emergency Operations Center.

It is also necessary to have direct radio communications between the Emergency Operations Center and each Relocation Shelter. Conditions can change quickly, especially when migrating vapor clouds are involved. The Relocation Shelters may have to be moved, or displaced people may temporarily need to take shelter (protect-in-place).

ALERTING AND NOTIFICATION

In order for evacuation to be successful, the IC must assure that people are quickly alerted that there is an emergency in progress. The methods of notification will vary depending on the location of the emergency, the type of plan and hardware in place, and the time of day.

The location of the general population and the time of day should always be a factor in the IC's decision-making process. For example, FEMA studies indicate that on any given weekday, 40% of the community are in their home while 60% are at work or in school. In contrast, the nighttime figures are roughly reversed. In either case, the majority of the population is under roof and would require relocation during an evacuation. One research study indicated that it takes 2.5 to 3 hours to warn 90% of the public through door-to-door contact but only 20 to 35 minutes with sirens or the Emergency Broadcast System.

In urban areas, up to 20% of the population could be in transit to or from their homes. Rush-hour traffic and the time of day are significant factors in deciding whether or not to evacuate. Some studies conducted by the nuclear power industry show that with good planning and traffic control assistance from police agencies, many urban highways are capable of handling large traffic flows created by a full-scale evacuation. The same studies, however, also point out that high-density traffic jams can be created at critical traffic arteries when large crowds attempt to evacuate locations like athletic stadiums and concert halls. Figure 5.15 provides a more detailed picture of principal locations of the population.

A variety of communication technologies exist to assist the IC with the warning process. There are many advantages and disadvantages of each of these systems, and it is important to recognize that one is not necessarily better than another. Each one is a different type of tool and must be selected based on local conditions. See Figure 5.16.

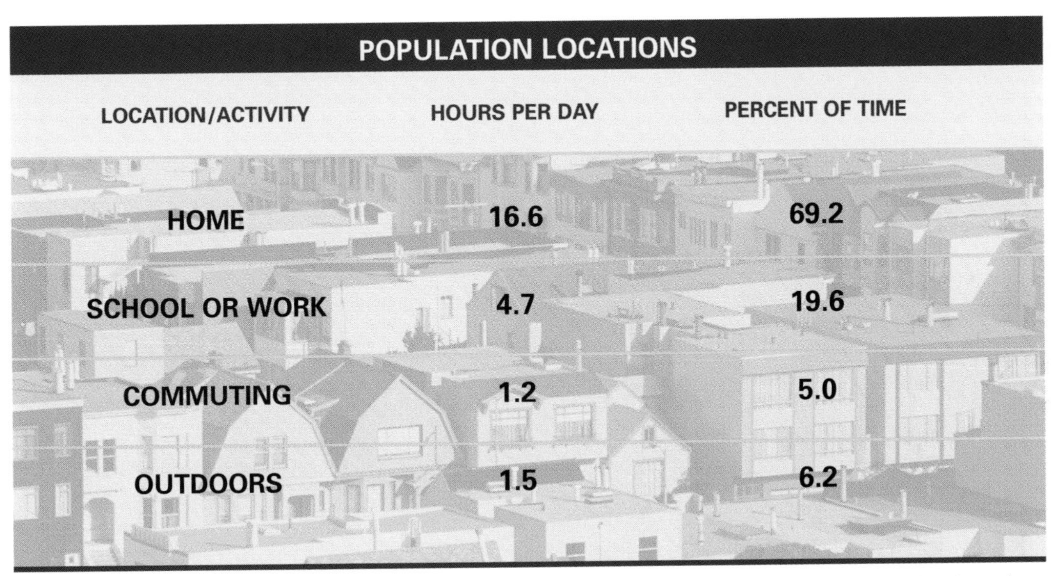

POPULATION LOCATIONS

LOCATION/ACTIVITY	HOURS PER DAY	PERCENT OF TIME
HOME	16.6	69.2
SCHOOL OR WORK	4.7	19.6
COMMUTING	1.2	5.0
OUTDOORS	1.5	6.2

Figure 5.15 Average 24 hour population location.

A brief summary of the different alerting methods is as follows:

- **Door-to-door**—This consists of a simple door-to-door visit to residents in the affected area. This is usually completed by police agencies. This initial contact has a multiplier effect as residents call neighbors and relatives, alerting them that there is a problem.
- **Loudspeakers/public address systems**—Loudspeakers on emergency vehicle siren systems are an effective way of alerting people outside in public areas such as parks. Sound from public address systems on emergency vehicles travel 500 feet, which means the vehicle has to stop every 1,000 feet and broadcast the message. This is a slow process. Public Address Systems inside shopping malls and public assembly buildings may also be used.

- **Tone alerted radios**—Some fixed facilities such as chemical plants and oil and gas plants have provided special Tone Activated Radios for residents living near their facility. These operate on the same principle as a volunteer firefighter's radio. A radio signal is sent from the control room at the plant that sets off an alerting tone inside each home's radio. A live, real-time message can then be broadcast. Special weather radios are also available that are activated by the National Weather Service (NWS) for severe storm warnings. This system can be used to issue special warnings if prior arrangements have been made with the NWS.
- **Emergency Alerting System**—The Emergency Alerting System, or EAS, is an effective method of alerting people in buildings and automobiles. The EAS is operated by the Federal Emergency Management Agency and the National Weather Service (NWS). The system allows emergency messages to be broadcast by cable, satellite, and other services through the more than 14,000 radio and television stations and 11,000 cable systems serving over 33,000 communities in the United States. By law, the EAS must be available to the President within 10 minutes using normal activation procedures from any area in the United States for national emergencies. State Emergency Management agencies can provide local emergency notification using the EAS communications backbone.
- **Scanner Radios and Weather Radios**—Scanner radios are widely used by fire and police buffs, boaters, and truck drivers. The National Oceanographic and Atmospheric Administration (NOAA) weather radio system covers a major portion of the population within the country. The station broadcasts continuously and can be used to warn people of special atmospheric emergencies such as migrating toxic vapor clouds. This is especially useful for boaters who may be downwind on lakes or rivers. The U.S. Coast Guard may also contact boaters directly by issuing special broadcasts on marine channel 16.
- **Commercial Television and Radio**—Television Capture Systems are becoming more popular as alerting tools in communities where cable television systems are used. Emergency services can "capture" the cable station and transmit a message called a "crawler" across the bottom of the viewer's screen. The media may also break into normal radio programming with a special broadcast. Some Emergency Management Agencies have special agreements with local radio stations that allow them to break into programming and broadcast an emergency message live from the Emergency Operations Center.
- **Sirens and Alarms**—These may include the community Emergency Preparedness Agency sirens or special sirens installed in areas around fixed facilities like chemical plants, refineries, and weapons depots. Some of these devices can also function as a public address system.
- **Aircraft**—Helicopter loudspeakers can be an effective method of alerting people in outdoor and remote areas (e.g., flying over parks, campgrounds, and hunting areas).
- **Electronic Billboards**—Many interstate highways and commuter routes have electronic message signs for alerting drivers of local traffic conditions, kidnappings, terrorist threats, and so on. Some bus and train stations have similar electronic bulletin boards.
- **Computerized Telephone Notification Systems**—CT/NS can reach a potentially large number of people by telephoning hundreds of phone numbers

simultaneously by computer. Prerecorded emergency messages provide instructions to residents.

- **Low-power AM Radio Systems**—These systems use a low-power radio transmitter in the AM band to broadcast traffic and weather information 24 hours a day. Some systems also broadcast continuous public safety messages. Telephone books, flashing roadside signs, and other public education literature directs the citizen to switch to the designated frequency whenever a public emergency occurs (e.g., "Switch your radio to AM 1600"). These systems are very effective when they regularly broadcast useful information on a daily basis like the current weather. People become used to going to the public safety radio program to get reliable information about what is going on in their community.

A good community alerting and notification system is based on a variety of systems that are described in the Community Emergency Response Plan. The plan should spell out who has the authority and responsibility to activate each system. Each system component should be tested on a regular basis.

ESTIMATES OF WARNING TIMES FOR ALTERNATIVE SYSTEMS

WARNING SYSTEM	PERCENT OF POPULATION WARNED			
	25	50	75	90
MEDIA	20–30	45–60	80–120	180–240
DOOR TO DOOR	40–45	60–80	100–120	150–180
ROUTE ALERT	25–35	40–50	60–70	90–150
TONE ALERT RADIO OR AUTO TELEPHONE	2–3	4–5	7–10	10–15
SIREN/MEDIA	5–10	12–15	15–20	20–35
SIREN/FIXED RESPONSE	1–2	2–3	14–15	10–15

Figure 5.16 Estimates of alerting times in minutes.

Many locations have improved their alerting systems by providing detailed Public Protective Action instructions in the local telephone directory. Alerting messages received by the Emergency Alerting System direct residents to a specific page number in their telephone book, where detailed instructions are provided.

ALERTING SYSTEMS IN FIXED FACILITIES

At a fixed facility such as a refinery or chemical plant, the notification process normally occurs by activation of sirens or by use of an on-site public address system.

One of the most frequent problems encountered in fixed facilities is the confusion created by a single warning tone that may also be used to indicate the beginning or ending of a work shift, a fire, toxic gas release, and so on. The same tone is used with one, two, or three blasts on the horn, which have different meanings. As a general rule, evacuation alarms should be unique and distinctly different from any other type of alarm in the facility.

Bill McEnroe

Figure 5.17 An activated evacuation alarm at an industrial facility does not mean that the occupants have either protected in-place or evacuated.

As part of the pre-incident planning process, fixed facility personnel should have special knowledge of the following:

- Methods by which their personnel are notified of an emergency evacuation, including sound of the alarm system
- Instructions on where personnel should report to and assemble when the evacuation alarm is sounded
- Facility evacuation routes and corridors
- Safe havens where personnel can seek refuge if evacuation is not possible or advisable
- Ability to communicate with facility personnel at evacuation assembly areas. This is critical in accounting for evacuees and initiating search and rescue operations
- Location of both primary and alternate assembly locations

From a tactical perspective, the IC should be aware that an activated evacuation alarm at a fixed facility does not necessarily mean that the occupants have either protected-in-place or evacuated. Additional direction and assistance from emergency response personnel may be required to complete the evacuation. See Figure 5.17.

Once notified and evacuated, a head count should be taken of facility personnel, contractors, and visitors to ensure that all personnel are accounted for. Supervisors are usually responsible for coordinating all personnel accountability activities. Information regarding any missing personnel and their previously known location within the facility should be relayed to the IC so that search and rescue operations can begin. However, the IC should be aware that initial reports of people missing based on head counts are usually not correct.

TRANSPORTATION

Experience in large-scale evacuations indicates that the majority of the affected population will leave the hazardous area by use of their own automobile or by catching a ride with a friend or neighbor. However, a significant portion of the population cannot drive or do not own an automobile. This is especially the case in urban areas, where many people use public transportation as their only means of getting around.

In the 1979 Mississauga, Ontario incident discussed earlier, 40% of the population took almost immediate action to leave, 65% evacuated after 30 minutes, and almost 90% evacuated 45 minutes after being warned. Figure 5.16 provides an overview of the time required to alert and evacuate people.

Good advanced planning is required to prepare for transporting people during an evacuation. The best sources of transportation in most areas are school buses and public transit buses. Many communities have designated specific buses for emergency response duty. These buses may be equipped with Emergency Preparedness radios, and their drivers are specially trained.

While school bus fleets are an excellent form of transportation, they may not be immediately available during the summer months. An MOU should be in place between emergency response organizations and the agency providing bus services. This should outline how many vehicles can be committed to emergency evacuation duty, who has the authority to request and assign vehicles, and how the drivers will be contacted on a 24-hour basis.

Special emergency extraction may be required to transport people from areas close to the hazardous materials release. Some organizations have addressed this problem by purchasing Emergency Escape Packs, also known as Emergency Breathing Apparatus (EBA). These devices typically have 5 to 10 minutes of breathing air in their cylinders. The facepiece is a simple plastic hood that is placed over the head to provide a fresh-air breathing supply for limited duration. This is usually adequate to move someone from inside their home and into a awaiting vehicle, where they will be transported to a safe area.

EBAs may be carried on Special Air Units operated by the fire department or be staged in strategic locations for use by Emergency Response Personnel. They may be found at or near special project sites, such as oil and gas wells, where hydrogen sulfide is present, or toxic waste dump remediation projects.

RELOCATION FACILITIES

Relocation Facilities are used to temporarily house people displaced during an evacuation. One study suggests that approximately 65% of evacuees do not stay at public relocation shelters. They may check into a hotel or stay with friends and relatives outside of the evacuated area. Using these numbers, that means that the remaining 35% of the population will require some form of public shelter. In a town of 25,000 people that would be about 8,700 people…still a lot of people that must be cared for!

Relocation facilities are typically located at schools, at National Guard armories, or at community centers. In order for relocation shelters to be effective, they need the following elements in place:

- **Safe Building**—Relocation facilities may be in service for just several hours or for several days. The building must be safe, have a food service area, adequate restrooms and bathing facilities, and an area large enough for temporary sleeping furniture such as cots. The building must be air conditioned or heated and have adequate security to protect residents. Relocation Shelters should always be energy efficient in case the occupants need to protected-in-place.

- **Shelter Manager**—An individual trained in Shelter Management techniques should be assigned to each Relocation Facility. The Shelter Manager organizes and supervises shelter activities and is the single point of contact between the shelter and the field command post. Many jurisdictions have a cooperative arrangement with the local Red Cross chapter or religious and civic organizations like the Salvation Army. The local Emergency Preparedness Agency usually coordinates the shelter program, provides the funding, and trains the Shelter Managers, while the cooperating organizations provide the personnel to staff the shelter.
- **Shelter Support Staff**—If the Relocation Shelter will be operating for an extended period, a shelter staff should be provided to assist with its operation. Examples include receptionists to document who arrives and departs the shelter, EMTs or nurses to attend to the sick and disabled, food service personnel to prepare meals, building engineers or maintenance personnel, and counselors. It is also advisable to assign a police officer at each shelter for security. Direct radio contact with the Emergency Operations Center is recommended.

If shelters will remain in operation for an extended period, arrangements must be made for around-the-clock staffing and provisions.

INFORMATION

The Incident Commander should assure that displaced people are kept informed of the actions being taken by emergency responders to mitigate the problem. Failure to keep people informed creates political and public affairs problems.

Relocation facility Shelter Managers should receive regular situation briefings from the ICP or the Emergency Operations Center. This can be accomplished by direct radio or telephone contact on an hourly basis or by faxing a brief status report to each shelter location.

When the incident covers a period of days, the IC should consider issuing a written progress report, which can be posted twice daily in the shelter. If the displaced are not kept informed, they will quickly form a negative opinion of your operation and make their feelings known to the rest of the world through the media.

The news media can be a powerful tool in confirming that the initial evacuation was well handled and is still necessary for public safety. It is important that the IC project to the media an image of professionalism and control during the evacuation. The IC should hold regularly scheduled joint press briefings with senior representatives of the media present. For example, the Emergency Management, Police, and Fire agencies should conduct their briefings as a team and project unity in their decision making.

TYPES OF WARNING SYSTEM MESSAGES

Emergency Broadcast System (EBS):

The national emergency notification system that uses commercial AM and FM radio stations for emergency broadcasts. The EBS is usually initiated and controlled by Emergency Preparedness agencies.

Sample EBS Messages:

Initial Warning: During the initial size-up stages of an incident, the IC often does not have enough information to make a decision whether to protect-in-place or evacuate. Alerting time is critical. An initial warning message can be transmitted to alert people that there is a problem.

Example:

"We interrupt your normally scheduled program for the following emergency broadcast. The County Fire Department has just issued a hazardous materials emergency warning for the Evanston, Wyoming area. You are instructed to take immediate shelter inside your home or any public building until the danger has passed. Close all doors and windows and turn off your heating or air conditioning system. Stay tuned to this station for more details. Additional emergency instructions for hazardous materials emergencies are provided on page 2 of the local telephone directory." See opposite page.

Protection-in-Place Warning: When the IC has decided to implement the Protection-in-Place option, the EBS message should be designed to get people indoors and to seal off sources of contaminated air that could filter into the home.

Example:

"This is the County Fire Protection District. Emergency response units have responded to a hazardous materials incident along the Interstate-80 area. The emergency is under control; however, you are advised to remain indoors with your doors and windows closed and the heating and air conditioning system off. This is only a precaution which will keep any contaminated air which may be in your area outside of your home. You are completely safe as long as you remain indoors. Emergency repairs are underway and we will advise you as soon as it is safe to go back outside. Stay tuned to this station for more information. If you have an emergency at your location, dial 911 for help."

Evacuation Warning: When the IC issues an evacuation order by EBS, the instructions must tell the public where to evacuate to. General instructions may cause residents to enter dangerous areas.

Example:

"This is the County Fire Protection District. Emergency response units have responded to a hazardous materials incident in the Interstate-80 area. Emergency response teams are on the scene at this time and are taking immediate action to control the hazardous materials release. As a precaution, we have ordered an immediate evacuation of the following areas. These include all buildings in the East Park and Quail Creek neighborhoods. If you are in one of these areas, you are instructed to evacuate immediately. Proceed to a designated emergency shelter at the Evanston High School. Representatives from the County will meet you at the relocation center. Police agencies will patrol your area until the emergency is over. If you have an emergency at your location dial 911. Please do not call 911 if you do not have an emergency. Additional information will be provided at the relocation center. Again, please leave immediately."

Figure 5.18 Warning system messages.

HAZARDOUS MATERIALS EMERGENCY PROCEDURES

HOW YOU WILL BE NOTIFIED OF A HAZARDOUS MATERIAL EMERGENCY

If there is a hazardous material emergency in the Evanston area, the community's warning and notification system will alert you. The system consists of eight sirens strategically placed around the community and a public address network that immediately captures the Emergency Broad- cast System (EBS) radio stations and the cable TV stations. The official EBS stations are KEVA (1240 AM) and J KOTB (106.3 FM) radio.

The sirens sound for three-minute periods several times to raise the alert at the beginning of an emergency and to indicate all clear at the end.

In areas not served by warning sirens, official cars with loudspeaker systems will alert residents throughout the neighborhoods.

WHAT TO DO IF YOU HEAR THE COMMUNITY ALERT SIREN

- Stay calm.
- Go indoors, close all windows, doors, fireplace dampers, and vents, and turn off fans, heating systems, or air conditioners.
- If you are in a vehicle, close all windows and vents and stop at the nearest building for shelter.
- Turn on an EBS radio station for information and instructions. The message will be repeated on the EBS stations as necessary until conditions change.
- If you are instructed to protect your breathing, cover your nose and mouth with a handkerchief or other cloth, wet if possible.
- Stay indoors until instructed to do otherwise.
- Do not use the telephone unless you have an emergency. The lines are needed for official business, and your call could delay emergency response.
- As soon as it can be determined that the hazardous condition has passed, local authorities will announce the emergency is over.
- If the emergency involves a toxic gas release, at the "All Clear" you may be instructed to open windows and doors, ventilate the building, and go outside.

INSTRUCTIONS

All EBS messages begin with a description of the emergency, including time and location. The instructions that follow depend on the severity of the emergency. Residents will hear directions for one of four emergency procedures: Warning Alert, Shelter in Place, Prepare to Evacuate, or Evacuate.

Hazardous Material Warning Alert

The WARNING ALERT indicates there is a problem that poses no present danger to the community. However, there could be a potential for escalation to a more serious situation. The WARNING ALERT informs residents to stand by.

The following are typical instructions for a WARNING ALERT:

1. Stay indoors.

2. Stay tuned to an EBS radio station for further instructions.

Shelter in Place

In certain hazardous material incidents it is safer to keep community residents indoors rather than to evacuate them.

The following are typical insructions for SHELTERING IN PLACE:

1. If you are outdoors, protect your breathing until you can reach a building.

2. Go to an inside room, preferably one with no or few windows.

3. Close all windows, doors, and vents and cover cracks with plastic wrap, tape, or wet rags.

4. Keep your pets inside.

5. Listen to an EBS radio station for further advice.

Prepare to Evacuate

You may be asked to PREPARE TO EVACUATE if a situation has the potential of escalating to the point where an evacuation is required.

During this time, authorities will act to alleviate the emergency and also will prepare for an orderly evacuation if it becomes necessary.

The following are typical instructions for PREPARING TO EVACUATE:

1. All person in (names of areas) should stay indoors and prepare to evacuate.

2. If you are in your home, gather all necessary belongings. Pack only the items you need most, such as clothing, medicine, baby supplies, portable radio, credit cards, flashlight, and checkbook.

3. Locate and review your Community Emergency Action Card. You need not evacuate at this time.

4. Stay tuned to your EBS radio station for further instructions.

Evacuate

An EVACUATION may be ordered if the community is threatened and there is time to evacuate safely. Make sure the EVACUATE order applies to your area.

The following are typical instructions for an EVACUATION:

1. All persons in (names of areas) should evacuate in an orderly manner.

2. In certain circumstances you may need special equipment for evacuation. Under those conditions you are required to remain in your home until fire department personnel arrive with the equipment and instruct you in its use.

3. Children in school will be taken to an evacuation shelter if necessary.

4. Lock your house and turn on the porch light as you leave. Your neighborhood will be guarded while you are away.

5. Drive or walk toward the main roadway in your-area.

Emergency personnel stationed along these routes will direct you away from the emergency area toward an evacuation shelter.

6. Use your own car if you can. Keep all vehicle windows, doors, and vents closed. If you have room, take passengers. Please observe traffic laws.

7. Turn on your car radio for information.

8. If you can't find transportation, walk to one of the pickup points along the nearest MAIN emergency route in your area. You can get a ride there.

9. Law enforcement personnel will be in place in the evacuated area to prevent looting, vandalism, etc.

10. You may return home as soon as the emergency is declared over and it is safe to return.

SPECIAL ARRANGEMENTS FOR MOBILITY-IMPAIRED PERSONS

If you are disabled or need assistance, send the Uinta County Fire department a postcard with your name, address and telephone number, a description of the extent of your difficulty, and any other pertinent information about the state of your health, such as diet, medicines, etc.

This information will be placed on file so you will receive the necessary assistance in the event of an emergency. To help the department maintain current files, it is important for you to mail them an update card if there are changes in your status.

Uinta County Fire Department P.O.Box 6401
Evanston, WY 82931
(307) 789-3013

WHERE TO GET MEDICAL, FIRE, OR POLICE HELP DURING AN EMERGENCY

If you need medical assistance, or if you need to report a fire or violation of the law, call 911. Give your name, address (including ZIP code), the nature of the emergency, and stay on the line.

INFORMATION

For more information about hazardous material emergencies contact the Uinta County Fire Department at 789-3013.

Information for the Evanston Area Hazardous Material Emergency Pages was provided courtesy of Union Pacific Resources Company.

Figure 5.19 Telephone directory information.

SUMMARY

Site Management is the first step in the Eight Step Incident Management Process©. Its major focus is on establishing control of the incident scene by assuming command of the incident and isolating people from the problem by establishing an Isolation Perimeter and Hazard Control Zones. Site management and control provide the foundation for the response. Responders cannot safely and effectively implement an incident action plan unless the playing field is clearly established and identified for both emergency responders and the public

REFERENCES AND SUGGESTED READINGS

Air Force Civil Engineer Support Agency (AFCESA) and PowerTrain, Inc., HAZARDOUS MATERIALS INCIDENT COMMANDER EMERGENCY RESPONSE TRAINING CD-ROM. Tyndall Air Force Base, FL: AFCESA (2002).

Air Force Civil Engineer Support Agency (AFCESA) and PowerTrain, Inc., HAZARDOUS MATERIALS TECHNICIAN EMERGENCY RESPONSE TRAINING CD-ROM. Tyndall Air Force Base, FL: AFCESA (1999).

Agency for Toxic Substances and Disease Registry, HAZARDOUS SUBSTANCES EMERGENCY EVENTS SURVEILLANCE 1999–2000 BIENNIAL REPORT. Atlanta, GA: ATSDR (2000)

Brown, Michael, J., and Gerald, E. Streit, "Emergency Responders' Rules-of-Thumb for Air Toxics Releases in Urban Environments." U.S. Department of Energy Publication LA-UR-98-4539. Los Alamos, NM: Los Alamos National Laboratory (1998).

Brunacini, Alan V., FIRE COMMAND (2nd edition). Quincy, MA: National Fire Protection Association (2002).

Carter, Harry, "Downtown HazMat," Pennsylvania Fireman (May 1993). pages 96–100.

Federal Emergency Management Agency, HAZARDOUS MATERIALS WORKSHOP FOR LAW ENFORCEMENT. Washington, DC (April 1992).

Federal Emergency Management Agency—U.S. Fire Administration, CSX TUNNEL FIRE, BALTIMORE, MD (USFA-TR-140). Washington, DC: FEMA (July 2001).

Emergency Film Group, Site Management and Control. Plymouth, MA: Emergency Film Group (2004).

Lindell, Michael, K., Patricia, A. Bolton, Ronald, W. Perry, et al., "Planning Concepts and Decision Criteria for Sheltering and Evacuation in a Nuclear Power Plant Emergency." A report prepared for the National Environmental Studies Project of the Atomic Industrial Forum, Inc. by the Battelle Human Affairs Research Centers, Seattle, WA (June 1985).

Maloney, Daniel, M., Anthony, J. Policastro and Larry Coke, "The Development of Initial Isolation and Protection Action Distances Table for the U.S. DOT Publication—1990 Emergency Response Guidebook." A paper presented at the 3rd Annual HazMat 90 Central Conference (March 1990).

Michigan State Police, Emergency Management Division WARNING, EVACUATION & IN-PLACE PROTECTION HANDBOOK (EMD PUB-304). Lansing, MI: Michigan State Police (January 1994).

National Fire Service Incident Management System Consortium Model Procedures Committee, IMS MODEL PROCEDURES GUIDE FOR HAZARDOUS MATERIALS INCIDENTS. Stillwater, OK: Fire Protection Publications, Oklahoma State University (2000).

National Fire Service Incident Management System Consortium Model Procedures Committee, IMS MODEL PROCEDURES GUIDE FOR HIGHWAY INCIDENTS. Stillwater, OK: Fire Protection Publications, Oklahoma State University (2003).

National Institute for Chemical Studies, SHELTERING IN PLACE AS A PUBLIC PROTECTIVE ACTION. Charleston, WV: National Institute for Chemical Studies (June 2001).

National Institute for Occupational Safety and Health, GUIDANCE FOR PROTECTING BUILDINGS FROM AIRBORNE CHEMICAL, BIOLOGICAL OR RADIOLOGICAL ATTACKS. Cincinnati, OH: NIOSH (May 2002).

National Regulatory Commission and U.S. Environmental Protection Agency Task Force Report on Emergency Planning, PLANING BASIS FOR THE DEVELOPMENT OF STATE AND LOCAL GOVERNMENT RADIOLOGICAL EMERGENCY RESPONSE PLANS IN SUPPORT OF LIGHT WATER NUCLEAR POWER PLANTS (NUREG-0396, EPA 520/1-78-016). Washington, DC: NRC.

National Transportation Safety Board, "Special Investigation Report: Survival in Hazardous Materials Transportation Accidents" (Report NTSB/HZM-79-4). Washington, DC: National Transportation Safety Board (December 1979).

Nordin, John, "Technical Dialogue—Evacuate or Shelter in Place," THE FIRST RESPONDER NEWSLETTER (May 2003).

Sherman, M. H., D. J. Wilson, and D. E. Kiel, "Variability in Residential Air Leakage." A technical paper presented at the ASTM Symposium on Measured Air Leakage Performance in Buildings, Philadelphia, Pennsylvania (April 1984).

Sorenson, John, H., "Evaluation of Warning and Protective Action Implementation Times for Chemical Weapons Accidents." A report prepared for the Aberdeen Proving Ground, MD by Oak Ridge National Laboratory (Report No. ORNL/TM-10437). Oak Ridge TN: Oak Ridge National Laboratory (April 1988).

U.S. Environmental Protection Agency, ESTABLISHING WORK CONTROL ZONES AT UNCONTROLLED HAZARDOUS WASTE SITES (EPA Publication, 9285.2-06FS), Washington, DC: EPA (April 1991).

U.S. Environmental Protection Agency, MANUAL OF PROTECTIVE ACTION GUIDES AND PROTECTIVE ACTIONS FOR NUCLEAR INCIDENTS (EPA-520/1-75-001), Washington, DC: EPA (September 1975).

Wagman, David, "Get Out of Town," Homeland Protection Professional, (September 2003). Pages 20-27.

Wheeler, W. H. and W. R. Byrd, "Emergency Planning for H_2S Releases: Utilizing Shelter in Place and Interagency Drills." A report prepared for presentation at the Society of Petroleum Engineers Conference held in San Antonio, Texas. SPE Report No. 25979 (March 1993), pages 373–77.

Wilson, D. J., "Accounting for Peak Concentrations in Atmospheric Dispersion for Worst Case Hazard Assessments," Department of Mechanical Engineering, University of Alberta, Edmonton, Alberta (May 1991).

Wilson, D. J., "Effectiveness of Indoor Sheltering during Long Duration Toxic Gas Releases," Department of Mechanical Engineering, University of Alberta, Edmonton, Alberta (May 1991).

Wilson, D. J., "Wind Shelter Effects on Air Infiltration for a Row of Houses," Department of Mechanical Engineering, University of Alberta, Edmonton, Alberta (September 1991).

Wilson, D. J., "Variation of Indoor Shelter Effectiveness Caused by Air Leakage Variability of Houses in Canada and the USA." Prepared by the Department of Mechanical Engineering, University of Alberta, Edmonton, Alberta. Presented at the U.S. EPA/FEMA Conference on the Effective Use of Sheltering-in-Place as a Potential Option to Evacuation During Chemical Release Emergencies, Emmitsburg, MD (December 1988).

CHAPTER 6

Houston, TX F.D.

IDENTIFY THE PROBLEM

OBJECTIVES

1. Describe the principles of recognition, identification, classification, and verification as they apply to hazardous materials emergencies.
2. List and describe the seven basic methods of identifying hazardous materials.
3. Given examples of the following railroad cars, identify each car by type and identify the material and its hazard class that is found in each car. [NFPA 472-6.2.1.1(A)].
   - Cryogenic liquid tank cars
   - High-pressure tube cars
   - Nonpressure tank cars
   - Pneumatically unloaded hopper cars
   - Pressure tank cars
4. Given examples of the following intermodal tanks, identify each intermodal tank by type and identify the material and its hazard class that is found in each tank [NFPA 472-6.2.1.1(B)].
   - Nonpressure intermodal tanks:
     - IM-101 (IMO Type 1 internationally) portable tank
     - IM-102 (IMO Type 2 internationally) portable tank
   - Pressure intermodal tanks—DOT Spec. 51 (IMO Type 5 internationally)
   - Specialized intermodal tanks:
     - Cryogenic intermodal tanks (IMO type 7 internationally)
     - Tube modules
5. Given examples of the following cargo tanks, identify each cargo tank by type [NFPA 472-6.2.1.1(C)].
   - Nonpressure liquid tanks
   - Low-pressure chemical tanks
   - Corrosive liquid tanks
   - High-pressure tanks
   - Cryogenic liquid tanks
   - Tube trailers
6. Given examples of the following tanks, identify at least one material and its hazard class that is typically found in each tank. [NFPA 472-6.2.1.1(D)].
   - Nonpressure tank
   - Pressure tank
   - Cryogenic liquid tank
7. Given examples of the following nonbulk containers, identify at least one material and its hazard class that is typically found in each container [NFPA 472-6.2.1.1(E)].

- Bags
- Carboys
- Cylinders
- Drums

8. Given examples of the following radioactive materials packages, identify each package by type and identify at least one typical material found in each package [NFPA 472-6.2.1.1(F)].
   - Type A
   - Type B
   - Industrial
   - Excepted
   - Strong, tight containers
9. Given examples of containers, DOT specification markings for nonbulk and bulk packaging, and the associated reference guide, identify the basic design and construction features of each container [NFPA 472-6.2.3.1(A) through (D)].
   - Cargo tanks
   - Fixed facility tanks
   - Intermodal tanks
   - One-ton containers
   - Pipelines
   - Railroad cars
   - Intermediate bulk containers (i.e., tote tanks)
   - Nonbulk containers (e.g., carboys, drums, cylinders)
   - Radioactive material containers
10. Given examples of facility and transportation containers, identify the approximate capacity of each container [NFPA 472-6.2.1.2].
11. Describe the types of specialized marking systems found at fixed facilities.
12. Describe how a pipeline can carry different products [NFPA 472-6.2.3.2].
13. Given an example of a pipeline, identify the following: [NFPA 472-6.2.3.3].
    - Ownership of the pipeline
    - Procedures for checking for gas migration
    - Procedures for shutting down the pipeline or controlling the leak
    - Type of product in the pipeline
14. Identify and describe the placards, labels, markings, and shipping documents used for the transportation of hazardous materials.
15. Given a label for a radioactive material, identify vertical bars, contents, activity, and transport index, and then describe the labeled items and its significance in surveying a hazardous materials incident [NFPA 472-6.2.1.4].

ABBREVIATIONS AND ACRONYMS

AAR	Association of American Railroads
ANSI	American National Standards Institute
API	American Petroleum Institute
ASME	American Society of Mechanical Engineers
ASTM	American Society of Testing and Materials
CAS	Chemical Abstract Service
CTC	Canadian Transport Commission
CHEMTREC	Chemical Transportation Emergency Response Center
COFC	Container on Flat Car
DOT	U.S. Department of Transportation
EPA	Environmental Protection Agency
ERG	Emergency Response Guidebook
HMT	Hazardous Materials Table
LC_{50}	Lethal Concentration, 50% Kill
LEPC	Local Emergency Planning Committee
LPG	Liquefied Petroleum Gas
LSA	Low Specific Activity
MAWP	Maximum Allowable Working Pressure
N.O.S.	Not Otherwise Specified
NTSB	National Transportation Safety Board
ORM	Other Regulated Material
PG	Packing Group
PIH	Poison—Inhalation Hazard
RQ	Reportable Quantity
SARA	Superfund Amendments and Reauthorization Act
SCADA	Supervisory Control and Data Acquisition System
SCO	Surface Contaminated Objects
SI	International System Units
STCC	Standard Transportation Commodity Code
TC	Transport Canada
TI	Transport Index
TIH	Toxic—Inhalation Hazard
TOFC	Trailer on Flat Car

INTRODUCTION

In 1971, a railroad derailment in the Houston, Texas metropolitan area caused a breach in a pressure tank car transporting propane, which subsequently ignited. The propane fire then impinged on an adjoining tank car of vinyl chloride. After approximately 45 minutes of exposure to fire, the vinyl chloride tank car violently ruptured. As a result, a fire department training officer photographing the incident was killed and a number of emergency response personnel and civilians were injured. Because of the nature of the incident, the National Transportation Safety Board (NTSB) conducted an investigation and concluded that the following factors contributed to the severity of the accident:

- The lack of adequate training, information, and documented procedures for on-scene identification
- The lack of adequate assessment of threats to safety
- Reliance on firefighting recommendations that did not take into consideration the full range of hazards

The American philosopher George Santayana (1863–1952) stated, "Those who do not remember the past are condemned to relive it." Although the Houston incident occurred over 30 years ago, the lessons learned are timeless. Timely identification and verification of the hazmats involved are critical to the safe and effective management of a hazmat incident. Failure to perform basic tasks of recognition, identification, and verification is like going to war without knowing who the enemy is.

This chapter discusses the second step in the Eight Step Process©—Identify the Problem. Although this text is directed toward the Hazardous Materials Technician and the On-Scene Incident Commander, the recognition and identification process actually starts as soon as emergency responders are notified. Remember the Eight Step Process©—Problem Identification cannot be safely and effectively accomplished if responders have not first controlled the incident scene. Likewise, strategic goals and tactical response objectives cannot be formulated if the nature of the problem is not defined. Remember—a problem well-defined is half-solved!

This chapter reviews the basic principles of problem identification and methods of hazmat recognition, identification, and classification. The authors would especially like to recognize Charles J. Wright, Manager of Hazardous Materials Training for the Union Pacific Railroad, for his life-long efforts in developing many of the recognition and identification tools, and training materials that are referenced in this chapter.

BASIC PRINCIPLES

KNOWING THE ENEMY

Managing a hazmat incident is much like trying to manage a war. Neither effort will be very successful unless you learn as much as possible about the "enemy," where it can be found, and its general tendencies and behaviors. Among the most critical tasks in managing a hazmat incident are surveying the incident scene to detect the presence of hazmats, identifying the nature of the problem and the materials involved, and identifying the type of hazmat container and the nature of its release. This effort is made more difficult by the number and variety of hazardous materials found in society and the increasing likelihood of hazmats being used for both criminal and terrorism events.

Historically, there have been numerous public and private sector studies that have evaluated hazmat transportation, hazmat incidents, the materials involved, and the source of their release. Although the specific numbers may vary, these studies clearly show some specific trends and patterns, including the following:

- Approximately 75% of releases occur in facilities that produce, store, manufacture, or use chemicals; the remaining 25% occurred during transportation.
- Approximately 50% of all hazmat emergencies involve flammable and combustible liquids (e.g., gasoline, diesel fuel, and fuel oils) and flammable gases (e.g., natural gas, propane, butane). The next most common hazard class is corrosive materials, specifically materials such as sulfuric acid and sodium hydroxide.
- The top hazardous materials transported by rail include trailers on flat cars (TOFC) and containers on flat cars (COFC), LPG, sodium hydroxide, sulfuric acid, anhydrous ammonia, chlorine, gasoline, and blended motor fuels.

While responders may deal with thousands of chemicals, a rather small list of hazmats accounts for the majority of our problems. Be practical and realistic—the routine establishes the foundation on which the non-routine must build. If you don't "have your act together" managing high-probability flammable liquid and gas emergencies, it's unlikely that you'll perform well trying to manage an incident involving a more exotic hazardous material.

This background information is no substitute for conducting a hazard analysis and developing an Emergency Response Plan for your plant or community. However, it does support the point that with proper analysis and planning, your hazmat emergency response program can become more effective and efficient.

TOP CHEMICALS INVOLVED IN HAZMAT RELEASES

CHEMICAL NAME	% OF RELEASES	% OF DEATHS AND INJURIES
PCB (Polychlorinated Biphenyls)	23.0%	2.8%
Sulfuric Acid	6.5%	4.7%
Anhydrous Ammonia	3.7%	6.8%
Chlorine	3.5%	9.6%
Hydrochloric Acid	3.1%	5.6%
Sodium Hydroxide	2.6%	1.9%
Methyl Alcohol	1.7%	0.4%
Nitric Acid	1.7%	1.5%
Toluene	1.4%	2.4%

U.S. Environmental Protection Agency

Figure 6.1 Top chemicals involved in hazmat releases.

SURVEYING THE INCIDENT

The identification process starts with a survey of the incident site and surrounding conditions. Responders should complete the following tasks: (1) Identify the hazmats involved; (2) identify the presence and condition of the containers involved; and (3) assess the conditions at the incident site, including exposures, topography, injuries, etc.

The identification process is built on the following basic elements:

1. *Recognition*—Recognize the presence of hazardous materials. Positive recognition automatically shifts the response into the hazmat mode. Basic recognition clues include occupancy and location, container shapes, markings and colors, placards and labels, shipping and facility documents, monitoring and detection instruments, and senses.

2. *Identification*—Identify the hazardous materials involved and the nature of the problem. Primary hazmat identification clues include markings and colors, shipping and facility documents, and monitoring and detection instruments. Regardless of the method of recognition or identification, always verify the information obtained. Don't take anybody's word at face value, including other responders. Always verify, verify, verify!
3. *Classification*—Determine the general hazard class or chemical family of the hazardous materials involved. A rule of thumb is, "If you can't identify, then try to classify." Basic classification clues include occupancy and location, container shapes, placards and labels, monitoring and detection instruments, and senses.

When dealing with unknown substances, responders should rely on monitoring instruments and chemical analytical kits, which use a systematic process to determine the unknown's identity and hazards. The most widely used system is the HazCat® Chemical Identification System. Although responders may not always be able to identify the hazmat(s) involved, they will usually be able to determine the unknown's hazard class or chemical family. Being able to determine what an unknown material *isn't* may be as important as identifying what a material *is*.

In recent years there has been increasing emphasis on the potential use of hazardous materials and weapons of mass destruction (WMDs) in terrorism or criminal events (e.g., clandestine labs). These scenarios bring a number of new issues, including the use of weapons and armed assaults, secondary devices, and booby traps. However, from a health and safety perspective, the "bad stuff" is still a hazardous materials and the basic concepts of recognition, identification, and verification still apply. You can't play in the WMD world if you don't first understand the hazmat response world!

CLUES AND IDENTIFICATION PROCESS

CLUE	HM RECOGNIZE	HM IDENTIFY	HM CLASSIFY	WEAPONS OF MASS DESTRUCTION
Occupancy & location	✓		✓	✓
Container shapes	✓		✓	✓
Markings & colors	✓	✓		
Labels & placards	✓		✓	
Shipping papers & facility documents	✓	✓		
Monitoring & detection equipment	✓	✓	✓	✓
Senses	✓		✓	✓

Figure 6.2 Clues and the identification process.

IDENTIFICATION METHODS AND PROCEDURES

In some situations, responders will initiate offensive response operations without immediately realizing that hazardous materials are involved. In other cases, they may initiate aggressive, offensive operations before the material(s) are positively identified, verified, and evaluated. In either case, responders have placed themselves at an unacceptable risk.

Emergency responders rely on seven basic clues as part of their identification process. Look for hazmats in every incident; then identify or at least classify the material. The seven clues are

- Occupancy and location
- Container shapes
- Markings and colors
- Placards and labels
- Shipping papers and facility documents
- Monitoring and detection equipment
- Senses

These clues are also listed in relation to their distance from the hazmat (see Figure 6.3). The closer you are to the release site, the greater the likelihood that you can accurately identify the material(s). The downside is that you also face a substantially higher risk of exposure and/or contamination. Your initial objective should be to learn as much as possible about the problem from as far away as possible.

To ensure personnel safety during size-up, many emergency responders rely on binoculars or spotting telescopes for identification. These have many applications, including surveying outdoor and indoor incidents to verify information such as container labels, markings and other clues from a distance. Although they provide a narrow field of vision, telescopes can also be a useful tool for the identification process.

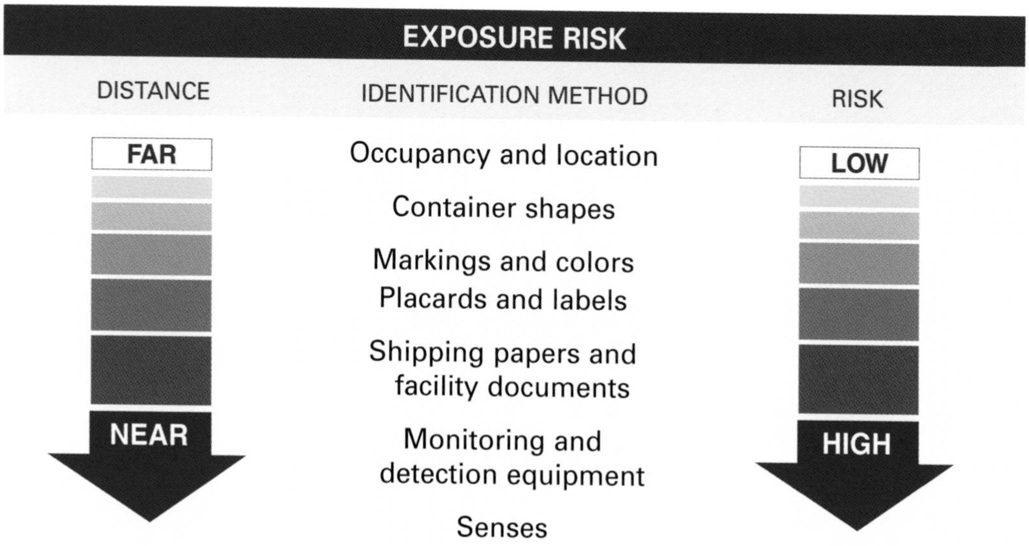

EXPOSURE RISK

DISTANCE — IDENTIFICATION METHOD — RISK

FAR — LOW

Occupancy and location

Container shapes

Markings and colors

Placards and labels

Shipping papers and facility documents

Monitoring and detection equipment

Senses

NEAR — HIGH

Figure 6.3 The closer you are to the problem when identifying the hazmat(s) involved, the greater your risk of exposure.

OCCUPANCY AND LOCATION

Hazardous materials surround us every day—not only in transportation and industrial facilities but in stores, hospitals, supermarkets, warehouses, garages, and even in our homes. These potential locations can be categorized into four basic areas—production, transportation, storage, and use. Occupancy and location is the first clue for the recognition and classification of the hazmats involved.

The key for determining these potential sites is through the hazard analysis process. SARA Title III and state/local "Right-to-Know" legislation also requires facilities to notify the fire department, the Local Emergency Planning Committee (LEPC), and other government agencies when on-site quantities exceed established threshold values (e.g., 500 pounds, 10,000 pounds, etc.). Hazard analysis information should include a list of the hazmats on site, their quantity and location, and hazards. In addition, material safety data sheets (MSDS) or comparable information should be available. Many local jurisdictions require fixed facilities to use Knox Boxes™ or similar devices at the front gate or main entrance for the storage of all pertinent facility information. In other instances, response information is provided on a CD-Rom for rapid access.

Experienced responders are able to associate certain hazmats with different types of occupancies. Do you know what's in your community? Check it out—you might be surprised! Consider the following examples:

- You respond for a person down in a metal plating firm. What types of hazmats could you encounter? If you said sodium cyanide, potassium cyanide, and strong inorganic acids (e.g., nitric acid, sulfuric acid, chromic acid, etc.), you've got the right idea.
- Water treatment plants are found in almost every community, and the far majority use chlorine in the treatment/disinfection process. Chlorine is typically transported and stored in these facilities in either 1-ton containers or 90-ton railroad tank cars, and represent both a hazmat and terrorist target hazard. In a worst-case scenario, a major release from a 90-ton chlorine tank car could generate a plume that could impact downwind areas over 10 miles away.
- You have a facility that produces the metal bases for light bulbs, a nonhazardous product. No hazmat problems, right? A facility walk-through would show that anhydrous ammonia is used as a fuel source—"you mean that same anhydrous ammonia which is non-flammable?" The anhydrous ammonia is piped to a dissociator, where it is split into hydrogen and nitrogen; the hydrogen is then used as a fuel source for burners and the nitrogen is released. Other hazmats found include sulfuric and hydrochloric acid, caustic cleaners, and oils, solvents and lubricants.
- You are involved in the planning process for a raid on a suspected methamphetamine lab. What types of hazards may be present? First, there are the obvious concerns with the perpetrator's use of weapons and booby traps. From a hazmat perspective, there are flammable liquids and solvents, corrosives, reactives, and possibly anhydrous ammonia depending on the cooking process. Would you want to take a lab down while the bad guys are cooking?

Responders must recognize that hazmat exposures are no longer limited to industrial or transportation emergencies. The use of hazardous materials for criminal or terrorist purposes is a real threat. Not every emergency is a hazmat emergency; however, responders have the potential of being exposed to hazmats at ANY emergency!

Figure 6.4 Hazardous materials may be found at any location.

CONTAINER SHAPES

All hazardous materials are controlled as long as they remain within their container. The size, shape, and construction features of a container/packaging are the second clue to the standard hazmat identification process and can be used as a clue for both hazmat recognition and classification. The U.S. Department of Transportation (DOT) defines packaging as a receptacle, which may require an outer packaging and any other components or materials necessary for the receptacle to perform its containment function and to ensure compliance with minimum packaging requirements.

Packaging used for transporting hazardous materials is regulated by the DOT. Other types of containers are used only at fixed facilities, such as process towers, piping systems, and reactors. Packaging used for production, storage, and use is usually built according to nationally recognized consensus standards, such as those provided by the American Society of Mechanical Engineers (ASME), the American Society of Testing and Materials (ASTM), and the American National Standards Institute (ANSI). For example, welded steel atmospheric pressure petroleum storage tanks are constructed according to specifications of the American Petroleum Institute (API), Standard 650. This standard is, in turn, referenced by the NFPA *Technical Standard on Flammable and Combustible Liquids (NFPA 30)*.

Packaging is divided into three general groups: nonbulk packaging, bulk packaging, and facility containment systems.

NONBULK PACKAGING

Nonbulk packaging will hold solid, liquid or gaseous materials. DOT provides the following definitions:

- Liquid—capacity of 119 gallons (450 liters) or less
- Solid—net mass of 882 pounds (400 kg) or less for solids, or capacity of 119 gallons (450 liters) or less
- Compressed Gas—water capacity of 1,001 pounds (454 kg) or less

Nonbulk packaging may consist of single packaging (e.g., drum, carboy, cylinder) or combination packaging—one or more inner packages inside of an outer packaging (e.g., glass bottles inside a fiberboard box, infectious disease sample containers). Nonbulk packaging may be palletized or placed in overpacks for transport in vehicles, vessels, and freight containers. Examples include bags, boxes, carboys, cylinders, and drums.

BULK PACKAGING

Any packaging, including transport vehicles, meeting the following DOT definition:

- Liquid—capacity greater than 119 gallons (450 liters).
- Solid—net mass greater than 882 pounds (400 kg) for solids, or capacity greater than 119 gallons (450 liters).
- Compressed Gas—water capacity greater than 1,001 pounds (454 kg).

Bulk packages can be an integral part of a transport vehicle (e.g., cargo tank trucks, tank cars, and barges) or packaging placed on or in a transport vehicle (e.g., intermodal portable tanks, ton containers).

FACILITY CONTAINMENT SYSTEMS

Packaging, containers, and containment systems that are part of a fixed facility's operations. Depending on the nature of the facility and its operations, the types of containment systems can vary greatly. Examples include pressurized and nonpressurized storage tanks, process towers, chemical and nuclear reactors, piping systems, pumps, storage bins and cabinets, dryers and degreasers, machinery, and so forth.

Responders should become familiar with container and packaging shapes and be able to relate them to potential contents and hazard classes (i.e., container versus contents). The military has used aircraft silhouettes for years as a training tool for recognizing both friendly and enemy aircraft. This same concept can be applied by emergency responders in recognizing and identifying hazmat containers and packaging.

NONBULK, BULK AND FACILITY CONTAINMENT SYSTEMS		
TRANSPORTATION CONTAINMENT SYSTEMS		FACILITY CONTAINMENT SYSTEMS
NONBULK	BULK	
Bags	Bulk bags	Buildings
Bottles	Bulk boxes	Machinery
Boxes	Cargo tanks	Open piles (outdoors & indoors)
Carboys	Covered hopper cars	Piping
Cylinders	Freight containers	Reactors (chemical & nuclear)
Drums	Gondolas	Storage bins, cabinets, or shelves
Jerricans	Pneumatic hopper trailers	
Multicell	Portable tanks and bins	
Wooden Barrels	Protective overpacks for radioactive materials	
	Tank cars	
	Ton containers	
	Van trailers	

Figure 6.5 Examples of nonbulk, bulk, and facility containment systems.

Source: Chemical Manufacturers Association and Association of American Railroads, *Technical Bulletin on Packaging for Hazardous and Non-Hazardous Materials.* Washington, DC: Chemical Manufacturers Association (1989).

NONBULK PACKAGING

Nonbulk packaging is constructed to performance or specification standards mandated by DOT. Performance tests for nonbulk packaging include a drop test, leakproof test, hydrostatic test, stacking test, and vibration standard test.

The type of material and the end use of the product will determine the type of packaging. For example, packaging for household or consumer commodities is usually different from industrial applications, even for the same material. Industrial grade materials are usually more concentrated, while the household version is often diluted with other materials, such as water or a solvent.

The type of container is typically a good clue to its hazards and contents. In general, the more substantial or durable the container, the greater the hazards of the material, especially when dealing with drums.

Figures 6.6–6.12 illustrates the design and construction features, and contents found in the common types of nonbulk packaging.

NONBULK PACKAGING

BAGS

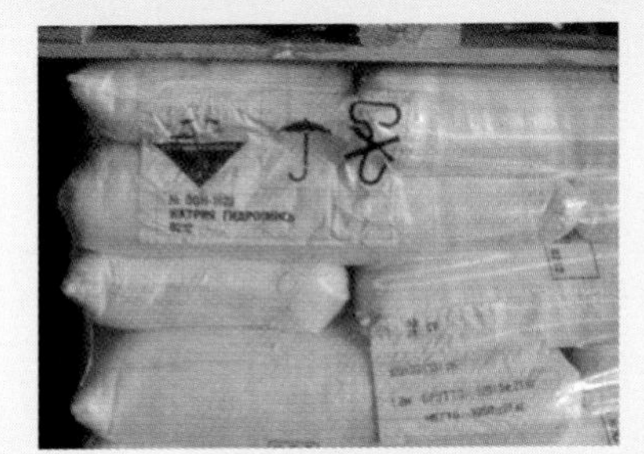

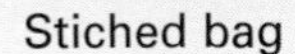

Stiched bag

Folded and glued bag

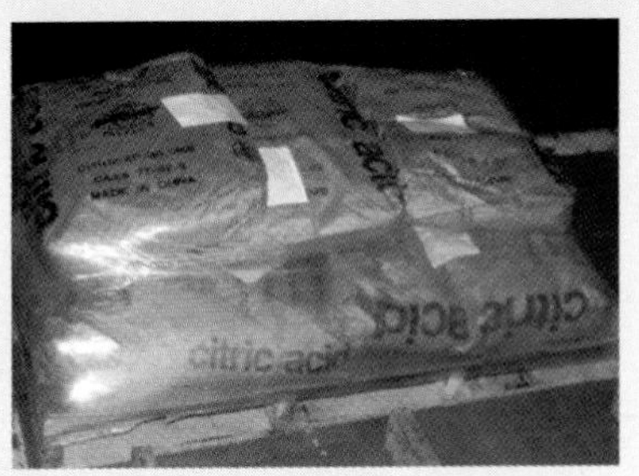

Shrunk wrapped bags

DESIGN AND CONSTRUCTION

- Flexible packaging constructed of cloth, burlap, kraft paper, plastic, or a combination of these materials.
- Closed by folding and gluing, heat sealing, stitching, crimping with metal, or twisting and tying.
- Typically contain up to 100 lbs. of material, although large tote bags weighing up to 500 lbs. may be found.
- Usually palletized when found in quantities.
- Tote bags may be palletized or hung inside of boxed vehicles.

CONTENTS / HAZARD CLASS

- Solid materials, including explosives, flammable solids, oxidizers and organic peroxides, poisons and corrosives.
- Common examples include fertilizers, pesticides, caustic powders.
- Large reinforced bladders may be used for the transport of non-hazardous materials.

Figure 6.6

BOTTLES

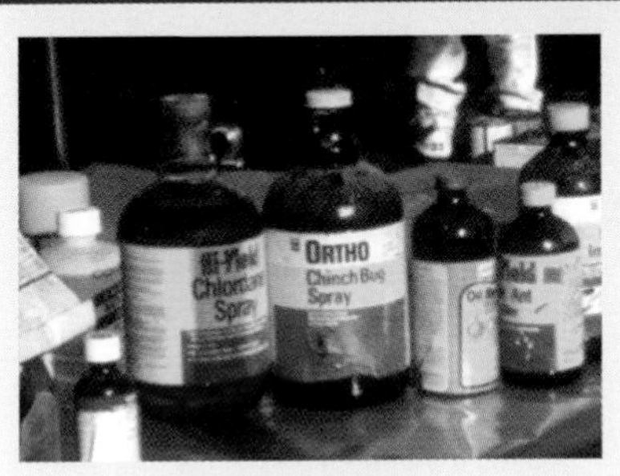

Protected bottles

Plastic bottles

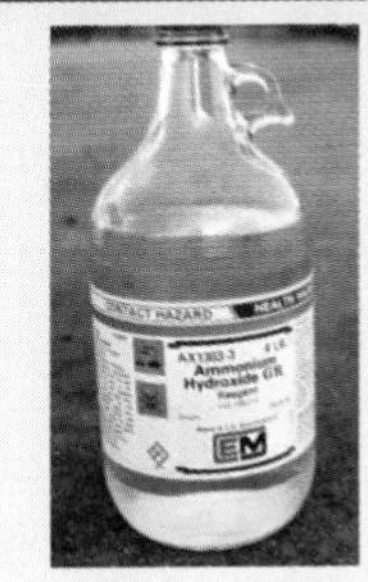

Glass bottle

DESIGN AND CONSTRUCTION

- Constructed of glass, plastic and metal, although ceramic may be found. Some newer generation of glass bottles may be encased in an outer plastic coating.
- Closed head with threaded caps or stoppers.
- Range from several ounces to >20 gal.
- Usually placed within an outer package for transport, such as a box.
- May be referred to as jugs or jars.

CONTENTS / HAZARD CLASS

- Liquid and solid materials, including flammable and combustible liquids, poison liquids, and corrosives.
- Common examples include laboratory reagents and corrosive liquids.
- Generally brown glass bottles are commonly used for light sensitive and reactive materials, such as ethers and organic peroxides.

Figure 6.7

DRUMS

5 gallon drum
(pail, bucket, can)

Closed head
stainless steel drum

Closed head
plastic drum

Open head
plastic drum

Open head
fiber (fibre) drums

DESIGN AND CONSTRUCTION

- A cylindrical package constructed of metal, plastic, fiberboard, or other suitable materials.
- Typical capacity is 55 gallons, although smaller (e.g., pails) and larger drums can be found.
- May be open-head (i.e., drum lid comes off) or closed-head (i.e., drum lid is fixed and cannot be removed). Closed head drums tend to be used for the more hazardous materials, and usually contain 2 openings—2-inch and 3/4-inch diameter plugs called "bungs." Bungs can be a source of leaks.
- Overpack drums are used for transporting damaged or leaking nonbulk containers of all types and sizes. Range from 5-gallon lab packs to 100-gallon overpacks.

CONTENTS / HAZARD CLASS

- Liquids, solids and mixtures, including flammable and combustible liquids, flammable solids, oxidizers, organic peroxides, poisons, corrosives, and radioactive materials.
- Steel Drums—Commonly used for flammable and combustible liquids, mild corrosives, and liquids used in food production.
- Stainless Steel Drums—Commonly used for more hazardous and reactive liquids, such as nitric acid or oleum (super concentrated sulfuric acid @ 120–160% concentrations).
- Aluminum Drums—Hold materials that would react with rust or with steel, and cannot be shipped in a poly drum. Contents are often combustible or toxic, such as some pesticides. Caustic corrosives would NOT be shipped in an aluminum drum.
- Plastic (Poly) Drums—Commonly used for corrosive liquids, some flammable or combustible liquids, and food production liquids.
- Fiberboard Drums—Commonly hold solid materials, such as powders, granules or pellets. May be toxic or corrosive, or may present little or no risk. Are usually lined with a poly liner; when lined, may also hold gels and some low hazard liquids.

Figure 6.8

CARGO TANK TRUCKS

Cargo tank trucks also have various markings that, when combined with container shapes, can be important clues in the overall identification process. These include company names and logos, vehicle identification numbers, and the manufacturers specification plate.

Manufacturers Specification Plate. Mounted on the front third of the tank on the left side. For tanks constructed prior to July 1, 1985, the plate will be found on the right side near the front. Key information on the plate includes:

- Name of the manufacturer, manufacturer's serial number, and date of manufacture.
- DOT container specification number (e.g., MC-306/DOT-406)
- Materials of construction, including head, shell, weld and lining material (if applicable).
- Maximum allowable working pressure (MAWP)
- Nominal compartments capacity (in gallons) of each compartment, front to rear.
- Water capacity of the tank in pounds. This will be found on pressure vessels, such as MC-331 and MC-338 cargo tanks. Remember—1 gallon of water weighs 8.35 lbs.
- Maximum product load (in pounds).
- Maximum temperature is the maximum temperature at which the cargo tank will safely carry material without failure.
- Date of manufacture.

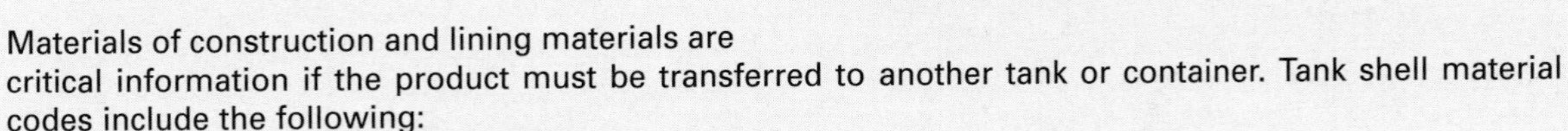
Materials of construction and lining materials are critical information if the product must be transferred to another tank or container. Tank shell material codes include the following:

AL	Aluminum
CS	Carbon Steel
HSLA	High Strength Low Alloy Steel
HSLA-QT	High Strength Low Alloy, Quenched and Tempered Steel
MS	Mild Steel
SS	Stainless Steel

For example, an MC-406 AL specification would indicate that the cargo tank is an MC-406 (low pressure cargo tank) constructed of aluminum.

Some cargo tanks are designed to multiple container specifications that allow the unit to transport more than one type of commodity. Common multi-purpose configurations include combination MC-306/DOT-406 and MC-307/DOT-407 units, or combination MC-307/DOT-407 and MC-312/DOT-412 units. In addition to the manufacturer's specification plate, these tanks have a second "multi-purpose" plate that identifies the specification under which the cargo tank is being operated. These plates are color-coded, as are the fittings that are added to make the cargo tank meet the respective specifications. A sliding shield exposes the specification plate currently in use. However, these plates may move as a result of an accident or may not be properly positioned by the vehicle driver. The recomended color codes are:

MC-306/DOT-406	Red Plate and Fittings
MC-307/DOT-407	Green Plate and Fittings
MC-312/DOT-412	Yellow Plate and Fittings
Non-Specification Tank	Blue Plate and Fittings

ASME Containers. Cargo tanks that are pressurized and that are built to the requirements of the American Society of Mechanical Engineers (ASME) Pressure Vessel Code will have this information stamped on the specification plate or will carry a separate plate outlining the ASME requirements. The ASME plate is easily identified by an embossed "U" in the upper left corner of the plate. Examples of tanks subject to the ASME requirements include MC-331 high pressure and MC-338 cryogenic liquid cargo tanks.

Inspection Markings. Cargo tanks are subject to various inspections, and inspection markings can be found on the front head or tank shell near the specification plate. These markings will indicate the date and type of inspection conducted. For example, MC-331 cargo tanks are required to be inspected annually and be tested every 5 years; an example of a marking and abbreviations are listed below.

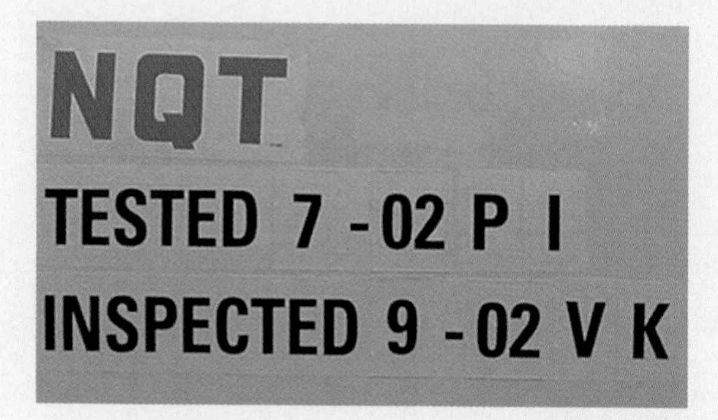

ABBREVIATION	TEST INSPECTION TYPE
V	External Visual Inspection
I	Internal Visual Inspection
K	Leakage Test
L	Lining Inspection
P	Pressure Test
T	Thickness Test

Figure 6.25

MC-331 Tank Shell Markings. MC-331 cargo tanks will also be stenciled with the letters "QT" or "NQT" near the specification plate. During manufacture, the tank shell steel is hardened through different heat treatment or quenching processes—quench tempering (QT) or non-quench tempering (NQT). Understanding these points can become a critical consideration if an MC-331 cargo tank is involved in a severe rollover incident.

Responders should understand the following basic principles regarding QT and NQT tanks:

- QT tanks are heated to a high temperature, then rapidly cooled. This produces a metal that is quite hard and strong, but can also be somewhat brittle. The brittleness of a metal can become an issue if the cargo tank is subjected to extensive mechanical stress, such as when the tank is involved in a serious rollover or strikes a concrete abutment.
- NQT tank metals are heated to specified temperature for a specified period of time, then slowly cooled through the use of some type of heated media (e.g., melted salt). This produces a metal that is hardened, but not as hard as QT metal. In addition, NQT metals are not as brittle as QT metals.

Tank Color. MC-331 specification tanks are required to have the upper two-thirds of the tank painted white. Corrosive tank trucks (MC-312/DOT-412) will often have a contrasting color band on the tank in line with the dome covers and overturn protection. This band is usually a corrosive-resistant paint or rubber material.

CARGO TANK TRUCKS

	DESIGN AND CONSTRUCTION	CONTENTS / HAZARD CLASS
MC-306 / DOT-406 **ATMOSPHERIC PRESSURE CARGO TANK TRUCK**	• Oval cross-section on most indicates an atmospheric pressure tank (less than 4 psi). • Normally constructed of aluminum. May also find round cross-section tanks constructed of stainless steel. • Maximum of 5 compartments per trailer. • Maximum capacity of approximately 9,000 gallons in U.S. transportation system; 13,800 gallons in Canada and border states. • Valves can be air or cable operated. • Safety features include internal safety valves with sheer protection, fusible links, nuts and plugs, emergency remote shutoff, vacuum relief protection, vapor emission controls, and overturn protection.	• Flammable and combustible liquids, including gasoline, diesel fuel, and fuel oils. • Round cross-section, stainless steel units often transport flammable solvents and chemicals.
MC-307 / DOT-407 **LOWPRESSURE CHEMICAL CARGO TANK TRUCK**	• Circular cross-section or noninsulated version with visible strengthening rings. Stainless steel shell is most common. • Horseshoe shape or insulated versions with an outer jacket (usually aluminum) that conceals rings and insulation. • Up to 5 compartments with overturn protection and crash boxes. • Maximum allowable working pressure (MAWP) up to 40 psi. • Total capacity of 5,000 to 8,000 gallons. • Valves can be air, hydraulic or cable operated. • Safety features include internal safety valves, fusible links, nuts and plugs, emergency remote shutoff, pressure and vacuum relief protection, and overturn protection.	• Flammable and combustible liquids, poisons, mild corrosives, and chemicals with a vapor pressure of 18 psi at 100°F (37.8°C) or greater, but not more than 40 psi at 170°F (4.4°C). • Workhorse of the chemical industry.

CARGO TANK TRUCKS

MC-312 / DOT-412 CORROSIVE CARGO TANK TRUCK

DESIGN AND CONSTRUCTION

- Narrow circular cross-section on noninsulated version with visible strengthening rings.
- Horseshoe shape on insulated versions with an outer jacket (usually aluminum) that conceals rings and insulation.
- Normally a single compartment cargo tank with spill box in the rear. Capacity of 3,300 to 6,500 gallons, depending upon the product.
- Multiple spill boxes may indicate that the tank is a top-unloading version; not necessarily an indicator of the total number of compartments.
- Normal maximum allowable working pressure (MAWP) between 40 and 50 psi; rare versions may operate at pressures up to 100 psi.
- Valves can be manual, air or hydraulic operated.
- Safety features include internal safety valves, pressure and vacuum relief protection, and overturn protection.
- Does not have an emergency shut-off device. However, cutting off the air supply during a transfer can reduce the flow of product.

CONTENTS / HAZARD CLASS

- Corrosives and high density liquids.

MC-330 / MC-331 HIGH PRESSURE CARGO TANK TRUCK

DESIGN AND CONSTRUCTION

- Single compartment unit constructed of top quality seamless or welded steel; circular cross-section with rounded ends or heads. Steel may be quench tempered (QT) or non-quenched tempered (NQT).
- Minimum design pressure of 100 psi and maximum of 500 psi.
- Single shell noninsulated tank.
- May be found as a bobtail or a transport. Bobtail capacities range from 2,300 to 3,000 gallons; transports range from 9,000 to 14,500 gallons.
- Transports as large as 17,500 gallons are allowed in some states.
- Valves can be air or cable operated.
- Safety features include pressure relief devices, internal safety valves with sheer protection, excess flow valves, fusible links and nuts, emergency remote shutoff, and off-truck emergency remote shutdown devices.

CONTENTS / HAZARD CLASS

- Liquefied gases, primarily propane, butane and anhydrous ammonia.

CARGO TANK TRUCKS

MC-338 CRYOGENIC LIQUID CARGO TANK TRUCK

DESIGN AND CONSTRUCTION

- Large, well insulated "thermos bottle" design. Inner container holds the product, and the vacuum-sealed outer shell is filled with insulation.
- Many versions include a work box at the rear of the unit and evaporator coils underneath the belly of the tank. Work box includes discharge and fill piping, valves and pump.
- Single compartment unit with working pressures ranging from 25 to 500 psi. Capacity of 8,000 to 10,000 gallons.
- Safety features include pressure relief devices on both the inner tank and the outer tank, excess flow valves, fusible links and nuts, and emergency remote shutoff.
- MC-338 units transporting gases that occur naturally in the atmosphere (e.g., nitrogen, oxygen) are designed to regularly vent off vapors during transport.

CONTENTS / HAZARD CLASS

- Cryogenic liquids, including liquid oxygen, liquid nitrogen, ethylene and liquid carbon dioxide.

COMPRESSED GAS TRAILER

DESIGN AND CONSTRUCTION

- Consist of 2 to approximately 20 seamless steel cylinders permanently mounted together. Often referred to as a "tube trailer."
- Service pressures range up from 2,000 to 5,000 psi.
- Cylinder valves, piping and controls are typically at the back of the unit.
- Safety features include pressure relief devices.

CONTENTS / HAZARD CLASS

- Pressurized gases, such as oxygen, hydrogen, nitrogen, and helium.
- Units are commonly found at construction and industrial sites. When empty, the entire unit is removed and replaced with a similar full unit.

CARGO TANK TRUCKS

	DESIGN AND CONSTRUCTION	CONTENTS / HAZARD CLASS
HEATED MATERIAL CARGO TANK TRUCKS	• Constructed of mild steel or aluminum. • Tank is covered with insulation and an outer jacket (usually aluminum) that conceals the insulation. • May be provided with steam coils to keep the product heated and fluid (e.g., molten sulfur). • Some asphalt tanks have burner tubes and may carry propane cylinders to fuel the burners. • Required to be placarded as a "HOT" material. Molten sulfur cargo tanks are required to have the contents stenciled on both sides and ends of the tank.	• Heated materials, such as tar, asphalt, molten sulfur, and #7 and #8 fuel oils.
DRY BULK TANK TRUCKS	• Bottom hopper design for off-loading the product. Air pressure is generally used for product transfer. • Carry very heavy loads; centrifugal force the cause of many rollovers. • Static charge build-up is a common hazard when transferring product.	• Solid products, such as fertilizers (e.g., ammonium nitrate), oxidizers, cement, dry caustic soda, plastic pellets, etc.

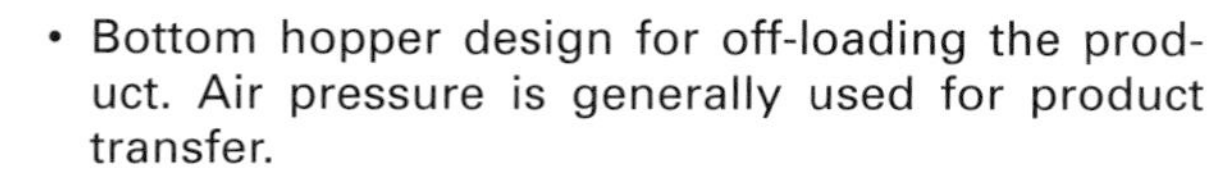

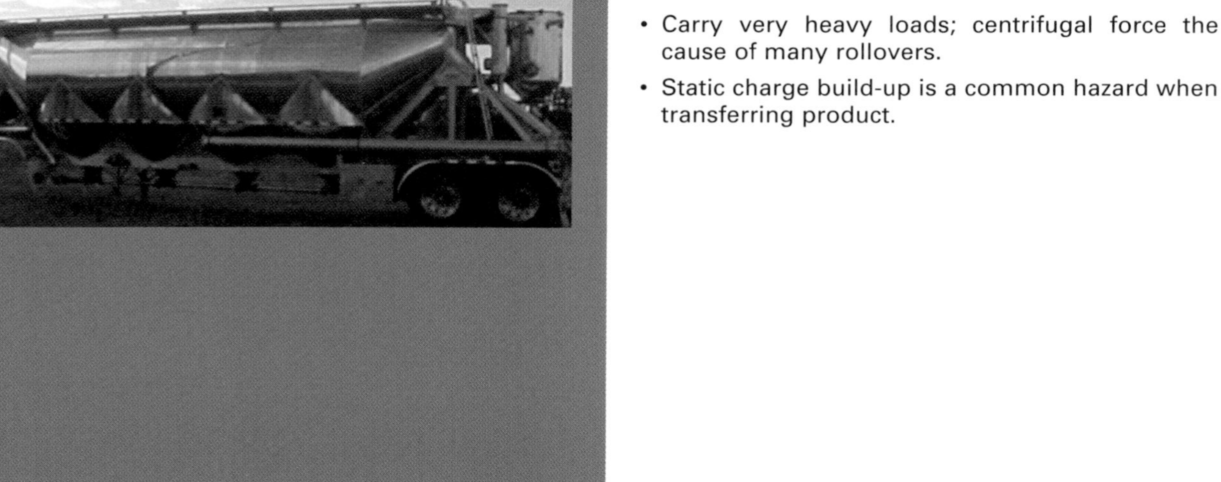

Figure 6.26 Examples of cargo tank trucks.

RAILROAD TANK CARS

Railroad Tank Car Nomenclature. The railroad industry uses tank car test pressure as the criterion for differentiating between pressurized and non-pressurized tank cars. Non-pressure tank cars have a test pressure of 100 psig or less, while pressure tank cars have a test pressure greater than 100 psig.

When describing a tank car, the "B-end" is used as the initial reference point. The B-end is where the hand brake wheel is located; numbers 1 through 4 indicate the wheels.

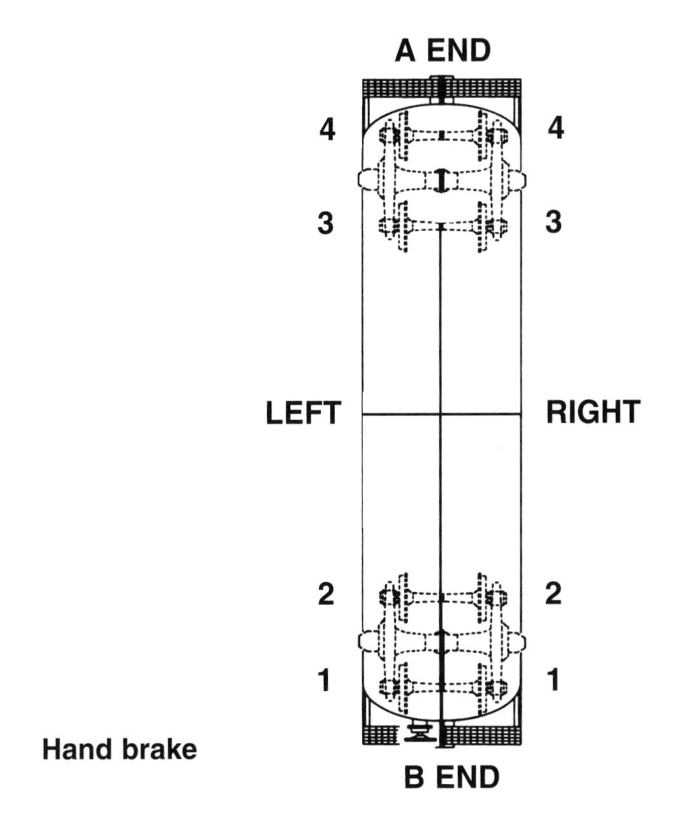

Railroad Tank Car Markings and Colors can be used to gain knowledge about the tank itself and its contents. This information would be useful in evaluating the condition of the container. These markings include the following:

- **Commodity Stencil.** Over 90 different commodities must have the name of the commodity marked on both sides of the tank in 4-inch (102 mm) minimum letters. Examples include tank cars transporting anhydrous ammonia, ammonia solutions with more than 50% ammonia, Division 2.1 material (flammable gas), and Division 2.3 material (poison gas)
- **Reporting Marks and Number.** Railroad cars are marked with a set of initials and a number (e.g., GATX 12345) stenciled on both sides (left end as one faces the tank) and both ends of the car. These markings can be used to obtain information about the contents of the car from the railroad, the shipper or CHEMTREC. The last letter in the reporting marks has special meaning:
  - "X" indicates a rail car is not owned by a railroad (for a rail car, the lack of an "X" indicates rail road ownership).
  - "Z" indicates a trailer.
  - "U" indicates a container.

  Some shippers and car owners also stencil these markings on top of tank cars to assist in identification in an accident or derailment scenario.

- **Capacity Stencil**. Shows the volume of a tank car in gallons (and sometimes liters), as well as in pounds (and sometimes kilograms). These markings are found on the sides of the car under the reporting marks. For certain tank cars (e.g., DOT-105, DOT-109, DOT-112, DOT-114 and DOT-111A100W4), the water capacity / water weight of the tank car is stenciled near the center of the car. The term load limit may also be used for capacity.

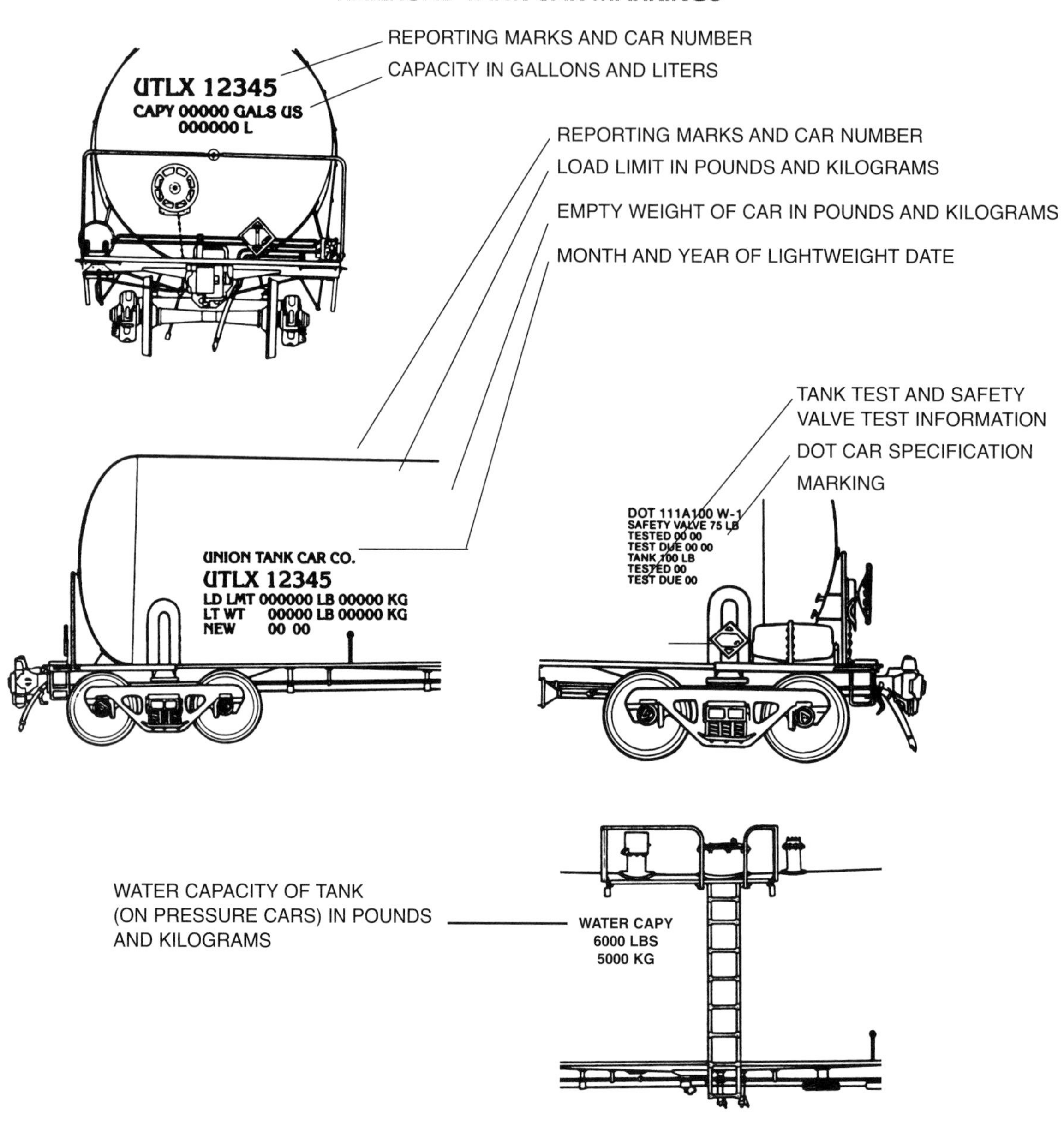

- **Specification Marking**. The specification marking indicates the standards to which a tank car was built. These markings will be on both sides of the tank car (right end as one faces the tank) and provide the following information:

Approving Authority	Tank test pressure
Class Number	Type of material used in construction
Separator / Delimiter Character (significant in certain tank cars)	Type of weld used
	Fittings/material/lining

- **Specification Plate.** Tank cars ordered after 2003 will have a plate on the A-end right bolster and the B-end left bolster (i.e., structural cross-member which cradles the tank) that provides information about the tank car's characteristics. Although not designed as an emergency response tool, the plate will provide the following information:

  Car Builder's Name
  Builder's Serial Number
  Certificate of Construction/Exemption
  Tank Specification
  Tank Shell Material/Head Material
  Insulation Materials
  Insulation Thickness
  Underframe/Stub Sill Type
  Date Built

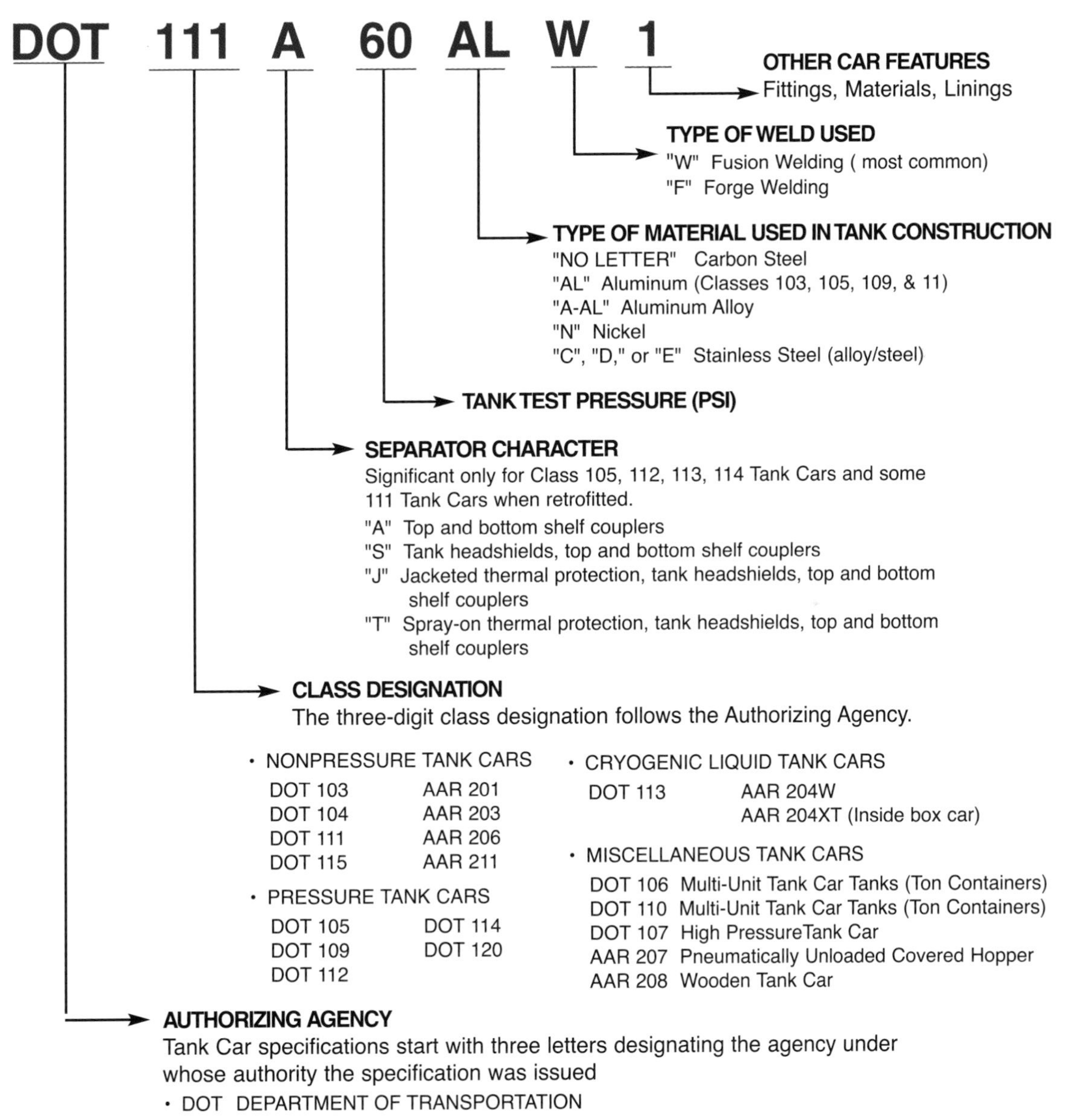

Figure 6.27 Railroad Tank Cars and Markings.

RAILROAD TANK CARS

NON-PRESSURE TANK CARS

DESIGN AND CONSTRUCTION

- Horizontal tank with flat or nearly flat ends (NOTE: Only jacketed tank cars will appear to have flat ends).
- Fittings and valves visible on top of tank car. However, some non-pressure cars may be found with a protective housing around all fittings.
- Older cars will have an expansion dome with visible fittings.
- Although classified as non-pressure, actually have tank test pressures at 60 and 100 psi.
- Capacities of 4,000 to 45,000 gallons. Capacity can be determined from tank car markings.
- May have up to 6 compartments, each having its own set of fittings.
- Often has bottom unloading valve.
- Non-pressure tank cars transporting strong corrosives may have a wide band of corrosion-resistant paint running vertically at the manway, or may be identified by staining or corrosion around the manway area.
- Safety features can include pressure and vacuum relief valves, shelf couplers, head shields, and thermal protection.
- Non-pressure car classes

DOT 103	DOT 111	AAR 201	AAR 206
DOT 104	DOT 115	AAR 203	AAR 211

CONTENTS / HAZARD CLASS

- Transports liquids and solids with vapor pressures below 25 psig at 105–115°F.
- Examples include flammable and combustible liquids, flammable solids, reactive liquids and solids, oxidizers, organic peroxides, liquid poisons, and corrosives.
- Non-hazardous liquids include food grade liquids, such as corn syrup and vegetable oils.

RAILROAD TANK CARS

PRESSURE TANK CARS

DESIGN AND CONSTRUCTION

- Cylindrical tank with rounded ends. May be insulated or thermally protected (either jacketed or sprayed on). Those without insulation or jacketed thermal protection must have the top two-thirds painted white.
- Fittings and valves enclosed in a protective housing with a cover.
- Tank test pressures range from 100 to 600 psi.
- Capacities range from 4,000 to 45,000 gallons. Capacity can be determined from tank car markings.
- Safety features can include pressure relief protection, excess flow valves, shelf couplers, and head shields. Tank cars transporting flammable liquefied gases must be provided with thermal protection. Other gases will require thermal protection by 2006.
- Pressure car classes

  DOT 105 DOT 114
  DOT 109 DOT 120
  DOT 112

CONTENTS / HAZARD CLASS

- Transports flammable, non-flammable and poisonous compressed gases. May also transport flammable liquids.
- Examples include LPG, chlorine, anhydrous ammonia, and anhydrous hydrogen fluoride.
- Hydrocyanic acid is typically transported in a car painted white with a red stripe running the length and around the ends of the car. Anhydrous hydrogen fluoride may be found in a tank car with an orange/red stripe running the middle of the car. These color and design schemes are an industry standard and are not DOT requirements. NOTE: Security concerns may cause these markings to be changed in the future; check with the railroad or shipper for the latest information

CRYOGENIC LIQUID TANK CARS

DESIGN AND CONSTRUCTION

- Well-insulated "Thermos Bottle" design. Inner stainless steel tank holds the product, and the space between the inner and outer tanks is filled with insulation and under a vacuum.
- Transports low pressure refrigerated liquids (pressures 25 psig or lower). Capacity can be determined from tank car markings.
- Tank test pressure can range up to175 psi.
- Absence of any top fittings or protective housing.
- Loading/unloading fittings and pressure relief device often found in cabinets on both sides or at one end of the tank car.
- Cryogenic tank car classes

  DOT 113 AAR 204

CONTENTS / HAZARD CLASS

- Gases that are liquefied through refrigeration rather than pressurization. Pressures usually below 25 psig at product temperatures < -130°F.
- Examples include liquid oxygen, liquid hydrogen, and liquid argon. LNG and ethylene may be found at slightly higher pressures.
- Combination of insulation and vacuum protects the contents from ambient temperatures for only 30 days. Shippers and railroads closely track these "time sensitive" shipments.

RAILROAD TANK CARS

	DESIGN AND CONSTRUCTION	CONTENTS / HAZARD CLASS
THE TANK TRAIN 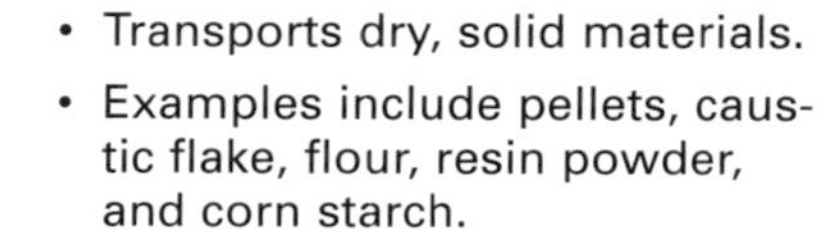	• Series of non-pressure tank cars interconnected with flexible hoses to allow for loading / unloading of the cars from one end. • Used when all tank cars are transporting the same product. • Consist of spring-loaded, butterfly valve arrangement on each car which is pneumatically controlled from the loading / unloading point. After loading, hoses are purged of liquid and valves are closed, thereby isolating each car.	• Transports liquids with low vapor pressures. • Examples include flammable and combustible liquids, crude oil, phosphoric acid
PNEUMATICALLY UNLOADED COVERED HOPPER CARS 	• Covered hopper car that is loaded through a series of hatches on top of the car and unloaded pneumatically. • Although non-pressure cars, tank test pressures range from 20 to 80 psi. Pressure is only on the car during unloading operations. • The movement of finely divided solids and air during transfer operations can generate static charges. Dust explosions may be a possibility if the solid is combustible. • Tank car class AAR 207.	• Transports dry, solid materials. • Examples include pellets, caustic flake, flour, resin powder, and corn starch.

Figure 6.28 Examples of railroad tank cars.

RADIOACTIVE MATERIAL PACKAGING

Federal regulations place strict controls on the transportation of radioactive materials. The transport of radioactive materials is based on the philosophy that (1) safety should be primarily focused on the package (i.e., packaging is the first line of defense), and (2) package integrity should be directly related to the degree of hazard of the material it contains. Radiological packaging may be made of any number of materials, depending on the level of radioactivity involved.

More than two thirds of radioactive material shipments are of human-made radioisotopes used in medicine, industry, agriculture, and scientific research. These are typically shipped in their most stable forms, which is usually as a solid material. The majority of radioactive material shipments are made by truck, while most larger shipments (e.g., Type B packaging) are transported by rail.

Responders should be familiar with four basic types of radioactive material packaging:

1. *Excepted Packaging*—Used to transport material with low levels of radioactivity. Excepted Packaging does not have to pass any performance tests but must meet specific design requirements spelled out in DOT regulations. Key design and construction features include the following:
   - Authorized for limited quantities of radioactive material that would pose a very low hazard if released in an accident. Examples include consumer goods, such as smoke detectors and lantern mantles.
   - Are exempted from specific radioactive material packaging, marking, labeling, and shipping paper requirements.

   Included under Excepted Packaging is Strong Tight Packaging. Strong Tight Packaging is allowed for shipping Low Specific Activity (LSA) and Surface Contaminated Objects (SCO) transported domestically in "exclusive use" shipments. These must meet only the general requirements for any hazardous materials package.

2 *Industrial Packaging*—Used in certain shipments of LSA material and Surface Contaminated Objects, which are typically categorized as radioactive waste. Most low-level radioactive waste, such as contaminated protective clothing and handling materials, is shipped in secured packaging of this type. DOT regulations require that these packagings allow no identifiable release of the material into the environment during normal transportation and handling.

3. *Type A Packaging*—Used to transport small quantities of radioactive material with higher concentrations of radioactivity than those shipped in Industrial Packaging. Key design and construction features include the following:
   - Typically constructed of steel, wood or fiberboard and have an inner containment vessel made of glass, plastic, or metal surrounded with packing material made of polyethylene, rubber or vermiculite.
   - Designed to ensure that the package retains its containment integrity and shielding under normal transport conditions. However, they are not designed to withstand the forces of an accident.
   - The consequences of a release of material from a Type A package would not be major since the quantity of radioactive material transported in this package is limited.

- Are only used to transport non-life-threatening amounts of radioactive material, such as radiopharmaceuticals, radioactive waste, and radioactive sources used in industrial applications.

4. *Type B Packaging*—Used to transport radioactive material with the highest levels of radioactivity, including potentially life-endangering amounts that could pose a significant risk if released during an accident. Key design and construction features include the following:
   - Must meet all of the Type A requirements, as well as a series of tests that simulate severe or worst-case accident conditions. Accident conditions are simulated by performance testing and engineering analysis.
   - Range from small, hand-held radiography cameras to heavily shielded steel casks that weigh up to 125 tons. Examples of materials transported in Type B packaging include spent nuclear fuel, high-level radioactive waste, and high concentrations of other radioactive materials such as cesium and cobalt.
   - In the 50+ year history of transporting radioactive material, there has never been a release from a certified Type B package or an injury or death resulting from the release of a radioactive material in a transportation incident.

Figure 6.29 Responders should be familiar with the four basic types of radioactive material packaging.

FACILITY CONTAINMENT SYSTEMS

Facility containment systems are packaging, containers, and/or associated systems that are part of a fixed facility's operations. Examples may include storage tanks, process towers, chemical and nuclear reactors, piping systems, pumps, storage bins and cabinets, dryers and degreasers, and machinery.

The following charts provide further information on shapes, descriptions, and contents of common facility storage tanks and vessels:

- Atmospheric and Low-Pressure Liquid Storage Tanks (Figure 6.30)
- Pressurized Storage Vessels (Figure 6.31 and 6.32)

ATMOSPHERIC PRESSURE LIQUID STORAGE TANKS

ATMOSPHERIC TANKS	DESIGN AND CONSTRUCTION	CONTENTS / HAZARD CLASS
CONE ROOF TANK 	• Welded steel tank with vertical cylindrical walls supporting a fixed bottom and a flat or conical roof. However, older riveted tanks still exist. • Typically 20 to 100 ft. diameter, but can be as large as 200+ ft. Depending upon tank diameter and height, capacities can range from 1,000 to 100,000+ barrels (NOTE: 1 barrel = 42 gallons). • Have a vapor space between the liquid level and the roof. Operates at atmospheric pressures, and uses a pressure-vacuum valve for normal tank "breathing" during transfer operations. • Emergency venting provided by a weak roof-to-shell seam designed to fail in case of fire or explosion (if tank is constructed to API 650 specifications). • May be insulated, particularly for fuel oil or asphalt service. • Tanks in chemical service may be lined; high vapor pressure solvents may have an inert blanket (e.g., nitrogen) over the product.	• Flammable and combustible liquids, solvents, oxidizers, and corrosives. • At petroleum storage facilities, will primarily store combustible liquids due to environmental emissions and economic considerations. • At chemical facilities, may store virtually any hazardous or non-hazardous liquid.
COVERED FLOATING ROOF TANK	• Welded steel tank with vertical cylindrical walls supporting a fixed flat or conical roof, with an internal floating roof or pan that floats on the surface of the product. • Is essentially a cone roof tank with an internal floating roof. • Have no vapor space, except when the product is at lowest levels and the floating roof / pan is sitting on its roof supports. • Identified by the large "eyebrow" vents at the top of the tank shell. • Typically 40 to 120 ft. diameter, but can be as large as 300+ ft. at refineries and crude oil terminals. Depending upon tank diameter and height, capacities can range from 1,000 to 500,000+ barrels (NOTE: 1 barrel = 42 gallons). • Fires can be difficult to extinguish through portable application devices.	• Flammable and combustible liquids. • At petroleum storage facilities, will primarily store flammable liquids and higher vapor pressure products due to environmental emissions and economic considerations.

ATMOSPHERIC PRESSURE LIQUID STORAGE TANKS

ATMOSPHERIC TANKS	DESIGN AND CONSTRUCTION	CONTENTS / HAZARD CLASS
OPEN FLOATING ROOF TANK	• Steel tank with vertical cylindrical walls and a roof that floats on the surface of the product. There is a seal area between the tank shell and floating roof. • Have no vapor space, except when the product is at lowest levels and the floating roof is sitting on its roof supports. • Identified by the wind girder around the top of the tank shell and the roof ladder. • Floating roof may be a pontoon or honeycomb design, and have a drainage system to carry normal rainwater off the roof and to the ground. • Typically 40 to 120 ft. diameter, but can be as large as 300+ ft. at refineries and crude oil terminals. Depending upon tank diameter and height, capacities can range from 1,000 to 500,000+ barrels (NOTE: 1 barrel = 42 gallons). • Most common storage tank found in the petroleum industry; most common scenario is a seal fire. • When installed, semi-fixed suppression systems are generally designed to protect only the seal area.	• Flammable and combustible liquids. • At petroleum storage facilities, will primarily store flammable liquids and higher vapor pressure products due to environmental emissions and economic considerations.
OPEN FLOATING ROOF TANK WITH GEODESIC DOME	• Is essentially an open floating roof tank with a lightweight aluminum geodesic dome installed over the top. • Geodesic roof structure is installed for environmental emissions and economic considerations.	• Flammable and combustible liquids. • At petroleum storage facilities, will primarily store flammable liquids and higher vapor pressure products due to environmental emissions and economic considerations.

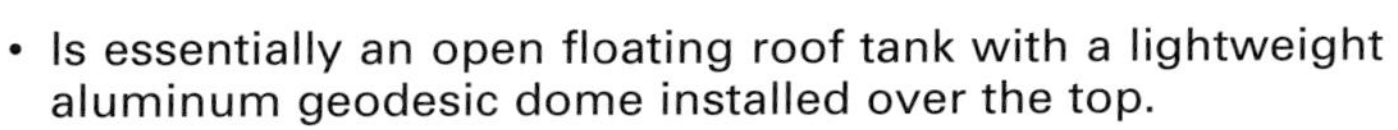

Figure 6.30 Examples of atmospheric liquid storage tanks.

LOW PRESSURE LIQUID STORAGE TANKS

LOW PRESSURE TANKS

VERTICAL STORAGE TANK

DESIGN AND CONSTRUCTION

- Steel tank with vertical cylindrical walls supporting a fixed bottom and a flat or conical roof. Typically welded construction, although older riveted tanks are still in-service. Also referred to as dome roof tanks.
- Smaller diameter (approximately 25 ft. and less). Capacities range from 100 to 10,000 barrels (NOTE: 1 barrel = 42 gallons).
- Working pressures range from 2.5 to 15 psi.
- Uses a pressure-vacuum valve for normal tank operations. If designed, emergency relief protection provided by a weak roof-to-shell seam. However, the roof may not always fail as designed.
- Tanks in chemical service may be lined; high vapor pressure solvents may have an inert blanket (e.g., nitrogen) over the product.
- Smaller fiberglass tanks with capacities up to 25,000 gallons may also be found in chemical service.

CONTENTS / HAZARD CLASS

- Flammable and combustible liquids, solvents, oxidizers, poison liquids, and corrosives.
- At petroleum storage facilities, will primarily store combustible liquids due to environmental emissions and economic considerations.
- At chemical facilities, may store virtually any hazardous or non-hazardous liquid.

HORIZONTAL STORAGE TANK

DESIGN AND CONSTRUCTION

- Wide range of horizontal tanks in service. Range from welded or riveted steel plate construction to double-walled steel tanks encased in fire retardant materials with its own containment box (i.e., tank within a tank design).
- Capacities typically range from 300 gallons to 10,000 gallons, although larger tanks may be found. Working pressures are at atmospheric pressure.
- Structural integrity of tank supports is critical in a fire scenario. Under NFPA standards, horizontal tanks storing flammable liquids must be protected by materials having a fire resistance of not less than 2 hours.
- Uses a pressure-vacuum valve for normal tank operations.

CONTENTS / HAZARD CLASS

- Flammable and combustible liquids, solvents, oxidizers, poison liquids, and corrosives.
- Newer generation of tank within a tank design is widely used for motor fuels applications (e.g., gasoline, diesel fuel).

ATMOSPHERIC AND LOW PRESSURE LIQUID STORAGE TANKS

LOW PRESSURE TANKS

UNDERGROUND STORAGE TANK

DESIGN AND CONSTRUCTION

- EPA defines an Underground Storage Tank System as a tank and any underground piping connected to the tank that has at least 10% of its combined volume underground.
- Horizontal tank constructed of steel, fiberglass, or steel with a fiberglass coating. Tank capacities can range up to 25,000+ gallons.
- Leak detection, overfill protection, and cathodic protection are required under EPA regulations. Tanks may be single or double-walled to meet requirements.
- Visible clues are vents, fill points, and potential occupancy and locations (e.g., service station, fleet maintenance).

CONTENTS / HAZARD CLASS

- Flammable and combustible liquids.
- At chemical facilities, may store virtually any hazardous or non-hazardous liquid.

Figure 6.31 Examples of low pressure liquid storage tanks.

HIGH PRESSURE STORAGE TANKS

HIGH PRESSURE HORIZONTAL TANK

DESIGN AND CONSTRUCTION

- Single shell, non-insulated welded storage tanks constructed to ASME standards.
- Pressures range from 100 to 500 psi, and are protected with pressure relief valves.
- Capacities can vary—residential tanks range from 100 to 1,000 gallons; industrial and distribution facilities range from 1,000 to 120,000 gallons.
- Majority are horizontal, although some smaller vertical tanks may be found. Some tanks may be mounded, where most of the tank surface is covered by a earthen surface.
- Piping systems will include both a liquid line and a vapor line. Liquid lines are generally larger than vapor lines, and may be painted different colors.
- Some storage tanks may be protected with water spray systems.

CONTENTS / HAZARD CLASS

- Flammable and non-flammable liquefied gases, including LPG, anhydrous ammonia, chlorine, and vinyl chloride.
- Bulk LPG storage facilities can also include cylinder and bobtail filling stations, and truck and rail car unloading racks.

HIGH PRESSURE SPHERICAL TANK

- Single shell, non-insulated welded storage tanks constructed to ASME standards. Can be up to 90+ ft. diameter.
- Pressures range from 100 to 500 psi, and are protected with pressure relief valves.
- Capacities can be as large as 600,000 gallons.
- Piping systems will include both a liquid line and a vapor line. Liquid lines are generally larger than vapor lines, and may be painted different colors.
- Some storage tanks may be protected with water spray and deluge systems.

- Flammable and non-flammable liquefied gases, including LPG, anhydrous ammonia, and vinyl chloride.
- Found at large gas distribution facilities and petrochemical manufacturing facilities.

HIGH PRESSURE UNDERGROUND TANK

- Single shell, non-insulated welded storage tanks constructed to ASME standards. Outer shell coated with a mastic material and provided with cathodic protection.
- Capacities can range from 500 to 10,000+ gallons, with pressures ranging from 100 to 500 psi.
- Most underground propane tanks under 2,000 gallons (water capacity) have a riser with a combination valve containing the filler valve, service valve, pressure relief valve and other appurtenances. The protective shroud covering the riser may be the only identification clue.
- Some areas store LPG gases in earth covered domes, underground caverns and salt mines.

- Liquefied petroleum gases (propane, butane and mixtures)

CRYOGENIC LIQUID AND REFRIGERATED STORAGE TANKS

	DESIGN AND CONSTRUCTION	CONTENTS / HAZARD CLASS
CRYOGENIC LIQUID STORAGE TANK	• Large, well insulated "thermos bottle" design. Inner container constructed of stainless steel or metal suitable for low temperature service, while the outer shell is typically a gastight carbon steel jacket, with the space filled with insulation and under vacuum. Constructed to ASME standards. • Normally have vaporizing unit and pumps to warm the product and convert it to a gas, as well as pressurize the container and "move" the gas. • Liquid oxygen tanks will always be over a concrete pad; never over asphalt or grass. • Capacities range from 500 to 20,000 gallons (most are in the range of 1,500 to 11,000 gallons), with design working pressures up to 250 psi. • Safety features include pressure relief devices on both the inner tank and the outer tank, excess flow valves, fusible links and nuts, and emergency remote shutoff.	• Cryogenic liquids, including liquid oxygen, liquid nitrogen, liquid carbon dioxide, liquid argon, and liquid helium. • Often find liquid oxygen (LOX) at hospitals and some industries, and liquid nitrogen at hospitals, universities, and industry.
REFRIGERATED STORAGE TANK 	• Insulated and refrigerated cylindrical bulk storage tank. Depending upon product and service, may be a combination of single steel wall, double steel wall, or a concrete exterior and double steel interior wall combined with insulation. • Products are stored at temperatures near their boiling point. • Design working pressures <15 psi. • Typically 50 to 150 ft. diameter, but can be as large as 200+ ft. Depending upon tank diameter and height, capacities can range from 20,000 to 200,000+ barrels (NOTE: 1 barrel = 42 gallons).	• Liquefied gases, including LPG, LNG, ethylene and anhydrous ammonia. • Found at pipeline terminals, LNG peak shaving plants, and utility plants.

Figure 6.32 Examples of high pressurized storage tanks.

MARKINGS AND COLORS

Markings and colors on hazmat packaging or containment systems are the third clue in the standard identification process. These clues may include color codes, container specification numbers, signal words or even the content's name and associated hazards. At facilities, clues may include Hazard Communication markings, piping color code systems, and specific signs and/or signal words (e.g., "Hydrofluoric Acid Area").

Markings and colors can be used as a clue for hazmat recognition, identification, and classification. We will evaluate these systems based on the nature and use of the package.

NONBULK PACKAGE MARKINGS

Agricultural Chemicals and Pesticide Labels. Individual nonbulk packages, particularly those storing pesticides and agricultural chemicals, will display useful information for identification and hazard assessment. These container markings include the following:

- *Toxicity Signal Word*—The signal word indicates the relative degree of acute toxicity. Located in the center of the front label panel, it is one of the most important label markings. The three toxic categories are high toxicity, moderate toxicity, and low toxicity (see Figure 6.13).

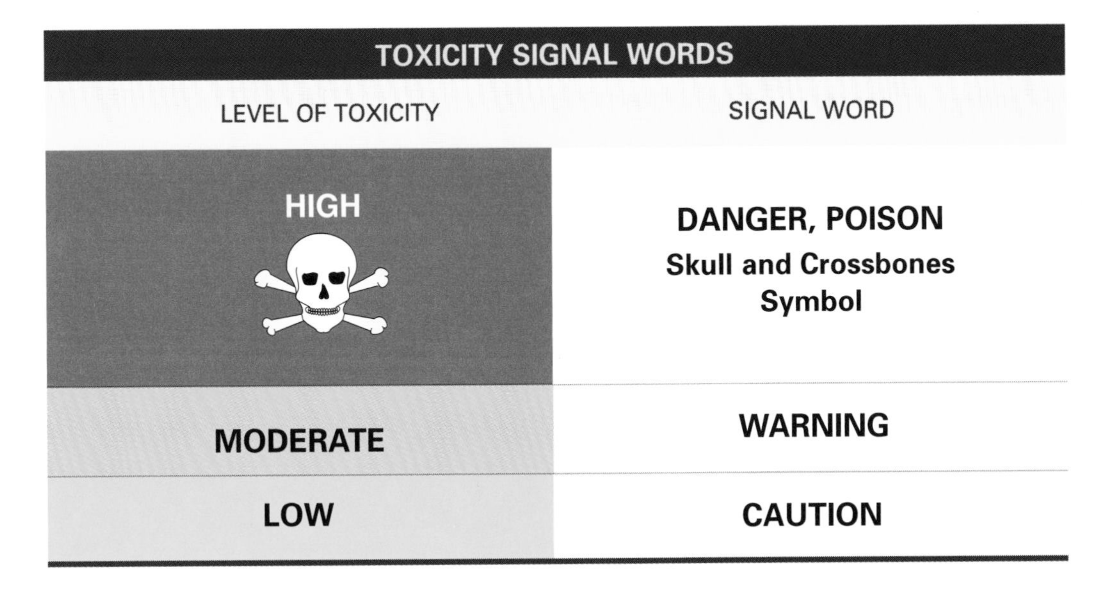

TOXICITY SIGNAL WORDS	
LEVEL OF TOXICITY	SIGNAL WORD
HIGH	DANGER, POISON Skull and Crossbones Symbol
MODERATE	WARNING
LOW	CAUTION

Figure 6.33 Toxicity signal words found on ag chem and pesticide containers indicate the relative degree of acute toxicity of the contents.

- *Statement of Practical Treatment*—Located near the signal word on the front panel, it is also referred to as the "first aid statement" or "note to physician." It may have precautionary information as well as emergency procedures for exposures. Antidote and treatment information may also be added.

- *Physical or Chemical Hazard Statement*—A statement displayed on a side panel, as necessary. It will list special flammability, explosion, or chemical hazards posed by the product.
- *Product Name*—The brand or trade name is printed on the front panel. If the product name includes the term *technical*, as in Parathion-Technical, it generally indicates a highly concentrated pesticide with 70% to 99% active ingredients.
- *Ingredient Statement*—All pesticide labels must have statements that break down the chemical ingredients by their relative percentages or as pounds per gallon of concentrate. Active ingredients are the active chemicals within the mixture. They must be listed by chemical name, and their common name may also be shown. Inert ingredients have no pesticide activity and are usually not broken into specific components, only total percentage. A number of agricultural chemical products, particularly those used in the home, use flammable products (e.g., propane used for aerosols and diesel fuel or solvents used with liquids) only as the inert ingredients and may not be easy to identify.
- *Environmental Information*—The label may provide information on both the storage and disposal of the product, as well as environmental or wildlife hazards that could occur. This information can be most useful when planning clean-up and disposal after a fire or spill.
- *EPA Registration Number*—This number is required for all ag chems and pesticide products marketed in the United States. It is one of three ways to positively identify a pesticide. The other ways are by the product name or by the chemical ingredient statement. The registration number will appear as a two- or three-section number (e.g., EPA Reg. No. 239-2491-AA) and indicates the manufacturer, specific product, and locations where the product may be used. When relaying this number to Chemtrec or other sources, make sure you include each dash. A U.S. Department of Agriculture number may appear on products registered before 1970.
- *EPA Establishment Number*—The location where the product was manufactured. This number has little significance to responders.

When faced with an incident involving poisonous or toxic containers, obtain an uncontaminated container or label for reference whenever possible. When victims are transported to a medical facility as a result of a suspected exposure, ensure proper identification and medical care by also forwarding an uncontaminated label or container.

Chemical Abstract Service (CAS) Number. Often found on hazardous materials containers, as well as MSDSs, the CAS number is often used by state and local right-to-know regulations for tracking chemicals in the community and workplace. Sometimes referred to as a chemical's "social security number," sequentially assigned CAS numbers identify specific chemicals and have no chemical significance.

Cylinder Color Codes. A number of gases are stored and transported in compressed gas cylinders. Although there are several voluntary color schemes, none are mandatory. The only reliable method to identify cylinder contents is to check the DOT label attached to the cylinder head. Unfortunately, if cylinders are exposed to fire, the labels may burn off and further hamper identification.

BULK PACKAGING AND TRANSPORTATION MARKINGS

Identification Number. Four digit identification numbers are assigned to a hazardous material or group of hazardous materials. They are used to determine the name of the material and to obtain hazard and response information from emergency response guidebooks. They are required on shipping papers and on the following:

- Non-bulk packages of hazardous materials (except limited quantities) printed adjacent to the required labels on the package
- Shipments of non-bulk packages when loaded at one location of more than 8,820 lbs. of a single material
- Bulk packages of hazardous materials, including cargo tanks, rail cars, portable tanks, and railroad hopper cars

Acceptable methods of displaying this marking are shown in Figure 6.34. Identification numbers are prohibited on DOT explosives, dangerous, radioactive, and subsidiary hazard placards. The identification number must be displayed on the supplemental orange panel for these materials.

When viewing shipping papers or MSDSs, the identification number may be found with the prefix "UN" or "NA." Those identification numbers preceded by the letters "UN (United Nations)" are appropriate for domestic and international transportation. Those preceded by the letters "NA (North America)" are not recognized for international transportation, except to and from Canada.

Figure 6.34 The four-digit identification number may be displayed on a bulk container in one of three ways.

Inhalation Hazard Markings. The words *Poison—Inhalation Hazard* or *Inhalation Hazard* indicate that the material is considered toxic by inhalation (e.g., chlorine).

Marine Pollutant Markings. These markings are displayed on both sides and both ends of bulk packages of materials designated on the shipping papers as a marine pollutant, except when the container is properly placarded (e.g., Poison, Flammable Liquid, etc.). The mark must appear when the package moves by water.

Elevated Temperature Materials. Elevated temperature materials are materials that, when offered for transportation in a bulk container, are:

- Liquids at or above 212°F (100°C).
- Liquids with a flash point at or above 100°F (37.8°C) that are intentionally heated and are transported at or above their flash point.
- Solids at a temperature at or above 464°F (240°C).

Except for a bulk container transporting molten aluminum or molten sulfur (which must be marked "MOLTEN ALUMINUM" or "MOLTEN SULFUR" respectively), the container must be marked on each side and each end with the word "HOT" in black or white lettering on a contrasting background. The word "HOT" may be displayed on the bulk packaging itself or in black lettering on a white square-on-point configuration similar in size to the placard.

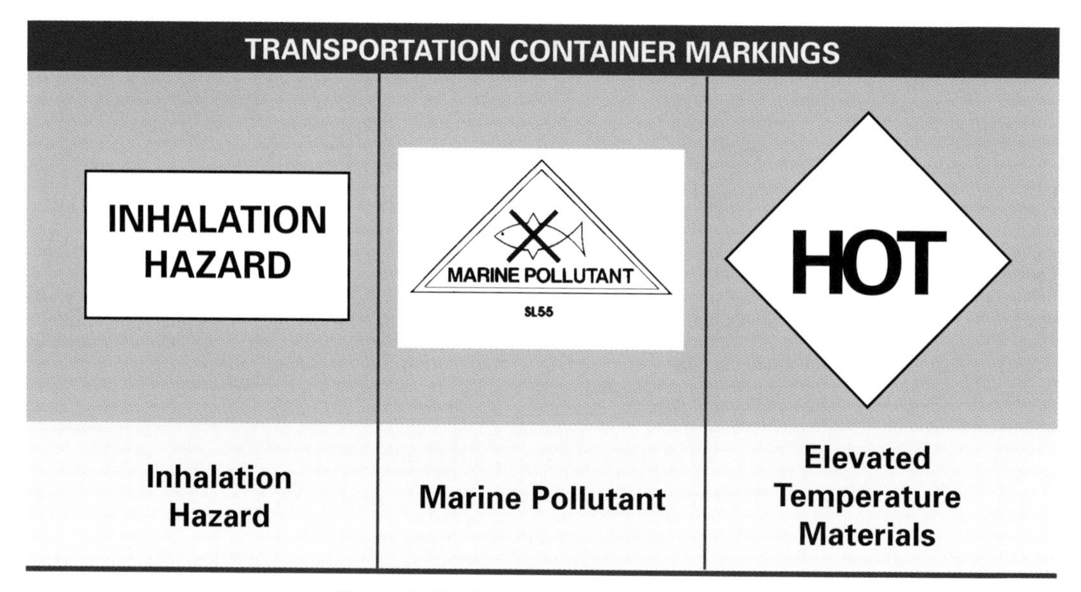

Figure 6.35 Transportation container markings for special situations.

Pipelines. Pipelines are the second largest hazmat transportation mode and often cross over or under roads, waterways, and railroads. At each of these crossover locations, a marker should identify the pipeline right-of-way. Although its format and design may vary, all markers are required to provide the pipeline contents (e.g., natural gas, propane, liquid petroleum products, etc.), the pipeline operator, and an emergency telephone number (see Figure 6.36).

The pipeline emergency telephone number goes to a control room, where an operator monitors pipeline operations and can start emergency shutdown procedures. It should be stressed that even when a ruptured pipeline is immediately shut down, product backflow may continue for several hours until the product drains to the point of release.

Most gas pipelines are dedicated to one product (e.g., natural gas, butadiene, anhydrous ammonia). However, liquid petroleum transmission pipelines may carry several different petroleum products simultaneously. Figure 6.37 shows a typical liquid products shipping cycle. Liquid pipeline personnel normally refer to product flows in terms of "barrels" rather than gallons (Note: 1 barrel equals 42 gallons).

Product flows through many transmission pipeline systems are monitored through a computerized pipeline SCADA System (Supervisory Control and Data Acquisition System). The exact injection date and time of the particular product into the pipeline is noted and its delivery date/time is projected. As the product gets close to its destination, a sensor in the line signals the arrival of the shipment. The SCADA System provides pipeline personnel with the ability to monitor pipeline flows and pressures and initiate emergency shutdown procedures in the event of a release.

For liquid petroleum pipelines, there is usually no physical separator (e.g., sphere or pig) between different products. Rather, the products are allowed to "co-mingle." This interface can range from a few barrels to several hundred, depending on the pipeline size and products involved. Verification of the shipment arrival is made by examining a sample of the incoming batch for color, appearance, and/or chemical characteristics.

Figure 6.36 Pipeline markers must provide the pipeline contents, the pipeline operator, and an emergency telephone number.

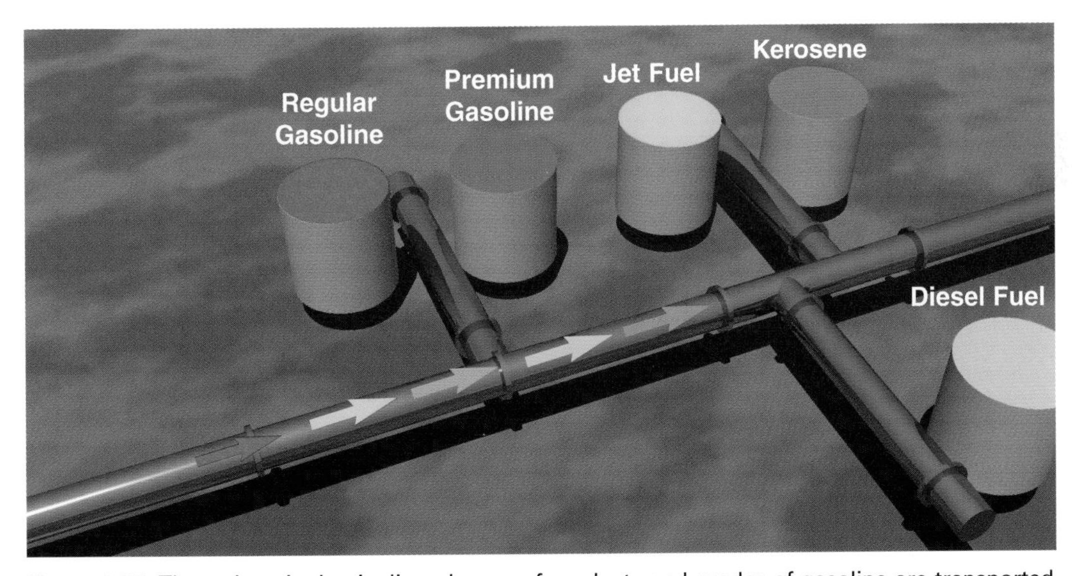

Figure 6.37 Through a single pipeline, dozens of products and grades of gasoline are transported at once through a process called "batching".

FACILITY MARKINGS

NFPA 704 System. Many state and local fire codes mandate the use of the NFPA 704 marking system at all fixed facilities in their jurisdiction, including tanks and storage areas. The NFPA 704 system, which is shown in Figure 6.38, is not used on transport vehicles.

THE NFPA 704 MARKING SYSTEM

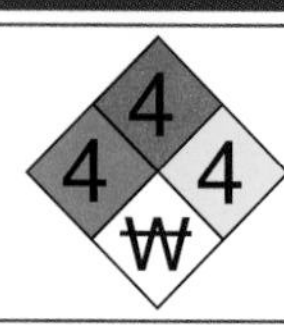

The NFPA 704 Marking System distinctively indicates the properties and potential dangers of hazardous materials. The following is an explanation of the meanings of the Quadrant Numerical Codes:

HEALTH (Blue)

IN GENERAL, HEALTH HAZARD IN FIREFIGHTING IS THAT OF A SINGLE EXPOSURE WHICH MAY VARY FROM A FEW SECONDS UP TO AN HOUR. THE PHYSICAL EXERTION DEMANDED IN FIREFIGHTING OR OTHER EMERGENCY CONDITIONS MAY BE EXPECTED TO INTENSIFY THE EFFECTS OF ANY EXPOSURE. ONLY HAZARDS ARISING OUT OF AN INHERENT PROPERTY OF THE MATERIAL ARE CONSIDERED. THE FOLLOWING EXPLANATION IS BASED UPON PROTECTIVE EQUIPMENT NORMALLY USED BY FIREFIGHTERS:

4 MATERIALS TOO DANGEROUS TO HEALTH TO EXPOSE FIREFIGHTERS. A FEW WHIFFS OF THE VAPOR COULD CAUSE DEATH OR THE VAPOR OF LIQUID COULD BE FATAL ON PENETRATING THE FIREFIGHTER'S NORMAL FULL PROTECTIVE CLOTHING. THE NORMAL, FULL-PROTECTIVE CLOTHING AND BREATHING APPARATUS AVAILABLE TO THE AVERAGE FIRE DEPARTMENT WILL NOT PROVIDE ADEQUATE PROTECTION AGAINST INHALATION OR SKIN CONTACT WITH THESE MATERIALS.

3 MATERIALS EXTREMELY HAZARDOUS TO HEALTH, BUT AREAS MAY BE ENTERED WITH EXTREME CARE. FULL-PROTECTIVE CLOTHING INCLUDING SELF-CONTAINED BREATHING APPARATUS, COAT, PANTS, GLOVES, BOOTS AND BANDS AROUND LEGS, ARMS, AND WAIST SHOULD BE PROVIDED. NO SKIN SURFACE SHOULD BE EXPOSED.

2 MATERIALS HAZARDOUS TO HEALTH, BUT AREAS MAY BE ENTERED FREELY WITH FULL-FACE MASK AND SELF-CONTAINED BREATHING APPARATUS WHICH PROVIDES EYE PROTECTION.

1 MATERIALS ONLY SLIGHTLY HAZARDOUS TO HEALTH. IT MAY BE DESIRABLE TO WEAR SELF-CONTAINED BREATHING APPARATUS.

0 MATERIALS WHICH WOULD OFFER NO HAZARD BEYOND THAT OF ORDINARY COMBUSTIBLE MATERIAL UPON EXPOSURE UNDER FIRE CONDITIONS.

FLAMMABILITY (Red)

SUSCEPTIBILITY TO BURNING IS THE BASIS FOR ASSIGNING DEGREES WITHIN THIS CATEGORY. THE METHOD OF ATTACKING THE FIRE IS INFLUENCED BY THIS SUSCEPTIBILITY FACTOR.

4 VERY FLAMMABLE GASES OR VERY VOLATILE FLAMMABLE LIQUIDS. SHUT OFF FLOW AND KEEP COOLING WATER STREAMS ON EXPOSED TANKS OR CONTAINERS.

3 MATERIALS WHICH CAN BE IGNITED UNDER ALMOST ALL NORMALTEMPERATURE CONDITIONS. WATER MAY BE INEFFECTIVE BECAUSE OF THE LOW FLASH POINT.

2 MATERIALS WHICH MUST BE MODERATELY HEATED BEFORE IGNITION WILL OCCUR. WATER SPRAY MUST BE USED TO EXTINGUISH THE FIRE BECAUSE THE MATERIAL CAN BE COOLED BELOW ITS FLASH POINT.

1 MATERIALS THAT MUST BE PREHEATED BEFORE IGNITION CAN OCCUR. WATER MAY CAUSE FROTHING IF IT GETS BELOW THE SURFACE OF THE LIQUID AND TURNS TO STEAM. HOWEVER, WATER FOG GENTLY APPLIED TO THE SURFACE WILL CAUSE A FROTHING WHICH WILL EXTINGUISH THE FIRE.

0 MATERIALS THAT WILL NOT BURN.

REACTIVITY (STABILITY) (Yellow)

THE ASSIGNMENT OF DEGREES IN THE REACTIVITY CATEGORY IS BASED UPON THE SUSCEPTIBILITY OF MATERIALS TO RELEASE ENERGY EITHER BY THEMSELVES OR IN COMBINATION WITH WATER. FIRE EXPOSURE WAS ONE OF THE FACTORS CONSIDERED ALONG WITH CONDITIONS OF SHOCK AND PRESSURE.

4 MATERIALS WHICH (IN THEMSELVES) ARE READILY CAPABLE OF DETONATION OR OF EXPLOSIVE DECOMPOSITION OR EXPLOSIVE REACTION AT NORMAL TEMPERATURES AND PRESSURES. INCLUDES MATERIALS WHICH ARE SENSITIVE TO MECHANICAL OR LOCALIZED THERMAL SHOCK. IF A CHEMICAL WITH THIS HAZARD RATING IS IN AN ADVANCED OR MASSIVE FIRE, THE AREA SHOULD BE EVACUATED.

3 MATERIALS WHICH (IN THEMSELVES) ARE CAPABLE OF DETONATION OR OF EXPLOSIVE DECOMPOSITION OR EXPLOSIVE REACTION BUT WHICH REQUIRE A STRONG INITIATING SOURCE OR WHICH MUST BE HEATED UNDER CONFINEMENT BEFORE INITIATION. INCLUDES MATERIALS WHICH ARE SENSITIVE TO THERMAL OR MECHANICAL SHOCK AT ELEVATED TEMPERATURES AND PRESSURES OR WHICH REACT EXPLOSIVELY WITH WATER WITHOUT REQUIRING HEAT OR CONFINEMENT. FIREFIGHTING SHOULD BE DONE FROM AN EXPLOSIVE-RESISTANT LOCATION.

2 MATERIALS WHICH (IN THEMSELVES) ARE NORMALLY UNSTABLE AND RAPIDLY UNDERGO VIOLENT CHEMICAL CHANGE BUT DO NOT DETONATE. INCLUDES MATERIALS WHICH CAN UNDERGO CHEMICAL CHANGE WITH RAPID RELEASE OF ENERGY AT NORMAL TEMPERATURES AND PRESSURES OR WHICH CAN UNDERGO VIOLENT CHEMICAL CHANGE AT ELEVATED TEMPERATURES AND PRESSURES. ALSO INCLUDES THOSE MATERIALS WHICH MAY REACT VIOLENTLY WITH WATER OR WHICH MAY FORM POTENTIALLY EXPLOSIVE MIXTURES WITH WATER. IN ADVANCE OR MASSIVE FIRES, FIREFIGHTING SHOULD BE DONE FROM A SAFE DISTANCE OR FROM A PROTECTED LOCATION.

1 MATERIALS WHICH (IN THEMSELVES) ARE NORMALLY STABLE BUT WHICH MAY BECOME UNSTABLE AT ELEVATED TEMPERATURES AND PRESSURES OR WHICH MAY REACT WITH WATER WITH SOME RELEASE OF ENERGY BUT NOT VIOLENTLY. CAUTION MUST BE USED IN APPROACHING THE FIRE AND APPLYING WATER.

0 MATERIALS WHICH (IN THEMSELVES) ARE NORMALLY STABLE EVEN UNDER FIRE EXPOSURE CONDITIONS AND WHICH ARE NOT REACTIVE WITH WATER. NORMAL FIREFIGHTING PROCEDURES MAY BE USED.

SPECIAL INFORMATION (White)

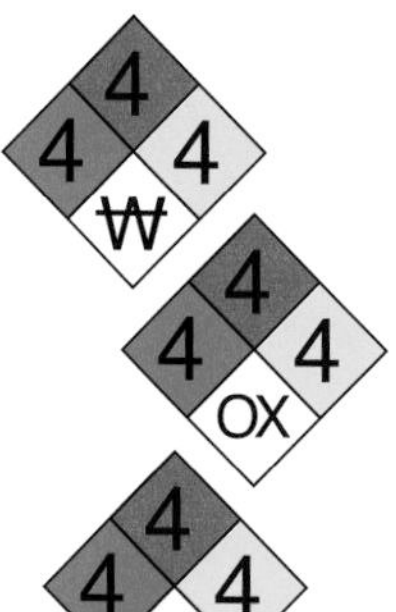

MATERIALS WHICH DEMONSTRATE UNUSUAL REACTIVITY WITH WATER SHALL BE IDENTIFIED WITH THE LETTER W WITH A HORIZONTAL LINE THROUGH THE CENTER (W).

MATERIALS WHICH POSSESS OXIDIZING PROPERTIES SHALL BE IDENTIFIED BY THE LETTERS OX.

MATERIALS POSSESSING RADIOACTIVITY HAZARDS SHALL BE IDENTIFIED BY THE STANDARD RADIOACTIVITY SYMBOL

Figure 6.38

DOT HAZARDOUS MATERIALS PLACARDING REQUIREMENTS

GENERAL GUIDELINES

- Placard transport vehicles, freight containers or rail cars containing any quantity of hazardous materials listed in Table 1.
- Transport vehicles and freight containers containing 1,001 lbs. (454 kg) or less of Table 2 hazardous materials in non-bulk packages are not required to be placarded.
- Placards may be displayed for a hazardous material, even when not required, if it conforms to DOT placard requirements.
- Text indicating the hazard (e.g., FLAMMABLE, CORROSIVE) is not required on a placard, except for Class 7 materials and the DANGEROUS placard.
- A bulk container that is placarded must remain placarded when empty unless it is cleaned and purged of vapors, or is refilled with another material that may or may not require placards.
- Placards are NOT required for the following:

  Infectious substances, ORM-D materials, and Limited Quantities.

  Combustible liquids in non-bulk packaging.

  Non-bulk packaging with residue of a Table 2 material.

TABLE 1 MATERIALS (PLACARD ANY QUANTITY)

PLACARD NAME	PLACARD	HAZARD CLASS OR DIVISION	EXAMPLES
EXPLOSIVES 1.1	EXPLOSIVES 1.1A 1	1.1 NOTE: Will have compatibility group letter when required	Detonators, Black Power, Dynamite
EXPLOSIVES 1.2	EXPLOSIVES 1.2B 1	1.2	Detonating Cord, Aerial Flares
EXPLOSIVES 1.3	EXPLOSIVES 1.3C 1	1.3	Liquid Fuel Rocket Motors

PLACARD NAME	PLACARD	HAZARD CLASS OR DIVISION	EXAMPLES
POISON GAS		2.3	Chlorine, Phosgene, Arsine
DANGEROUS WHEN WET		4.3	Sodium, Magnesium, Calcium Carbide
ORGANIC PEROXIDE		5.2 Note: Organic Peroxide, Type B, liquid or solid, temperature controlled	
POISON INHALATION HAZARD		6.1 Note: Inhalation Hazard, Zone A or B	Aniline, Arsenic
RADIOACTIVE		7	Cobalt, Uranium Hexaflourisde

TABLE 2 MATERIALS (REQUIRES 1,001LBS. OR MORE)			
PLACARD NAME	**PLACARD**	**HAZARD CLASS OR DIVISION**	**EXAMPLES**
EXPLOSIVES 1.4	1.4 EXPLOSIVES B 1	1.4 Note: Will have compatibility group letter when required	Common Fireworks Small Arms Ammunition
EXPLOSIVES 1.5	1.5 BLASTING AGENTS D 1	1.5	Ammonium Nitrate Fuel oil (ANFO) mixtures
EXPLOSIVES 1.6	1.6 EXPLOSIVES N 1	1.6	Very insensitive material with no explosion or fire hazard
FLAMMABLE GAS	FLAMMABLE GAS 2	2.1	Propane, Hydrogen, Acetylene
NONFLAMMABLE GAS	NON-FLAMMABLE GAS 2	2.2	Carbon Dioxide, Nitrogen, Argon, Anhydrous Ammonia
FLAMMABLE	FLAMMABLE 3	3	Gasoline, Acetone Methyl Alcohol

PLACARD NAME	PLACARD	HAZARD CLASS OR DIVISION	EXAMPLES
COMBUSTIBLE	COMBUSTIBLE 3	3	Diesel Fuel, Kerosene
FLAMMABLE SOLID	FLAMMABLE SOLID 4	4.1	Magnesium ribbons road flares
SPONTANEOUSLY COMBUSTIBLE	SPONTANEOUSLY COMBUSTIBLE 4	4.2	Air reactive materials or self-heating materials
OXIDIZER	OXIDIZER 5.1	5.1	Calcium hypochlorite, Calcium perchlorate, Ammonium nitrate
ORGANIC PEROXIDE	ORGANIC PEROXIDE 5.2	5.2 Note: Other than organic peroxide, Type B, liquid or solid, temperature controlled	
POISON	POISON 6	6.1 Note: Other than Inhalation Hazard, Zone A or B	Carbon tetrachloride

PLACARD NAME	PLACARD	HAZARD CLASS OR DIVISION	EXAMPLES
CORROSIVE	CORROSIVE 8	8	Sulfuric Acid, Hydrochloric Acid, Sodium Hydroxide, Potassium Hydroxide
COMBUSTIBLE	COMBUSTIBLE 3	9	Dry ice, lead, hazardous waste, molten sulfur

OTHER PLACARDS, LABELS AND MARKINGS

PLACARD NAME	PLACARD	APPLICATION AND USE
DANGEROUS PLACARD	DANGEROUS	• Transport container with 2 or more categories of HM that require different placards specified in Table 2 MAY be placarded with DANGEROUS placard instead of the specific placards required for each of the Table 2 materials. • If 2,205 lbs. (1,000 kg) or more of one category of material is loaded at one loading facility, the placard specified in Table 2 must be applied. • May not be placed on a cargo tank, portable tank, or rail tank car.
PLACARD WITH SPECIAL WHITE BACKGROUND	POISON 6	• Rail shipments of Class 1.1, 1.2, 2.3—Hazard Zone A, and 6.1, PG 1, Zone A materials. • Shipments of Class 2.1 materials in a DOT 113 tank car. • Highway route controlled quantity radioactive material.
SUBSIDIARY HAZARD LABELS OR PLACARDS	 	• Used for HM that meet the definition of one hazard class, but have subsidiary or multiple hazards. • Before October 1, 2005, the subsidiary hazard label or placard may not display the hazard class or the four-digit identification number. • After October 1, 2005, all subsidiary hazard labels and placards must show the hazard class or division number.

PLACARD NAME	PLACARD	APPLICATION AND USE
POISON INHALATION HAZARD	INHALATION HAZARD 6	Materials that meet DOT inhalation toxicity criteria have additional requirements: • Packaging must be marked as "Inhalation Hazard," or have a placard or label marked as such. • This requirement is in addition to any other required label or placard.
ELEVATED TEMPERATURE MATERIALS	HOT	The "HOT" marking will be used if the material meets one of the following criteria: • Liquid at temperatures >212°F. (100°C). • Liquid intentionally heated and has a flash point above 100°F. (37.7°C). • Solids at 464°F. (240°C) or above.
INFECTIOUS SUBSTANCES	INFECTIOUS SUBSTANCE IN CASE OF DAMAGE OR LEAKAGE IMMEDIATELY NOTIFY PUBLIC HEALTH AUTHORITY 6	• Label that marks packages within infectious substances. • Examples include medical waste, infectious substances, anthrax.

FIGURE 6.43 DOT hazardous materials placarding requirements.

Radioactive Material Labels and Placards. Radioactive packages will be labeled based on the type and quantity of material being shipped and associated levels of radiation. Three different labels are used on radioactive material packaging—Radioactive White–I, Radioactive Yellow–II, and Radioactive Yellow–III.

Radioactive label markings are essential in assessing the integrity of the packaging during an incident, and will include:

- *Contents*—The radioactive content of the packaging (e.g., Am-241 Americium, Cs-129 Cesium).
- *Activity*—The rate of disintegration or decay of a radioactive material. Listed in the appropriate SI units [e.g., Becquerels (Bq), Terabecquerels (Tbq), etc.] or in both SI units and the appropriate customary units [Curies (Ci), millicuries (mCi) microcuries (uCi), etc.]. Activity indicates how much radioactivity is present and not how much material is present.
- *Transport Index (TI)*—In most cases, this is the maximum radiation level in millirems per hour (mrem/hour) at 1 meter (3 ft) from an undamaged package. The TI can be an indicator for determining the external radiation hazard of an undamaged package and can be a starting point for determining whether or not damage has occurred.

One exception may be fissile material packages. For these, the TI is often an assigned number, and there may be little correlation between the TI and the measurement taken at 1 meter. Fissile material shipments will be identified as such on the shipping papers, and the TI may be higher than a reading taken at 1 meter.

LABEL	EXTERNAL RADIATION LEVELS	TRANSPORT INDEX (Dose at 1 Meter)
RADIOACTIVE WHITE I LABEL	• Packages with extremely low or almost no levels of radiation. • Maximum contact radiation level of 0.5 mrem/hour. No TI on this label.	No TI on this label.
RADIOACTIVE YELLOW II LABEL	• Packages with low radiation levels. • Maximum contact radiation level ranging from 0.5 mrem/hour to 50 mrem/hour. Maximum Allowable TI = 1.	Maximum allowable TI=1.
RADIOACTIVE YELLOW III LABEL	• Packages with higher radiation levels. • Maximum contact radiation level ranging from 50 mrem/hour to 200 mrem/hour. • Also required for Fissile Class III or large quantity shipments, regardless of radiation level. Maximum Allowable TI = 10.	Maximum allowable TI=10.

FIGURE 6.44

Placards and labels are simply another clue and should not be considered as a definitive source of hazmat information. Although enforcement measures are constantly improving, experience shows that a number of vehicles and containers are still improperly labeled or placarded.

SHIPPING PAPERS AND FACILITY DOCUMENTS

SHIPPING PAPER REQUIREMENTS

Shipping papers are required to accompany each hazmat shipment and are the fifth clue in the standard identification process. Responders must be familiar with the information noted on shipping papers, their location on each transport vehicle, and the individual responsible for them. Figure 6.45 summarizes this information for each transportation mode.

Basic Description. Each transport mode has its own terms for shipping papers. All shipping papers are required to contain the following entries, known as the hazardous material's basic description. All information shall be printed in English:

- *Proper Shipping Name.* Identifies the name of the hazmat as found in the Hazardous Materials Table. The word "WASTE" will precede the proper shipping name for those shipments that are classified as hazardous wastes.

SHIPPING PAPERS INFORMATION			
MODE OF TRANSPORTATION	TITLE OF PAPER	LOCATION OF PAPERS	RESPONSIBLE PERSON(S)
Highway	Bill of lading or freight bill	Cab of vehicle	Driver
Railroad	Waybill and/or consist, switch list, training list, track list	With train crew (e.g., conductor or engineer)	Conductor
Water	Dangerous cargo manifest	Wheelhouse or pipe-like container on barge	Captain or master
Air	Airbill with shipper's declaration of dangerous goods	Cockpit (may also be found attached to the outside of packages)	Pilot

Figure 6.45 Types and sources of shipping papers by transportation.

- *DOT Hazard Class/Division Number*. Indicates the material's primary and secondary (as appropriate) hazard as listed in the DOT Hazardous Materials Regulations. A division is a subset of a hazard class. Note: A hazmat may meet the criteria for more than one hazard class, but is assigned to only one hazard class.
- *Identification Number(s)*. The four-digit identification number assigned to each hazardous material. The identification number may be found with the prefix "UN"—United Nations or "NA"—North America.
- *Packing Group*. DOT regulations require that all shipments meet basic DOT requirements. Packing group further classifies hazardous materials based on the degree of danger represented by the material. There are three groups:

  1. Packing Group I indicates great danger.
  2. Packing Group II indicates medium danger.
  3. Packing Group III indicates minor danger.

  Packing Groups may be shown as "PG I", etc. Packing Groups are not assigned to Class 2 materials (compressed gases), Class 7 materials (radioactive materials), some Division 6.2 materials (infectious substances), and ORM-D materials.
- *Total Quantity*. Indicates the quantity by net or gross mass, capacity, and so on. May also indicate the type of packaging. The number and type of packaging (e.g., 1 TC, 7 DRM) may be entered on the beginning line of the shipping description. Carriers often use abbreviations to indicate the type of packaging. The following are examples of packaging abbreviations used by the Union Pacific Railroad:

BA = Bale	CH = Covered Hopper	KIT = Kit
BG = Bag	CL = Carload	KL = Container Load
BOX = Box	CY = Cubic Yard	PA = Pail
BC = Bucket	CYL = Cylinder	PKG = Package
CA = Case	DRM = Drum	SAK = Sack
CAN = Can	JAR = Jar	TB = Tube
CR = Crate	JUG = Jug	TC = Tank Car
CTN = Carton	KEG = Keg	TL = Trailer Load

- *Emergency Contact.* All shipping papers must also contain an emergency response telephone number. This is a telephone number for the shipper or shipper's representative that may be accessed 24 hours a day, 7 days a week in the event of an accident. The contact person must be either knowledgeable of the hazardous characteristics and emergency response information for the hazmat(s) listed on the shipping paper or have immediate access to someone who has that knowledge. If the shipper is registered with CHEMTREC or comparable service, that phone number may be displayed as the emergency contact.

SHIPPING PAPERS—ADDITIONAL ENTRIES

Additional shipping paper entries may be required for some hazardous materials. They include the following:

- *Compartment Notation.* Identifies the specific compartment of a multi-compartmented rail car or cargo tank truck in which the material is located. On rail cars, compartments are numbered sequentially from the "B" end (the end where the hand brake wheel is located), while cargo tank trucks are numbered sequentially from the front.
- *Residue (Empty Packaging).* Identifies packaging that contains a hazmat residue and has not been cleaned and purged or reloaded with a material that is not subject to the DOT Hazardous Materials Regulations. Residue is indicated by the words "Residue: Last Contained" before the proper shipping name. It is only used in rail transportation.
- *HOT.* Identifies elevated temperature materials other than molten sulfur and molten aluminum.
- *Technical Name.* Identifies the recognized chemical name currently used in scientific and technical handbooks, journals, and texts. Generic descriptions may be found provided that they identify the general chemical group. With some exceptions, trade names may not be used as technical names. Examples of acceptable generic descriptions are organic phosphate compounds, tertiary amines, and petroleum aliphatic compounds.
- *Not Otherwise Specified (NOS) Notations.* If the proper shipping name of a material is an "N.O.S." notation, the technical name of the hazardous material must be entered in parentheses with the basic description. If the material is a mixture or solution of two or more hazardous materials, the technical names of at least two components that most predominantly contribute to the hazards of the mixture/solution must be entered on the shipping paper. For example, Flammable Liquid, Corrosive Liquid, NOS, (contains methanol, potassium hydroxide), 3, UN 2924, PGII.

- *Subsidiary Hazard Class.* Indicates a hazard of a material other than the primary hazard assigned.
- *Reportable Quantity (RQ) Notation.* Indicates that the material is a hazardous substance by the EPA. The letters "RQ" (reportable quantity) must be shown either before or after the basic shipping description entries. This designation indicates that any leakage of the substance above its RQ value must be reported to the proper agencies (e.g., National Response Center). Regardless of which agencies are involved, the legal responsibility for notification still remains with the spiller.
- *Marine Pollutant.* Indicates that the material meets the definition of a marine pollutant. If the basic description does not identify the component that makes the material a marine pollutant, the name of the component(s) must appear in parentheses.
- *EPA Waste Stream Number.* Indicates the number assigned to a hazardous waste stream by the U.S. EPA to identify that waste stream. Note: For all hazardous waste shipments, a Uniform Hazardous Waste Manifest must be prepared in accordance with both DOT and EPA regulations.
- *EPA Waste Characteristic Number.* Indicates the general hazard characteristics assigned to a hazardous waste by the U.S. EPA. Waste characteristics include EPA corrosivity, EPA toxicity, EPA ignitability, and EPA reactivity.
- *Radioactive Material Information.* Should provide the following:
  - "Radioactive Material"—if not part of the proper shipping name
  - Name of each radionuclicide
  - Physical/chemical form
  - Activity in curies
  - Label applied
  - Transport Index (if applicable)
  - U.S. Department of Energy Approval Number (if applicable)
  - Fissile Exempt (if applicable)
  - Fissile Class (if applicable)
- *Poison Notation.* Indicates that a liquid or solid material is poisonous when the fact is not disclosed in the shipping name.
- *Poison-Inhalation Hazard (PIH) or Toxic-Inhalation Hazard (TIH) Notation.* Indicates gases and liquids that are poisonous by inhalation.
- *Hazard Zone.* Indicates relative degree of hazard in terms of toxicity (only appears for gases and liquids that are poisonous by inhalation):
  - Zone A—LC_{50} less than or equal to 200 ppm (most toxic)
  - Zone B—LC_{50} greater than 200 ppm and less than or equal to 1000 ppm
  - Zone C—LC_{50} greater than 1000 ppm and less than or equal to 3000 ppm
  - Zone D—LC_{50} greater than 3000 ppm and less than or equal to 5000 ppm (least toxic)
- *Dangerous When Wet Notation.* Indicates a material that, by contact with water, is liable to become spontaneously flammable or give off flammable or toxic gas at a rate greater than 1 liter per kilogram of the material, per hour.

- *Limited Quantity (LTD QTY).* Indicates a material being transported in a quantity for which there is a specific labeling and packaging exception.
- *Canadian Information.* Indicates information required for hazardous materials entering or exiting Canada in addition to that required in the United States (e.g., ERG reference number, the 24-hour Canadian emergency telephone number, Canadian class).
- *Placard Notation.* Indicates the placard applied to the container. Where placards are not required (e.g., ORM commodities or nonflammable cryogenic gases), the notation "MARKED" is followed by the four-digit identification number.
- *Trade Name.* A name that enables organizations, such as emergency responders and CHEMTREC®, to access the MSDS for additional information.
- *DOT Exemption Notation.* If a hazmat shipment is made under a DOT exemption for specific packaging or shipping procedures, the shipping papers must include the letters "DOT—E" followed by the assigned exemption number. The exemption number must be placed so that it is clearly associated with the description to which the exemption applies.
- *Hazardous Materials STCC Number.* A seven-digit Standard Transportation Commodity Code (STCC) number will be found on all shipping papers accompanying rail shipments of hazmats. It will also be found when intermodal containers are changed from rail to highway movement. Look for the first two digits—"49"—as the key identifier for a hazmat. The STCC number will follow the notation "HAZMAT STCC ."
- *Shipper Contact.* Indicates the identity of the producer or consolidator of the materials described.

SHIPPING PAPERS—EMERGENCY RESPONSE INFORMATION

Emergency response information must also be included with shipping papers. An emergency telephone number for the shipper or shipper's representative is required on the shipping paper. Emergency response information must provide the following:

- Brief product description
- Emergency actions involving fire
- Emergency actions involving release only
- Personnel protective measures
- Environmental considerations, as appropriate
- First aid measures

Several common sources of emergency response information requirements are a MSDS, a copy of the *Emergency Response Guidebook* (ERG), a copy of the specific page from the ERG for the hazmat being transported, or railroad emergency response information sheets that are cross-referenced with the train consist. With most major railroads, the railroad emergency response information sheets will be part of the train consist.

FACILITY DOCUMENTS

Various types of facility documents are available to assist in the information process. They can be a source for hazmat recognition, identification, and classification at an emergency.

The specific type and nature of information provided will vary based on pertinent federal, state, and / or local reporting requirements. Examples include hazmat inventory forms, shipping and receiving forms, Risk Management Plans and supporting documentation, MSDSs, and Tier II reporting forms required to be submitted to the LEPC and the fire department under SARA Title III.

Both the Risk Management Plans and the Tier II reporting forms can be used as part of the hazards analysis process. For example, the Tier II reporting forms provide information such as chemicals on-site that exceed the reporting thresholds, physical and chemical hazards, average and maximum amounts on site, and types of storage containers and location.

MONITORING AND DETECTION EQUIPMENT

If the hazardous material cannot be identified from the previously discussed methods, monitoring and detection equipment can often provide data and information concerning the overall nature of the problem you face as well as the specific materials involved. Monitoring and detection equipment is essential for identifying, verifying, or classifying the hazmat(s) involved and is the sixth clue in the standard identification process.

Although considered here as an identification tool, monitoring and detection equipment are also critical tools for evaluating real-time data and developing a risk-based response. Monitoring helps responders to:

- Determine the appropriate levels of personal protective clothing and equipment.
- Determine the size and location of hazard control zones.
- Develop protective action recommendations and corridors.
- Assess the potential health effects of exposure.

The selection, application, and use of monitoring instruments are addressed in Chapter 7—Hazard and Risk Evaluation.

SENSES

Sometimes clues are not as obvious as a building occupancy, container shapes, or markings, and your senses must come into play. Senses are not a primary identification tool and are the final clue in the standard identification process. The only senses that offer some protection are visual and hearing; the use of other senses means you are at potential risk from the hazmat. In most cases, if you are close enough to smell, feel, or hear the problem, you are probably too close to operate safely.

Nonetheless, senses can be valuable assets and can offer immediate clues to the presence of hazardous materials. For our purposes, "senses" refers to any personal physiological reaction to or visual observation of hazmat release. Smells, dizziness, unusual noises (i.e., relief valve actuations), and destroyed vegetation are some examples.

The inhalation (or smelling) of chemicals should always be avoided, but there are times when you may enter a situation and an odor will be present. Being able to characterize those odors is important to your safety so that you can move to a safe area. Unfortunately, being able to detect the odor effectively also means that you have most likely been exposed to potentially dangerous concentrations of the gas or vapors at some point.

In Chapter 2 the concept of *Immediately Dangerous to Life or Health (IDLH)* atmospheres was discussed. As a review, there are some basic street smart clues of IDLH atmospheres that you should always remember:

- *Visible vapor clouds*—Avoid entering any vapor clouds, smoke, or mists. Fires are considered IDLH conditions. Vapor clouds with colors (e.g., green, orange, brown) are not good!
- *Releases*—Releases from bulk containers or a pressurized container are extremely hazardous because the more product that is available, the greater the risk. Most pressurized containers hold liquefied gases with large liquid to vapor expansion ratios, thereby presenting greater risk.
- *Large liquid leaks*—Avoid contact with any amount of released materials. The larger the leak, the more vapors that may be produced.
- *Below grade or confined spaces*—These can include artificial barriers, all of which can be oxygen deficient or accumulate toxic gases.
- *Dead birds, foliage, sick animals and sick humans*—These are all biological indicators of a chemical release.
- *Physical senses and "street smarts"*—If a situation doesn't seem right, it probably isn't. Be aware for odors and sources of energy—mechanical, electrical and kinetic. Remember—dealing with unknowns is always dangerous!

> **NOTE:** An excellent textbook on basic street smart hazmat response issues is ***Street Smart HazMat Response*** by Michael Callan. The book is available through Oklahoma State University—Fire Protection Publications at (800) 654-4055 or Red Hat Publishing at (800) 603-7700.

SUMMARY

The evaluation of hazards and the assessment of the risks build on the timely identification and verification of the hazardous materials involved. A problem well defined is half-solved. Identification and verification of the hazmats involved are critical to the safe and effective management of a hazmat incident.

The seven basic clues for recognition, identification, and classification are

- Occupancy and location
- Container shapes
- Markings and colors
- Placards and labels
- Shipping papers and facility documents
- Monitoring and detection equipment
- Senses

REFERENCES AND SUGGESTED READINGS

Air Force Civil Engineer Support Agency (AFCESA) and PowerTrain, Inc, HAZARDOUS MATERIALS INCIDENT COMMANDER EMERGENCY RESPONSE TRAINING CD-ROM, Tyndall Air Force Base, FL: AFCESA (2002).

Air Force Civil Engineer Support Agency (AFCESA) and PowerTrain, Inc, HAZARDOUS MATERIALS TECHNICIAN EMERGENCY RESPONSE TRAINING CD-ROM, Tyndall Air Force Base, FL: AFCESA (1999).

Bowen, John, "PCB"—An Update." AMERICAN FIRE JOURNAL (June, 1987), pages 32–37.

Callan, Michael. STREET SMART HAZMAT RESPONSE. Chester, MD: Red Hat Publishing, Inc. (2001).

Carr, John and Les Omans, "Responding to Commercial Aircraft Haz-Mat Incidents." FIRE ENGINEERING (November, 2001), pages 47–52.

Chemical Manufacturers Association and the Association of American Railroads, PACKAGING FOR TRANSPORTING HAZARDOUS AND NON-HAZARDOUS MATERIALS, Washington, DC: Chemical Manufacturers Association (June, 1989).

Code of Federal Regulations, TITLE 49 CFR PARTS 100–199 (TRANSPORTATION), Washington, DC: U.S. Government Printing Office.

Compressed Gas Association, HANDBOOK OF COMPRESSED GASES (4th Edition). Boston, MA: Kluwer Academic Publishers (2000).

Emergency Film Group, CYLINDERS—CONTAINER EMERGENCIES (videotape), Edgartown, MA: Emergency Film Group (2002).

Emergency Film Group, THE EIGHT STEP PROCESS: STEP 2—IDENTIFYING THE PROBLEM (videotape), Edgartown, MA: Emergency Film Group (2004).

General American Transportation Corporation, GATX TANK CAR MANUAL (6th edition), Chicago, IL: General American Transportation Corporation (1994).

Hawley, Chris, HAZARDOUS MATERIALS AIR MONITORING AND DETECTION INSTRUMENTS. Albany, NY: Delmar – Thomson Learning (2002).

Hawley, Chris, HAZARDOUS MATERIALS INCIDENTS. Albany, NY: Delmar —Thomson Learning (2002).

Hawley, Chris, Gregory G. Noll and Michael S. Hildebrand, SPECIAL OPERATIONS FOR TERRORISM AND HAZMAT CRIMES. Chester, MD: Red Hat Publishing, Inc. (2002).

Hildebrand, Michael S. and Gregory G. Noll, GASOLINE TANK TRUCK EMERGENCIES: GUIDELINES AND PROCEDURES (2nd Edition), Stillwater, OK: Fire Protection Publications – Oklahoma State University (1996).

Hildebrand, Michael S., Gregory G. Noll and Michael Donahue, HAZARDOUS MATERIALS EMERGENCIES INVOLVING INTERMODAL CONTAINERS, Stillwater, OK: Fire Protection Publications—Oklahoma State University (1995).

Hildebrand, Michael S. and Gregory G. Noll, PROPANE EMERGENCIES (2nd Edition), Washington, Dc: National Propane Gas Association (2002).

Hildebrand, Michael S. and Gregory G. Noll, STORAGE TANK EMERGENCIES, Chester, MD: Red Hat Publishing, Inc. (1997).

Hazardous Materials Advisory Council, AN OVERVIEW OF INTERMODAL PORTABLE TANKS, Washington, DC: Hazardous Materials Advisory Council (1986).

Hazmat World, "Safety Upgrades Plug Tank Car Leaks." HAZMAT WORLD (August, 1993), pages 42–43.

International Association of Fire Fighters, TRAINING FOR HAZARDOUS MATERIALS RESPONSE: TECHNICIAN, Washington, DC: IAFF (2002).

Isman, Warren E. and Gene P. Carlson, HAZARDOUS MATERIALS, Encino, CA: Glencoe Publishing Company (1980).

Lesak, David M. HAZARDOUS MATERIALS STRATEGIES AND TACTICS, Upper Saddle River, NJ: Prentice-Hall Inc. (1999).

Maslansky, Carol J. and Steven P. Maslansky, AIR MONITORING INSTRUMENTATION. New York, NY: Van Nostrand Reinhold (1993).

National Fire Protection Association, FIRE PROTECTION HANDBOOK (19th edition), Quincy, MA: National Fire Protection Association (2003).

National Fire Protection Association, HAZARDOUS MATERIALS RESPONSE HANDBOOK (4th Edition), Quincy, MA: National Fire Protection Association (2002).

National Fire Protection Association, LIQUEFIED PETROLEUM GASES HANDBOOK (6th edition), Quincy, MA: National Fire Protection Association (2001).

National Fire Protection Association, NFPA 30—NATIONAL FLAMMABLE AND COMBUSTIBLE LIQUIDS CODE, Quincy, MA: National Fire Protection Association (2000).

NIOSH/OSHA/USCG/EPA, OCCUPATIONAL SAFETY AND HEALTH GUIDANCE MANUAL FOR HAZARDOUS WASTE SITE ACTIVITIES, Washington, DC: U.S. Government Printing Office (1985).

Safe Transportation Training Specialists, "Cargo Tank Truck Construction And Emergency Response." Student handout presented at International Association of Fire Chiefs (IAFC) Hazardous Materials Conference, Hunt Valley, MD (June, 2003).

Sea Containers Limited, GENERAL GUIDE TO TANK CONTAINER OPERATION, London, England: Sea Containers Group (1983).

Sea Containers Limited, INSPECTION, REPAIR AND TEST REQUIREMENTS FOR TANK CONTAINERS, London, England: Sea Containers Group (undated).

Thomas, Charles R., "Hazardous Materials in Air Transport." Student handout presented at International Association of Fire Chiefs (IAFC) Hazardous Materials Conference, Hunt Valley, MD (June, 2000).

Union Pacific Railroad Company, A GENERAL GUIDE TO TANK CARS, Omaha, NE: Union Pacific Railroad Company, Technical Training (April, 2003).

Union Pacific Railroad Company, A GENERAL GUIDE TO TANK CONTAINERS, Omaha, NE: Union Pacific Railroad Company, Technical Training (January, 1999).

Union Pacific Railroad Company, RECOGNIZING AND IDENTIFYING HAZARDOUS MATERIALS, Omaha, NE: Union Pacific Railroad Company, Technical Training (December, 2001).

U.S. Department of Energy – Office of Transportation and Emergency Management, MODULE EMERGENCY RESPONSE RADIOLOGICAL TRANSPORTATION TRAINING (MERRTT), WASHINGTON, DC: DOE (2000).

U.S. Department of Justice—Office of Justice Programs, Office for Domestic Preparedness (OJP/ODP), WEAPONS OF MASS DESTRUCTION (WMD) RADIATION/NUCLEAR COURSE FOR HAZARDOUS MATERIALS TECHNICIANS, Mercury, NV: Department of Energy (DOE) and Bechtel Nevada (2002).

U.S. Department of Transportation—Research and Special Programs Administration, HAZARDOUS MATERIALS TRANSPORTATION TRAINING MODULES, Washington, DC: DOT/RSPA (2002).

U.S. Environmental Protection Agency, "Review of the EPA Rule Regulating Polychlorinated Biphenyl (PCB) Transformer Fires" (Report E1E57-11-0024-80780), Washington, DC: EPA, Office of the Inspector General (March, 1988).

Wright, Charles J., "Handling Rail Emergencies Involving Hazardous Materials." Student handout presented at the Kansas Hazardous Materials Symposium, Wichita, KS (November 7, 1993).

Wright, Charles J. and William T. Hand, "Intermodal Tank Containers—Parts I and II," FIRE COMMAND, (June and July, 1988), pages 17, 36–37.

York, Kenneth J. and Gerald L. Grey, HAZARDOUS MATERIALS/WASTE HANDLING FOR THE EMERGENCY RESPONDER, Tulsa, OK: Fire Engineering Books and Videos (1989).

ACME CHEMICAL
HEY O.T. = MY COMPUTER SAYS TO USE COPIOUS QUANTITIES OF WATER...
KID, MINE SAYS: "BEND OVER AND KISS YOUR GOODBYE"

CHAPTER 7

FDNY

HAZARD ASSESSMENT AND RISK EVALUATION

OBJECTIVES

1. Describe the concept of hazard assessment and risk evaluation.
2. Describe the following terms and explain their significance in the risk assessment process [NFPA 472-6.2.2(B)].

   - Acid, caustic
   - Biological agents and toxins
   - Catalyst
   - Chemical reactivity
   - Concentration
   - Critical temperature and pressure
   - Dose rate
   - Expansion ratio
   - Flammable (explosive) range
   - Flash point
   - Halogenated hydrocarbon
   - Inhibitor
   - Ionic and covalent bonding
   - Maximum safe storage temperature (MSST)
   - Miscibility
   - Organic and inorganic
   - pH
   - Physical state (solid, liquid, gas)
   - Radioactivity
   - Self-accelerating decomposition temperature (SADT)
   - Specific gravity
   - Sublimation
   - Toxic products of combustion
   - Vapor pressure
   - Viscosity
   - Water reactivity
   - Air reactivity
   - Boiling point
   - Chemical interactions
   - Compound, mixture
   - Corrosivity
   - Dose
   - Evaporation rate
   - Fire point
   - Half-life
   - Ignition (autoignition) temperature
   - Instability
   - Irritants (riot control agents)
   - Melting point/freezing point
   - Nerve agents
   - Oxidation potential
   - Persistence
   - Polymerization
   - Saturated, unsaturated, and aromatic hydrocarbons
   - Solution, slurry
   - Strength
   - Temperature of product
   - Vapor density
   - Vesicants (blister agents)
   - Volatility
   - Water solubility

3. Describe the heat transfer processes that occur as a result of a cryogenic liquid spill [NFPA 472-6.2.2(C)].
4. Identify and interpret the types of hazard and response information available from each of the following resources, and explain the advantages and disadvantages of each resource [NFPA 472-6.2.2(A)].

   - Hazardous materials databases
   - Maps and diagrams

- Monitoring equipment
- Reference manuals
- Technical information centers
- Technical information specialists

5. Identify the steps in an analysis process for identifying unknown solid and liquid materials [NFPA 472-6.2.1.3(A)].
6. Identify the steps in an analysis process for identifying an unknown atmosphere [NFPA 472-6.2.1.3(B)].
7. Identify the types of monitoring equipment, test strips, and reagents used to determine the following hazards [NFPA 472-6.2.1.3(C)]:
   a. Corrosivity
   b. Flammability
   c. Oxidation potential
   d. Oxygen deficiency
   e. Radioactivity
   f. Toxic levels
8. Identify the capabilities and limiting factors associated with the selection and use of the following monitoring equipment, test strips, and reagents [NFPA 472-6.2.1.3(D)]:
   a. Carbon monoxide meter
   b. Colorimetric tubes
   c. Combustible gas indicator
   d. Oxygen meter
   e. Passive dosimeter
   f. Photoionization detector
   g. pH indicators and/or pH meters
   h. Radiation detection and measurement instruments
   i. Reagents
   j. Test strips
9. Describe the basic identification tools and detection devices for each of the following [NFPA 472-6.2.1.1(G)]:
   a. Nerve agents
   b. Vesicants (blister agents)
   c. Biological agents and toxins
   d. Irritants (riot control agents)
10. Identify two methods for determining the pressure in bulk packaging or facility containers [NFPA 472-6.2.2(F)].
11. Identify one method for determining the amount of lading remaining in damaged bulk packaging or facility containers [NFPA 472-6.2.2(G)].
12. Identify and describe the components of the General Hazardous Materials Behavior Model (GEBMO).

13. Identify the types of damage that a pressure container could incur [NFPA 472-6.2.3.4].
14. Identify at least three resources available that indicate the effects of mixing various hazardous materials [NFPA 472-6.2.4.1].
15. Identify the steps for determining the extent of the physical, safety, and health hazards within the endangered area of a hazardous materials incident [NFPA 472-6.2.5.2].
16. Identify two methods for predicting the areas of potential harm within the endangered area of a hazardous materials incident [NFPA 472-6.2.5.2(C)].
17. Describe the steps for estimating the outcomes within an endangered area at a hazardous materials incident [NFPA 472-6.2.5.3].
18. Describe the steps for determining response objectives (defensive, offensive, and nonintervention) given an analysis of a hazardous materials incident. [NFPA 472-6.3.1.2].
19. Identify the possible action options to accomplish a given response objective [NFPA 472- 6.3.2.2].
20. Describe the factors that influence the underground movement of hazardous materials in soil and through groundwater.
21. Identify the hazards associated with the movement of hazardous materials in the following types of sewer collection systems:
    a. Storm sewers
    b. Sanitary sewers
    c. Combination sewers
22. List five site safety procedures for handling an emergency involving a hydrocarbon spill into a sewer collection system.

ABBREVIATIONS AND ACRONYMS

ACC	American Chemistry Council
ALOHA	Aerial Locations of Hazardous Atmospheres
ATSDR	Agency for Toxic Substances and Disease Registry
CAMEO	Computer Assisted Management of Emergency Operations
CANUTEC	Canadian Transport Emergency Centre
CEPPO	Chemical Emergency Prevention and Preparedness Office (EPA)
CHEMNET	Chemical Industry Mutual Aid Network
CHEMTREC	Chemical Transportation Emergency Center
cpm	Counts per Minute
eV	Electron Volts

FT-IR	Fourier Transform Infrared Spectrometry
FOG	Field Operations Guide
GC	Gas Chromatograph
GEBMO	General Hazardous Materials Emergency Behavior Model
IP	Ionization Potential
LEPC	Local Emergency Planning Committee
MARPLOT	Mapping Applications for Response, Planning and Local Operational Tasks
MS	Mass Spectrometer
MSST	Maximum Safe Storage Temperature
NAPCC	National Animal Poison Control Center
NEC	National Electrical Code
NPIC	National Pesticide Information Center
NTSB	National Transportation Safety Board
pH	Power of Hydrogen
PID	Photo-Ionization Detector
RQ	Reportable Quantity
SADT	Self-Accelerating Decomposition Temperature
SETIQ	Mexican Emergency Transportation System for the Chemical Industry
UV	Ultraviolet

INTRODUCTION

The evaluation of hazard information and the assessment of risks is the most critical decision-making point in the successful management of a hazardous materials incident. The decision to intervene, or more often to not intervene, is not easy. While most responders recognize the initial need for isolating the area, denying entry, and identifying the hazardous materials involved, failure to develop effective analytical and problem-solving skills can lead to injury, and in a worst-case scenario, death.

This chapter discusses the third step in the Eight Step Process©—Hazard and Risk Evaluation. The chapter is based on the premise that responders have (1) successfully implemented site management procedures, and (2) identified the nature of the problem and the materials potentially involved. Topics include understanding hazardous materials behavior, outlining the common sources of hazard information, evaluating risks, and determining response objectives.

Case Study

HAZARDOUS MATERIALS TRAIN DERAILMENT AKRON, OHIO FEBRUARY 26, 1989

On February 26, 1989, a CSX Transportation, Inc. freight train derailed in Akron, OH. Twenty-one freight cars, including nine butane tank cars, derailed. The nine butane tank cars came to rest adjacent to the B. F. Goodrich Chemical Company plant, and butane released from two breached tank cars immediately caught fire. The fire required the evacuation of approximately 1,750 residents.

After the Akron Fire Department (AFD) received and verified the shipping information from the train crew, on-scene activities were accomplished in a timely and professional manner. These activities included controlling the butane tank car fires, controlling the fire at the B. F. Goodrich plant.

Both the fire department and the city of Akron depended on the expertise of CSX for the removal of wreckage from the derailment site. The AFD Operations Section Chief considered it unsafe to offload the butane tank cars because of a continuing fire from one of the derailed tank cars. After discussions with CSX, he agreed with an action plan proposed by CSX to rerail the tank cars, move them to the Akron Junction Yard where the tank cars would be more permanently secured, and finally transport them to Canton, Ohio, where the product would be offloaded. During the course of these conversations, however, the railroad did not discuss alternatives with the fire department; nor did the railroad advise the fire department of the possible risks associated with rerailing the tank cars.

On February 28, while the rerailed tank cars were being moved from the accident site, a butane tank car rolled off its tracks and forced the evacuation of approximately 25 families from the area. Only after this second derailment were alternative plans and the risks associated with each potential course of action thoroughly discussed with the fire department and city of Akron officials.

In their accident investigation final report, the National Transportation Safety Board (NTSB) stated that it recognized the limited technical resources that may be available to local communities regarding train wreckage clearing operations and the communities' reliance on the railroad to take the appropriate course of action. Thus, NTSB believed that it is necessary for the railroad to discuss with local emergency response personnel (1) the severity of known damage to hazmat tank cars; (2) the relative dangers posed to public safety; (3) all possible courses of action; and (4) any associated risks.

NTSB also recognized the need for the Incident Commander to play an active role during the hazard and risk assessment process. Tasks should include searching out information on the severity of known tank car damage and the dangers posed, potential alternatives and solutions, and the risks involved.

Hazard and risk evaluation is the cornerstone of decision making at a hazmat incident. The NTSB Accident Investigation Report was one of the first to address the issue of hazard and risk evaluation from the perspective of emergency responders.

BASIC PRINCIPLES

The concept of hazard and risk evaluation is recognized as a critical benchmark in safe and successful emergency response operations. Although our focus is directed toward hazmat response, the reality is that the risk management and evaluation process influences all aspects of emergency response.

If we review incidents and case studies where emergency responders have been injured or killed, in most instances it is *not* due to their failure to assess and understand the hazard. In contrast, one of the most common root causes is our failure to adequately evaluate and understand the level of risk involved.

What are hazards and risks?

Hazards refer to a danger or peril. In hazardous materials response operations, hazards generally refer to the physical and chemical properties of a material. You can obtain hard data on hazards through emergency response guidebooks, computer databases, and the like. Examples include flash point, toxicity levels (LC50, LD50), exposure values (TLVs), protective clothing requirements, and compatibility.

Risks refer to the probability of suffering harm or loss. Risks can't be determined from books or pulled from computerized data bases—they are those intangibles that are different at every hazmat incident and must be evaluated by a knowledgeable Incident Commander. Although the risks associated with hazmat response will never be completely eliminated, they can be successfully managed. The objective of response operations is to minimize the level of risk to responders, the community, and the environment. Hazmat responders must see their role as risk evaluators, not risk takers.

Risk levels are variable and change from incident to incident. Factors that influence the level of risk include the following:

- **Hazardous nature of the material(s) involved**. For example, toxicity, flammability, and reactivity.
- **Quantity of the material involved**. Risks will often be greater when dealing with bulk quantities of hazardous materials as compared to limited-quantity, individual containers. However, quantity must also be balanced against the hazardous nature of the material(s) involved—small quantities of highly toxic or reactive materials can create significant risks.
- **Containment system and type of stress applied to the container**. Containers may be either pressurized or nonpressurized. Risks are inherently higher for pressurized containers as compared to low-pressure and atmospheric-pressure containers. In addition, the type of stressors involved (thermal, chemical, mechanical, or combination) and the ability of the container to tolerate that stress will influence the level of risk.
- **Proximity of exposures**. This would include both distance and the rate of dispersion of any chemical release. Exposures include emergency response personnel, the community, property, and the environment.
- **Level of available resources**. The availability of resources and their response time will influence the level of risk. This includes both the training and knowledge level of responders.

This hazard and risk evaluation process will ultimately lead to the development of an Incident Action Plan (IAP) that should be designed to favorably change or influence the outcome.

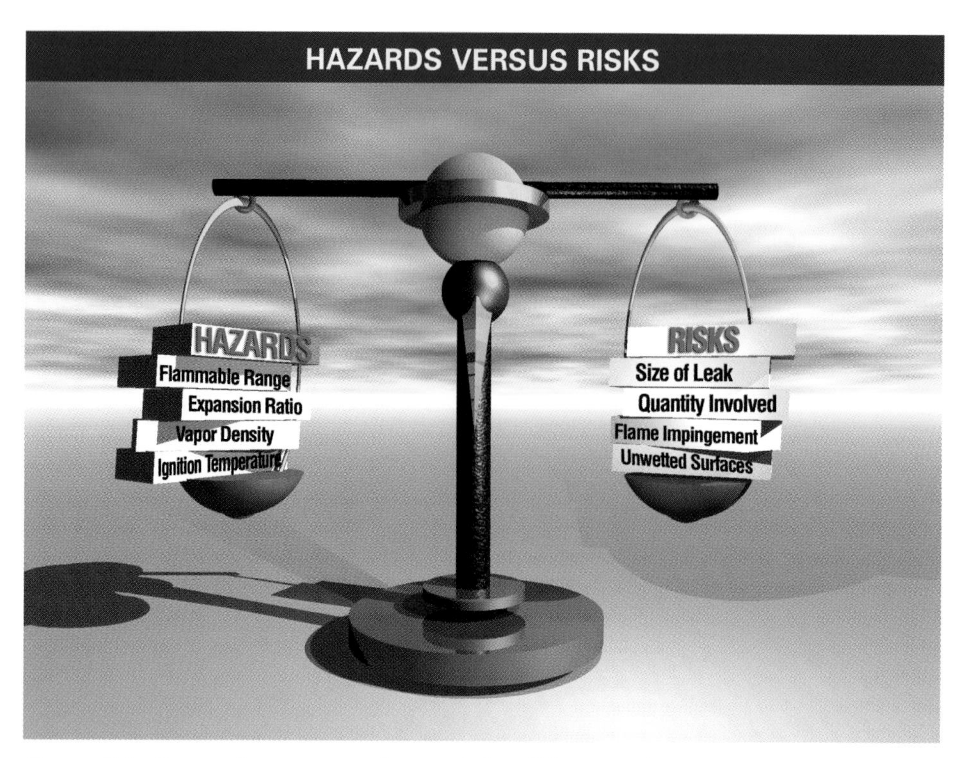

Figure 7.1 Responders must understand the differences between hazards and risks. Failure to understand these differences can result in responder injuries and death!

In this chapter, the hazard and risk evaluation process will be viewed as three distinct yet inter-related tasks:

1. *Hazard Assessment*—Assessment of the hazards that may be involved in the incident, including the collection and interpretation of hazard and response information. Information pertaining to this task will be covered in the "Understanding the Enemy" and "Sources of Hazard Information" sections that follow.
2. *Risk Evaluation* —This is the process where responders analyze the problem and assess potential outcomes. Factors that will influence the level of risk include the hazardous materials involved, type of container and its integrity, and the environment where the incident has occurred. This task will be covered in the "Evaluating Risks" section.
3. *Development of the IAP*—The output of the risk evaluation process is the implementation of strategies and tactics that will produce a favorable outcome. This will be covered in the section titled "Developing an Incident Action Plan." Detailed information on the implementation of response objectives will be covered in Chapter 9, Implementing Response Objectives.

UNDERSTANDING THE ENEMY: PHYSICAL AND CHEMICAL PROPERTIES

To evaluate risks effectively, responders must be able to identify and verify the materials involved, and determine their hazards and behavior characteristics. This includes collecting and interpreting available hazard data.

To be effective in combat, one must understand the tendencies and behavior of your enemy. In hazmat response, our enemy is the hazardous material. To mount a safe and effective hazmat response, responders must understand (1) how the enemy will behave (i.e., its physical properties) and (2) how it can harm (i.e., its chemical properties). In this section we review the key physical and chemical properties of hazardous materials and their role in the risk assessment process.

GENERAL CHEMICAL TERMS AND DEFINITIONS

The following terms are commonly found on MSDSs and in various emergency response references as part of a material's description or basic chemical make-up. Understanding these terms and definitions is critical in evaluating and predicting the behavior of hazardous materials and their containers during an incident.

- **Element**—Pure substance that cannot be broken down into simpler substances by chemical means.
- **Compound**—Chemical combination of two or more elements, either the same elements or different ones, that is electrically neutral. Compounds have a tendency to break down into their component parts, sometimes explosively (e.g., organic peroxides).
- **Mixture**—Substance made up of two or more elements or compounds, physically mixed together. Each element or compound in the mixture keeps its own properties.
- **Solution**—Mixture in which all of the ingredients are completely dissolved. Solutions are composed of a solvent (water or another liquid) and a dissolved substance (known as the solute).
- **Slurry**—Pourable mixture of a solid and a liquid.
- **Cryogenic liquid**—Gases that have been transformed into extremely cold liquids that are stored at temperatures below -130°F (-90°C). Cryogenic liquids will have the following hazards: (1) extremely cold temperature; (2) tremendous liquid-to-vapor expansion ratio; and (3) hazards of the respective material (e.g., liquid oxygen is an oxidizer, while liquid hydrogen is flammable). When released from their container, cryogenic liquids will start to boil off rapidly due to their extremely low boiling point.
- **Ionic bonding**—The electrostatic attraction of oppositely charged particles. Atoms or groups of atoms can form ions or complex ions.
- **Covalent bonding**—The force holding together atoms that share electrons.
- **Organic materials**—Materials that contain carbon atoms. Organic materials are derived from materials that are living or were once living, such as plants or decayed products. Most organic materials are flammable. Examples include methane (CH_4) and propane (C_3H_8).
- **Inorganic materials**—Compounds derived from other than vegetable or animal sources which lack carbon chains but may contain a carbon atom (e.g., sulfur dioxide—SO_2).
- **Hydrocarbons**—Compounds primarily made up of hydrogen and carbon. Examples include LPG, gasoline, and fuel oils.
- **Saturated hydrocarbons**—A hydrocarbon possessing only single covalent bonds,

and all of the carbon atoms are saturated with hydrogen. May also be referred to as alkanes. Examples include methane (CH_4), propane (C_3H_8), and butane (C_4H_{10}).

- **Unsaturated hydrocarbons**—A hydrocarbon with at least one multiple bond between two carbon atoms somewhere in the molecule. Generally, unsaturated hydrocarbons are more active chemically than saturated hydrocarbons and are considered more hazardous. May also be referred to as the alkenes and alkynes. Examples include ethylene (C_2H_4), butadiene (C_4H_6)., and acetylene (C_2H_8).
- **Aromatic hydrocarbons**—A hydrocarbon containing the benzene "ring," which is formed by six carbon atoms and contains resonant bonds. Examples include benzene (C_6H_6) and toluene (C_7H_8).
- **Halogenated hydrocarbons**—A hydrocarbon with halogen atom (e.g., chlorine, fluorine, bromine, etc.) substituted for a hydrogen atom. They are often more toxic than naturally occurring organic chemicals, and they decompose into smaller, more harmful elements when exposed to high temperatures for a sustained period of time.

PHYSICAL PROPERTIES

Physical properties provide information on the behavior of a material. These properties or characteristics of a material can be observed and measured and will provide responders with an understanding of how a material will behave both inside and after being released from its container.

- **Normal physical state**—The physical state or form (solid, liquid, gas) of a material at normal temperatures [68°F (20°C) to 77°F (25°C)]. Determining the physical state of a material can allow responders to assess potential harm. Consider the following points:
  - Solids will generally cause limited harm, as the typical route of exposure is physical contact or ingestion. Exceptions to this are dusts (depending on the size of the particles) and radioactive substances.
  - Liquids present additional risks, as they are not as easily controlled, and may have the ability to evaporate, thereby creating an inhalation hazard. In addition, they have the ability to damage skin, and some are toxic through skin absorption. Of the 700 chemicals listed in the *NIOSH Pocket Guide to Chemical Hazards*, 85 are listed as toxic through skin absorption.
  - Gases present the greatest risk, as they may be odorless, colorless, toxic, corrosive, and/or flammable. Gases and liquids with high vapor pressures pose among the greatest risks to emergency responders.
- **Temperature of product**—The temperature of the material within its container. A material's temperature will influence both the range of hazards and potential counter-measures. Temperatures are usually measured in Fahrenheit (°F) or Centigrade (°C).
- **Specific gravity**—The weight of a solid or liquid material as compared with the weight of an equal volume of water. If the specific gravity is less than one, the material is lighter than water and will float. If the specific gravity is greater than one, the material is heavier than water and will sink. Most insoluble hydrocarbons are lighter than water and will float on the surface. This is a significant property for evaluating spill control options and clean-up procedures for waterborne releases.

- **Vapor density**—The weight of a pure vapor or gas compared with the weight of an equal volume of dry air at the same temperature and pressure. If the vapor density of a gas is less than one, the material is lighter than air and may rise. If the vapor density is greater than one, the material is heavier than air and will collect in low or enclosed areas. Materials with a vapor density close to 1.0 (e.g., 0.8 to 1.2) will likely hang at midlevel and will not travel unless moved by wind or ventilation drafts. In the *NIOSH Packet Guide to Chemical Hazards,* vapor density is identified as RgasD or relative gas density. Vapor density is a significant property for evaluating exposures and determining where gases and vapors will travel.

  If a reference source does not provide a vapor density, you can calculate it by using the molecular weight of the material. The molecular weight of air is 29; materials with a molecular weight of <29 will rise, and those with a molecular weight >29 will sink. For example, anhydrous ammonia has a molecular weight of 17 and a vapor density of 0.59.

  An easy way to remember common hazardous gases and simple asphyxiants that are lighter than air is the acronym 4H MEDIC ANNA:

  H = Hydrogen (VD = 0.069)
  H = Helium (VD = 0.14)
  H = Hydrogen Cyanide (VD = 1.0)
  H = Hydrogen Fluoride (VD = 0.70)

  M = Methane (VD = 0.554)
  E = Ethylene (VD = 0.97)
  D = Diborane (VD = 0.96)
  I = Illuminating gas (10% Ethane and 90% methane mixture –VD = 0.6)
  C = Carbon Monoxide (VD = 0.97)

  A = Anhydrous Ammonia (VD = 0.588)
  N = Neon (VD = 0.7)
  N = Nitrogen (VD = 0.96)
  A = Acetylene (VD = 0.90)

Note that there are situations where some of these gases, if sufficiently chilled, could initially sink and be considered heavier than air. Examples include liquefied gases such as anhydrous ammonia, anhydrous hydrogen fluoride, and ethylene.

- **Boiling point**—The temperature at which a liquid changes its phase to a vapor or gas. The temperature at which the vapor pressure of the liquid equals atmospheric pressure. The lower the boiling point, the more vapors that are produced at a given temperature. The closer a material is to its boiling point, the more vapors that are produced.

  When evaluating flammable liquids, remember that flash point and boiling point are directly related. A low flash point flammable liquid will also have a low boiling point, which translates into greater amounts of vapors being given off

- **Melting point**—The temperature at which a solid changes its phase to a liquid. This temperature is also the freezing point depending on the direction of the change. For mixtures, a melting point range may be given. This is a significant property in evaluating the hazards of a material, as well as the integrity of a container (e.g., frozen material may cause its container to fail).

- **Sublimation**—The ability of a substance to change from the solid to the vapor phase without passing through the liquid phase (e.g., dry ice, naphthalene or moth balls). An increase in temperature can increase the rate of sublimation. Significant in evaluating the flammability or toxicity of any released materials that sublime. The opposite of sublimation is deposition (changes from vapor to solid).
- **Critical temperature and pressure**—Critical temperature is the temperature above which a gas cannot be liquefied no matter how much pressure is applied. Critical pressure is the pressure that must be applied to liquefy a gas at its critical temperature. Both terms relate to the process of liquefying gases. A gas cannot be liquefied above its critical temperature. The lower the critical temperature, the less pressure required to bring a gas to its liquid state.
- **Volatility**—The ease with which a liquid or solid can pass into the vapor state. The higher a material's volatility, the greater its rate of evaporation. Vapor pressure is a measure of a liquid's propensity to evaporate; the higher a liquid's vapor pressure, the more volatile the material. This is a significant property in that volatile materials will readily disperse and increase the hazard area.
- **Evaporation rate**—The rate at which a material will vaporize or change from liquid to vapor, as compared to the rate of vaporization of a specific known material—*n*-butyl acetate. Is useful in evaluating the health and flammability hazards of a material. The relative evaporation rate of butyl acetate is 1.0, and other materials are then classified as follows:

SPEED	EVAPORATION RATE (Butyl Acetate = 1.0)	EXAMPLES
Fast	>3.0	Methy ethyl ketone = 3.8 Acetone = 5.6 Hexane = 8.3
Medium	0.8 to 3.0	Naptha = 1.4 Xylene = 0.6 Isobutyl alcohol = 0.6
Slow	<0.8	Water = 0.3 Mineral spirits = 0.1

Figure 7.2

- **Expansion ratio**—The amount of gas produced by the evaporation of one volume of liquid at a given temperature. This is a significant property when evaluating liquid and vapor releases of liquefied gases and cryogenic liquids. The greater the expansion ratio, the more gas that is produced and the larger the hazard area.
- **Vapor pressure**—The pressure exerted by the vapor within the container against the sides of a container. This pressure is temperature dependent; as the temperature increases, so does the vapor pressure. Consider the following three points:
  1. The vapor pressure of a substance at 100°F (37.7°C) is always higher than the vapor pressure at 68°F (20°C).
  2. Vapor pressures reported in millimeters of mercury (mm Hg) are usually very low pressures. 760 mm Hg is equivalent to 14.7 psi or 1 atmosphere. Materials with vapor pressures greater than 760 mm Hg are usually found as gases.
  3. The lower the boiling point of a liquid, the greater vapor pressure at a given temperature.

Water has a vapor pressure of 25 mm Hg; materials with a vapor pressure above 25 mm Hg are producing vapors and can present a significant inhalation risk. Materials with a vapor pressure over 760 mm Hg will be gases under normal conditions.

- **Solubility**—The ability of a solid, liquid, gas, or vapor to dissolve in water or other specified medium. The ability of one material to blend uniformly with another, as in a solid in liquid, liquid in liquid, gas in liquid, or gas in gas. This is a significant property in evaluating the selection of control and extinguishing agents, including the use of water and firefighting foams.
- **Miscibility**—The ability of materials to dissolve into a uniform mixture. If a material is miscible in water, we mean it is infinitely dissolvable in water.
- **Degree of solubility**—An indication of the solubility and/or miscibility of the material.

  Negligible—less than 0.1%

  Slight—0.1 to 1.0%

  Moderate—1 to 10%

  Appreciable—greater than 10%

  Complete—soluble at all proportions
- **Viscosity**—Measurement of the thickness of a liquid and its ability to flow. High-viscosity liquids, such as heavy oils, must first be heated to increase their fluidity. A low-viscosity liquid will spread like water and increase the size of the hazard area.

CHEMICAL PROPERTIES

Chemical properties are the intrinsic characteristics or properties of a substance described by its tendency to undergo chemical change. In simple terms, the true identity of the material is changed as a result of a chemical reaction, such as reactivity and the heat of combustion. Chemical properties typically provide responders with an understanding of how a material may harm.

FLAMMABILITY HAZARDS

- **Flash point**—Minimum temperature at which a liquid gives off sufficient vapors that will ignite and flash over but will not continue to burn without the addition of more heat. Significant in determining the temperature at which the vapors from a flammable liquid are readily available and may ignite. Flash point is also linked to boiling point and vapor pressure; low flash point materials will typically have low boiling points and increasing vapor pressures.
- **Fire point**—Minimum temperature at which a liquid gives off sufficient vapors that will ignite and sustain combustion. It is typically several degrees higher than the flash point. In assessing the risk posed by a flammable liquids release, greater emphasis is placed on the flash point, since it is a lower temperature and sustained combustion is not necessary for significant injuries or damage to occur.
- **Ignition (autoignition) temperature**—The minimum temperature required to ignite gas or vapor without a spark or flame being present. Significant in evaluating the ease at which a flammable material may ignite. Materials with lower ignition temperatures generally have a greater risk of ignition.

- **Flammable (explosive) range**—The range of gas or vapor concentration (percentage by volume in air) that will burn or explode if an ignition source is present. Limiting concentrations are commonly called the lower flammable (explosive) limit and the upper flammable (explosive) limit. Below the lower flammable limit, the mixture is too lean to burn; above the upper flammable limit, the mixture is too rich to burn. If the gas or vapor is released into an oxygen-enriched atmosphere, the flammable range will expand. Likewise, if the gas or vapor is released into an oxygen-deficient atmosphere, the flammable range will contract. Chemical families with wide flammable ranges include alcohols, aldehydes, and ethers.
- **Toxic products of combustion**—The byproducts of the combustion process that are harmful to humans. Based on the burning material(s), the byproducts of combustion will vary as follows:
  - If there is incomplete combustion = carbon monoxide
  - If nitrogen is present = nitrogen oxide gases. Incomplete combustion can produce hydrogen cyanide and ammonia.
  - Organic materials containing a halogenated material (e.g., chlorine, bromine, fluorine) = hydrogen chloride, hydrogen bromide or hydrogen fluoride in a fire.
  - If sulfur is present = sulfur dioxide gas. Incomplete combustion = hydrogen sulfide and carbonyl sulfide.

REACTIVITY HAZARDS

- **Reactivity/instability**—The ability of a material to undergo a chemical reaction with the release of energy. Reactivity can be initiated by mixing or reacting with other materials, the application of heat, physical shock and so on. Reactive materials include materials that decompose spontaneously, polymerize, or otherwise self-react. Examples include oxidizers and organic peroxides, corrosives, pyrophoric (i.e., air reactive) materials, and water reactive materials.
- **Oxidation ability**—The ability of a material to (1) either give up its oxygen molecule to stimulate the oxidation of organic materials (e.g., chlorate, permanganate, and nitrate compounds), or (2) receive electrons being transferred from the substance undergoing oxidation (e.g., chlorine and fluorine).
- **Water reactivity**—Materials that react with water and release a flammable gas or present a health hazard. Some materials may also react explosively with water (e.g., sodium).
- **Air reactivity (pyrophoric materials)**—Materials that ignite spontaneously in air without an ignition source (e.g., aluminum alkyls, white phosphorous).
- **Chemical reactivity**—A process involving the bonding, unbonding and rebonding of atoms, that can chemically change substances into other substances. The interaction of materials in a container may result in a build-up of heat and pressure, and may cause container failure. Similarly, the combined materials may be more corrosive than the container was originally designed to withstand and lead to container failure.
- **Polymerization**—A reaction during which a monomer is induced to polymerize by the addition of a catalyst or other unintentional influences, such as excessive heat, friction, contamination. If the reaction is not controlled, it is possible to have an excessive amount of energy released.

- **Catalyst**—Used to control the rate of a chemical reaction by either speeding it up or slowing it down. If used improperly, catalysts can speed up a reaction and cause a container failure due to pressure or heat build-up.
- **Inhibitor**—Added to products to control their chemical reaction with other products. If the inhibitor is not added or escapes during an incident, the material will begin to polymerize, possibly resulting in container failure.
- **Maximum safe storage temperature (MSST)**—The maximum storage temperature that an organic peroxide may be maintained, above which a reaction and explosion may occur.
- **Self-Accelerating decomposition temperature (SADT)**—The temperature at which an organic peroxide or synthetic compound will react to heat, light, or other chemicals and release oxygen, energy, and fuel in the form of an explosion or rapid oxidation. When this temperature is reached by some portion of the mass of an organic peroxide, irreversible decomposition will begin.

CORROSIVITY HAZARDS

- **Corrosivity**—A material that causes visible destruction of, or irreversible alterations to, living tissue by chemical action at the point of contact. Corrosive materials include acids and caustics or bases.
- **Acids**—Compound that forms hydrogen ions in water. These compounds have a pH < 7, and acidic aqueous solutions will turn litmus paper red. Materials with a pH < 2.0 are considered a strong acid.
- **Caustics (base, alkaline)**—Compound that forms hydroxides ions in water. These compounds have a pH > 7, and caustic solutions will turn litmus paper blue. Materials with a pH >12 are considered a strong base. Also known as alkali, alkaline, or base.
- **pH**—Acidic or basic corrosives are measured to one another by their ability to dissociate in solution. Those that form the greatest number of hydrogen ions are the strongest acids, while those that form the hydroxide ion are the strongest bases. The measurement of the hydrogen ion concentration in solution is called the pH (power of hydrogen) of the compound in solution. The pH scale ranges from O to 14, with strong acids having low pH values and strong bases or alkaline materials having high pH values. A neutral substance would have a value of 7.
- **Strength**—The degree to which a corrosive ionizes in water. Those that form the greatest number of hydrogen ions are the strongest acids (e.g., pH < 2), while those that form the greatest number of hydroxide ions are the strongest bases (pH > 12).
- **Concentration**—The percentage of an acid or base dissolved in water. Concentration is not the same as strength.

RADIOACTIVE MATERIALS

- **Radioactivity**—The ability of a material to emit any form of radioactive energy.
- **Activity**—The rate of disintegration or decay of a radioactive material. Measured in curies (1 curie = 37 billion disintegrations per second), although it is usually expressed in either millicuries or microcuries. Activity indicates how much radioactivity is present and not how much material is present.

- **Dose**—A quantity of radiation or energy absorbed by the body, usually measured in millirems (mrem).
- **Dose rate**—The radiation dose delivered per unit of time (e.g., mrem/hour).
- **Half-life**—The time it takes for the activity of a radioactive material to decrease to one half of its initial value through radioactive decay. The half-life of known materials can range from a fraction of a second to millions of years.

CHEMICAL AND BIOLOGICAL AGENTS/WEAPONS

- **Biological agents and toxins**—Biological threat agents consist of pathogens and toxins. Pathogens are disease-producing organisms and include bacteria (e.g., anthrax, cholera, plague, e coli), and viruses (e.g., small pox, viral hemorrhagic fever). Toxins are produced by a biological source and include ricin, botulinum toxins, and T2 mycotoxins.
- **Chemical agents**—Chemical agents are classified in military terms based upon their effects on the enemy. The intent of using chemical weapons is to incapacitate and to kill. Categories of chemical agents are
  - Nerve agents (neurotoxins)
  - Choking agents (respiratory irritants)
  - Blood agents (chemical asphyxiants)
  - Vesicants or blister agents (skin irritants)
  - Antipersonnel agents (riot control agents)

  Nerve agents—Chemical warfare agents that are the most toxic of the known chemical agents. Primarily consist of organophosphate agents that attack the central nervous system. Nerve agents are hazardous in both their liquid and vapor state, and can cause death within minutes after exposure. Examples of chemical warfare agents include tabun, sarin, soman, and VX.

  Choking agents—Chemical agent that can damage the membranes of the lung. Examples include phosgene and chlorine.

  Blood agents—Chemical agents that consist of a cyanide compound, such as hydrogen cyanide (hydrocyanic acid) and cyanogens chloride. These agents are identical to their civilian counterpart used in industry.

  Vesicants (blister agents)—Chemical agents that pose both a liquid and vapor threat to all exposed skin and mucous membranes. These are exceptionally strong irritants capable of causing extreme pain and large blisters upon contact. Examples include mustard, lewisite, and phosgene oxime.

  Riot control agents—Usually solid materials that are dispersed in a liquid spray and cause pain or burning on exposed mucous membranes and skin. Common examples include "Mace" (CN) and pepper spray (i.e., capsaicin).

  Persistence—Refers to the length of time a chemical agent remains as a liquid. A chemical agent is said to be "persistent" if it remains as a liquid for longer than 24 hours and nonpersistent if it evaporates within that time. Among the most persistent chemical agents are VX, tabun, mustard, and lewisite.

TECHNICAL INFORMATION CENTERS

A number of private and public sector hazardous materials emergency "hotlines" exist. Their functions include (1) providing immediate chemical hazard information; (2) accessing secondary forms of expertise for additional action and information; and (3) acting as a clearinghouse for spill notifications. They include both public and subscription-based systems.

In the United States, the most recognized emergency information center is CHEMTREC (Chemical Transportation Emergency Center). Operated by the American Chemistry Council (ACC) in Arlington, Virginia, CHEMTREC is a free public service that can be contacted 24 hours daily at (800) 424-9300 from anywhere within the United States, as well as Puerto Rico, the Virgin Islands, and Canada.

Callers outside of the United States and ships at sea can contact the Center using CHEMTREC's international and maritime number at (703) 527-3887. (Note: Collect calls are accepted.) Non-English-speaking callers to CHEMTREC are handled through the use of an interpreter service.

The CHEMTREC™ Center provides a number of emergency and non-emergency services, including the following:

1. *Emergency response information.* CHEMTREC provides immediate advice to callers anywhere on how to cope with chemicals involved in a transportation or fixed facility emergency. In addition to its extensive database, it can also access shippers, manufacturers, or other forms of expertise for additional and appropriate follow-up action and information. CHEMTREC also has immediate access to medical professionals and toxicologists, who can provide medical information for incidents involving chemical exposures.

   CHEMTREC emergency specialists have the capability to initiate conference phone calls between on-scene responders and company representatives as well as fax MSDSs from their database directly to on-scene personnel.

2. *Emergency Communications.* CHEMTREC helps hazmat shippers comply with DOT regulations (49 CFR 172.604) that require shippers to provide a 24-hour emergency point-of-contact. Manufacturers and shippers that register with CHEMTREC must provide MSDS's for their products, as well as emergency and administrative points of contact in the event additional information or assistance is required.

3. *Chemical Industry Mutual Aid Network.* CHEMTREC is the coordination point for the chemical industry mutual aid network known as CHEMNET. The mutual aid network consists of response teams from commercial contractors under contract to ACC. The primary objective of the network is to provide a chemical industry presence on scene as soon as possible. Chemical mutual aid teams cannot be activated by emergency responders; they must be requested by chemical companies that are members of the network.

   The network works like this—If a mutual aid member is unable to respond to an incident involving one of its products in a timely manner, CHEMTREC links the company with the member response team or contractor closest to the scene and able to provide assistance. The member team would then respond to the incident scene and render assistance until the shipper's personnel arrive on-scene.

CHEMTREC also serves as the point of contact for several product-specific industry mutual aid programs. These include products such as chlorine, sulfur dioxide, hydrogen fluoride, hydrogen cyanide, phosphorus, vinyl chloride, compressed gases, and swimming pool chemicals.

4. *Participation in drills and exercises.* CHEMTREC is available to participate in local exercises and provide assistance as in an actual emergency. Requests for this service should be coordinated with CHEMTREC at least 48 hours before the exercise. Further information on this program and CHEMTREC operations can be found at www.chemtrec.org.

Other emergency response and general information numbers that may be useful for both emergency planning and response purposes include the following:

- CANUTEC (Canadian Transport Emergency Centre) is operated by Transport Canada and can be contacted at (613) 996-6666. The general information number is (613) 992-4624. The Canadian counterpart of CHEMTREC, it provides assistance in identification and establishing contact with shippers and manufacturers of hazardous materials that originate in Canada. Additional information on CANUTEC can be found at www.tc.gc.ca/canutec.
- SETIQ (Emergency Transportation System for the Chemical Industry) is a service of the Mexico National Association of Chemical Industries and can be contacted at 01-800-00-214-00 in the Mexican Republic. For calls originating in Mexico City and its metro area, SETIQ can be contacted at 5559-1588. For calls originating elsewhere, call 0-11-52-5-559-1588. Additional information on SETIQ can be found at www.aniq.org.mx/setiq/setiq.htm.
- U.S. Coast Guard and the Department of Transportation National Response Center (NRC) at (800) 424-8802, or at (202) 267-2675 for those without 800 access. The NRC is the federal government's central reporting point for all oil, chemical, radiological, biological, and etiological releases into the environment within the United States and its territories. In addition, the NRC receives reports via the toll-free number on potential or actual domestic terrorism and coordinates notifications and response with the FBI and the Soldier and Biological Chemical Command (SBCCOM).

  The NRC must be notified by the responsible party (i.e., the spiller) if a hazardous materials release exceeds the reportable quantity (RQ) provisions of CERCLA. Additional information on the NRC can be found at www.nrc.uscg.mil.
- The Agency for Toxic Substances and Disease Registry (ATSDR) at (404) 498-0120. ATSDR is the leading federal public health agency for hazmat incidents and operates a 24-hour emergency number for providing advice on health issues involving hazmat releases. If necessary, ATSDR can deploy an Emergency Response Team to provide on-scene assistance.

  ATSDR also has developed Medical Management Guidelines (MMGs) for Acute Chemical Exposures to aid emergency department physicians and other emergency healthcare professionals who manage acute exposures resulting from chemical incidents. Additional information on ATSDR can be found at www.atsdr.cdc.gov/atsdrhome.html.
- National Animal Poison Control Center (NAPCC) at (900) 680-0000 or (800) 548-2423. Operated by the University of Illinois at Urbana-Champaign, this number

provides 24-hour consultation in the diagnosis and treatment of suspected or actual animal poisonings or chemical contamination. In addition, it staffs an emergency response team to investigate such incidents in North America and performs laboratory analysis of feeds/animal specimens/environmental materials for toxicants and chemical contaminants. Additional information on NAPCC can be found at www.workingdogs.com/doc0002.htm

- National Pesticide Information Center (NPIC) at (800) 858-7378. Operated by Oregon State University in cooperation with EPA, NPIC provides information on pesticide-related health/toxicity questions, properties, and minor clean-up to physicians, veterinarians, responders, and the general public. Hours of operation are 6:30 A.M. to 4:30 P.M. (Pacific Time), seven-days a week. Additional information on NPIC can be found at www.npic.orst.edu.

Emergency responders should develop a telephone roster of those individuals and agencies at the state and local level that can offer technical assistance. Examples include environmental and health departments, local chemical industry personnel, local hazmat spill cooperatives and clean-up contractors, and regional poison control centers. Your Local Emergency Planning Committee (LEPC) and Regional Counter-Terrorism Task Force can be a good resource in this process.

HAZARDOUS MATERIALS WEB SITES AND COMPUTER DATABASES

Portable computers, personal data assistants (PDAs), smart phones, CD-ROMs, and Internet access have literally revolutionized the ability of emergency responders to search and access hazard information from the field. Virtually all of the major emergency response guidebooks previously cited have an electronic equivalent that can be referenced through third-party software products, CD-Rom, or Web site. In addition, there are numerous public and commercial web-based tools that can facilitate the risk evaluation process.

Examples of some computer-based and electronic tools include:

- CAMEO® (Computer Assisted Management of Emergency Operations) is the most widely used computer-based software tool used by hazmat responders. Developed by the EPA Chemical Emergency Preparedness and Prevention Office (CEPPO) and the National Oceanic and Atmospheric Administration (NOAA) Office of Response and Restoration, CAMEO has application as both a planning and response tool. It consists of three major elements:
  - The *CAMEO database* contains over 6,000 hazardous chemicals, 80,000 synonyms, and product trade names. CAMEO provides a powerful search engine that is linked to chemical-specific information on fire and explosive hazards, health hazards, firefighting techniques, clean-up procedures, and protective clothing. CAMEO also contains basic information on facilities that store chemicals, on the inventory of chemicals at the facility (Tier II), and on emergency planning resources. Additionally, there are templates where users can store local planning information.
  - *MARPLOT (Mapping Applications for Response, Planning, and Local Operational Tasks)* is the mapping application that allows users to "see" their data (e.g., roads, facilities, schools, response assets), display this information on computer maps, and print the information on area maps. Areas contaminated by potential or actual chemical release scenarios also can be overlaid on the maps

to determine potential impacts. These maps are created from the U.S. Bureau of Census TIGER/Line files and can be manipulated quickly to show possible hazard areas.

- *ALOHA (Aerial Locations of Hazardous Atmospheres)* is an atmospheric plume dispersion model used for evaluating releases of hazardous materials vapors. ALOHA allows the user to estimate the downwind dispersion of a chemical cloud based on its physical and toxicological properties, atmospheric conditions, and specific circumstances of the release. Graphical outputs include a "cloud footprint" that can be plotted on maps with MARPLOT to display the location of other hazmat facilities and vulnerable locations (e.g., hospitals, schools). Specific information on these locations can be extracted from CAMEO information modules to help make decisions about the degree of risk posed by a release scenario.

❑ The CHEMTREC and EPA Chemical Emergency Preparedness and Prevention Office (CEPPO) Web sites provide good starting points for gathering hazard information. With their links, responders have a free, regularly updated starting point for accessing a number of both public and private Web sites that can provide both chemical or container information. The CHEMTREC Web site can be found at www.chemtrec.org, while the EPA Web site can be found at www.epa.gov.

❑ The NOAA Chemical Reactivity Worksheet is an excellent tool that can be downloaded and used for determining the effects of various chemical mixtures. It includes reactivity information for more than 6,000 chemicals, and can be referenced at http://response.restoration.noaa.gov/chemaids/react.html

❑ The Operation Respond Institute's OREIS provides responders with the ability to access real-time hazmat shipping information for participating railroads and trucking companies directly from carrier databases. Participating carriers include all of the Class I railroads in the United States and Canada. In addition, chemical databases and other hazard information are available. A Web-based version of OREIS is available at www.oreis.org.

❑ Virtually all of the major federal agencies with an emergency planning or response mission (e.g., DOT, OSHA, EPA, NRC) have Web sites with a range and variety of reference sources and links.]

Web-based tools provide a wide range of options for accessing technical response information. In addition, the Internet provides response organizations with the capability to create an incident-specific Web site where response partners and parties not on scene can have access to incident-specific information, including status reports, digital photos, video clips, and so forth.

When evaluating electronic-based information sources, consider the following criteria:

- How will the tool complement or improve your response operations and decision making?
- Costs, including initial subscription and user fees.
- Hardware and software requirements, including communications technology.
- Communications security (COMSEC), as appropriate.
- Ease of use and user friendliness.
- Technical support.

TECHNICAL INFORMATION SPECIALISTS

A common source of hazard information are personnel who either work with the hazardous material(s) or their processing, or who have some specialized knowledge, such as container design, toxicology, or chemistry. When evaluating these product and container specialists and the information they provide, consider these observations and lessons learned:

- Many individuals who are specialists in a narrow, specific technical area may not have an understanding of the broad, multi-disciplined nature of hazmat emergency response. For example, some information sources will provide extensive data on container design yet may be unfamiliar with basic emergency response principles.
- Each information specialist has their own strengths and limitations. It's a good idea to remove the term *expert* from your vocabulary; be wary of self-proclaimed experts without first verifying their background and knowledge.
- You will often interact with individuals with whom you have had no previous contact. Before relying on their recommendations, verify their level of expertise and job classification by asking specific questions. More than one responder has been disappointed or embarrassed to find that the "expert" they have been waiting for is actually a truck driver or a product marketing representative.
- When questioning outside information sources, consider yourself as playing the role of a detective. Remember, final accountability always rests with the IC. While this is certainly not an interrogation process, you must be confident of the specialist's expertise and authority. In some cases, responders will ask questions for which they already know the answer in order to evaluate that person's competency and knowledge level.
- Local responders and facility personnel must get out into their communities and establish personal contacts and relationships with your response partners. These include state, regional, and federal environmental response personnel, law enforcement, clean-up contractors, industry representatives, wrecking and rigging companies, and so on.
- Investigate the existence of local and state "Good Samaritan" legislation that may cover outside representatives as they assist you on the scene.

HAZARD COMMUNICATION AND RIGHT-TO-KNOW REGULATIONS

Numerous state and local worker and community right-to-know laws exist across the country. In addition, the OSHA Hazard Communication Standard (OSHA 1910.1200) has specific requirements pertaining to hazard markings, worker access to hazmat information, and worker exposures to chemicals in the workplace.

While the scope of these regulations may vary, most right-to-know laws provide emergency responders with access to MSDSs and have specific requirements mandating the development of facility preincident plans and community hazardous materials plans.

Responders must be able to both read and interpret material safety data sheets. OSHA requires that certain basic data and information be provided on each MSDS, including the following:

- *General information*—Includes manufacturer's name, address and emergency phone number, chemical name and family, and all synonyms.
- *Hazardous ingredient statement*—Breaks out the active ingredients by percentage. Trade secret restrictions may sometimes minimize the amount of information available on an MSDS, although responders should have access to this data during an emergency.
- *Physical data* —Includes physical properties.
- *Fire and explosion data*—Includes control and extinguishment measures, proper extinguishing agents, and so on.
- *Health and reactivity hazard data (as necessary)*—Includes toxicology information, signs and symptoms of exposure, emergency care, chemical incompatibilities, decomposition products, and so on.
- *Spill and leak control procedures*—Include procedures and precautions for handling chemical releases, as well as waste disposal methods.
- *Special protection information*—Includes protective clothing and respiratory protection requirements.
- Other special precautions (as necessary).

Although MSDSs can provide responders with a significant amount of data and information, you should realize that

- MSDSs have no uniform or consistent format or layout. The only regulatory requirements are that the specified data be provided.
- Computer-generated MSDSs may be difficult to initially use and interpret because of their layout.
- There are no regulatory requirements concerning the language and terminology used. MSDSs for the same chemical that are produced by different manufacturers may appear different and, in some instances, may use different terminology.

MONITORING INSTRUMENTS

Monitoring and detection equipment are critical tools for evaluating real-time incident data to

- Determine if anything (i.e., a hazmat) is present.
- Classify or identify unknown hazards.
- Determine the appropriate levels of personal protective clothing and equipment.
- Determine the size and location of hazard control zones.
- Develop protective action recommendations and corridors.
- Assess the potential health effects of exposure.
- Determine when the incident scene is safe so that the public and/or facility personnel may be allowed to return.

Monitoring is an integral part of site safety operations and a cornerstone of a risk-based emergency response philosophy. Numerous response organizations have been issued regulatory citations for their failure to identify hazardous and IDLH conditions, to evaluate constantly the incident site for changes, and to verify the accuracy of hazard control zone locations.

Hazardous materials concentrations can be identified, quantified, and/or verified in two ways: (1) on-site use of direct-reading instruments, which provide readings at the same time that monitoring is being performed, and (2) laboratory analysis of samples obtained through several collection methods. Both tools are discussed in this section.

Figure 7.5 A variety of monitoring and detection instruments are required. No single instrument does it all.

SELECTING DIRECT-READING INSTRUMENTS

Direct-reading instruments provide information at the time of sampling, thereby allowing for rapid, on-scene risk evaluation and decision making. They are use to detect and monitor flammable or explosive atmospheres, oxygen-enriched and -deficient atmospheres, certain toxic and hazardous gases and vapors, chemical agents, certain biological agents, and ionizing radiation. The selection of types of monitoring instruments should be based on local/facility hazards and anticipated response scenarios.

When evaluating survey instruments for emergency response use in the field, consider the following criteria:

- *Portability and user friendliness*—Ease to carry, weight, etc. Consider who will be using the instrument and their ability to consistently use it safely and correctly.
- *Instrument Response Time*—Also known as lag time, this is the period of time between when the instrument senses a product and when a monitor reading is produced. Depending on the instrument, lag times can range from several seconds to minutes. Variables will include the following:
  - Does the instrument have a pump? Monitors with a pump typically have a response time of 3–5 seconds; monitors without a pump and operating in a diffusion mode have response times of 30–60 seconds. Consult your manufacturer's operation manual for specific information.
  - Use of sampling tubing—add 1 to 2 seconds of lag time for each foot of hose.

  Monitors also have a recovery time, which is the amount of time it takes the monitor to clear itself of the sample. Recovery time is influenced by the properties of the sample, the amount of sampling hose, and the amount absorbed by the monitor.

- *Sensitivity and selectivity*—The ability of the instrument to select slight changes in product concentrations and select a specific chemical or group of chemicals that react similarly. Monitoring instruments are calibrated on specific materials (e.g., methane, pentane, isobutylene). Increased selectivity widens the relative response of an instrument and can increase its accuracy; however, it may not be possible to determine the exact contaminant present.

  Note: Amplifiers are used in some monitoring instruments to widen their response to more hazmats and increase accuracy. However, other electrical equipment in the area, including radios, other types of monitoring instruments, power lines, and transformers, may interfere with the amplifier in the instrument being used.

- *Lower detection limit (LDL)*—The lowest concentration to which a monitoring instrument will respond. The lower the LDL, the quicker contaminant concentrations can be evaluated. Many instruments have several scales of operation for monitoring both very low and very high concentrations.

- *Calibration*—The process of adjusting a monitoring instrument so that its readings correspond to actual, known concentrations of a given material. If the readings differ, the monitoring instrument can then be adjusted so that readings are the same as the calibrant gas. Conditions that will affect calibration include atmosphere, humidity, temperature, and atmospheric pressure.

  There are four types of calibration:

  1. *Factory calibration*—Instrument is returned to a certified factory/facility for testing and adjustment by certified instrument technicians.
  2. *Full calibration*—Instrument is shown a calibration gas and the readings are adjusted (automatically or manually) to the certified calibration gas values.
  3. *Field calibration*—Instrument is exposed to a known calibration gas and the user verifies that the readings correspond to +10% of the calibration gas. Field adjustments to the monitor may then be made, as appropriate.
  4. *Bump test*—Instrument is exposing to a known calibration gas and the sensors show a response or alarm. If the instrument does not respond appropriately, then it should undergo a field or full calibration.

- *Correction factors (i.e., relative response curves)*—Relative response is the difference between a calibrated response and a response from a product for which the meter is not calibrated. This is a common issue with combustible gas indicators (CGIs) and photoionization detectors when dealing with known materials. The manufacturer should provide a comparison or response curve table to adjust the readings for the product being evaluated.

- *Inherent safety*—The inherent safety and the ability of the device to operate in hazardous atmospheres must be evaluated (see Scan Sheet 7-A). In addition, instruments should be certified by an approved testing laboratory for expected operating conditions.

Scan 7-A

NATIONAL ELECTRICAL CODE HAZARDOUS CLASSIFICATIONS

When evaluating direct-reading monitoring instruments, responders must have a basic understanding of certain terms and classifications that relate to the ignition potential of electrical instruments in hazardous environments. Hazardous locations are defined in Article 500 of *NFPA 70—The National Electrical Code (NEC).*

Three simultaneous conditions can create a hazardous location:

1. Vapors, dusts, or fibers are present in sufficient quantity to ignite (i.e., within the flammable range).
2. Source of ignition may be present (e.g., the instrument).
3. The resulting exothermic reaction could propagate beyond where it started.

HAZARDOUS CLASSIFICATIONS

NEC hazardous classifications are divided into a system of three classes, two divisions, and various subgroups.

Classes describe the type of flammable materials that produce the hazardous atmosphere.

- *Class I locations*—Flammable gases or vapors may be present in quantities sufficient to produce explosive or ignitable mixtures.
- *Class II locations*—Concentrations of combustible dusts may be present (e.g., coal or grain dust).
- *Class III locations*—Areas concerned with the presence of easily ignitable fibers or flyings (e.g., cotton milling).

Groups are products within a Class. Class I is divided into four groups (Groups A–D) on the basis of similar flammability characteristics. Class II is divided into three groups (Groups E–G). There are no groups for Class III materials.]

- *Group A Atmospheres*
  - Acetylene
- *Group B Atmospheres (not sealed in conduit 1/2-inch or larger)*
  - 1,3-butadiene
  - Carbon monoxide
  - Formaldehyde (gas)
  - Hydrogen
  - Manufactured gas (containing > 30% hydrogen by volume)
- *Group C Atmospheres (selected chemicals)*
  - Acetaldehyde
  - Ethylene oxide
  - Dicyclopentadiene
  - Diethyl ether
  - Epichlorohydrin
  - Ether acetate
  - Ethylene
  - Ethylene glycol
  - Ethyl mercaptan
  - Hydrazine
  - Hydrogen cyanide
  - Hydrogen selenide
  - Hydrogen sulfide
  - Methylacetylene
  - Monoethyl ether
  - Nitropropane
  - Tetraethyl lead
  - Tetrahydrofuran

- *Group D Atmospheres (selected chemicals)*

Acetone	Acetonitrile	Acrylonitrile
Ammonia	Aniline	Benzene
Butane	Chlorobenzene	Cyclohexane
Dichloroethane	Ethane	Ethyl alcohol
Etylene glycol	Fuel oils	Gasoline
Hexane	LPG	Methane
Methyl alcohol	Methyl ethyl ketone	Monomethyl ether
Naptha	Propane	Styrene
Vinyl chloride	Xylene	

- *Group E Conductive Dusts*—Atmospheres containing metal dusts, including aluminum, magnesium, and their commercial alloys, and other metals of similarly hazardous characteristics.
- *Group F Semi-Volatile Dusts*—Atmospheres containing carbon black, coal, or coke dust with more than 8% volatile material.
- *Group G Nonconductive Dusts*—Atmospheres containing flour, starch, grain, carbonaceous, chemical thermoplastic, thermosetting, and molding compounds.

Divisions describe the type of location that may generate or release a flammable material.

- *Division 1* are locations where the vapors, dust, or fibers are continuously generated and released. The only element necessary for a hazardous situation is a source of ignition.
- *Division 2* are locations where the vapors, dusts, or fibers are generated and released as a result of an emergency or failure in the containment system.

Division 1 areas have a greater probability of generating a hazardous atmosphere than Division 2 areas. Although Division 1 devices are permitted for use in Division 2 areas, instruments approved for Division 2 areas are not usable in Division 1 areas. At a minimum, it is recommended that direct-reading instruments be rated for operations in Class I, Division 2 hazardous classifications.

REDUCING THE IGNITION POTENTIAL OF ELECTRICAL EQUIPMENT

To reduce the ignition potential of both fixed and portable monitoring instruments, instrument manufacturers can use several different methods.

Explosion-proof construction encases the electrical equipment (i.e., potential ignition source) in a rigidly built container so that (1) it withstands the internal explosion of a flammable mixture, and (2) it prevents propagation to the surrounding flammable atmosphere. Used in Class I, Division 1 atmospheres at fixed installations.

Intrinsically safe equipment or wiring is incapable of releasing sufficient electrical energy under both normal and abnormal conditions to cause the ignition of a flammable mixture. Commonly used in portable direct-reading instruments for operations in Class I, Division 2 hazardous classifications.

Purging is used for protecting totally enclosed electrical equipment with an inert gas under a slight positive pressure from a reliable source. The inert gas provides positive pressure within the enclosure and minimizes the development of a flammable atmosphere. Used in Class I, Division 1 atmospheres at fixed installations.

Explosion-proof construction and purging are primarily used at fixed facilities and for protecting stationary instrumentation. In contrast, direct-reading instruments used for field applications will rely on intrinsically safe construction. Users should ensure that instruments are certified by an approved testing laboratory (e.g., Underwriters' Laboratories or Factory Mutual) for operations within the required class, group, and division.

In addition to the previous criteria, the following operational, storage, and use considerations should be evaluated:

- Where and in what type of storage container will the instruments be stored? This is especially critical when placing monitoring instruments inside or outside of vehicles. In addition, some storage containers may off-gas and contaminate sensors.
- Can field maintenance be easily performed? For example, are field calibration kits available and can sensors be easily changed in the field?
- Can buttons, switches, and so on be easily manipulated while wearing chemical gloves?
- How long does it take for the monitoring instruments to "warm up" before they can be used in the field?
- What types of alarms does the instrument have? Can meters be read easily while wearing respiratory protection? Is there a glare problem during daytime operations and a lighting problem for operations at night?
- What types of batteries are required for the instrument—off-the-shelf batteries or rechargeable batteries? How long will the unit operate with a full charge?

TYPES OF DIRECT-READING INSTRUMENTS

All direct-reading instruments have inherent limitations. Many detect and/or measure only specific classes of chemicals. As a general rule, they are not designed to measure and/or detect airborne concentrations below 1 ppm. Also, many direct-reading instruments designed to detect one particular substance may detect other substances (interference) and give false readings.

Figure 7.6 outlines the common types of direct-reading instruments used in the emergency response field.

When using direct-reading instruments, interpret instrument readings conservatively and consider the following guidelines:

- Conduct a daily check of your instruments, as well as before use.
- Use chemical correction factors when dealing with known materials, as appropriate.
- Remember that instrument readings have some limitations when dealing with unknown substances. Report readings of unknown contaminants as positive instrument response rather than specific concentrations (i.e., ppm). Conduct additional monitoring at any location where a positive response occurs.
- A reading of zero should be reported as no instrument response rather than clean, since quantities of chemicals may be present that cannot be detected by that particular instrument technology.
- Remember the Rule of Threes when dealing with unknowns and suspected criminal scenarios involving hazardous materials; use several types of detection technologies to classify or identify the hazard.
- After the initial survey, continue frequent monitoring throughout the incident.

DIRECT-READING MONITORING INSTRUMENTS

CORROSIVE MONITOR INSTRUMENTS

HAZARD MONITORED	APPLICATION	METHOD OF OPERATION	GENERAL COMMENTS
Corrosivity—Acidity or Alkalinity 7.6 A 7.6 B	Measures the acidity or alkalinity of a corrosive material	• pH Paper—chemical reaction changes the color of the detection paper. Can detect both corrosive liquids and vapors in air, with a range of 0 to 14. Acids are normally shades of red and purple, while bases/caustics are shades of blue. • pH meters use a probe that is inserted into the liquid.	• Readings less than 2 or greater than 12 present a significant risk for injury. • Neither pH paper or pH meters will provide the specific concentration of the corrosive. • Using neutral water to wet pH paper will make it easier to detect corrosive vapors in air. • pH meters must be calibrated before each use. Probes must be thoroughly rinsed with distilled water before and after each calibration and use. They are typically not used extensively for emergency response purposes.

RADIATION SURVEY MONITORS

HAZARD MONITORED	APPLICATION	METHOD OF OPERATION	GENERAL COMMENTS
Ionizing Radiation (alpha and beta particles, gamma rays) 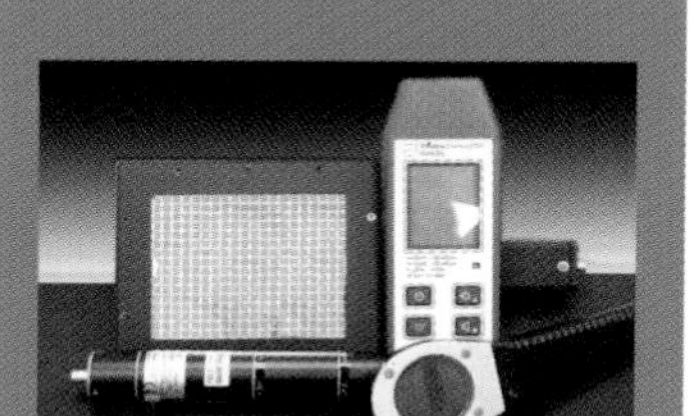 7.6 C	Measure accumulated radiation exposure	• Ionization detectors that collect and count ions electronically. • Three types of radiation detectors: scintillation, Geiger Mueller tubes, and ion chambers. Different probes can be used to measure for alpha, beta or gamma sources. – Scintillation—range of 0.02 to 20 mR/hr. – GM Tubes—range of 0.2 to 20 mR/hr or 800 to 80,000 cpm (most common detection technology found in emergency response). – Ion Chamber—range of 1 mR/hr to 500 R/hr.	• Radiation instruments should be selected based upon: – Type of radiation to be detected – Energy of the radiation to be detected – Range high enough to measure the intensity of the source. • An instrument reliability check should be performed before each use. If the reading is not with + 20% of the initial reading listed on the calibration sticker, the instrument should be recalibrated. • Meters using the International System (SI) for measuring radiation (e.g., Gray, Sievart) may be found at fixed facilities.

HAZARD MONITORED	APPLICATION	METHOD OF OPERATION	GENERAL COMMENTS
Ionizing Radiation (alpha and beta particles, gamma rays) 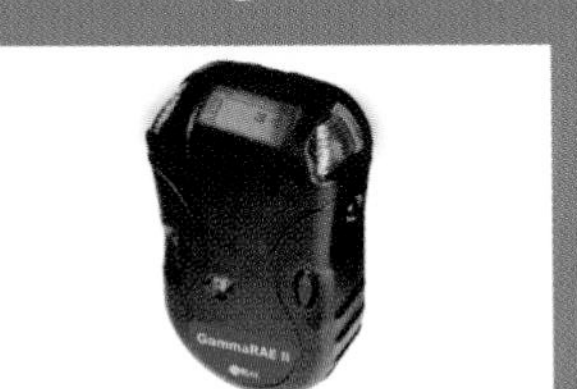  7.6 D   7.6 E		• Meter may read as counts per minute (CPM), and mR/Hr to R/hr. • Radiation pagers are used for detecting gamma and x-ray radiation, but may also detect high energy beta particles. Excellent tool when dealing with terrorism or criminal events when radioactive materials are suspected. – Units are activated when radiation levels exceed background levels. – Can provide an audible, visual (LED light), or vibra alarm. • Personal dosimeters monitor the accumulated dose received by an individual. Types include older style personal dosimeters that were visually read in the field, to sensors that must be read in a laboratory, to newer generation electronic dosimeters integrated into radiation pagers.	• Electromagnetic fields can give "false positive" readings.
OXYGEN MONITORS			
Oxygen Deficient and Enriched Atmospheres 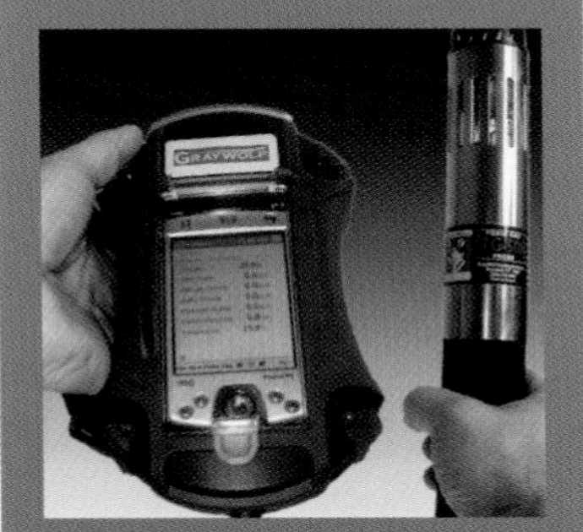 7.6 F	Measures the percentage of oxygen in air. Should measure both oxygen deficient (<19.5%)and enriched (>23.5%) atmospheres.	• Operate by a diffusion process, in which air diffuses into the sensor. Oxygen reacts with electrolytes in a cell, thereby generating a current flow in the meter. • May be either a passive sensor, or have an internal pump that draws in the air sample. • Meter will normally read percent of oxygen in the sample (e.g., 21% O_2).	• Most O_2 sensors are combined into a multi-gas meter (e.g., LEL, x O_2 and 1 or 2 toxicity sensors). • Some materials (e.g., chlorine, fluorine) will indicate a high or normal level of O_2, when the actual atmosphere may be oxygen deficient. • Extreme cold temperatures can often result in sluggish, delayed movement of the meter. • O_2 sensors are adversely affected by materials with oxidation potential in their chemical structure (e.g., chlorine, ozone, carbon dioxide). • A drop of 0.1% of O_2 means that 5,000 ppm of "something else" is in the atmosphere. • Must be calibrated prior to use to compensate for altitude and barometric pressure.

HAZARD MONITORED	APPLICATION	METHOD OF OPERATION	GENERAL COMMENTS
COMBUSTIBLE GAS INDICATORS (LEL METERS)			
Flammable Gases and Vapors 7.6 G 7.6 H	Measures the concentration of a flammable gas or vapor in air	• Basic principle is that a stream of sampled air passes through the senor housing, causing a heat increase thereby increasing the resistance in the electrical circuit, and subsequently showing an instrument reading (except for infrared sensor). • Three types of LEL sensors: catalytic bead (most common), metal oxide semi-conductors, and infrared. • Meter may read in either percent of the LEL, ppm, or percent of gas by volume.	• Although commonly known as CGI's, they are actually used for detecting flammable gases and vapors. • Readings are relative to a calibrant gas (e.g., methane, pentane, hexane). Response curves required. • Catalytic bead sensors may have problems including: – Intended for use in normal atmospheres (i.e., not oxygen deficient or enriched). Will affect LEL readings if O_2 levels are deficient (will lower readings). – Sensor can be damaged by certain materials, including silicone, tetraethyl lead, and acid gases. Some problems can be reduced through use of filters and water traps between instrument and sampling tube. – Chronic exposure through high levels may saturate the sensor and cause it to be useless for a long period of time until purged and recalibrated. • 1% of the atmosphere = 10,000 ppm. Just because you don't have a fire problem does not mean you don't have a health problem!
COLORIMETRIC INDICATOR TUBES (DETECTOR TUBES)			
Specific Gases and Vapors 7.6I	Measures the concentration of specific gases and vapors in air. Used in hazard categorization systems for testing for unknowns; enable the user to classify the hazard class or chemical family of the unknown.	• Glass tubes are filled with different reagents that react with the chemical being tested. When that chemical is present, the reagent may change color, or produce a colored stain that must be evaluated or measured using the reference points on the tube (ppm or % of material). • Ends of glass tube are first broken off; arrow on the tube should face TOWARDS the pump. Sample is then drawn through the tube using a bellows, piston or thumb pump.	• Primarily used to determine if a specific chemical is present (or not present), as compared to specific quantitative results. • May also be found in pre-packaged Hazmat Kits with a sampling matrix for the identification of unknowns or chemical warfare agents. • Operational issues include: – Greatest sources of error are (1) how the user judges the color stain or stain's endpoint, and (2) the tubes accuracy. Tubes may have error margin up to 35%.

HAZARD MONITORED	APPLICATION	METHOD OF OPERATION	GENERAL COMMENTS
Specific Gases and Vapors 7.6 J 7.6 K		• Readings can be directly read, or must be interpreted based upon the number of strokes or pumps. Wrong number of strokes/pumps can lead to incorrect readings. • Read the instructions for each tube prior to use! • Draeger Chip Measurement System (CMS) uses a bar-coded chip that is inserted into a pump. Sampling is performed automatically and readings are provided on an LCD screen in ppm.	– Tubes have a limited shelf life; can be affected by high humidity (tubes are calibrated at 50% humidity) and temperature extremes (most operate from 32°F [0°C] to 122°F [50°C]). – Response times may vary greatly from chemical to chemical. – Tubes can have cross-sensitivities—many similar chemicals can interfere with the sampling and give "false positive" readings. – Measured concentration of the same compound may vary between different manufacturers tubes. In addition, different manufacturer's tubes should not be used with another manufacturer's pump.
TOXIC GAS SENSORS			
Specific Gases and Vapors 7.6 L	Designed to detect a specific chemical Common toxic gas sensors include carbon monoxide, hydrogen sulfide, chlorine, ammonia, sulfur dioxide, etc.	• Sensor technology includes electrochemical sensors with two or ore electrodes and a chemical mixture sealed in a sensor housing, and metal oxide semi-conductors (MOS). • Meter will normally read ppm of specific gas or vapor.	• More accurate than detector tubes, but are limited to significantly fewer chemicals. • Hydrogen sulfide and carbon monoxide sensors are often combined into a multi-gas meter (e.g., LEL, x O_2 and toxicity sensors). • Toxic sensors can have a range of shelf-lives and a maximum exposure limit. High concentration exposures will shorten the life of the sensor. • Chemicals in the same chemical family as the sensor may give a reading that may be misinterpreted by the user.
Passive Dosimeters 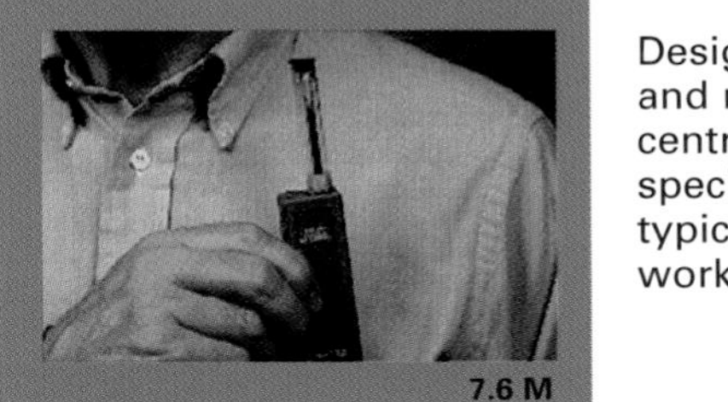 7.6 M	Designed to detect and measure concentrations of a specific chemical, typically in the workplace	• Passive monitor that measures an individual's exposure to a specific chemical. • Provides either immediate results or is required to be sent to a laboratory for analysis.	• Must acquire the specific dosimeter for the materials in question (e.g., organic vapors, mercury, ethylene oxide, etc.). • Commonly used to monitor for TLV/TWA and TLV/STEL exposures in non-emergency scenarios.

IONIZATION DETECTORS

NOTE: There are two primary ionizing detectors used in the field: the photo-ionization detector (PID) and the flame ionization detector (FID). These are discussed below. Other types of ionizing detectors used in the weapons of mass destruction area will be covered later in this section under terrorism agent detection.

PHOTO-IONIZATION DETECTORS (PID)

HAZARD MONITORED	APPLICATION	METHOD OF OPERATION	GENERAL COMMENTS
Organic and Some Inorganic Gases and Vapors 7.6 N 7.6 O 7.6 P	Detects total concentration of many organic and some inorganic gases and vapors (e.g., ammonia, arsine, phosphine, hydrogen sulfide)	• Gas/vapor sample is exposed to an ultraviolet (UV) lamp, which ionizes the sample. Ions are collected, amplified and produce a current which is read as total meter units. • PID reads from 0.1 to 2,000 ppm, although some units may read up to 10,000 ppm. There are also PID's available which will read in the ppb range. All readings are expressed in terms of meter units for the calibration gas—iso-butylene. • The vapor or gas must be able to be ionized, known as its Ionization Potential (IP). IP is measured in electron volts (eV). • There are various UV lamps, with their intensity measured also measured in eV. Most common bulbs are 10.2, 10.6 and 11.7 eV. The 10.2 or 10.6 bulbs are typically used for emergency response purposes. • In order for the PID to read a gas/vapor, the IP of the sample must be less than the eV rating of the bulb. Response times are usually less than 5 seconds.	• General survey instrument that does NOT tell the user what is there; only that something is there! • IP's for gases and vapors are provided by the PID manufacturer, and can also be referenced from the NIOSH Pocket Guide to Chemical Hazards. • Operational issues include: – Lamps are affected dirt and dust, as well as high humidity and fog. – Higher levels of methane may suppress some of the ionizing potential of the lamp. Methane has an IP of 13 eV. – PID's cannot separate out mixed gases.

FLAME IONIZATION DETECTORS (FID)

HAZARD MONITORED	APPLICATION	METHOD OF OPERATION	GENERAL COMMENTS
Organic Gases and Vapors 7.6 Q 7.6 R 7.6 S	Detects total concentration of many organic gases and vapors Operates in either the survey mode or gas chromatograph mode	• Gases / vapor sample is exposed to a hydrogen flame to ionize the sample. • Operates in two modes: – Survey Mode—detects total concentration of many organic gases and vapors. All organic compounds are ionized and detected at the same time. – Gas Chromatograph (GC) Mode—identifies and measures specific organic compounds; volatile compounds are separated. • Survey Mode Method of Operation—ionizes any chemical with an IP < 15.4 eV. Sample is exposed to a hydrogen flame and is ionized; ions are then collected, amplified, and produce a current which is read on the display as total organic vapors present (in ppm). • GC Method of Operation – Sample is drawn into a column with an inert material. Hydrogen gas is then passed through the tube and picks up various components of the sample. Each chemical takes a period of time to exit the tube. – The mass spectrometer (MS) is usually coupled with a GC and is the identifying portion of the device. As the sample passes into the detector, the energy is measured on a strip recorder as a peak. – The MS measures the relative mass of the molecular fragments of the sample and compares it to materials within the MS library, as each molecular fragment has a different weight.	• Operational issues include: – Not all FID's are intrinsically safe. Must first establish that the operating area is safe (i.e., not in the explosive range). – FID is more accurate than the PID. – FID has the ability to read methane. – Response affected by temperatures below 40°F. (4.4°C), as gases may condense in the pump and column. – There is a high learning curve for using GC's and mass spectrometers in field settings for emergency response purposes. • The more extensive the MS library, the more likely one will be able to identify an unknown substance. • Many police department crime labs will have GC/MS capabilities.

FOURIER-TRANSFORM INFRARED SPECTROMETRY (FT-IR)

HAZARD MONITORED	APPLICATION	METHOD OF OPERATION	GENERAL COMMENTS
Organic and Some Inorganic Gases and Vapors 7.6 T 7.6 U	Identification of unknown solid and liquid substances	• Sample is obtained and taken to the unit. FT-IR records the interaction of infrared radiation with the chemical sample, measuring the frequencies at which a sample absorbs radiation and the intensities of the absorptions. • As chemical functional groups are known to absorb light at specific frequencies, the FT-IR technology can allow for the specific identification of liquid and solid samples. • Reading is compared against a library of various hazardous and non-hazardous substances. If the material is in the library, identification is confirmed in <1 minute.	• Libraries include hazardous materials, chemical agents, white powders, explosives, clan lab precursors, drugs, and other common liquid and solid materials. • Although they cannot identify biological materials, they can determine if a compound is of biological or non-biological origin.

OTHER MISCELLANEOUS DETECTION DEVICES

HAZARD MONITORED	APPLICATION	METHOD OF OPERATION	GENERAL COMMENTS
HazCat™ Chemical Identification System 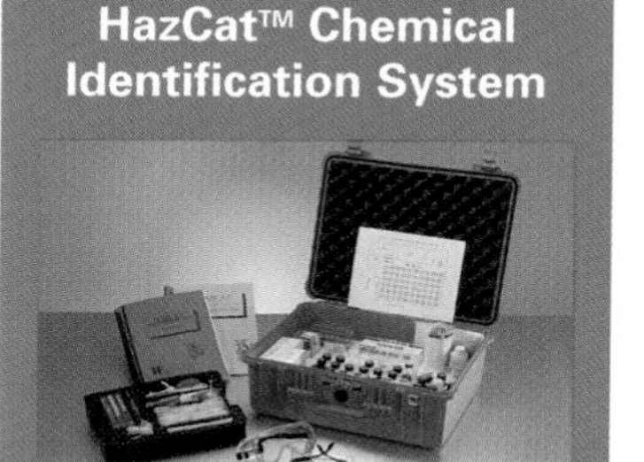 7.6 V	On-site identification or characterization of hazardous and non-hazardous materials	• Field chemistry testing kit that provides a logic system for conducting specific chemical tests that will result in either the identification or characterization of solids, liquids and gases. • System is based upon the concept that chemicals within the same family react in the same manner.	• Requires specific training; can typically provide characterization of unknowns within 10–30 minutes, depending upon the substance. • Kit options are offered for methamphetamine labs and chemical / biological / radiological applications.

HAZARD MONITORED	APPLICATION	METHOD OF OPERATION	GENERAL COMMENTS
Chemical Test Strips 7.6 W 7.6 X	Detects the presence of specific chemicals or hazards	• Requires a sample of the contaminant to come in contact with the test strip or chemical indicator paper. If the contaminant is present, it will trigger a change in color on the test strip or indicator paper.	• May be part of a hazmat identification system, such as the HazCat™ Chemical Identification System. • Indicator papers include pH paper, and M-8 / M-9 tape for chemical agents. • Chemical Classifier™ Kits can test liquid spills for corrosivity, oxidizers, fluoride, organic solvents/petroleum distillates, and iodine/bromine/chlorine. • HazMat Smart-Strip™ can test liquids and aerosols for chlorine, pH, sulfides, oxidizer, fluorides, nerve agents, cyanide, and arsenic. • Readings should be confirmed by other instruments, as appropriate (e.g., colorimetric detector tubes, chemical agent monitors, etc.).
Mercury Detector 7.6 Y	Detects mercury vapors in air	• Mercury vapors collect on a gold foil, changing the electrical resistance and causing a corresponding display reading.	• Unit can be susceptible to false readings if hand carried or moved around; suggest placing the unit on a stable surface and bringing the contaminant to the unit, if possible. • If the unit must be moved around, take three readings and then develop an average. • If clothing is possibly contaminated, place clothing in a plastic bag and insert the instrument probe into the bag to sample the atmosphere. • ACGIH TLV/TWA is 0.025 mg/m^3; ACGIH recommends that women of childbearing age should not be exposed to concentrations > 0.010 mg/m^3.

Figure 7.6 Direct-Reading Monitoring Instruments.

Initial air monitoring and reconnaissance operations pose the greatest threat to emergency responders. Always remember these basic safety considerations:

- Air monitoring personnel have the greatest risk of exposure—protective clothing must be sufficient for the expected hazards. The situation should determine the level of protective clothing used.
- The air monitoring team should consist of at least two personnel, with a backup team wearing an equal level of protection.
- Protect the instruments as appropriate. If they have the potential to become contaminated, the instrument may be wrapped in clear plastic to protect against contamination. However, make sure that the intake and exhaust ports are open.
- Approach the hazard area from upwind whenever possible. The initial site survey should begin upwind and then move to the flanks of the release. As soon as any positive indication is received, proceed with caution.
- Priority areas should include confined spaces, low-lying areas, and behind natural or artificial barriers (e.g., hills, structures, etc.), where heavier-than-air vapors can accumulate.

MONITORING STRATEGIES

The Incident Commander and/or the Hazmat Group Supervisor must establish a monitoring strategy. In developing this strategy, the following operational issues must be considered:

- Establish monitoring priorities based on whether the incident is in open air or in an enclosed or confined space environment.
- Always use the appropriate monitoring instrument(s) based on dealing with known or unknown materials. The instrument(s) should be able to detect the anticipated hazard(s), measure appropriate concentrations, and operate under the given field conditions.
- Monitoring personnel should have a good idea of what readings to expect. In the event that abnormal or unusual readings are encountered, the possibility of instrument failure should be considered. Try to confirm the initial reading with another instrument.
- The absence of a positive response or reading does not necessarily mean that contaminants are not present. A number of factors can affect contaminant concentrations, including wind, temperature, and moisture.
- Never assume that only one hazard is present.
- Remember the Rule of Threes—when dealing with unknowns and suspected criminal scenarios involving hazardous materials, use several types of detection technologies to classify or identify the hazard.
- Interpret the instrument readings in more than one manner (i.e., always play devil's advocate).
- Establish action levels based on instrument readings.

Monitoring operations at long-term incidents may be performed by multiple responders, using different monitoring instruments at various locations. Reliable monitoring results can only be obtained by using consistent monitoring procedures.

MONITORING FOR TERRORISM AGENTS

The terrorism problem has created new challenges for the emergency response community. As a result, hazmat trained personnel are now being required to broaden their skills and knowledge-base as it related to chemical, biological and radiological hazards.

Monitoring and detection instruments will play a key role in assessing the credibility of a threat. Among the clues in making the transition that an incident may be terrorist in nature are:

- Is the incident at a target occupancy or event?
- Has there been a history of any previous threats?
- Are multiple casualties involved? Are the reason(s) known?
- Have there been reports of any unusual odors? Explosions? Hazardous materials?
- Are initial emergency responders down?
- Have any secondary events occurred?

These clues will ultimately determine your initial monitoring strategy at a suspected terrorism incident. If a suspected explosive device is involved, the explosives hazard will take precedent over any other threats. If responders have arrived at an incident where an explosion has already occurred, the potential for an radiological dispersal device (RDD) or "dirty bomb" should be initially considered. The radiological threat can be quickly assessed in most instances through the use of a radiation pager, or monitor.

When conducting monitoring and sampling at a suspected terrorism incident where hazardous materials or weapons of mass destruction may be involved, consider the following:

- Avoid contact with the suspected material(s).
- When sampling for chemical warfare agents, be aware of the potential for false positives.
- Remember the "Rule of Three's"—use several types of detection technologies to classify, identify or verify the hazard.
- Grab a sample and do the testing away from the hazard area.

Remember your product and container specialists for terrorism and WMD agents. Initial sources may include your local FBI WMD Coordinator, Terrorism Task Force, or the FBI Hazardous Materials Response Unit (HMRU). Additional sources are available through agencies such as the U.S. Army Soldiers Biological and Chemical Command (SBCCOM), the U.S. Army Medical Research Institute of Infectious Diseases (USAMRID), and the Centers for Disease Control (CDC).

MONITORING INSTRUMENT TECHNOLOGY	HAZARD MONITORED	GENERAL COMMENTS
Photo-Ionization Detector 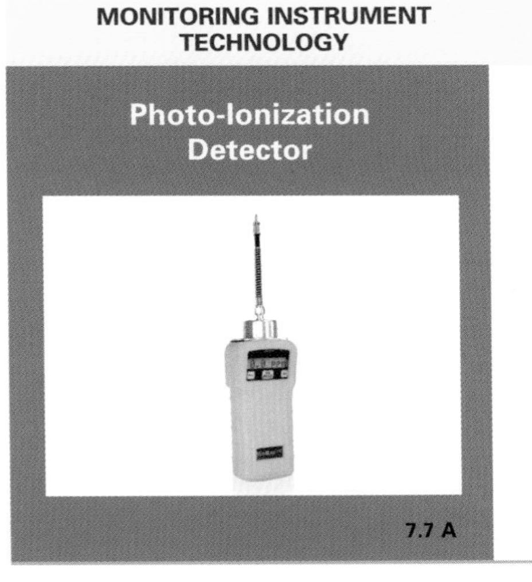 7.7 A	• Toxic Industrial Materials • Chemical—Nerve Agents (tabun, sarin, soman)	• The 10.6 eV lamp will see most chemical agents, but the 11.7 eV lamp is the better to see the nerve agents. Given their low vapor pressures, none of the chemical agents will likely read very high on the PID. • Always use more than one detection technology when using a PID. • Will not tell you WHAT is there; only that something is present. • Available for monitoring point sources or for ambient air monitoring of a given area connected to a host controller unit. • Difficult to calibrate after agent exposure.

MONITORING INSTRUMENT TECHNOLOGY	HAZARD MONITORED	GENERAL COMMENTS
Ion-Mobility Spectrometry (IMS) 7.7 B	• Chemical—Nerve and Blister Agents • Irritants and Riot Control Agents (APD-2000) • Explosives	• May use a radioactive source to ionize contaminants in the air. • Can be used for point or area monitoring. • Extremely sensitive to false positives (e.g., cleaning agents, floor wax). • Examples include: – APD 2000 – Improved Chemical Agent Monitor (I-CAM)
Flame Spectro-Photometry 7.7 C	• Toxic Industrial Materials • Chemical—Nerve Agents and Hydrogen Mustard	• Burns agent with hydrogen gas, which decomposes any chemical agents or TIM's in the sample. Decomposed agent turns into hydrogen phosphorous or sulfur which emit light and then read by optical fibers. • Is generally used in conjunction with gas chromatograph. • Provides a very quick response time, but cannot be used in an explosive environment. • Primarily designed for military battlefield applications. • Examples include: – Miniature CAM Monitor – AP2C Detector
Colorimetric Detector Tubes and Color Change Chemistry 7.7 D	• Many Toxic Industrial Materials • Most Chemical Agents—Nerve, Blister, Blood and Choking Agents • Some Riot Control Agents	• Color changes alerts user to the presence or absence of a suspected agent. Can sample both liquids and vapors. • WMD colorimetric tubes may be used differently than some "regular" detector tubes. Read the manufacturer's instructions! • Examples include: – M8 / M9 chemical agent detector papers. NOTE: can get false positives from pesticides and some petroleum products. – M18A2 chemical agent detector kit – Colorimeteric Detector Kits (Draeger, MSA, etc.) – M256A1 chemical agent detector kit – Hazmat Smart Strips™

MONITORING INSTRUMENT TECHNOLOGY	HAZARD MONITORED	GENERAL COMMENTS
Surface Acoustic Wave (SAW) 7.7 E	• Chemical—Nerve, Blister Blood and Choking Agents • Some Toxic Industrial Materials	• Uses surface acoustic wave technology to identify the agent. Capability to detect more than one agent at the same time. • Can be used for point or area monitoring. • Subject to electro-magnetic radiation interference and false positives from solvents, household chemicals and aftershave. • Examples include: – Hazmat CAD Plus – JCAD Chem Sentry
Infrared Spectrometry (FT-IR) 7.7 F	• Toxic Industrial Materials • Chemical Agents • Explosives • Indicate the Presence of a Biological	• Reading is compared against a library of various hazardous and non-hazardous substances. If the material is in the library, identification is confirmed in <1 minute. • Libraries include hazardous materials, chemical agents, white powders, explosives, clan lab precursors, drugs, and other common liquid and solid materials. • Although they cannot identify biological materials, they can determine if a compound is of biological or non-biological origin. • Examples include: – TravelIR HCI™ – HazMat™ID – Avatar™ ESP
Hand-Held Immunoassays (HHA) 7.7 G	• Biological Agents and Toxins	• Similar to pregnancy tests which tag / identify the antibodies present and provide a color change. Color change is read either visually or through use of an instrument. • May be subject to false-positives and false negatives. • Are not a definitive tool to determine the presence of any bio pathogen or toxin. Should not be used to make decisions on patient management or prophylaxis. However, can be useful in assessing credibility of the threat. • If threat is credible, sample should be sent to an accredited lab for analysis. • Examples include: – RAMP (Rapid Analyte Measurement Platform) – Bio Threat Alert™ Test Strips and Guardian Reader System™ – Smart Tickets

MONITORING INSTRUMENT TECHNOLOGY	HAZARD MONITORED	GENERAL COMMENTS
Polymerase Chain Reaction (PCR) Technology  7.7 H	• Biological Agents	• DNA sample of a suspected bio-agent is put into a reaction and will go through a chain reaction process to amplify a specific DNA sequence. • Cannot differentiate between live and dead organisms; for definitive results, test must be confirmed through lab analysis. • Not widely found in hazmat response operations at the present time. • Example include: – R.A.P.I.D. System™ (Ruggedized Advance Pathogen Identification Device) – Bio Seeq™

Figure 7.7 Monitoring For Terrorism Agents.

Monitoring locations must be identified and described so that subsequent responders will conduct air monitoring at the same location and height. Various systems can be used to identify monitoring and sampling locations, including systematically dividing the incident scene into quadrants (i.e., grid system) or using GPS devices at potential crime scenes.

Monitoring results should be documented as follows:

- Instrument—the type of monitoring instrument being used
- Location—specific location where the monitoring is conducted
- Time—time at which the monitoring is conducted
- Level—level where the monitoring reading is taken (e.g., foot, waist, head)
- Reading—the actual reading given by the monitoring instrument

Monitoring priorities will be dependent on whether responders have identified the hazmat(s) involved. Priorities should be systematic and continuously evaluated throughout the course of the emergency.

Unknowns will create the greatest challenge for responders. The nature of the incident (e.g., credible threat scenario involving WMD agents), the location of the emergency (e.g., outdoors, indoors, confined space), and the suspected physical state of the unknown (i.e., solid, liquid, or gas) will influence the monitoring strategy. In scenarios involving unknowns, the role of hazmat responders is much like that of a detective. At the conclusion of the testing process, responders may still be unable to specifically identify the material(s) involved; however, they should be able to rule out a number of hazard classes and shorten the list of possibilities.

The following monitoring priority is used by many hazmat responders when dealing with scenarios involving unknown substances in an open-air environments. If a corrosive liquid is suspected, responders should use pH paper to determine if a corrosive atmosphere is present, as these vapors may adversely affect some meters. Initial efforts should be toward determining if IDLH concentrations are present and providing an initial base to confirm or refute the existence of specific hazards.

1. *Radiation*—If there is any doubt, radiation detection should be the first priority. Remember—gamma rays travel the greatest distance and are the primary hazard for external exposure. A positive reading twice above background levels would confirm the existence of a radiation hazard. If initial gamma radiation readings are negative but clues are present indicating the possible presence of radioactive materials, additional testing for beta and alpha sources should be conducted.
2. *Flammability*—Since flammability and oxygen levels are directly related, monitoring for flammability and oxygen is usually implemented simultaneously through combination CGI/O_2 meters. Remember that an oxygen-deficient atmosphere will shorten the flammable range, while an oxygen-enriched atmosphere will expand the flammable range.
3. *Oxygen*—Monitoring should evaluate the presence of both oxygen-deficient and -enriched atmospheres, particularly when dealing with confined spaces. Responders must consider that oxygen levels may also be influenced by the level of contaminants (i.e., an increasing level of contaminants displacing available oxygen).

4. *Toxicity*—The level and sophistication of toxicity monitoring will depend upon available instrumentation. Resources range from the simple to the sophisticated and may include the following:
   - Indicator papers, such as pH paper, and M-8 / M-9 tape for chemical agents. Chemical Classifier™ kits can test liquid spills for corrosivity, oxidizers, fluoride, organic solvents/petroleum distillates, and iodine/bromine/chlorine, while the HazMat Smart-Strip™ can test liquids and aerosols for chlorine, pH, sulfides, oxidizer, fluorides, nerve agents, cyanide, and arsenic.
   - Specific or combination air monitors, which detect toxic gases such as hydrogen sulfide or carbon monoxide. Carbon monoxide monitors are useful in fire and post-fire situations. Hydrogen sulfide monitors are useful when dealing with confined spaces and when working around petroleum facilities where "sour" gas is handled.
   - Colorimetric detector tubes can be used for both known and unknown substances. Commercial detector tube kits are available for identifying unknown airborne hazards. By conducting a series of measurements with various pre-established detector tubes, responders can often determine the chemical class (e.g., organic gases, alcohols, acidic gases) that is present. Detector tube kits are also available for determining the presence of chemical agents. These kits only indicate the presence of several classes of gases and vapors; they provide a gross estimate of concentration and may not differentiate between specific gases within a certain class.
   - Survey instruments, such as flame ionization detectors (FIDs) and photoionization detectors (PIDs).

If the incident involves a confined space scenario, *OSHA 1910.126—Permit-Required Confined Space Standard*, clearly outlines the required monitoring priority as follows:

- Oxygen deficiency and enrichment
- Flammability
- Toxicity

From a practical perspective, the use of multi-sensor instruments (e.g., three gas and four gas sensors that allow one instrument to provide monitoring for all three hazards) can make this initial monitoring priority discussion a moot point.

EVALUATING MONITORING RESULTS—ACTIONS LEVELS AND GUIDELINES

Initial air monitoring efforts should be directed toward determining if IDLH concentrations are present. Decisions regarding protective clothing recommendations, establishing hazard control zones, and evaluating any related public protective actions should be based on the following parameters:

1. *Radioactivity*—Any positive reading twice above background levels or alpha and/or beta particles that are 200 to 300 counts per minute (cpm) above background would confirm the existence of a radiation hazard and should be used as the basis for initial actions.
2. *Flammability*—The IDLH action level is 10% of the lower explosive limit (LEL).
3. *Oxygen*—An IDLH oxygen-deficient atmosphere is 19.5% oxygen or lower, while an oxygen-enriched atmosphere contains 23.5% oxygen or higher. In eval-

The overall objective of emergency responders at any emergency is to favorably change or influence the outcome. Direct outcomes are typically stated as fatalities, injuries, property and environmental damage. Indirect outcomes include systems disruptions (e.g., water, transportation, utility), damaged reputations, and residual fears.

To determine whether or not to intervene, responders must first estimate the likely harm that will occur without intervention. Simply, what will happen if you do nothing? This requires you to (1) visualize the likely behavior of the hazardous material and/or its container, along with the likely harm associated with that behavior; and (2) describe the outcome of that behavior.

To visualize likely behavior, five basic questions must be addressed:

1. Where will the hazardous material and/or its container go when released?
2. How will the hazardous material and/or its container get there?
3. Why are the hazardous material and/or its container likely to go there?
4. What harm will the hazardous material and/or its container do when it gets there?
5. When will the hazardous material and/or its container get there?

When answering these questions, recognize and understand the factors that will affect hazmat behavior, including the following:

- Inherent properties and quantities of the materials involved (i.e., toxicity, flammability, reactivity, etc.)
- Built-in design and construction features of the container (thermal insulation, pressure relief devices, fixed water spray systems, etc.)
- Natural laws of physics and chemistry, as these will influence dispersion patterns and where the product will go once it is released from its container
- Pertinent environmental factors (i.e., terrain, weather and atmospheric conditions, wind direction and speed, and the physical surroundings [i.e., rural location vs. downtown]).

BEHAVIOR OF HAZMATS AND CONTAINERS

All hazmat releases will follow a logical sequence of events, regardless of the hazard class involved. The concept of events analysis is a useful tool to visualize hazmat behavior and estimate what is likely to occur. Events analysis is defined as the process of breaking down complex actions into smaller, more easily understood parts. It helps responders to (1) understand, track, and predict a given sequence of events; and (2) decide when and how to change that sequence.

An easy way to visualize hazmat behavior is by using the General Hazardous Materials Behavior Model or GHBMO, pronounced "gebmo." Originally developed by Ludwig Benner of the National Transportation Safety Board (NTSB) and published in 1978, the GHBMO is an excellent tool for understanding and predicting the behavior of the container and its contents at a hazmat incident. See Figure 7.9.

GENERAL HAZARDOUS MATERIALS BEHAVIOR MODEL©					
EVENT					
Stress	**Breach**	**Release**	**Engulf**	**Impinge**	**Harm**
EVENT CATEGORIES					
Thermal	Disintegration	Detonation	Cloud	Short term	Thermal
Radiation	Runaway cracking	Violent rupture	Plume	Medium term	Radiation reactive
	Attachments opening	Rapid relief	Cone	Long term	Asphyxiation
Chemical	Punctures	Leak	Stream		Chemical
Mechanical	Splits or Tears	Spill	Irregular deposit		Etiologic Mechanical
EVENT INTERRUPTION PRINCIPLES					
Influence Applied Stresses	**Influence Breach Size**	**Influence Quantity Released**	**Influence Size of Danger Zone**	**Influence Exposure Impinged**	**Influence Severity of Injury**
Redirect impingement	Chill contents	Change container position	Initiate controlled ignition	Provide shielding	Rinse off contamination
Shield stressed system	Limit stress level	Minimize pressure differential	Erect dikes or dams	Begin evacuation	Increase distance from source
Move stressed system	Activate venting devices	Cap off breach	Dilute		Provide shielding

Source:
Ludwig Benner, Jr.

Figure 7.9 The General Hazardous Materials Behavior Model (GHBMO) provides a framework for visualizing and predicting hazardous materials behavior.

STRESS EVENT

Under normal conditions, hazardous materials are controlled within some type of container or containment system. Containment systems can range from nonbulk containers such as bags, bottles and drums, to bulk containers, including cargo tank trucks, pressurized storage vessels, and chemical reactors. For an emergency to occur, either the container or its contents must first be disturbed or stressed in some fashion.

Stress is defined as an applied force or system of forces that tend to either strain or deform a container (external action) or trigger a change in the condition of the contents (internal action). It is important to recognize that this stress can affect the container and/or its contents.

Three types of stress—thermal, mechanical, and chemical—are common. Less likely, though still possible, are etiological and radiation stresses. These stressors may be present alone or in combination with each other.

- *Thermal Stress*—Generally associated with hot or cold temperatures and their effects upon the container or its contents. Examples include fire, sparks, friction

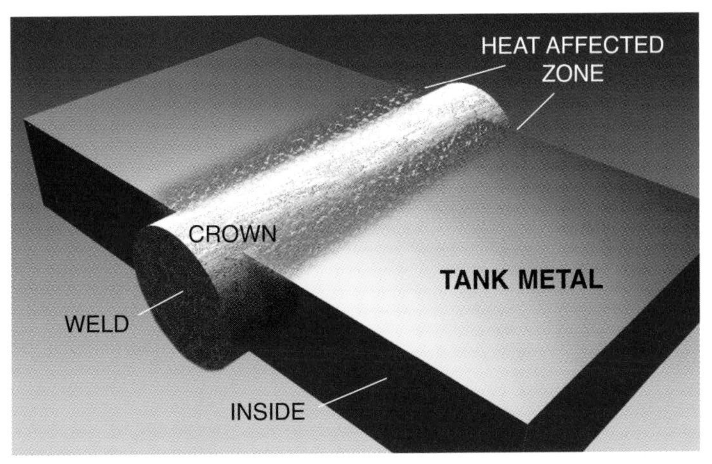

Figure 7.13 The heat-affected zone is the tank metal next to a weld. It may be vulnerable to container stress as cracks are likely to start there.

MECHANICAL STRESS AND DAMAGE

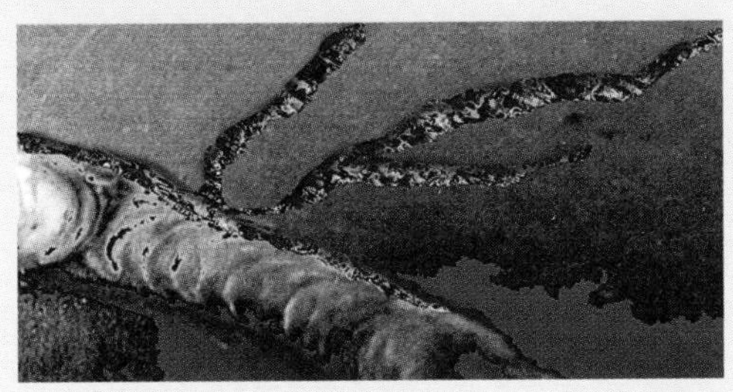

7.14 A

Crack—Narrow split or break in the container metal that may penetrate through the container metal (may also be caused by fatigue). It is a major mechanism that could cause catastrophic failure.

7.14 B

Score—Reduction in the thickness of the container shell. It is an indentation in the shell made by a relatively blunt object. A score is characterized by the reduction of the container or weld material so that the metal is pushed aside along the track of contact with the blunt object. Scores caused by prolonged contact with a tank car wheel are called "wheel burns."

7.14 C

Gouge—Reduction in the thickness of the tank shell. It is an indentation in the shell made by a sharp, chisel-like object. A gouge is characterized by the cutting and complete removal of the container or weld material along the track of contact.

7.14 D

Wheel burn—Reduction in the thickness of a railroad tank shell. It is similar to a score but is caused by prolonged contact with a turning railcar wheel.

7.14 E

Dent—A deformation of the tank head or shell. It is caused from impact with a relatively blunt object (e.g., railroad coupler, vehicle, concrete abutment). The sharper the radius of curvature of the dent, the greater the chance of cracking.

7.14 F

Rail burn—A deformation in the shell of a railroad tank car. It is actually a long dent with a gouge at the bottom of the inward dent. A rail burn can be oriented circumferentially or longitudinally in relation to the tank shell. The longitudinal rail burns are the more serious because they have a tendency to cross a weld. A rail burn is generally caused by the tank car passing over a stationary object, such as a wheel flange or rail.

7.14 G

Street burn—A deformation in the shell of a highway cargo tank. It is actually a long dent that is inherently flat. A street burn is generally caused by a container overturning and sliding some distance along a cement or asphalt road.

Figure 7.14 Examples of mechanical stress and damage to pressurized containers.

Guidelines for damage assessment of pressurized containers include the following:

1. Gather information concerning the type of container (e.g., DOT specification number), material of construction (e.g., aluminum, steel), and internal pressure. Methods for determining the internal pressure include
   - Using pressure gauges, attached to sample lines, gauging device, fittings, and so on.

- Use of temperature gauges with vapor pressure/temperature conversion charts. However, tank contents may stratify into layers having different temperatures due to external temperature changes; as a result, the pressure estimated from product temperature readings may be inaccurate (i.e., may be lower than the actual pressure).
- Using ambient temperature, recognizing that the temperature of the tank's contents may lag ambient temperatures up to 6 hours.

  The internal pressure in empty railroad tank cars that contain residual vapors may be equal to that in loaded cars (or greater than that in loaded tank cars if some inert gas is used for unloading). Vapor pressure/temperature graphs are available from the *Compressed Gas Handbook*, as well as the shipper or manufacturer of the material.

2. Determine the amount of material in the container.
3. Determine the type of stress applied to the container (e.g., thermal, mechanical, or combination).
4. Evaluate the stability of the container. Take caution when inspecting an unstable container, as it may move or shift during the inspection process. It may be necessary to stabilize the container with blocks, cribbing, or other means.
5. Examine all accessible surfaces of the container, paying attention to the types of damage and the radius (i.e., sharpness) of all dents. Railroad personnel will often use a tank car dent gauge as a "go/no go" device for comparing the radius of curvature of a tank car dent to accepted standards to determine the severity of damage. However, the tank car dent gauge cannot be used for assessing dents on cargo tank trucks due to differences in shell metal and thickness.

Experience shows that the most dangerous situations will include the following:

- Cracks in the base metal of a tank or cracks in conjunction with a dent, score, or gouge. Both of these situations justify offloading or reducing container pressure as soon as safely possible.
- Sharply curved dents or abrupt dents in the cylindrical shell section that are parallel to the long axis of the container. If a dent is considered critical, the container should be first offloaded or pressure reduced before moving it.
- Dents accompanied with scores and gouges.
- Scores and gouges across a container's seam weld or in the heat affected zone of the weld. If the score crosses a welded seam and removes no more than the weld reinforcement (i.e., that part of the weld which sticks above the base metal), the stress is considered noncritical. However, if the score removes base metal at the welded seam, the stress is considered critical.

6. If you are unsure of the container damage or how the container is likely to breach, get assistance from product or container specialists. This may include railroad personnel, gas industry representatives, and cargo tank truck specialists.

MOVEMENT AND BEHAVIOR OF HAZMATS UNDERGROUND

When petroleum products or chemicals are released into the ground, their behavior will depend on their physical and chemical properties (e.g., liquid versus gas, hydrocarbon versus polar solvent), the type of soil (e.g., clay versus gravel versus sand), and the underground water conditions (e.g., location and movement of the water table). While such incidents may involve any hazmat class, flammable liquids, gases, and vapors are the most common.

As with hazmat containers and their behavior, responders should have a basic understanding of geology, groundwater, and groundwater movement to evaluate the underground dispersion of hazmat releases and potential exposures and to determine potential outcomes. However, the ultimate clean-up and remediation of underground spills and releases is not the responsibility of emergency responders.

GEOLOGY AND GROUNDWATER

Soil consists of loose, unconsolidated surface materials, such as sand, gravel, silt, and clay. Bedrock is the hard, consolidated material that lies under the soil, such as sandstone, limestone, or shale. Most areas have a soil cover ranging from a few feet to several hundred feet.

Generally, rocks and soils consist of small fragments or sand grains. When compressed together, they may form small voids or pores. Measurement of the total volume of these voids is called the porosity of the rock or soil. If these pores are interconnected, the rock or soil is permeable (i.e., fluid can pass through it). The size

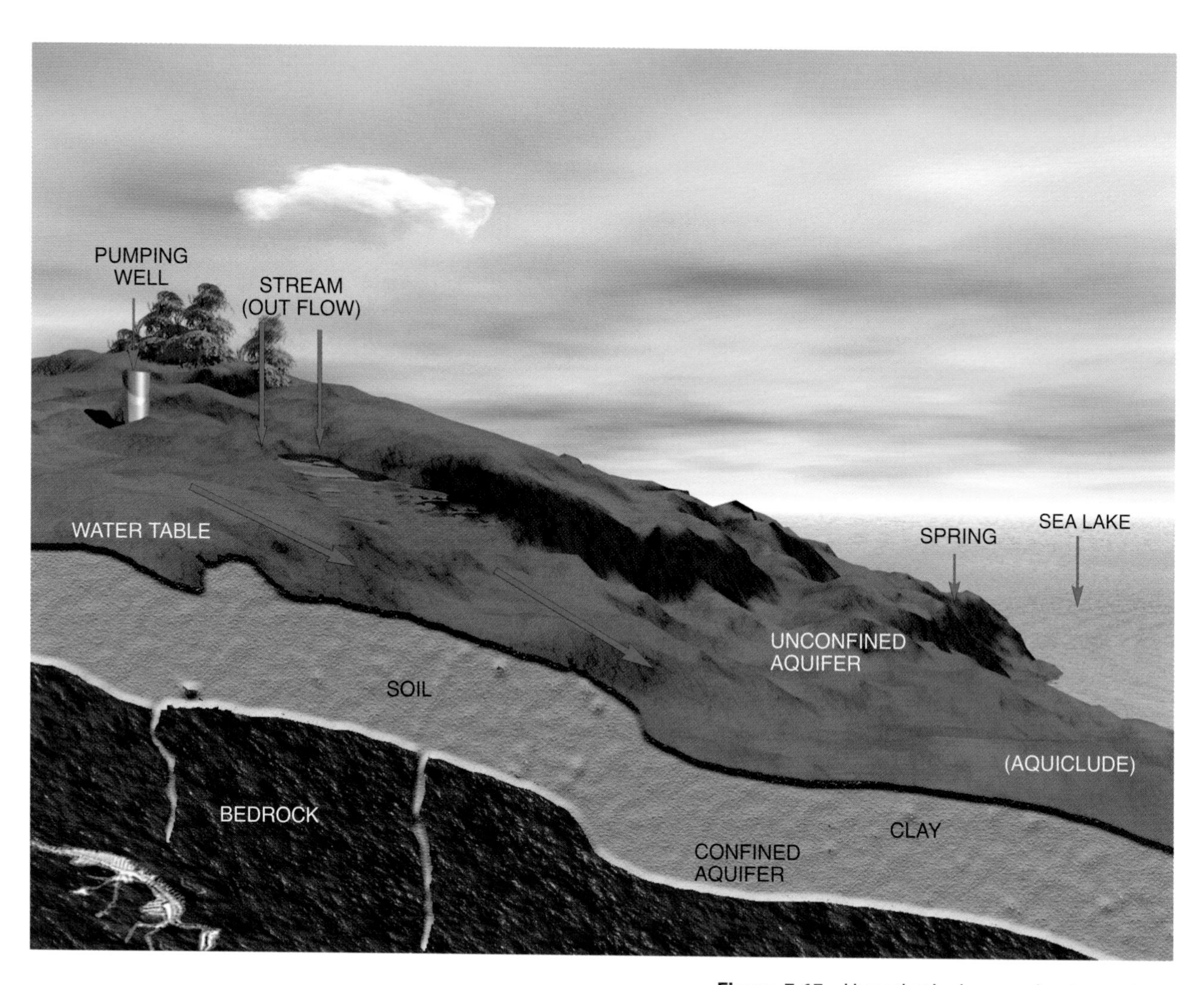

Figure 7.15 Hypothetical groundwater system.

of these voids will vary from large (e.g., gravel) to small (e.g., sand and topsoil) to essentially zero (e.g., dense clay). Rock almost never has large voids, but sandstone and limestone have voids that are similar to a fine sand. Aquifers are permeable sections of soil or rock capable of transmitting water. In contrast, silt and shale have many, but extremely small, pores that are poorly interconnected. Since fluids cannot readily pass through such materials, they are known as impermeable materials.

In most areas, water exists at some depth in the ground. The source of most groundwater is precipitation over land, which percolates into porous soils and rocks at the surface. Rivers and streams that seep water into the subsurface are a second source of groundwater. Groundwater accounts for the majority of the drinking water supply for the United States.

In most areas, groundwater moves extremely slowly. The rate of flow depends on the permeability of the underground aquifer and the slope or "hydraulic gradient" of the water table. Flows can range from 6 feet per year in fine clays to 6 feet per day in gravels. In addition, the location and relative production rates of groundwater wells can significantly disrupt normal flow patterns. Figure 7.15 shows a hypothetical groundwater system.

BEHAVIOR OF HAZMATS IN SOIL AND GROUNDWATER

Hazardous materials may be absorbed into the soil through either surface spills or leaks from underground pipelines or storage tanks. Flammable liquids and toxic solvents can create significant problems if they migrate into confined areas or are allowed to flow into waterways.

Flammable and toxic gases, such as natural gas, propane, or hydrogen sulfide, can also accumulate in underground pockets or confined areas. These can occur naturally or may be released from a storage tank or pipeline failure. When dealing with natural gas, recognize that some soils (e.g., clay) can "scrub" or remove the methyl mercaptan odorant commonly used on natural gas distribution lines, thereby removing any physical clues of smell or odor.

The underground movement of hazmats follows the most permeable, least resistant path. For example, the backfill in trenches carrying utility conduits, sewers, or other piping is often much more permeable than the undisturbed native soil. In urban areas, this can facilitate the rapid and easy movement of liquids and gases to nearby basements, sewers, or other below-grade structures. Identifying these conduits is critical in identifying potential exposures. Utilities and public works agencies, including regional "Miss Utility" and "One Call Systems," can often provide assistance in locating these underground conduits. Responders should also seek technical information specialists, such as geologists, who have either maps or knowledge of the local water tables, soil types and densities, underground rock formations, and so forth.

Liquid hazmats which are spilled into soil will tend to flow downward with some lateral spreading (see Figure 7.16). The rate of hazmat movement in soil will depend upon the viscosity of the liquid, soil properties, and the rate of release. For example, light hydrocarbon products, such as gasoline, will penetrate rapidly while heavier oils, such as #4 fuel oil, will move more slowly. If the soil near the surface has a high clay content and very low permeability, the hazmat may penetrate very little or not at all. However, a porous, sandy soil may quickly absorb the product. Eventually, the downward movement of the hazmat through soil will be interrupted by one of three events: (1) it will be absorbed by the soil; (2) it will encounter an impermeable bed; or (3) it will reach the water table.

Figure 7.16 The movement of hazardous materials through soil will be dependent on the viscosity of the liquid, properties of the soil, and the rate of the release.

Hazmats that are absorbed by the soil may move again at some later time as the water table is elevated. For example, responders will often receive a report of hydrocarbon or gasoline vapors in an area, with the source of the odor being unknown. In other situations, recent rains may cause the water table to rise and bring hydrocarbon liquid and vapors to the surface. With these scenarios, the suspected area should be divided into quadrants and monitoring readings "mapped" in a systematic manner to assist emergency responders in identifying the location of the problem and its direction of movement.

Although combustible gas indicators (CGIs) are excellent tools for evaluating flammable atmospheres, they may not be very effective for assessing low-level flammable concentrations such as found with subsurface and sewer spills. To register a positive reading, many CGIs require a concentration of up to 10,000 ppm (1% in air). While this may not be a flammable concentration, it often represents a significant environmental problem. Photoionization detectors (PIDs) are better suited for these scenarios.

Hazmats that encounter an impermeable layer will spread laterally until becoming immobile or until the hazmat comes to the surface where the impermeable layer outcrops. If the hazmat comes in contact with the water table, there is a high potential for the water supply to become contaminated and for the hazmat to move and accumulate in an underground structure (e.g., basement, sewer system). Remember that groundwater supplies can become contaminated by concentrations as small as 200 parts per billion (ppb). Preventing spills and releases from entering the soil is a critical element in many areas. While responders can do little to influence the underground movement of the material once it has entered the soil, underground aquifers and exposures must still be identified and protected.

Hydrocarbon liquids will not mix with water and will simply float on the surface of the water table. Many oils and refined petroleum products, however, contain certain components that are slightly soluble in water. Gasoline is high in water-soluble components which, when dissolved in water, produce odors and taste that can be detected at levels of only a few parts per million (e.g., ethanol, methanol, methyl tertiary butyl ether [MTBE]).

MOVEMENT AND BEHAVIOR OF SPILLS INTO SEWER COLLECTION SYSTEMS

Sewers, manholes, electrical vaults, french drains, and other similar underground structures and conduits can be critical exposures in the event of a hazmat spill. If the hazmat release penetrates the substructure walls or enters through surface sewers and manholes, significant quantities of liquid can be expected to flow into the sewer collection system. Depending on the hazmat involved, this situation can pose an immediate fire problem, as well as significant environmental concerns regardless of whether ignition occurs or not.

Most sewer emergencies involve flammable and combustible liquids. The probability of an explosion within an underground space will depend on two factors: (1) that a flammable atmosphere exists, and (2) that an ignition source is present. The severity of an explosion and its consequences will depend on the type of sewer collection system, the process and speed at which the hazmat moves through the system, and the ability of emergency responders to confine the release and implement fire and spill control procedures.

TYPES OF SEWER SYSTEMS

Sewer systems can be categorized based on their application:

- *Sanitary sewers.* This is a "closed" system that carries liquids and water-carried wastes from residences, commercial buildings, industrial plants and institutions, as well as minor quantities of storm water, surface water, and groundwater that are not admitted intentionally. The collection and pumping system will transport the wastewater to a treatment plant, where various liquid and solid treatment systems are employed to process the wastewater. Sewer diameters from 8 inches to 60 inches are common.
- *Storm sewers.* This is an "open" system that collects storm water, surface water, and street wash and other drainage from throughout a community but excludes domestic wastewater and industrial wastes. A storm sewer system may dump runoff directly into a retention area that is normally dry or into a stream, river, or waterway without treatment. However, large manufacturing facilities that use petroleum or chemicals in their process are required by EPA regulations to collect and treat all on-site surface runoff before it can be discharged. Storm sewers are generally much larger than sanitary sewers, with diameters ranging from 2-foot pipes to greater than 20-foot tunnels. Storm sewers are sometimes used by "illegal dumpers" as a means of hazardous waste disposal.
- *Combined sewers.* Carries domestic and industrial wastewater, as well as storm or surface water. Although separate sanitary sewers are being constructed today, combined sewers may be found in older cities and metro areas. Combined sewers are often very large and can be as much as 20 feet in diameter. Combined sewers may also have regulators or diversion structures that allow any sewer overflow to be discharged directly to rivers or streams during major storm events.

WASTEWATER SYSTEM OPERATIONS

There are four primary elements of a wastewater system: (1) collection and pumping, (2) filtering systems, (3) liquid treatment systems, and (4) solid treatment systems. The highest risks of a fire or explosion are associated with collection and pumping operations and with the early stages of liquids and solid processing. Similarly, the greatest potential for either environmental damage or shutdown of a wastewater treatment plant operation will take place at the liquid and solid stream treatment processes.

Wastewater, storm water, and surface water initially enter the collection and pumping system through a series of collectors and branch lines that tie together small geographic areas. These collectors and branch lines are eventually tied into a trunk sewer (also known as main sewer), which then carries the wastewater to its final destination for either treatment or disposal.

Where the terrain is flat, the collection system may consist solely of gravity piping. However, in most areas the collection system will require pumping or lift stations. Most pumping stations will have two parts—a wet well and a dry well. The wet well receives and temporarily stores the wastewater. Wet wells often contain electrical equipment such as fans, pumps, motors, and other accessories. In some instances, proper management of the wet well may provide an opportunity for the collection and removal of a flammable liquid. The dry well provides isolation and shelter for the controls and equipment associated with pumping the wastewater. They are designed to completely exclude wastewater and wastewater-derived atmospheres, although there may be accidental leakage from pumpshafts or occasional spills.

Depending on the type of sewer system and the specific location, most areas are classified by the National Electrical Code as Class I, Division 2 areas. However, pumping stations should be viewed as potential ignition sources when hydrocarbons are released into the sewer collection system. Pumping stations are sometimes equipped with hydrogen sulfide or fixed combustible gas detectors to detect the presence of flammable vapors and gases.

PRIMARY HAZARDS AND CONCERNS

There are two basic scenarios involving releases into a sewer collection system. The first scenario is an aboveground release where a spill flows into the sewer collection system through catch basins, manholes, and so on. The second scenario involves underground tank and pipeline leaks where the product migrates through the subsurface structure into the sewer collection system. This type of emergency is usually not obvious from the surface and presents a greater challenge in identifying the source of the problem and controlling the release within the sewer system.

With the subsurface scenario, responders will often receive a report of hydrocarbon or gasoline vapors in an area, with the source of the odor being unknown. In other situations, recent rains may cause the water table to rise and bring hydrocarbon liquid and vapors to the surface. Monitoring readings should be "mapped," as they can assist emergency responders in identifying the location of the problem and its direction of movement.

Some rules of thumb for evaluating monitoring readings are as follows:

1. If readings are high and then drop off or dissipate in a relatively short period of time, the source of the problem is often a spill or dumping directly into the sewer collection system.

2. If readings are consistent over a period of time, the source of the problem is often a subsurface release, such as an underground storage tank or pipeline.

Spills and releases into the sewer collection system will create both fire and environmental concerns.

FIRE CONCERNS

Flammable liquids, such as gasoline and other low flash point, high vapor pressure liquids, will create the greatest risk of a fire or explosion. The potential for ignition within a sewer collection system will be greatest at points where liquids may enter or where entry is possible. Manholes, storm sewers, catch basins, and pumping station wet wells are likely to present the greatest areas of concern.

If a flammable liquid enters a sanitary or combined sewer, the probability of ignition will be high. If floor traps are not filled with water, flammable liquid in either a sanitary or combined sewer collection system will back up vapors into building basements and other low-lying areas where there are multiple ignition sources, such as pilot lights, hot water heaters, electrical equipment, sparking and arcing, and so on. Ignition sources should be isolated and controlled and the area ventilated using positive or negative ventilation tactics, as appropriate.

When a flammable liquid enters a storm sewer collection system, manholes, catch basins, and pumping station wet wells are likely points of ignition. In some instances, vapors have flowed out of the sewer collection system and accumulated in low lying areas, only to be ignited by an ignition source completely outside of the sewer system (e.g., passing vehicle).

In the event of a fire or explosion, secondary and tertiary problems will likely be created by the emergency as the explosion will affect all utilities that occupy the same or nearby utility corridors. Natural gas leaks, electric and telephone utility outages, and a loss of both water and water pressure should be anticipated in areas which suffer a major fire or explosion.

ENVIRONMENTAL CONCERNS

Environmental concerns will be greatest when dealing with poisons and environmentally sensitive materials or when there is no ignition of flammable liquids following their release into the sewer system. Depending on the type of sewer collection and treatment system, environmental impacts may range from a shutdown of the wastewater treatment facility and/or destruction of microorganisms necessary for the treatment process, to a spill impacting environmentally sensitive areas (e.g., wetlands, wildlife refuge, etc.) or threatening both potable and aquifer drinking water supplies.

The selection of control agents, such as dispersants and firefighting foams, to control a fire or spill may also have secondary environmental impacts. Both sewer department and environmental personnel from the respective on-scene governmental agencies should be consulted on any decision to apply firefighting foams or dispersants either into a sewer collection system or onto a waterway.

COORDINATION WITH SEWER DEPARTMENT

Preplanning with the sewer department is critical. Responders should have a basic knowledge and understanding of the sewer system and its operations and a good working relationship with sewer department personnel. Responders should identify areas where there is a probability of hazmats entering the sewer collection system and discuss procedures and tactical options for handling such an emergency with the sewer department as part of its preincident planning and training activities.

When an emergency occurs, a sewer department representative should be requested on scene as soon as possible. The evaluation and selection of tactical control options should be based upon input from the sewer department and the respective governmental environmental agency. Maps of the local sewer system will be a key element in identifying the direction in which a spill may potentially head and in identifying likely exposures. Effective use of sewer maps will require a sewer department representative who is familiar with the unique aspects of the local system, local construction techniques, and so on. However, sewer maps may not always be accurate and up to date, particularly in identifying all lateral/domestic connections and branch connections on older combined sewer systems.

SITE SAFETY PROCEDURES FOR HYDROCARBON SPILLS INTO SEWER COLLECTION SYSTEMS

- Verify that the sewer department has been notified and enroute. Identifying the direction of flow is critical in identifying exposures and establishing evacuation zones.
- Continuous air monitoring must be provided, particularly in low-lying areas. Responders may have difficulty obtaining a sufficient number of monitoring instruments and trained personnel to perform monitoring over a relatively large geographic area.
- Monitoring readings should be mapped as this will assist responders in identifying the location of the spill within the sewer system and its general speed and direction of movement. This information, in turn, will assist in establishing response priorities.
- Control all ignition sources in the area, including vehicles, traffic flares, and smoking materials. Depending on the nature of the emergency, large spills into a sanitary or combined sewer collection system may require the shutdown of gas and electric utilities until the situation is under control.
- If liquids, gases, or vapors are found in tunnels or subways, traffic should be stopped until responders can further investigate and assess the level of risk.
- If the spill is in a sanitary or combined sewer collection system and its speed and direction of movement is known, responders may be able to notify homeowners and facilities ahead of the spill to pour water into their basement floor traps as a quick preventive measure to minimize hydrocarbon vapor migration and build-up.
- Do not allow any personnel to stand on or near manholes and catch basins. In the event of a fire or explosion, manhole lids can be blown into the air and fire can quickly emanate from catch basins and other sewer openings. Manhole lids can be blown into adjoining buildings and vehicles and represent a significant life safety hazard.

- Responders should not enter a sewer collection system unless advised by representatives of the sewer system. In addition to the obvious fire and health hazards, sewer collection systems consist of piping and collection areas of various diameters and depths and pose significant physical hazards.
- There are many confined spaces within a wastewater collection and pumping system. Responders should also consider the presence of oxygen deficient atmospheres and other toxic and flammable gases, including hydrogen sulfide, methane, and sewer and sludge gases.
- When either flushing or applying control agents into a sewer collection system, the agent must be applied at multiple points along the projected flow path. If control agents are only applied at the source of the release, the agent will never "catch up" with the head of the flow, and there will be a continuous flammable atmosphere within the sewer system.
- The injection of some control agents into a sewer collection system, such as firefighting foams, dispersants, and water, may also introduce air and possibly move the environment into the flammable range.

Figure 7.17

SUMMARY

Hazard and risk assessment is the most critical function in the successful management of a hazardous materials incident. The key tasks in this analytical process are (1) identifying the materials involved; (2) gathering hazard information; (3) visualizing hazmat behavior and predicting outcomes; and (4) based on the evaluation process, establishing response objectives. The system that ties these elements together is the General Hazardous Materials Behavior Model.

An accurate evaluation of the real and potential problems will enable response personnel to develop accurate and informed strategical response objectives and tactical decisions. Remember—your job is to be a risk evaluator, not a risk taker. Bad risk takers get buried; effective risk evaluators come home.

REFERENCES AND SUGGESTED READINGS

Air Force Civil Engineer Support Agency (AFCESA) and PowerTrain, Inc, HAZARDOUS MATERIALS INCIDENT COMMANDER EMERGENCY RESPONSE TRAINING CD-ROM, Tyndall Air Force Base, FL: AFCESA (2002).

Air Force Civil Engineer Support Agency (AFCESA) and PowerTrain, Inc, HAZARDOUS MATERIALS TECHNICIAN EMERGENCY RESPONSE TRAINING CD-ROM, Tyndall Air Force Base, FL: AFCESA (1999).

American Petroleum Institute. API 1628—GUIDE TO THE ASSESSMENT AND REMEDIATION OF UNDERGROUND PETROLEUM RELEASES (3rd Edition), Washington, DC: American Petroleum Institute (1996).

Andrews, Robert C., Jr., "The Environmental Impact of Firefighting Foam." INDUSTRIAL FIRE SAFETY (November/December, 1992), pages 26–31.

Armed Forces Radiobiology Research Institute,—Military Medical Operations, MEDICAL MANAGEMENT OF RADIOGICAL CASUALTIES HANDBOOK (2nd Edition), Bethesda, MD: AFRRI (April, 2003).

Bachman, Eric G. "Preplanning for Emergencies at Water-Treatment Facilities." FIRE ENGINEERING (August, 2003), pages 120–130.

Berger, M., W. Byrd, C. M. West, and R. C. Ricks, TRANSPORT OF RADIOACTIVE MATERIALS: Q & A ABOUT INCIDENT RESPONSE, Oak Ridge, TN: Oak Ridge Associated Universities (1992).

Bevelacqua, Armando, HAZARDOUS MATERIALS CHEMISTRY. Albany, NY: Delmar—Thomson Learning (2001).

Bevelacqua, Armando and Richard Stilp, TERRORISM HANDBOOK FOR OPERATIONAL RESPONDERS, Albany, NY: Delmar – Thomson Learning (1998).

Brunacini, Alan V., FIRE COMMAND (2nd Edition), Quincy, MA: National Fire Protection Association (2002).

Docimo, Frank, "METERS – Monitoring the Environment to Ensure Responders." Student Handout—Stamford, CT: Docimo and Associates (2003).

Emergency Film Group, Air Monitoring (two videotape series), Plymouth, MA: Emergency Film Group (2003).

Emergency Film Group, Detecting Weapons of Mass Destruction, (video-tape), Plymouth, MA: Emergency Film Group (2003).

Fender, David L. "Controlling Risk Taking Among Firefighters." PROFESSIONAL SAFETY (July, 2003), pages 14–18.

Fingas, Merv F. et. al. "The Behavior of Dispersed and Nondispersed Fuels in a Sewer System." AMERICAN SOCIETY OF TESTING AND MATERIALS—SPECIAL TECHNICAL PUBLICATION 1018 (1989).

Fingas, Merv F. et. al., "Fuels in Sewers: Behavior and Countermeasures." JOURNAL OF HAZARDOUS MATERIALS, 19 (1988), pages 289–302.

Ghormely, David M. "Emergency Response to Polymerizable Materials." Student Handout —Houston, TX: Rohm & Haas Company, Inc. (2003).

Hawley, Chris. HAZARDOUS MATERIALS AIR MONITORING AND DETECTION INSTRUMENTS. Albany, NY: Delmar – Thomson Learning (2002).

International Association of Fire Fighters, TRAINING FOR HAZARDOUS MATERIALS RESPONSE: TECHNICIAN, Washington, DC: IAFF (2002).

Larrañaga, Michael D., David L. Volz and Fred N. Bolton, "Pressure Effects on and Deformation of Hazardous Waste Containers." FIRE ENGINEERING (July, 1999).

Maslansky, Carol J. and Steven P. Maslansky, AIR MONITORING INSTRUMENTATION, New York, NY: Van Nostrand Reinhold (1993).

National Fire Protection Association, FIRE PROTECTION HANDBOOK (19th Edition), Section 7 – Managing Response to Hazardous Materials Incidents, Quincy, MA: National Fire Protection Association (2003).

National Fire Protection Association, HAZARDOUS MATERIALS RESPONSE HANDBOOK (4th Edition), Quincy, MA: National Fire Protection Association (2003).

National Fire Protection Association, NATIONAL ELECTRICAL CODE HANDBOOK, Quincy, MA: National Fire Protection Association (2002).

National Fire Protection Association, RECOMMENDED PRACTICE FOR HANDLING RELEASES OF FLAMMABLE AND COMBUSTIBLE LIQUIDS AND GASES—NFPA 329, Quincy, MA: National Fire Protection Association (1999).

National Fire Protection Association, STANDARD FOR FIRE PROTECTION IN WASTEWATER TREATMENT AND COLLECTION FACILITIES—NFPA 820, Boston, MA: National Fire Protection Association (2003).

National Transportation Safety Board, "Derailment of a CSX Transportation Freight Train and Fire Involving Butane in Akron, OH." (Report NTSB/HZM-90/2). Washington, DC: National Transportation Safety Board (February 26, 1989).

Office of Domestic Preparedness—Domestic Preparedness Equipment Technical Assistance Program (DPETAP), ADVANCED RADIOLOGICAL SURVEY TECHNIQUES COURSE, Pine Bluff, AR: DPETAP (2003).

Plog, Barbara A. and Patricia J. Quinlan, FUNDAMENTALS OF INDUSTRIAL HYGIENE (5th Edition), Chicago, IL: National Safety Council (2002).

RAE Systems, PROFESSIONAL GAS DETECTION TRAINER'S PACKAGE, Sunnyvale, CA: Rae Systems (2003).

Sidell, Frederick, M.D., William Patrick and Thomas Dashiell, JANE'S CHEM-BIO HANDBOOK (2nd Edition), Alexandria, VA: Jane's Information Group (2003).

Union Pacific Railroad Company, "Assessing Tank Car Damage," Participant's Manual—Tank Car Safety Course. Omaha, NE: Union Pacific Railroad Company (April, 2003).

U.S. Army Medical Research Institute of Chemical Defense, MEDICAL MANAGEMENT OF CHEMICAL CASUALTIES (3rd Edition), Aberdeen Proving Ground, MD: USAMIC—Chemical Casualty Care Office (July, 2000).

U.S. Army Medical Research Institute of Infectious Diseases (USAMRID), MEDICAL MANAGEMENT OF BIOLOGICAL CASUALTIES (4th Edition), Fort Detrick, MD: USAMRID (February, 2001).

Water Pollution Control Federation, EMERGENCY PLANNING FOR MUNICIPAL WASTEWATER FACILITIES (MOP SM-8), Arlington, VA: Water Pollution Control Federation (1989).

Wright, Charles, "Predicting Behavior and Estimating Outcomes." Student Handout—Omaha, NE: Union Pacific Railroad Company (2003).

York, Kenneth J. and Gerald L. Grey., HAZARDOUS MATERIALS/WASTE HANDLING FOR THE EMERGENCY RESPONDER, Tulsa, OK: Fire Engineering Books and Videos (1989).

HEY O.T. =
the CHEMTREC GUY
SAYS: WOW! THAT'S
A NEW ONE!

CHAPTER 8

SELECT PERSONAL PROTECTIVE CLOTHING AND EQUIPMENT

OBJECTIVES

1. Define the following terms and their impact and significance on the selection of chemical protective clothing: [NFPA 472-6.3.3.3(A)]
   a. Degradation
   b. Penetration
   c. Permeation
   d. Breakthrough time
   e. Permeation rate
2. Identify at least three indications of material degradation of chemical protective clothing. [NFPA 472-6.3.3.3(B)]
3. Describe the differences between limited-use and multiuse chemical protective clothing materials.
4. Identify the factors to be considered in selecting chemical-protective clothing for a specified action option. [NFPA 472-6.3.3.3]
5. Identify the four levels of personal protective equipment as specified by the Environmental Protection Agency (EPA) and the National Institute for Occupational Safety and Health (NIOSH). [NFPA 472-6.3.3.1]
6. Describe the advantages, limitations, and proper use of the following types of respiratory protection at hazmat incidents: [NFPA 472-6.3.3.2(A)]
   a. Air purifying respirators (APR)
   b. Powered air purifying respirators (PAPR)
   c. Positive pressure self-contained breathing apparatus (SCBA)
   d. Positive pressure air-line respirators with required escape unit (SAR)
7. Identify the factors to be considered in selecting respiratory protection at hazardous materials incidents. [NFPA 472-6.3.3.2(B)]
8. Identify the operational components of air-purifying respirators and airline respirators by name and describe their functions. [NFPA 472-6.3.3.2(C)]
9. Identify the procedures for donning, working in, and doffing the following types of respiratory protection: [NFPA 472-6.4.2(3)]
   a. Air-purifying respirators (APR)
   b. Powered air purifying respirators (PAPR)
   c. Positive pressure air-line respirators with required escape unit
10. Describe the advantages, limitations, and proper use of structural firefighting clothing at a hazmat incident.
11. Identify three types of vapor-protective and splash-protective clothing and describe the advantages and disadvantages of each type. [NFPA 472-6.3.3.2(C)]
12. Identify two types of high-temperature protective clothing and describe the advantages and disadvantages of each type.
13. Identify the process for selecting protective clothing at hazardous materials incidents. [NFPA 472-6.3.3.3(E)]
14. Identify the physical and psychological stresses that can affect users of specialized protective clothing. [NFPA 472-6.3.3.3(G)]

2. Diffusion of the chemical through the clothing material
3. Desorption of the chemical from the inner surface of the clothing material (toward the wearer)

Breakthrough time is defined as the time from the initial chemical attack on the outside of the material until its desorption and detection inside. The units of time are usually expressed in minutes or hours, and a typical test runs up to a maximum of 8 hours. If no measurable breakthrough is detected after 8 hours, the result is often reported as a breakthrough time of >480 minutes or >8 hours.

Permeation rate is the rate at which the chemical passes through the CPC material and is generally expressed as micrograms per square centimeter per minute ($\mu g/cm^2/min$). For reference purposes, .9 $\mu g/cm^2/min$ is equal to approximately 1 drop/hour. The higher the rate, the faster the chemical passes through the suit material. Comprehensive chemical compatibility charts will contain both the breakthrough time and the permeation rate.

Measured breakthrough times and permeation rates are determined by laboratory permeation testing procedures against a list of chemicals outlined in *ASTM F 1001, Standard Guide for Chemicals to Evaluate Protective Clothing Materials*. The ASTM F 1001 list is also used as the *NFPA 1991, Standard of Vapor Protective Suits for Emergency Response* battery of chemicals for determining CPC material permeation resistance.

PERMEATION THEORY

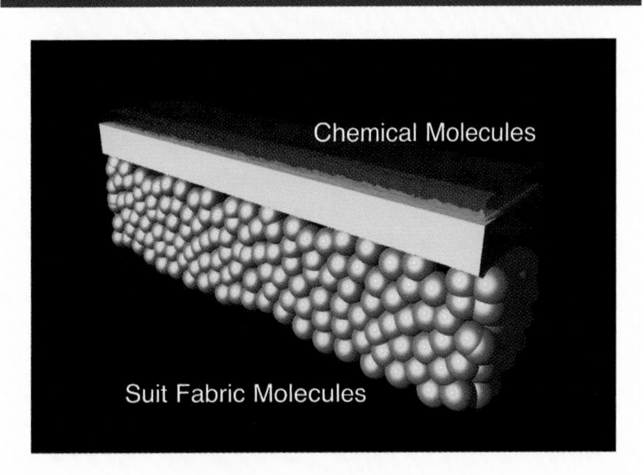

PHASE I: ADSORPTION

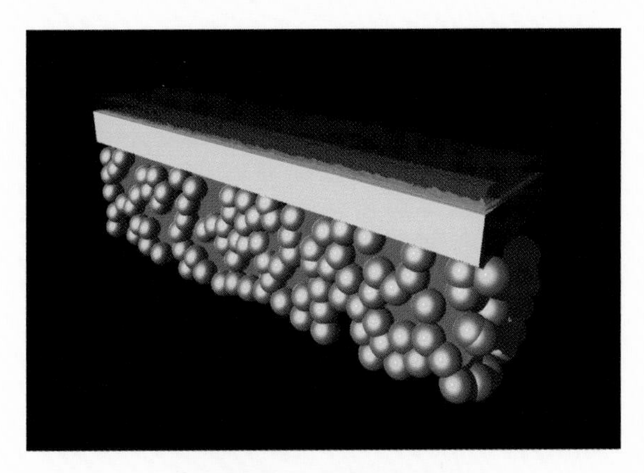

PHASE II: DIFFUSION

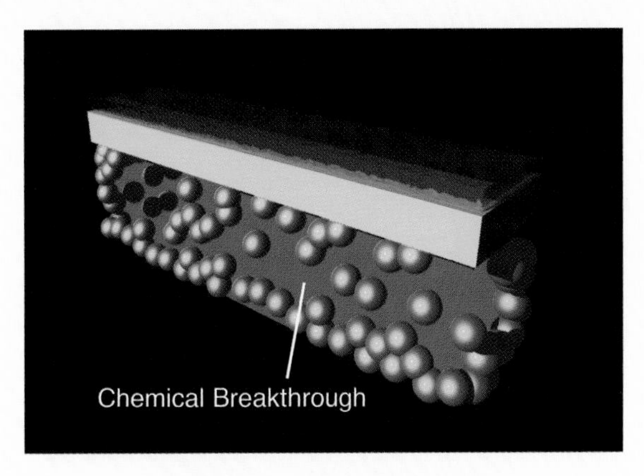

PHASE III: DESORPTION

Figure 8.2 Chemical permeation through protective clothing is a three-step process: absorption, diffusion, and desorption.

Permeation testing is conducted using pure, undiluted test chemicals on clean, uncontaminated swatches over a pre-established period of time (usually 2 to 8 hours). Virtually all testing is conducted at ambient room temperatures (70°F/21°C). Use of the breakthrough times from this testing process then allows responders to estimate the duration of maximum protection under a worst-case scenario of continuous chemical contact.

In evaluating permeation resistance and breakthrough times, several other terms may be found in the CPC manufacturer's compatibility information:

- *Actual breakthrough time*—Breakthrough time as previously defined.
- *Normalized breakthrough time*—A calculation, using actual permeation results, to determine the time at which the permeation rate reaches 0.1 μg/cm2/min. Normalized breakthrough times are useful for comparing the performance of several different protective clothing materials. Note that in Europe, breakthrough times are normalized at 1.0 μg/cm2/min., a full order of magnitude less sensitive.
- *Minimum detectable permeation rate (MDPR)*—The minimum permeation rate that can be detected by the laboratory analytical system being used for the permeation test.
- *System detection limit (SDL)*—The minimum amount of chemical breakthrough that can be detected by the laboratory analytical system being used for the permeation test. Lower SDL's result in lower (or earlier) breakthrough times.

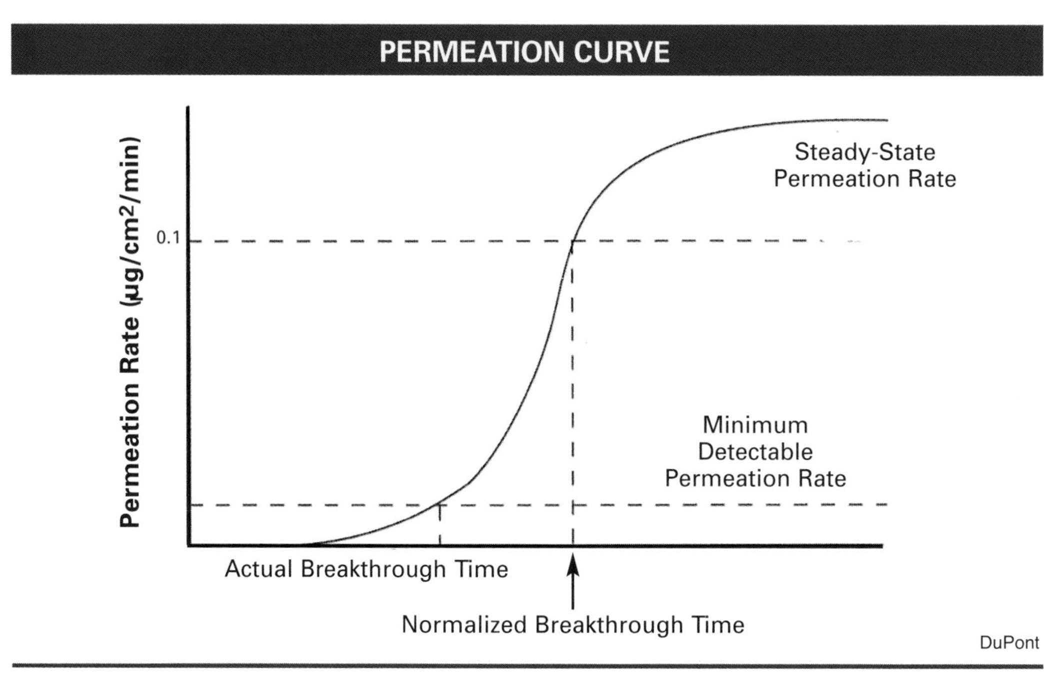

Figure 8.3 Permeation curves can visually illustrate the relationship between permeation rate and time. Breakthrough time is the initial point at which a chemical is detected on the inside of a CPC material.

Chemical permeation rates are a function of many factors, including the following:

- *Temperature.* Most chemical compatibility tests are conducted at ambient temperatures (68° to 72°F/20° to 22.2°C). However, as temperature increases, permeation rates increase and breakthrough times shorten.

- *Thickness*. Permeation is inversely proportional to the thickness of the clothing material. In other words, doubling its thickness will theoretically cut the permeation rate in half. The breakthrough time will, therefore, become longer.
- *Chemical mixtures and their effects upon chemical resistance are relatively unknown.* It is impossible to test for every possible chemical combination. Tests have also shown that combining data concerning exposures to individual chemicals has limited value. For example, Viton®/chlorobutyl laminate resists hexane for over three hours and acetone for one hour. However, any mixture of hexane and acetone will permeate Viton®/chlorobutyl laminate in under 10 minutes. Chemical mixtures may result in a stronger attack on protective clothing than with individual exposures.
- *Previous exposures*. Once a chemical has begun the diffusion process, it may continue to diffuse even after the chemical itself has been removed from the outside surface of the material. This is significant when considering the re-use of any protective clothing. **Decontamination is no assurance that permeation has stopped**. Although it is possible to test for permeation, the testing process also destroys the clothing material.

PROTECTIVE CLOTHING MATERIALS

Parameters that should be considered when evaluating and choosing chemical protective clothing materials include the following:

1. **Chemical resistance**. This is the most critical factor, as the clothing material must maintain its integrity and protection qualities when it comes in contact with a hazardous material. Chemical resistance (either permeation or penetration) must be evaluated. Permeation resistance test data should be used for vapor chemical protective clothing, particularly for chemicals that are toxic by skin absorption or present other hazards in vapor form.

   Chemical penetration data should be used for liquid splash chemical protective clothing. However, penetration data should not be used when chemicals are highly toxic or give off hazardous vapors.
2. **Flammability**. The CPC garment should not contribute to the fire hazard and should maintain its protective capabilities when exposed to elevated temperatures. When burning, the clothing material should not melt or drip.

   Be aware that chemical protective clothing is not appropriate for firefighting operations, or for protection in flammable or explosive environments. While some clothing manufacturers offer "flash" protective overgarments for use with CPC, they offer only limited protection for rapid, short-duration fires and should not be considered for firefighting applications.
3. **Strength and durability.** These characteristics reflect a material's ability to resist cuts, tears, punctures, abrasions, and other physical hazards found at the incident. The breaking strength, seam strength, and closure strength also relate to how well clothing materials withstand repeated use or the stresses of use.
4. **Overall integrity**. Chemical protective clothing should provide complete protection to the wearer. Each vapor protective garment should be tested for integrity with a "pressure" or inflation test. Similarly, liquid splash protective clothing designs should be tested for liquid-tight integrity to verify that the suit design does not allow liquids onto the wearer's skin.

5. **Flexibility**. This is the ability of the user to move and work in protective clothing and, generally, a factor of how the ensemble is fabricated. Emphasis is on the garment weight, fabrication of seams (i.e., bonded, sewn, glued, sealed, taped), and their resistance to chemical and wear exposures. Flexibility and dexterity are particularly important with gloves and chemical protective ensembles.

   When dealing with laminated fabrics, microcracking may become a concern. Continuous flexing of a laminate material may cause microcracks to develop on the inner laminate layers and lead to a CPC failure with little indication of potential failure.

6. **Temperature characteristics**. Temperature characteristics affect the ability of protective clothing to maintain its protective capacity in temperature extremes. Higher temperatures increase the effects of all chemicals upon polymers. A material suitable for chemical exposures at ambient temperatures may fail at elevated temperatures. Exposure to low temperatures may cause materials to stiffen, crack, flake, or separate.

7. **Shelf life**. Long-term exposure to sunlight or excessive heat (104°F/>40°C) causes many CPC materials to age and deteriorate. Some materials, particularly rubber-like materials, deteriorate over time, much like automobile tires. These changes may also occur without use. Shelf life information should be obtained from the manufacturer for the specific CPC products and materials being used. For example, DuPont suggests that garments older than 5 years old be downgraded for training use only.

8. **Decontamination and disposal**. The ability of the protective clothing material to be cleaned and decontaminated must be evaluated against potential chemical exposures. Limited-use garments often represent cost-effective options. Although decon may reduce the level of contamination, in many instances it will not be completely eliminated.

CATEGORIES OF CHEMICAL PROTECTIVE CLOTHING

Protective clothing materials can be classified by use into two broad categories – limited-use garments and reusable garments.

Limited-use garments and materials are protective clothing materials that are used and then discarded. They are engineered for one or a low number of wearings, and are discarded when damaged or contaminated. They eliminate many health and safety concerns regarding CPC decontamination and their return to service.

Most limited-use materials are usually constructed of a nonwoven fiber or a nonwoven fabric with a laminated film (e.g., Tychem® fabrics). In general, limited-use garments typically provide a broader range of chemical protection than reusable garments and are lighter in weight. However, they can be less durable and not as strong as reusable rubber or thermoplastic-based materials. The chemical resistance of these garments can sometimes be compromised by the physical breakdown of the material which may occur with improper environmental or storage conditions.

Limited-use garments are generally suitable for a single use and should be disposed of in accordance with local, state, and federal environmental regulations. Examples of limited-use garments used in hazmat emergency response include:

- Tychem® QC—polyethylene coated Tyvek®
- Tychem® SL—Saranex® 23P -laminated Tyvek®

- Tychem® F—barrier film laminated to Tyvek®
- Tychem® CPF® 2, 3 and 4—barrier film laminates
- Tychem® Responder®—multiple barrier film laminates
- Tychem® CSM®—multiple barrier film laminates
- Trellchem® TLU—polyamide fabric laminated on each side with a barrier film laminate
- Hazard-Gard™ I—nonwoven fabric laminated with a polyethylene film
- Hazard-Gard™ II—Saranex® 23P film laminated to polypropylene fabric

Advantages of limited-use materials include lower costs, ability to stock a larger and more varied protective clothing inventory, and reduced inspection and maintenance requirements. They are often used for support functions, including decontamination, remedial clean-up of identified chemicals, and training.

Reusable garments are designed and fabricated to allow for decontamination and reuse. Generally thicker and more durable than limited-use garments, they are used for liquid chemical splash and vapor protective suits, gloves, aprons, boots, and thermal protective clothing. Reusable garment materials are usually made from chlorinated polyethylene (CPE), vinyl (plasticized polyvinyl chloride—PVC), fluorinated polymers (Teflon®), and rubberlike fabrics, such as butyl rubber, neoprene rubber, and Viton®.

Although these garments are considered reusable, certain chemical exposures may require the disposal of this clothing as well. Disposal must be in accordance with local, state, and federal environmental regulations.

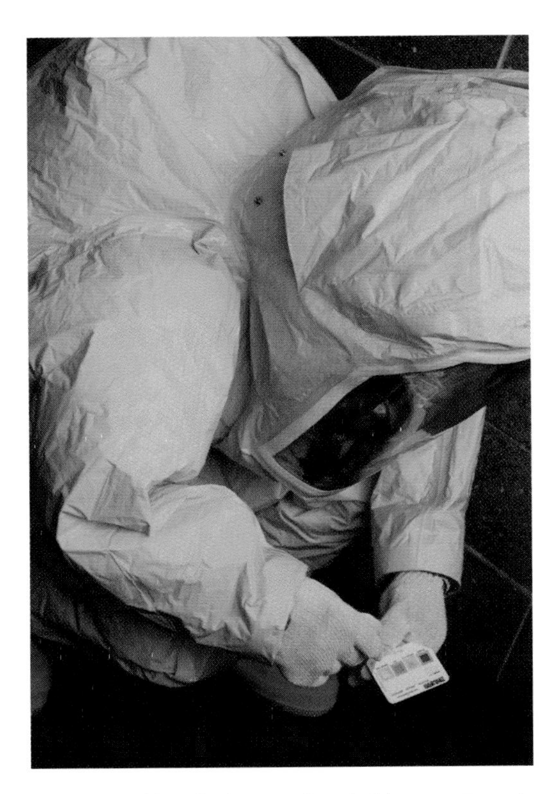

Figure 8.4 Chemical protective clothing must meet national standards in compliance with NFPA 1991, 1992 or 1994.

NFPA CHEMICAL PROTECTIVE CLOTHING STANDARDS

The NFPA Technical Committee on Hazardous Materials Protective Clothing and Equipment has developed three consensus standards that specify minimum documentation, design and performance criteria, and test methods for chemical protective clothing. These standards are often referenced as minimum requirements in purchase specifications and cover the following:

- NFPA 1991—Vapor Protective Ensembles for Hazardous Materials Emergencies
- NFPA 1992—Liquid Splash Protective Ensembles for Hazardous Materials Emergencies
- NFPA 1994—Protective Ensemble for Chemical/Biological Terrorism Incidents

Each standard requires independent, third party certification to ensure that the protective clothing meets the respective design, performance, and documentation requirements. Certification agencies, such as Underwriter's Labs (UL) or the Safety Equipment Institute (SEI), certify the garment performance, not NFPA. Compliant products must carry a product label indicating compliance with the NFPA standard, a technical data package, and user instructions. A chemical protective ensemble can be certified to several NFPA standards.

NFPA 1991 and 1992 performance requirements can be summarized as follows:

PERFORMANCE REQUIREMENT	NFPA 1991 CHEMICAL VAPOR PROTECTIVE CLOTHING	NFPA 1992 LIQUID SPLASH PROTECTIVE CLOTHING
SUIT INTEGRITY	Pressure test	Shower test
BARRIER	Permeation resistance	Penetration resistance
CHEMICAL BATTERIES	ASTM F1001 test battery or permeation resistance 21 chemicals (15 liquids & 6 gases)	ASTM F1001 test battery for penetration resistance (7 liquid chemicals)
PHYSICAL HAZARD RESISTANCE	Puncture / tear, burst resistance, seam strength, closure strength. Specific tests for footwear and gloves.	
DURABILITY	Abrasion resistance, flex fatigue	
FUNCTIONAL PERFORMANCE	User mobility, visor clarity, glove dexterity	
COMPONENT PERFORMANCE	Valve leakage, pass-thru strength	Pass-thru strength

PERFORMANCE REQUIREMENT	NFPA 1991 CHEMICAL VAPOR PROTECTIVE CLOTHING	NFPA 1992 LIQUID SPLASH PROTECTIVE CLOTHING
FLAME RESISTANCE	Primary materials exposed to flame in 2 stages: 3 seconds to determine ease of ignition; and 12 seconds to determine flame extinguishment. Fabric cannot melt or drip.	No flame impingement requirements in base specifications; part of the optional flame fire escape requirements
COMPONENT TESTING	Examples may include visibility through a visor; glove dexterity/ gripping ability; footwear slip resistance; exhaust valve operation.	
OTHER SUIT PERFORMANCE TESTS	Functionality test —User demonstrates ability to perform tasks, manipulate tools, etc. Airflow test—Assesses ability of CPC to exhaust air if SCBA goes into the bypass mode.	
OPTIONAL PERFORMANCE AREAS	Liquefied gases Flash fire escape protection Chem/bio terrorism	Flash fire escape protection

NFPA 1994—PROTECTIVE ENSEMBLE FOR CHEMICAL/BIOLOGICAL TERRORISM INCIDENTS

NFPA 1994 was originally enacted in 2001 as a result of the growing terrorism problem. It sets performance requirements for protective clothing used at chemical and biological terrorism incidents, and defines three classes of ensembles based on the perceived threats at an incident. Ensemble differences are based on (1) the ability of the design to resist inward leakage of chemical and biological contaminants; (2) resistance of the suit to chemical warfare and toxic industrial chemicals; and (3) the strength and durability of these materials. All NFPA 1994 ensembles (i.e., garment, gloves and footwear) are designed for a single exposure.

Many of the NFPA 1994 testing requirements are similar to those found in both NFPA 1991 and 1992. In addition, the standard permits dual certification

Class 1 ensembles offer the highest level of protection and are intended for use in worst-case circumstances, where the substance creates an immediate threat and is unidentified and of unknown concentrations. Scenarios for use may include ongoing release with likely gas/vapor exposures, the responder is close to the point of release, and most victims in the area appear to be unconscious or dead from exposure. SCBA and air supplied units would be used for respiratory protection.

Class 2 ensembles offer an intermediate level of protection and are intended for circumstances where the agent or threat may be identified, when the actual release has subsided, or in an area where live victims may be rescued. Possible exposure may exist with residual gases or vapors as well as highly contaminated surfaces. SCBA or powered air-purifying respirators (PAPR) may be used for respiratory protection.

Class 3 ensembles offer the lowest level of protection and are intended for use long after the initial release has occurred, at relatively large distances from the point of release, or for response activities such as decontamination, patient care, crowd control, traffic control, and clean-up operations. PAPRs be used for respiratory protection.

The performance classes can be summarized as follows:

CLASS	CHALLENGE	SKIN CONTACT	VAPOR THREAT	LIQUID THREAT	VICTIM'S CONDITION
1	Vapors Aerosols Pathogens	Not permitted	Unknown or Not Verified	High	Unconscious, not symptomatic and not ambulatory
2	Limited vapors Liquid splash Aerosols Pathogens	Not probable	IDLH	Moderate	Mostly alive, but not ambulatory
3	Liquid drops Pathogens	Not likely	TLV/STEL	Low to none	Self-ambulatory

Test chemicals currently used in NFPA 1994 are as follows:

TYPE OF MATERIAL	CHEMICAL
Chemical Agent	• Distilled Sulfur Mustard (HD) • Lewisite (L) • Sarin (GB) • V-Agent (VX)
Industrial Chemical (Liquid)	• Dimethyl Sulfate (DMA)
Industrial Chemical (Gas)	• Ammonia • Chlorine • Cyanogen Chloride (CK) • Carbonyl Chloride (CG) • Hydrogen Cyanide (AC)

NFPA protective clothing standards are reviewed on a five-year basis. The process is open to public participation. Proposed changes to any NFPA standard can be submitted at any time, and if of an emergency nature, can be implemented within several months. Otherwise, proposed changes are included in the normal revision cycle.

CHEMICAL COMPATIBILITY AND SELECTION CONSIDERATIONS

No single protective clothing material offers total chemical protection. The initial selection of protective clothing and equipment should be based on a hazard assessment of those chemicals found in the community or the facility. Unfortunately, there may be some chemicals for which there is no adequate protection.

Chemical protective clothing will often be constructed of a combination of several materials or laminates. Be aware of which of these materials form the basis for chemical compatibility recommendations. Recognize that it may not be possible to determine the specific material(s) used in some limited-use laminate garments due to product proprietary reasons.

The manufacturer should provide technical test data that reflects the chemical compatibility of both the primary suit material and all secondary components (e.g., gloves, boots, closure assemblies, visors, and exhaust valves). When evaluating chemical vapor protective suits, acquire a complete inventory of all suit components and their construction materials. The performance of chemical protective clothing is only as strong as its weakest material.

Manufacturers will publish quantitative chemical resistance data for particular chemicals. This data will normally be based on standardized laboratory tests such as those established by the ASTM F 1001 committee. Standard permeation and penetration tests often incorporate a very large safety factor in predicting failures. The size of this safety factor depends on the established testing criteria.

Chemical resistance data is described in terms of chemical permeation/breakthrough times and rates, or as "pass/fail" chemical penetration testing results. Remember that the longer the breakthrough time, the better the level of protection. Breakthrough rates are an indication of how quickly a chemical will permeate through a CPC material. If two CPC materials have comparable breakthrough times, the CPC with the lowest reported permeation rate should normally represent the better option.

Chemical compatibility recommendations for boots, gloves, and some garments may also be provided in the form of qualitative chemical resistance ratings or use recommendations for a specific protective clothing material and particular chemicals. These ratings are often in the form of a four- to six-grade scale (e.g., excellent, good, fair, and poor/not recommended) or a color code (e.g., green, yellow, red). They often will not include performance specifications or quantitative data such as breakthrough times. Degradation resistance data may be misleading and should be avoided when selecting CPC materials.

When evaluating chemical compatibility recommendations, consider the following guidelines:

- The primary reference source for chemical compatibility recommendations should be the CPC manufacturer's technical documentation. Other credible sources will include CPC reference manuals and computer databases, such as Forsberg and Keith's *Chemical Protective Clothing Permeation and Degradation Compendium*, and the National Toxicology Program's GlovES+™ computer expert system.
- Determine the basis of chemical compatibility recommendations. Degradation or immersion testing is not sufficient for compatibility assessment. Permeation or penetration test data should be sought since permeation of rubber or plastic fabrics can occur with little or no physical effect on the clothing material.

Remember, permeation is an insidious process that can occur with no sign of degradation.

- Compatibility recommendations based upon immersion testing data may be quite old, and they may also be based on the subjective evaluations rather than quantitative measurements for swelling, weight, or strength changes. In some cases, the testing criteria and qualitative descriptions for defining *good, excellent*, and other key words may not be documented.
- Materials constructed of the same primary fabric or material (e.g., butyl rubber, PVC) are not necessarily equal in performance. Variations in formulations, thickness, and coating and backing materials influence chemical exposure times.
- There may be a conflict in compatibility recommendations between sources. Responders should initially rely on the protective clothing manufacturer's chemical resistance recommendations. Always select the most conservative data.

RESPIRATORY PROTECTION

The respiratory system is the most direct and critical exposure route. Inhalation is the most common exposure route and is often the most damaging. Remember that a material does not have to be a gas in order to be inhaled—solid materials may generate fumes or dusts in a dry powdered form, while high-vapor-pressure liquid chemicals can generate vapors, mists, or aerosols that can be inhaled.

The selection of respiratory protection at a hazmat incident should be based on a number of factors, including the following:

- What is the physical form of the contaminant (i.e., solid, liquid or gas)?
- Has the contaminant been identified?
- Are concentrations known or unknown?
- What is the purpose of response operations?
- What will be the duration of response operations?
- What is the operating environment and operating conditions (e.g., indoors, outdoors, heat, cold, precipitation, etc.)?
- What type and level of skin protection will be required?

Respiratory protection can be provided by either air purification devices or by atmosphere supplying respiratory equipment.

AIR PURIFICATION DEVICES

Air purification devices are respirators that remove particulate matter, gases, or vapors from the ambient air before inhalation. When used for gases or vapors, they are commonly equipped with a sorbent material that absorbs or reacts with the hazardous gas. Particle-removing respirators use a mechanical filter to separate the contaminants from the air. Some cartridges are combined sorbent and mechanical filters. The proper cartridge must be used for the expected contaminants (e.g., acid gas, organic vapor, nuclear/bio/chemical agent, etc.). There is a uniform NIOSH color code system for the identification of cartridges.

The NIOSH Respirator Certification Requirements (42 CFR 84) outline the requirements for particulate respirators. Emergency responders may use particulate filters for white powder scenarios, and at structural collapse incidents. Part 84 defines nine classes of filters—three levels of filter efficiency (95%, 97%, and 99.7%), each with three categories of resistance to filter efficiency degradation (N, R, and P). The selection of N-, R- and P- filters depends on the presence or absence of oil particles, where N = Not resistant to oil; R = Resistant to oil, and P = oil Proof.

- If no oil particles are present, use any series (N, R, or P).
- If oil particles are present, use only R or P series.
- If oil particles are present and the filter is to be used for more than one work shift, use only P series.

The selection of filter efficiency (i.e., 95%, 99%, or 99.97%) will depend on how much filter leakage can be accepted. For example, high efficiency particulate air filters (HEPA) are at least 99.97% efficient in removing particles 0.3 micrometers in diameter and larger (e.g., N100, R100, and P100 filters).

Two basic types of air purification devices may be used for emergency response purposes:

Air Purification Respirators (APRs) are respirators with an air-purifying filter, cartridge, or canister that removes specific air contaminants by passing ambient air through the air-purifying element. These are negative pressure respirators and can be found with in either a full-face and half-face configuration with sorbent, mechanical or combination cartridges attached. They are commonly used in controlled industrial and workplace environments where the contaminants are known and concentrations measured. For emergency response applications, full-face respirators are typically the respirator of choice. If half-face respirators are used, eye protection must be provided.

Powered-Air Purification Respirators (PAPRs) are air-purifying respirators that use a blower to force the ambient air through air-purifying elements to a full-face mask. As a result, there is a slight positive pressure in the facepiece that results in an increased protection factor. Where an APR has a protection factor of 50 : 1, a PAPR will have a protection factor of 1000: 1 (Note: A protection factor of 1 = no respiratory protection in place. A protection factor of 1,000 means that the concentration of a breathed contaminant is reduced by a factor of one-thousand from the ambient concentration.) PAPR's are being used in a wide range of emergency response and post-emergency response applications, including decon, patient handling in medical facilities, and investigation of hazmat and terrorism crimes.

Operational considerations when using APRs and PAPRs include the following:

- **Air purification devices should not be used at hazmat releases unless qualified personnel have first monitored the environment and determined that such devices can be safely used (per OSHA 1910.120(q)(3)(iv). As a general rule, they should not be used for initial response operations at hazmat incidents and for emergency response operations involving unknown substances.**
- Cannot be used in IDLH environments or in oxygen-deficient atmospheres containing less than 19.5% oxygen. When used, both the contaminant and oxygen levels must be constantly monitored.

- Should not be used in the presence or potential presence of unidentified contaminants. Not recommended for areas where contaminant concentrations are unknown or exceed the designated use concentrations. "Designated use concentrations" are based on testing at a given temperature (usually room temperature) over a narrow range of flow rates and relative humidity. Therefore, the level of protection may be compromised in nonstandard conditions.
- Respiratory protection can be downgraded from air supplied to air-purifying respirators if (1) the contaminants have been identified; (2) the atmosphere is being monitored and contaminant levels are within acceptable limits; and (3) the IC approves.
- May present logistical problems for storage and maintenance because of the variety of filters and cartridges required. The shelf life of filters and cartridges will vary depending on the type of cartridge (i.e., sorbent versus mechanical filter), and its packaging. Always consult manufacturer instructions for guidance.
- Have a limited-protection duration. Once opened, sorbent canisters begin to absorb humidity and air contaminants whether in use or not, and their efficiency and service life will decrease dramatically. Where possible, cartridges should have an end-of-service-life indicator (ESLI) that warns the user of the approach of the end of adequate respiratory protection (e.g., the sorbent is approaching saturation or is no longer effective). If the cartridge doesn't have an ESLI indicator, a schedule must be established to ensure that the cartridges are changed before the end of their service life.
- APRs and PAPRs only protect against specific chemicals and only to specific concentrations. Their effectiveness against two or more chemicals simultaneously is highly questionable. They are well suited for operations involving solids, dusts, powders, and many biopathogens and toxins.
- Individuals must meet the fit testing and medical requirements as outlined by *OSHA 1910.134—Respiratory Protection.*

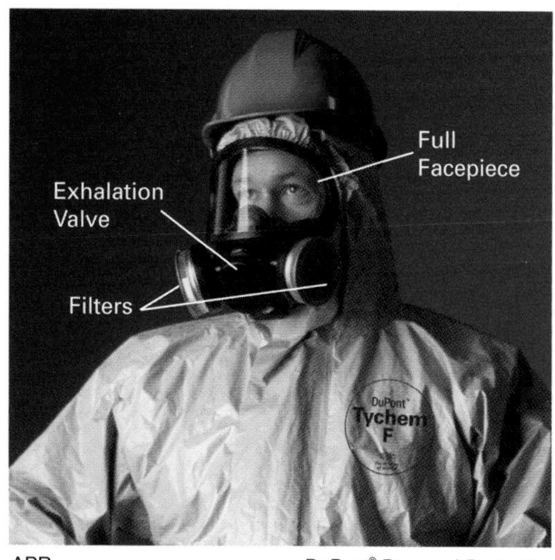

APR DuPont® Personal Protection

PAPR Chris Hawley

Figure 8.5 A and B Air-purifying respirators (APRs) and powered air purifying respirators (PAPRs) are respirators that remove particulate matter, gases, or vapors from the ambient air before inhalation.

Advantages of APRs and PAPRs include their light weight and lack of physical stress on the user. Limitations include many of the operational considerations in the preceding list, including the need for air monitoring to ensure that IDLH or oxygen-deficient conditions are not present, logistical requirements, and the fact that they only offer protection against specific chemicals and in specific concentrations. Negative-pressure respirators also carry a greater risk of leakage than positive pressure respirators.

ATMOSPHERE SUPPLYING DEVICES

Respiratory protection devices with an air source are referred to as atmosphere-supplying devices. There are two basic types: self-contained breathing apparatus (SCBA) and supplied air respirators (SAR), which supply air from a source away from the scene connected to the user by an airline hose.

These devices provide the highest available level of protection against airborne contaminants and in oxygen-deficient atmospheres. Only positive-pressure devices that maintain positive pressure in the facepiece during both inhalation and exhalation should be used for emergency response applications. Positive-pressure respirators will provide a protection factor of 10,000 : 1.

Self-Contained Breathing Apparatus. There are two basic types of SCBA: open-circuit and closed-circuit. *Open-circuit SCBA* are those where air is exhaled directly into the ambient atmosphere. It is the predominant type of SCBA used in the emergency response community. Closed-circuit SCBA are those where exhaled air is recycled by removing the carbon dioxide with an alkaline scrubber and replenishing the consumed oxygen from a solid, liquid, or gaseous oxygen source. Closed-circuit SCBA are used for specialized response scenarios where long, extended operations may be required, such as tunnel, subway, and mine rescue operations. However, they are not commonly used in conjunction with chemical vapor clothing and may generate heat which may add to the heat stress encountered in totally encapsulating suits.

Open-circuit SCBA used for firefighting applications should meet the requirements of *NFPA 1981—Standard for Open-Circuit SCBA for Fire Fighters*. As a result of the growing terrorism threat, NIOSH has also initiated a testing process to approve open-circuit SCBA that already meet the NFPA 1981 requirements against chemical, biological, radiological, and nuclear agents. At the time of publication, NIOSH was focusing on SCBA, but plans to develop similar criteria for approving other types of respirators, including APRs and PAPRs.

Advantages of using SCBA include that they are readily available in the emergency response community, most responders are proficient in their use, and they provide the highest level of respiratory protection. Limitations include their size, weight, bulkiness, and limited duration of air supply, overall resistance of the SCBA and its components to chemical exposures, and size restrictions when used in confined spaces.

Operational considerations when using SCBA include the following:

- Atmosphere-supplying units are required for initial response operations until the hazards and concentration of air contaminants can be fully assessed.
- Duration of the operation. Although 30-minute air cylinders are commonly used in the fire service, most active hazmat teams will use 45- to 60-minute air cylinders to provide a sufficient backup air supply for entry, exit, and decon operations.

- Depending on the type of cylinder, certain chemicals may attack the outer shell of an air cylinder. In 1996, a fiberglass-wrapped composite aluminum cylinder that was accidentally exposed to a commercial cleaning fluid containing hydrofluoric acid, phosphoric acid and sulfuric acid failed explosively approximately 6 days after exposure to the fluid.

Supplied Air Respirators. Although SCBA are most common, SARs may be used when extended working times are required for entry, decon or remedial clean-up operations. Components of a SAR include (1) source of breathing air—usually a cylinder, a cylinder cart or a cascade system; (2) airline hose; (3) positive-pressure respirator; and (4) emergency air supply, such as a small escape cylinder.

Operational considerations when using supplied air respirators include the following:

- Atmosphere-supplying units are required for initial response operations until the hazards and concentration of air contaminants can be fully assessed.
- NIOSH certification limits the maximum hose length from the source to 300 feet (91.4 meters).
- Use of airlines in IDLH or oxygen-deficient atmosphere requires a secondary emergency air supply, such as an escape pack for immediate backup protection in case of airline failure. In addition, use of a SAR will require personnel to monitor the air supply source.
- Using airline hose will probably impair user mobility and slow the operation. The user must retrace his or her entry path when leaving the work area.
- The airline hose is vulnerable to physical damage, chemical contamination, and degradation. Airline sleeves constructed of disposable materials can provide additional protection. Decontamination may be difficult.
- Dual flow SCBAs that have the capability of being supplied by either an SCBA or an airline may provide additional flexibility for both entry and decon operations. The user can operate in either the SCBA or airline hose modes by operating a manual or automatic switch.

Figure 8.6 Supplied air respirators being used with chemical vapor protective clothing.

Advantages of using SAR units include lower profile and weight, increased work durations, and their ability to provide the highest level of respiratory protection. Limitations include a number of the operational considerations listed above.

Air purification devices may be appropriate for operations involving volatile solids and for remedial clean-up and recovery operations where the type and concentration of contaminants is verifiable. However, air-supplied devices such as airline hose units will offer the greatest protection for exposures to gases and vapors.

U.S. ENVIRONMENTAL PROTECTION AGENCY LEVELS OF CHEMICAL PROTECTION

LEVEL A PROTECTION SHOULD BE WORN WHEN THE HIGHEST LEVEL OF RESPIRATORY, SKIN, EYE, AND MUCOUS MEMBRANE PROTECTION IS NEEDED.

- PERSONAL PROTECTIVE EQUIPMENT
  - ☐ POSITIVE-PRESSURE (PRESSURE-DEMAND), SELF-CONTAINED BREATHING APPARATUS (MSHA/NIOSH APPROVED).
  - ☐ FULLY-ENCAPSULATING CHEMICAL RESISTANT SUIT.
  - ☐ GLOVES, INNER, CHEMICAL RESISTANT.
  - ☐ GLOVES, OUTER, CHEMICAL RESISTANT.
  - ☐ BOOTS, CHEMICAL RESISTANT, STEEL TOE AND SHANK (DEPENDING ON SUIT BOOT CONSTRUCTION, WORN OVER OR UNDER SUIT BOOT). UNDERWEAR, COTTON, LONG-JOHN TYPE.*
  - ☐ HARD HAT (UNDER SUIT).*
  - ☐ COVERALLS (UNDER SUIT).*
  - ☐ TWO-WAY RADIO COMMUNICATIONS (INTRINSICALLY SAFE).

LEVEL B PROTECTION SHOULD BE SELECTED WHEN THE HIGHEST LEVEL OF RESPIRATORY PROTECTION IS NEEDED, BUT A LESSER LEVEL OF SKIN AND EYE PROTECTION. LEVEL B PROTECTION IS THE MINIMUM LEVEL RECOMMENDED ON INITIAL SITE ENTRIES UNTIL THE HAZARDS HAVE BEEN FURTHER IDENTIFIED AND DEFINED BY MONITORING, SAMPLING, AND OTHER RELIABLE METHODS OF ANALYSIS, AND PERSONNEL EQUIPMENT CORRESPONDING WITH THOSE FINDINGS UTILIZED.

- PERSONAL PROTECTIVE EQUIPMENT
  - ☐ POSITIVE-PRESSURE (PRESSURE-DEMAND), SELF-CONTAINED BREATHING APPARATUS (MSHA/NIOSH APPROVED).
  - ☐ CHEMICAL RESISTANT CLOTHING (OVERALLS AND LONG SLEEVED JACKET, COVERALLS, HOODED TWO PIECE CHEMICAL SPLASH SUIT, DISPOSABLE. CHEMICAL RESISTANT COVERALLS).*
  - ☐ COVERALLS (UNDER SPLASH SUIT).*
  - ☐ GLOVES, OUTER, CHEMICAL RESISTANT.
  - ☐ GLOVES, INNER, CHEMICAL RESISTANT.
  - ☐ BOOTS, OUTER, CHEMICAL RESISTANT, STEEL TOE AND SHANK.
  - ☐ BOOTS, OUTER, CHEMICAL RESISTANT.*
  - ☐ TWO-WAY RADIO COMMUNICATIONS (INTRINSICALLY SAFE).
  - ☐ HARD HAT.*

LEVEL C PROTECTION SHOULD BE SELECTED WHEN THE TYPE OF AIRBORNE SUBSTANCE IS KNOWN, CONCENTRATION MEASURED, CRITERIA FOR USING AIR-PURIFYING RESPIRATORS MET, AND SKIN AND EYE EXPOSURE IS UNLIKELY. PERIODIC MONITORING OF THE AIR MUST BE PERFORMED.

- PERSONAL PROTECTIVE EQUIPMENT
  - ☐ FULL-FACE, AIR-PURIFYING RESPIRATOR (MSHA/NIOSH APPROVED). CHEMICAL RESISTANT CLOTHING (ONE PIECE COVERALL, HOODED TWO PIECE CHEMICAL SPLASH SUIT,
  - ☐ CHEMICAL RESISTANT HOOD AND APRON, DISPOSABLE CHEMICAL RESISTANT COVERALLS).
  - ☐ GLOVES, OUTER, CHEMICAL RESISTANT.
  - ☐ GLOVES, INNER, CHEMICAL RESISTANT.*
  - ☐ BOOTS, STEEL TOE AND SHANK, CHEMICAL RESISTANT.
  - ☐ CLOTH COVERALLS (INSIDE CHEMICAL PROTECTIVE CLOTHING).*
  - ☐ TWO-WAY RADIO COMMUNICATIONS (INTRINSICALLY SAFE).*
  - ☐ HARD HAT.*
  - ☐ ESCAPE MASK.*

LEVEL D IS PRIMARILY A WORK UNIFORM. IT SHOULD NOT BE WORN ON ANY SITE WHERE RESPIRATORY OR SKIN HAZARDS EXIST.

* Optional

Figure 8.7

LEVELS OF PROTECTION

The need for proper protective clothing and equipment in a hostile environment is obvious. Unfortunately, there is no one type of PPE that satisfies our protection needs under all conditions. For example, chemical protection and thermal protection are very difficult to combine into one protective clothing material. The IC and the Hazmat Group Supervisor must be familiar with the various types and levels of protective clothing available.

Three basic types of protective clothing may be used at hazmat incidents:

1. Structural firefighting clothing is designed to protect against extremes of temperature, steam, hot water, hot particles, and the typical hazards of firefighting.
2. Chemical protective clothing is designed to protect skin and eyes from direct chemical contact. There are two basic types of CPC used: chemical splash protective clothing and chemical vapor protective clothing.
3. High temperature protective clothing is designed to protect against short-term exposures to high temperatures, such as proximity and fire entry suits.

The EPA has developed a classification scheme for the various levels of chemical protective clothing. Figure 8.7 outlines the equipment and its associated protection level. For our purposes in this chapter, protective clothing and equipment will be discussed in terms of its use—respiratory protection, structural firefighting clothing, chemical protective clothing, and high-temperature protective clothing.

STRUCTURAL FIREFIGHTING CLOTHING

While structural firefighting clothing (SFC) is the most common type of PPE used by emergency responders, it has a number of vulnerabilities when worn in hazmat environments. Although SFC may offer sufficient protection to the wearer who is fully aware of the hazards being encountered and the limitations of the protective clothing, it is normally not the first PPE choice for most hazmat response scenarios. An exception to this statement would be flammable gas and liquid fire incidents where SFC and SCBA will provide sufficient protection for most response scenarios.

For our purposes, SFC includes a helmet, positive-pressure SCBA, PASS device, turnout coat and pants, gloves and boots, and a hood made of a fire-resistant material. The ensemble should meet *NFPA 1971—Standard on Protective Ensemble for Structural Firefighting* requirements and is shown in Figure 8.8.

Figure 8.8 Structural firefighting clothing.

SFC provides limited protection from heat and cold but may not provide adequate protection from hazardous vapors and liquids. SFC may be used when the following conditions are met:

- Contact with splashes of extremely hazardous materials is unlikely.
- Total atmospheric concentrations do not contain high levels of chemicals toxic to the skin. In addition, there are no adverse effects from chemical exposure to small areas of unprotected skin.
- Live victims who are in need of rescue, such as those found in a terrorist attack. See Scan Sheet 8B for additional information in this area.

The increased presence of plastics and other toxic or carcinogenic synthetic materials found in structural fires has also led to increased concerns with the contamination and decontamination of SFC. Products of combustion include inorganic gases (e.g., hydrogen sulfide, nitrogen oxides), acid gases (e.g., hydrochloric acid, sulfuric acid), hydrocarbons (e.g., benzene), metals, and polynuclear aromatic compounds (PNAs). The inspection, cleaning and maintenance of SFC should be in accordance with *NFPA 1851—Structural Fire Fighting Protective Ensembles.*

Hazardous chemicals can both penetrate and permeate firefighting protective fabrics. Certain areas are more likely to absorb materials than others. Consider the following points:

- Clothing and equipment materials are porous and are easily contaminated by chemical penetration, such as:
  - Turnout clothing outer shells, thermal liners, collars, and wristlets
  - Station/work uniforms
  - Glove shells and liners
  - Fire retardant hoods
  - Boot linings
  - Helmet straps and linings
  - SCBA straps
- Coated or rubberlike materials are more likely to be affected by chemical permeation, such as:
  - Moisture barriers
  - Reflective trim
  - Boot outer layers
  - SCBA masks
  - Hard plastics used in the helmet and SCBA components
- Ash, resins, and other smoke particles can easily become trapped within the protective clothing fibers.
- Infectious bloodborne diseases, including the HIV, hepatitis B, and hepatitis C viruses, can be readily absorbed into the protective clothing fibers. SFC should be certified as protective against bloodborne pathogens.

Body protection. The outer shell that provides thermal protection is constructed of materials such as Kevlar™, PBI™, or Nomex™. Turnout coats and pants should be constructed with a moisture barrier, usually neoprene, Goretex™, or similar materials. Note that the manufacturers of Goretex™ recommend that their fabric not be worn in any type of chemical atmosphere since it will not stop the passage of chemical vapors.

The most serious problem faced when using SFC for hazmat operations is ensuring that all exposed skin surfaces are covered and protected. A hood made of fire-resistant materials such as Nomex® or Kevlar® will provide some protection for the head, ears, neck, and throat. When worn properly, they do not interfere with the SCBA face seal. A disadvantage is that any chemical splashed onto the hood may be absorbed and remain in direct contact with the skin.

In some situations, duct tape or elastic bands may be applied around the neck, waist, forearms, and ankles for additional vapor protection. These materials provide a false sense of security and do not increase the ability of SFC to provide either chemical splash or vapor protection. In addition, bleach materials that may be used for decon operations will also weaken several of the fire retardant materials used in SFC. **Remember—structural firefighting clothing is not designed to offer chemical protection!**

Gloves must be selected in reference to the tasks to be performed and the specific chemicals they will be exposed to. Because of the likelihood of physical contact, protective gloves should be considered as a critical element in the protective clothing ensemble. Factors to evaluate include chemical resistance, physical resistance, and temperature resistance.

Cotton, synthetic fiber, leather, firefighting, and rescue gloves will absorb liquids, keeping these chemicals in contact with the skin when working around any hazardous liquids. They will also deteriorate when exposed to corrosive liquids. In comparison, synthetic rubber and plastic gloves may melt when exposed to high temperatures associated with firefighting or may deteriorate on contact with certain petrochemical products. They also may not provide adequate protection against many corrosives and agricultural chemicals. Polyvinyl alcohol (PVA) gloves have excellent compatibility against certain petroleum solvents but break down on exposure to water.

Products that penetrate natural rubber and silicone may also create serious exposure problems for gloves, boots, and SCBA facepieces. Examples include methyl bromide, dichloropropene, and some chemical agents.

Respiratory protection. Since toxic, corrosive, and flammable vapors along with the products of combustion are present, air supplied respiratory protection devices are required. Positive-pressure SCBA is the minimum level of respiratory protection. Because of problems with decontamination before refilling, additional SCBA and air supply units are almost always required at hazmat incidents.

It is not uncommon for exposure to a specific chemical or hazmat environment to require the complete discarding of all SFC. Leather and fibrous materials are easily permeated by many chemicals and make decontamination difficult at best. Polycarbonate helmets may be affected by solvents. Pesticides, PCB-related fires, and radioactive materials incidents may make any decon impossible. Disposal should be done in accordance with local, state, and federal environmental regulations.

SCAN 8-B

STRUCTURAL PROTECTIVE CLOTHING AND CHEMICAL AGENTS

The U.S. Army Research, Development, and Engineering Command (RDECOM) in Edgewood, Maryland has conducted a series of research projects on the use of SFC and SCBA in a chemical agent environment where rescue operations may be required. In June 2003, RDECOM released an update to its initial August 1999 research report titled *Risk Assessment of Using Firefighter Protective Ensemble with SCBA for Rescue Operations during a terrorist Chemical Incident.*

The goal of the RDECOM report is to provide Incident Commanders with an understanding of the protection afforded by SFC and the risks involved if SFC is worn while performing rescue operations at the scene of a terrorist incident involving the use of military chemical warfare agents.

The following basic operational considerations summarize the contents of the RDECOM report:

- The presence of living victims inside the potential hazard area provides the basic indicator for firefighters to assess the level of nerve agent contamination.
- Rescue entry occurs after vapor concentration has peaked (assumed 10 minutes after the release of the agent).
- Firefighters using standard turnout gear and SCBA to perform rescue of known live victims can operate in a nerve agent vapor hazard for up to 30 minutes with minimal risks associated with nerve agent exposure.
- The risks associated with these 30-minute operations are that 50% of firefighters may experience increased sweating and muscle weakness 1–18 hours after exposure.
- Firefighters entering a nerve agent environment without known live victims using standard turnout gear and SCBA should limit their potential exposure to 3 minutes.
- Firefighters searching an enclosed area for victims should immediately exit the area and undergo decontamination if they encounter evidence of chemical contamination and cannot identify any living victims.
- If firefighters encounter oily liquid contamination (puddles/drops) and victims report signs of mustard agent (i.e., garlic odor), firefighters and victims should immediately exit the area and undergo decontamination.

The following points should be recognized in reviewing the preceding operational considerations:

1. SFC is not a substitute for chemical protective clothing. However, in certain response situations the Incident Commander may find himself or herself faced with a casualty rescue mission without having immediate access to CPC. Responders should recognize that there is a higher degree of risk to firefighters using SFC and SCBA than using chemical protective clothing.
2. SFC was tested against chemical vapor agent only. Chemicals that simulate the known characteristics of specific chemical agents were used for the testing. Future RDECOM research will address liquid hazard testing and analysis.
3. Rescue operations are based on the presence of live, viable victims being in the hazard area once responders arrive on scene. This is estimated to be approximately 10 minutes after the release of the chemical agent.
4. All responders and victims exiting the hazard area will immediately undergo a water-based (high volume, low pressure) decontamination.
5. Testing was performed on a limited number of various SFC gear configurations and styles used by the fire service. While thought to be representative, other configurations and styles may offer different amounts of protection.
6. Using SFC and SCBA while rescuing known live, viable victims in a chemical agent environment, does not justify using SFC for hazmat response operations (e.g., initial recon, leak control).

LIQUID CHEMICAL SPLASH PROTECTIVE CLOTHING

There are many hazmat incidents where the hazards and potential harm of the released material may require that specialized clothing be worn. Liquid chemical splash protective clothing consists of several pieces of clothing and equipment designed to provide skin and eye protection from chemical splashes. It does not provide total body protection from gases or vapors and should not be used for protection against liquids that give off vapors known to affect or be absorbed through the skin. Depending on the materials involved and the nature of response operations, respiratory protection may be initially be provided through SCBA or SAR, or by using APRs or PAPRs once air-monitoring operations have determined that the atmosphere allows for their use (i.e., EPA Level B or Level C configuration).

Liquid chemical splash protective clothing may be used under the following conditions:

- The vapors or gases present are not suspected of containing high concentrations of chemicals that are harmful to, or can be absorbed by, the skin.
- It is highly unlikely that the user will be exposed to high concentrations of vapors, gases, or liquid chemicals that will affect any exposed skin areas.
- Operations will not be conducted in a flammable atmosphere. If operations must be conducted in an environment with the potential for combined thermal and chemical hazards, responders should consider the use of flash protection overgarments available from some CPC manufacturers. Many response agencies also regularly use fire-retardant coveralls underneath the CPC.

Skin and body protection. Liquid chemical splash protective clothing is routinely used in hospitals and laboratories where various chemicals, biologicals, and infectious diseases are handled. They are also found in nuclear facilities and installations that handle or process radioactive materials and are used in handling mild corrosives and PCBs, in protecting against asbestos fibers and lead dust, and in formulating and applying agricultural chemicals.

In emergency response, liquid chemical splash protective clothing is often used for initial response operations, to protect decon personnel, and for postemergency response investigation and clean-up operations. Depending upon the application (e.g., emergency response versus law enforcement), CPC fabrics may be available in various colors.

Several common types include the following:

- **Single-piece suits**. Usually coveralls, a splash suit, or an encapsulating suit that is not vapor tight. Hoods and booties may be attached to coveralls and splash suits. Both limited-use and multiuse garments are available. Considering their low cost, absence of decon problems, and the varied needs of responders, single-piece limited-use suits are an excellent alternative used by many hazmat responders.
- **Two-piece suits**. Usually consist of bib overalls or pants worn with a jacket. Some ensembles include an expanded back or "humpback" design that covers an SCBA. They may encapsulate the user but do not provide vapor-tight protection. Accessories such as gloves, boots, and hoods are available.

Depending on the use of the CPC, seams may be sewn, bound, thermally welded, taped, or double taped.

Head protection. Hard hat, helmet, or hood. Some form of hard hat protection is recommended when using a hood or encapsulating suit. This gear is designed in various configurations and materials. Some manufacturers offer a respirator-fit hood that will provide a tighter fit around the facepiece and completely cover the neck area.

Gloves. Some coveralls and jackets have a sleeve mounted splash guard that prevents wrist exposure. Some manufacturers have also developed an O-ring glove and cuff assembly that ensures a leak-proof glove/cuff assembly. These designs allow gloves to be easily interchanged according to the hazard.

For maximum hand protection, overgloving and *doublegloving* should be used. Doublegloving involves the use of surgical gloves under a work glove. It permits doffing of the work glove without compromising exposure protection and also provides an additional barrier for hand protection. Doublegloving also reduces the potential for hand contamination when removing protective clothing during decon procedures. *Overgloving* is the wearing of a second glove over the work glove for additional chemical and abrasion protection during lifting and moving operations.

Gloves should be sized for use by individual responders. In general, size 9 gloves are commonly used for single- or doublegloves, while sizes 11 and 12 are typically used as overgloves for additional strength and chemical protection. Common glove materials include Viton®, neoprene, butyl rubber, Chloropel™, polylaminates such as SilverShield®, and polyvinyl chloride (PVC).

Footwear and shoe covers. Foot protection may be chemical boots, separate shoe covers, or booties that are part of the CPC ensemble. Boots should provide both chemical and mechanical protection (e.g., cuts, punctures, etc.). Chemical shoe boots are usually commonly used by responders, but work boots over footwear are also found. Common boot materials include PVC, neoprene, PVC/nitrile, and nitrile rubber.

Many liquid chemical splash suits constructed from limited-use materials have integral or connected sock booties. These attached "socks" are designed to be worn inside boots. They are not sufficiently durable or slip resistant to be worn as outer footwear. Specially designed shoe covers and booties are most often used in those areas where radioactive materials, etiological agents, and agricultural chemicals are handled or where cleanliness is a concern. There are chemical protective suits with attached boots specifically designed for hazmat response operations.

Aprons and body coverings. Aprons, lab coats, sleeve guards, and other body coverings are designed for protection against spills and splashes that occur when physically handling chemicals and other hazardous materials. They are primarily used for routine chemical handling operations rather than emergency response. Aprons and similar clothing may also be worn by responders while handling samples and performing HazCat tests.

Duct tape is sometimes used to ensure that zippers remain closed, to accommodate size differences, secure gloves and boots, and to secure the hood to the SCBA mask on some liquid chemical splash suits. Duct tape provides a false sense of security and does not increase the ability of the garment to provide liquid chemical splash protection. Duct tape is not a substitute for a properly designed garment. While it may be used to secure a glove over a glove gauntlet and cuff over a boot, duct tape does not provide a totally leak-proof seal. In some instances, duct tape may actually damage the garment material. There is a commercially available chemical tape (Chem-Tape® 2) that is specifically designed for protective clothing applications.

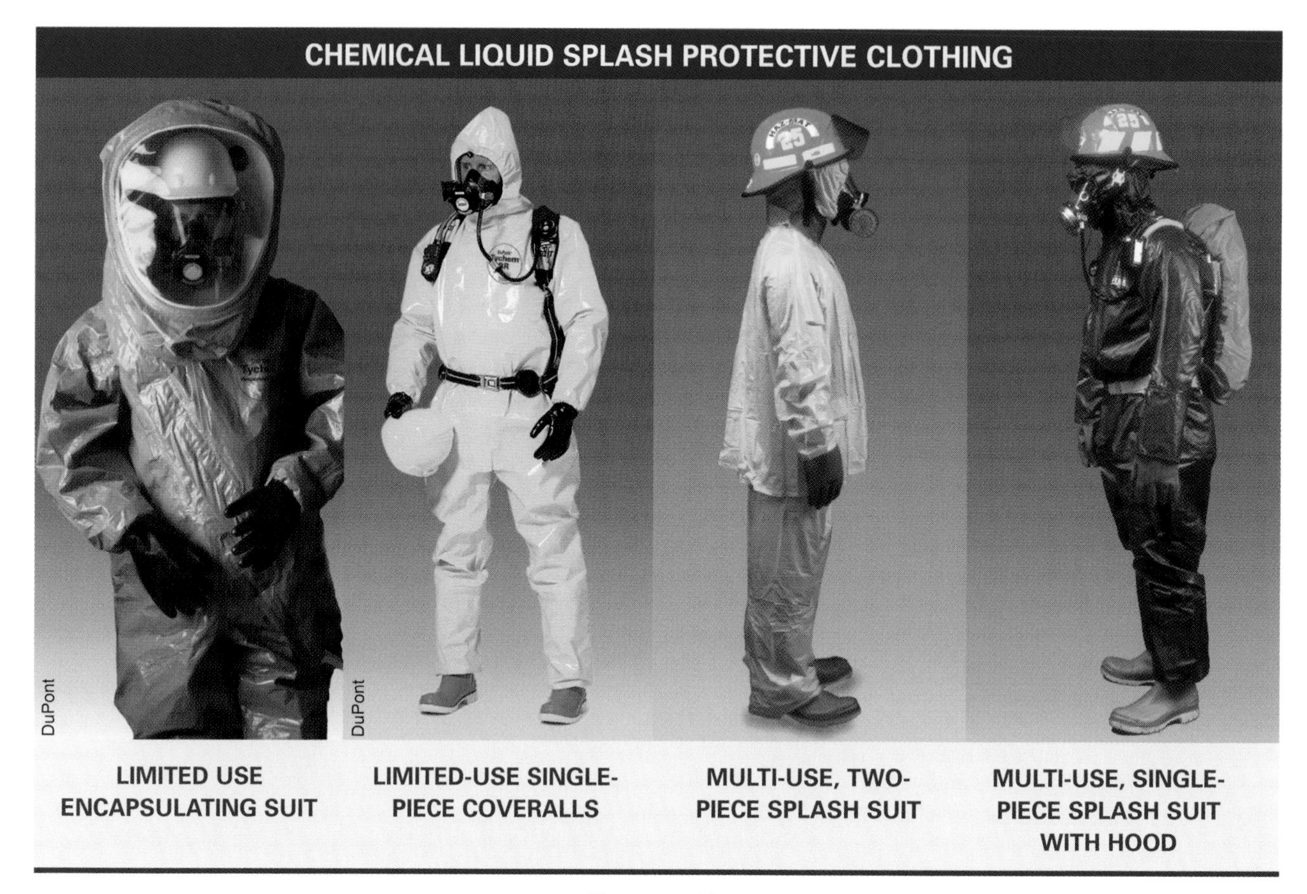

Figure 8.9 Examples of chemical liquid splash protective clothing.

CHEMICAL VAPOR PROTECTIVE CLOTHING

Chemical vapor protective clothing (i.e., EPA level A clothing) provides full-body protection with vapor-tight integrity. When used with air-supplied respiratory devices, it provides a gastight envelope around the wearer.

Chemical vapor protective clothing should be used when the following conditions exist:

- Extremely hazardous substances are known or suspected to be present, and skin contact is possible (e.g., cyanide compounds, toxic and infectious substances).
- There is potential contact with substances that harm or destroy skin (e.g., corrosives).
- Anticipated operations involve a potential for splash or exposure to vapors, gases, or particulates capable of being absorbed through the skin.
- Anticipated operations involve unknown or unidentified substances and the scenario dictates that vapor-tight skin protection is required.

Skin and body protection. Chemical vapor protective clothing is manufactured in several configurations. The most common is where the SCBA is worn underneath the ensemble, thereby providing total vapor protection by encapsulating the wearer. This configuration is easily identified by its "humpback" expanded-back design. CPC manufacturers will also incorporate an airline hose bulkhead connection onto the suit if a supplied air respirator will be used. These connections vary by respirator

- Emergency responders have the right resources (people, equipment, and supplies) to get the job done safely and rapidly.

Coordinating the information and resources required to resolve a working hazmat emergency may seem like an overwhelming problem (and it often is), but it isn't that difficult if you use a system. In this chapter, we will review several structured and systematic ways that responders can use to organize information and resources so they can be used to their fullest advantage.

Before we get into the chapter in detail, let's review several key points that we discussed in previous chapters that relate to information management and resource coordination:

- Failure to get the right information and resources to the right people at the right time can jeopardize the safety of responders and the overall success of the emergency response effort.
- Information and resources cannot be coordinated effectively if an incident management organization is not in place. The Information Officer and the Liaison Officer are the key players for moving information within the command structure and to external agencies. The Logistics Section Chief and the Staging Officer are the key players for obtaining and moving resources at the incident.
- Information that is poorly coordinated among the players at the emergency scene can politically damage the Incident Commander's credibility and ultimately undermine the response operation. When responders don't understand the plan or perceive that one doesn't exist, they improvise and start to freelance. Freelancing (also known as running around without a plan and doing your own thing) can get people hurt.
- The IC's Incident Action Plan (IAP) must have a solid technical basis (e.g., the actions proposed must be within the limits of science and technology and respect the basic laws of engineering, chemistry, and physics). A heavier-than-air gas will always be heavier than air, and it takes a 50-ton crane to lift 50 tons. If you buy into an action plan that has Magic Foo Foo Dust sprinkled onto the scientific facts you will eventually end up regretting it.

MANAGING INFORMATION IN THE FIELD

THE BASICS OF INFORMATION MANAGEMENT

Decisions cannot be made without reliable information. But how do we know the information we get is reliable, and how much information do we really need to make good decisions?

The reliability of the information used in decision making depends on the quality of the data and facts used to compile that information [Data + Facts = Information]. The IC must make decisions based on reliable data and facts, not assumptions.

- **Data**—Individual data elements that are gathered and organized for analysis. At a hazmat emergency the data we are most interested in concern data that describe the material's physical (i.e., how it behaves) and chemical (i.e., how it harms) properties, such as specific gravity, flash point, exposure values, and vapor density. The reliability of most published data on the various characteristics and hazards of hazardous materials is pretty high. Some common mistakes made at the emergency scene regarding data include looking up the wrong chemical in the database, not copying the information down correctly, and failure to validate the data using another reference source.

Data is no substitute for thinking! If there is any doubt about the data you are looking at, it's okay to question the source or authenticity. Just because it is written down on paper or shows up on a computer screen does not mean that you have to buy into it, subscribe to it, or believe what it says. Be inquisitive and challenge information you may have to rely on in the field.

- **Facts**—Statements made or observations about something that has occurred and has been verified and validated as being true. In emergency response work, facts are typically based upon objective observations made by trained and experienced personnel. For example, a digital photo taken by your RECON team of a score across a welded seam on an MC-331 propane transport is a pretty reliable fact. Comments by the RECON team that the metal below the surface of the weld seam may be cracked is just an opinion; it is not a fact. The RECON team's observation that the flammable gas instrument did not detect any flammable vapors near the damaged weld seam is a fact. When these two facts are combined, you have credible information that can help you make some decisions, or at least allow you to ask the right questions of product and container specialists.

Figure 9.1

INFORMATION MANAGEMENT REQUIREMENTS

Imagine that you have responded to a freight train derailment at 3:00 AM. There are 80 cars in the train and 20 contain different types of hazardous materials. The railroad presents you with a computer consist profiling how the train is made up, including emergency response information sheets on each hazmat carried in the train. You have many different documents to look at. This doesn't even include other sources of information you may have available through CHEMTREC®, technical reference manuals, computer databases, or pre-plans. Sorting, evaluating, interpreting, and communicating this information to people who need it can be a perplexing problem if you don't use a system to manage it.

Information management must begin well before the incident. Important decisions that must be made include the following:

1. What type of information will be needed at the emergency scene? How should the information be compiled?
2. What is the priority of the information that is needed? What do you need immediately versus an hour into the incident?
3. How will the information be stored for quick recovery at the incident scene—manually or electronically?
4. Are the information and retrieval systems suitable for field applications?
5. Who will be responsible for managing and coordinating information at the incident scene? Are they properly trained and equipped for the job?

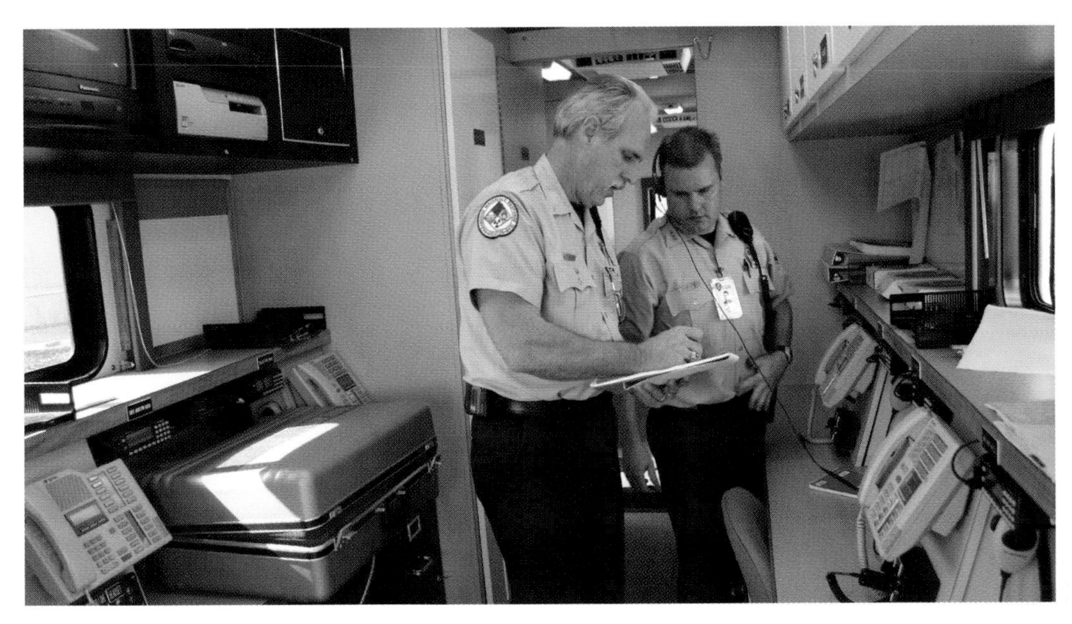

Figure 9.2 Information management must begin well before an incident.

WHAT DO YOU REALLY NEED TO KNOW?

In the old days responders lacked reliable and accessible hazard data to make decisions. Computers and the Internet solved that problem. Now we have access to what seems be unlimited information. What do we really need to know at a hazmat emergency? The process involves identifying what you need to know versus what is nice to know. Example: Is it more important to know the time or how the watch works and who made it? In the hazmat world your need to know information list should read something like this:

1. What are the hazards of the materials (e.g., flammability, toxicity, and reactivity)?
2. What are the PPE requirements (or what must I do to protect myself)?
3. What are the health concerns (e.g., exposure values, signs and symptoms of exposure, antidotes, etc.)?
4. What is the container type and condition (integrity, size, orientations etc.)? Hazmat incidents are a two-part problem—the product and the container.

You can have the worst stuff in the world but the situation is usually under control if the container is stable and the product is still in the container. Conversely, even corn syrup can present a serious problem if the container is breached and a lot of the product is on the ground in the wrong place at the wrong time.

5. What are the initial tactical recommendations (e.g., spill control, leak control, fire control, public protective actions)?
6. What type of decontamination procedures and methods will be required?

Trying to collect and carry a hard copy of every source of hazard and response information in an emergency response vehicle just isn't practical. Responder information needs will be different based on local or facility conditions.

There are three basic groups of information sources that should be immediately accessible from the incident scene:

- Facility Emergency Response Plans
- Pre-Incident Tactical Plans
- Published Emergency Response References
- Shipping Documents

FACILITY EMERGENCY RESPONSE PLANS

Refineries, chemical plants, and facilities that manufacture or store hazardous materials on site are required by OSHA to have a Facility Emergency Response Plan. These plans usually have a hazard analysis section, which identifies special problems that may exist within the plant or community, the potential risks and consequences of a hazmat incident, and the available emergency response resources in the facility.

While an Emergency Response Plan does not typically have much use in the field ("If we need the Plan to figure out what to do at 3 o'clock in the morning, we've got bigger problems than the Plan can fix!"), it is a good beginning point to identify target hazards for which a more detailed Pre-Incident Plan should be developed. This process may allow you to determine the types of information to be stored in a database.

Process safety and risk management documents can also be good sources of hazard analysis information. For example, the hazard analysis section may identify that poison gas is used in a particular process at a chemical plant. Having specific toxicity/exposure data and evacuation information would be very useful for immediate retrieval at an actual incident.

Chapter 1 describes several types of hazard analysis tools and sources of planning information that are available. Familiarize yourself with these different techniques so that you know what to ask for when you write or visit a facility for information.

PRE-INCIDENT TACTICAL PLANS

Pre-Incident Tactical Plans (or pre-plans) are like the quarterbacks play book on a football team. The plan explains exactly who does what and where they are supposed to be to execute the play. Pre-plans focus on a specific problem or location, such as a rail yard or a bulk storage facility. In fixed facilities with chemical manufacturing operations, pre-plans may concentrate on a particular process or tank(s) with special hazards.

PRE-PLANS SHOULD PROVIDE SPECIFIC TACTICAL INFORMATION

Pre-Determined Emergency Response Procedures

West Plant - Tank 142
Koch Refining Company, L.P.

Pre-Plan Section	Pre-Planned Event	Incident Level
A	Rim Seal Fire	Level I or Level II
B	Dike Fire	Level II
C	Fully Involved Tank	Level III

RTFC Pre-Plan

FACILITY ACCESS

Looking north up west-side access road.

BFG potential placement area at corner of H & 2nd.

Fire Boat / aux. pump access at Dock 9

Tank 142 foam manifold, SE corner along "H" street

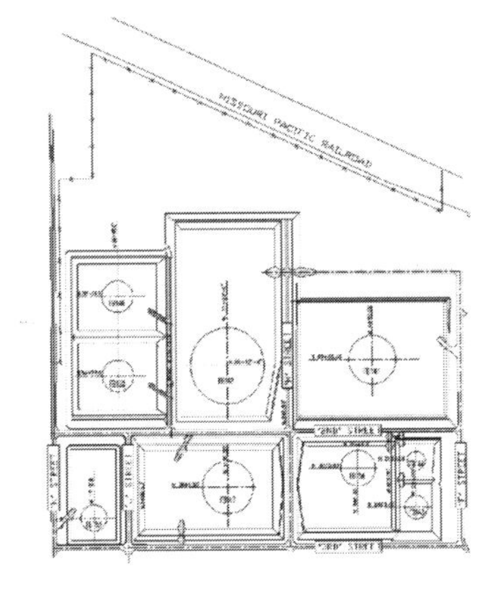

Looking west along 2nd St. - view of SE corner of tank 142

Fire Protection Equipment — Fixed and Tactical

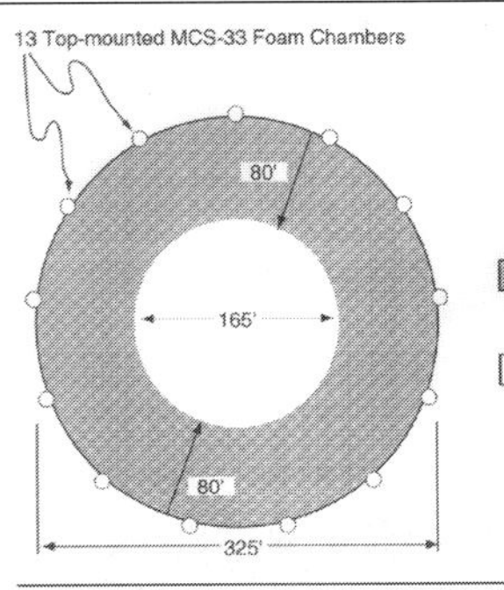

Area to be extinguished by fixed foam system

Area to be extinguished by BFG nozzle system

Foam Chamber Sector

Fixed Fire Protection:
(13) Top-mounted MCS-33 Foam Chambers, each rated at 638 gpm

Total Solution Flow:	8,296 gpm
Total Flow Time:	65 mins
Concentrate required:	16,177 gal.

Tactical Approach

Koch's pumper	@ 4,000 gpm
RTFC F-5	@ 4,000 gpm
Water required:	8,000 gpm
Concentrate required:	15,600 gal
Alcoseal transport	5,000 gal
F-5 Alcoseal tank	2,000 gal
RTFC XL-3 transport 1 - RTFC tractor	5,000 gal
RTFC XL-3 transport 1 - Koch tractor	5,000 gal
Concentrate available	**17,000 gal**

BFG Sector

165 ft diameter

@ 0.25 gpm / sq ft =	5,343 gpm
Operate BFG @	8,000 gpm
Total Flow Time =	65 mins
Concentrate Required:	15,600 gal

RTFC BFG	8,000 gpm
RTFC PT-1	4,000 gpm
RTFC PT-2	4,000 gpm
RTFC Hose Tender (H-1)	
RTFC Tractor 2	
RTFC ATC Transport 1	5,000 gal
RTFC ATC Transport 2	5,000 gal
RTFC ATC Transport 3	5,000 gal
Koch ATC Transport	3,100 gal
Concentrate available	**18,000 gal**

DISPATCH PROTOCOL

Equipment Housing Locations

RTFC Main Station
- RTFC Hose Tender H-1
- RTFC Pump Trailer PT-2
- RTFC Tractor 1
- Angus Alcoseal Transport
- RTFC ATC Transport 1
- RTFC Special Unit 1
- RTFC Rescue 1
- RTFC Foam 1 / Utility 1
- RTFC Foam 3 / Utility 3
- RTFC Foam 5
- RTFC Command Post CP-1
- RTFC Reserve Command Suburban C-6
- RTFC Forklift

Training Academy Station
- RTFC Special Unit 2
- RTFC BFG Trailer
- RTFC Tractor 2
- RTFC Pump Trailer PT-1
- RTFC Tractor 3
- RTFC ATC Transport 2
- RTFC Tractor 4
- RTFC ATC Transport 3
- RTFC Foam 4
- RTFC Foam 2 / Utility 2

Koch West Plant Fire Barn
- Koch Tractor
- Koch ATC Transport (3,100 gal)
- RTFC XL-3 Transport 4
- RTFC XL-3 Transport 5

The larger the facility or jurisdiction, the more essential that the pre-planning process be practical and prioritize problem areas based on the hazard analysis process. Criteria for developing special pre-plans in these situations may include the following:

- Type of hazards and risks present. Facilities that present high risks to the community should be pre-planned. Facilities that have a low probability of something going wrong, but a high consequence if it does go wrong, are good candidates for a pre-plan. When we respond to these facilities we have to get it right the first time. Nuclear plants, chemical plants, petroleum bulk storage facilities, and refineries are good examples.
- Environmentally sensitive exposures. Facilities in close proximity to waterways or sole source aquifers present high financial and environmental risks that need to be well thought out. For example, if the high-risk storage paint warehouse that is in close proximity to the community's primary aquifer catches on fire, should an offensive fire attack be initiated? If the answer is yes, what objectives and tactics should be implemented to control any runoff?
- Unusual or poor water supply requirements. Examples include facilities located in an area with a reputation for water supply problems or installations that may require unusually high fire flows (e.g., large diameter flammable liquid storage tank). You probably know where these areas are. Do you have a water supply plan for them?
- Locations that will require large quantities of foam concentrate such as bulk petroleum storage facilities and pipelines. The quantity requirements need to be known as well as application rates, availability of foam supplies, and so on.
- Restricted or delayed response routes. (for example, single approach and access corridors and railroad tracks or draw bridges that are frequently blocked).
- Poor accessibility. Examples include restricted entrance corridors, secure government installations with strong antiterrorism force protection countermeasures in place, obstacles and unusual ground slope.

Pre-plans can provide valuable information if the right type of field survey form is used to record key information during the site visit. Although informal site visits may be instructive for the personnel participating in them, they often provide little long-term benefit if there is no mechanism for compiling, maintaining, and disseminating key response information.

To be useful, pre-planning documents must be completed using a standardized format. A well-designed survey form assures that essential information is gathered and consistently recorded. Field testing a survey form before adopting a standard format will pay big dividends for your planning, training, and response programs. Bigger is not necessarily better; a simple form can reduce the maintenance time required to keep pre-plans current.

A well-prepared pre-plan should include a simple plot plan that shows the basic details of the facility but is not cluttered with extraneous information. The pre-plan should also indicate the availability of any special plans prepared by the facility that may be referenced during an emergency (for example, foam calculations, tactical checklist). One special note regarding pre-plans is Operations Security or OPSEC. We must apply good OPSEC to how we store pre-plans and control who has access to this information. In the wrong hands, pre-planning information could be used by criminals and terrorists to commit a crime or plan an attack against the facility we are trying to protect.

Key transportation areas should be pre-planned as well. Obvious areas include railyards and trucking terminals but should also include high-traffic-density highways, particularly interchanges with a history of trucks overturning, bridges, tunnels and toll plazas. These areas can be pre-planned for access, topography, environmental exposures, sensitive receptors, high-density populations and so on.

PUBLISHED EMERGENCY RESPONSE REFERENCES

Most emergency responders have some capability to access hazard information from laptops, on-board computer systems, or by tapping into the Internet through cellular, wireless or hard-line connections. There are a wide variety of published emergency response references available on the Internet, in hard copy, or through 24/7 watch offices. These references are generally divided into the following categories:

- Reference manuals and guidebooks (e.g., DOT ERG, AAR emergency action guides). You need to be selective in what you decide to purchase and carry to the scene of the emergency. It is much better to have a few published references that you have trained on and are comfortable with than to have a large and unfamiliar reference library.
- Technical information centers accessible by telephone (e.g., CHEMTREC®, CANUTEC).
- Hazardous materials databases accessed through the Internet either as public domain or subscription databases (e.g., TOXNET, TOMES).

Chapter 7 provides a good overview of the different types of widely recognized published references. Familiarize yourself with these sources of information and know how to interpret them before you need them.

INFORMATION STORAGE AND RECOVERY

Most HMRTs have field access to portable computer systems or a Personal Data Assistant (PDA). Having access to a lot of information on a computer is a big advantage, but don't overlook the simplicity of a three-ring binder, reference books, and a small on-board filing cabinet. Site drawings and process flow diagrams are especially well suited for hard-copy use.

No matter how good your on-board computer system is, it does not necessarily guarantee an improvement to your information management system at the incident scene. Don't expect much from your portable computer system if you haven't read the manual and had some realistic training that simulates field conditions.

The key to successfully managing and retrieving hazmat information under emergency conditions is good organization and simplicity. The "acid test" for deciding whether one type of information management system is better than another should be, "Will it work in the field in a consistent and reliable manner?" As previously noted, there will always be some situations where manual information management systems may be better suited than computerized systems.

When evaluating systems, consider the following:

- **User friendly**—Stressful situations call for simple solutions. Beware of hardware and software that requires a lot of training and experience to operate. Almost every HMRT has an in-house computer wizard. You know the type... you ask them a simple question in plain English and you get a high tech answer: "Hey, Chip Head, why won't this thing boot up?" Well, your hard drive probably has inadequate RAM capacity to synchronize with your main power feed boot-up, and therefore, you have generated a reverse C Drive power fluctuation that has incapacitated your Java script, and blah, blah, blah, blah." Don't build your information system around the expertise of one individual! The same case can be made for every piece of equipment in your inventory. If you rely on one person to work your gee wiz toys, you are "people dependent" and not system dependent. The best teams out there we have seen are system dependent not people dependent.
- **Durability**—We don't operate in a nice clean office environment. We operate in a dirty, hot, cold, humid, and dry environment. Mongo the Hazmat Tech, and a little spilled coffee on the keyboard often equals a dead computer. If you rely on a computer in the field, (1) the information should be backed up; and (2) it should be Mongo-proof.

Chemical databases are great for researching data on chemicals you have never dealt with before, but why spend a lot of time doing database research on common products like sulfuric acid? Commonly encountered products should be pre-planned. The research should be done ahead of time and kept in a computer file or hard copy.

Remember—a computer's outstanding attributes of speed, storage capacity, consistency, and the ability to process complex logical instructions are of no value unless they are applied within a good management process and the computer works 24/7 when you need it.

COORDINATING INFORMATION AMONG THE PLAYERS

Coordinating information in the field becomes particularly important as the Incident Commander, Hazmat Group Supervisor, and others evaluate options concerning protective clothing, decontamination requirements, and public protective actions.

Coordinating information is a dynamic process that must adjust its scale over time to provide the correct and credible information to the right people at the right time. The larger and more complex the incident, the larger the command organization needed to manage the incident. The larger the command organization, the more need there is for a formal structure to manage the data and facts that will flow between Command and the various individuals and organizations at the emergency scene. Information must also flow freely to and from the incident scene to off-site support facilities, such as the Emergency Operations Center (EOC), CHEMTREC®, and elected officials.

HAZMAT GROUP FUNCTIONS

If your organization has some depth in resources, an effective way to organize and manage a working hazardous materials incident is for Command to form a Hazmat Group and then delegate the responsibility for hazmat information to the Hazmat Group Supervisor. The Hazmat Group then sub-divides its functions into specific

Units, as described in the Hazmat Group Operations section of Chapter 3. Following this general approach, the Hazmat Group Supervisor assigns different functions to the first available and qualified person as the situation requires.

Primary functions and tasks assigned to the Hazmat Group include the following:

- **Safety function**—Primarily the responsibility of the Hazmat Group Safety Officer. Responsible for ensuring that safe and accepted practices and procedures are followed throughout the course of the incident. Possesses the authority and responsibility to stop any unsafe actions and correct unsafe practices.
- **Entry/backup function**—Responsible for all entry and backup operations within the hot zone, including reconnaissance, monitoring, sampling, and mitigation.
- **Decontamination function**—Responsible for the research and development of the decon plan and set-up and operation of an effective decontamination area capable of handling all potential exposures, including entry personnel, contaminated patients, and equipment. If necessary, will include the coordination of a Safe Refuge Area.
- **Site access control function**—Establish hazard control zones, establish and monitor egress routes at the incident site, and ensure that contaminants are not being spread. Monitor the movement of all personnel and equipment between the hazard control zones. Manage the Safe Refuge Area, if established.
- **Information/research function**—Responsible for gathering, compiling, coordinating and disseminating all data and information relative to the incident. This data and information will be used within the Hazmat Group for assessing hazards and evaluating risks, evaluating public protective options, the selection of PPE, and development of the incident action plan.

Secondary support functions and tasks that may be assigned to the Hazmat Group will include the following:

- **Medical Function**—Responsible for pre- and post-entry medical monitoring and evaluation of all entry personnel, and provides technical medical guidance to the Hazmat Group, as requested.
- **Resource Function**—Responsible for control and tracking of all supplies and equipment used by the Hazmat Group during the course of an emergency, including documenting the use of expendable supplies and materials. Coordinates, as necessary, with the Logistics Section Chief (if activated).

See Figure 3.7 (Chapter 3) for an example of how each of these functions is organized within the Hazmat Group.

CHECKLIST SYSTEM

The most simple and reliable method of coordinating information between the various Hazmat Group functions is to use the checklist system. Formal checklists have several distinct advantages as they relate to information management in the field. These include the following:

- Checklists don't panic. We all have bad days now and then, and when you are having a really bad day and are falling apart, a well-thought-out checklist can get you back with the program. If you are confused or lack the field experience, get the checklist out and start working through your assigned tasks.

- Checklists have institutional memory. Pilots routinely use checklists to take off and land their aircraft and to work through in flight emergencies. Newer generation of aircraft have these checklists integrated into cockpit information systems when an emergency occurs. Years of good and bad experience go into developing a checklist so that the lessons learned from the past are "institutionalized" and remembered from one pilot to the next. If you routinely critique your incidents and exercises, part of the feedback of the lessons learned should go into making your checklists smarter.
- Identify the tasks assigned to each Hazmat Group function. A good checklist assigns duties and responsibilities to each member of the team so that work is not duplicated and important information gathering activities are not overlooked.
- List critical activities and action items required for each function. As elements are addressed, they are formally "checked off" as a method of verifying that the activity has been completed. Old hands in the business pretty much have the checklist in their heads and work through what they need to do mentally, but an experienced professional eventually gets out the checklist and runs through it to make sure that something has not been overlooked.
- Prioritize actions so that important activities are completed early in the incident.
- Provide a framework for development of the IAP.
- Identify which Hazmat Group functions or individuals need to be formally contacted to coordinate information.
- Provide the required documentation of the incident for the post-incident analysis and the critique. See Chapter 12 for a discussion on these two activities.

In order for the Checklist System to be effective, checklists must be updated on a regular basis. It is important that responders take ownership of the checklists and make gradual improvements to the system over time.

While there is nothing wrong with "borrowing" an established checklist system from another organization as a beginning point, checklists should be adapted and customized to suit local needs.

HAZMAT INFORMATION OFFICER

The person designated as the HazMat Information Officer (INFO) will play a key role in the successful mitigation of any hazmat incident. This individual should be chosen because of his or her ability to communicate, comprehend and manage information, work effectively under stress, and coordinate activities with individuals from different backgrounds.

Extended incidents or incidents involving multiple chemicals (e.g., train derailment, building explosion with multiple chemicals) may require that an Information Unit be formed. This allows the workload to be split up into areas such as on-scene library research, contacting CHEMTREC®, the manufacturer, computer operations, and on-scene data collection. CHEMTREC® is a great resource early in the incident, but for a serious campaign incident you will usually be dealing with the product manufacturer or facility personnel after the first two hours.

In situations where there is insufficient room for everyone to function comfortably,

Prince George's County Fire/EMS Department Hazardous Materials Checklist

Location:			**Date:**
Incident Commander (IC):		**Operations:**	
Team Leader (TL):		**HAZMAT OPS:**	
HAZMAT Safety:		**Information (INFO):** /	
Decon:	**Entry:**	**Resources:**	**EMS:**

INFO Functions

The **Information** (**INFO**) Officer shall assume responsibility for developing, documenting, and coordinating all data relevant to the incident.

All data gathered shall be coordinated with **HAZMAT OPS**, **SAFETY**, **DECON**, **ENTRY**, **RESOURCE**, and **EMS** as appropriate.

A minimum of **three** (**3**) information sources shall be used to develop hazard and risk information.

INFO Operations

- ☐ **INFO** identified by vest.
- ☐ Collect all Personnel Accountability Tags.
- ☐ Identify hazardous material(s) involved.
  - **1.** ____
  - **2.** ____
  - **3.** ____
  - **4.** ____
  - **5.** ____
- ☐ Identify isolation and protective action distances.
  - Isolate: ____
  - Protective Action: ____
- ☐ Identify Control Zones.
  - Hot: ____
  - Warm: ____
- ☐ Is evacuation of exposures required based on situation? ☐ Yes ☐ No (Advise **HAZMAT OPS**.)
- ☐ Identify facility representatives (get business cards).

Name	Company	Phone #

- ☐ Obtain MSDSs and preplan, if available.
- ☐ Coordinate with **ENTRY** to obtain shipping papers, when applicable.
- ☐ Complete *Hazardous Materials Data Worksheet* for each hazardous material involved or potentially exposed.
  - ☐ Identify data sources used for each worksheet (*minimum of 3*).
- ☐ Coordinate with **ENTRY** to identify hazardous material container(s) involved or exposed.
  - Number: ____
  - Type: ____
  - Condition: ____
- ☐ Identify signs and symptoms of exposure.
  - ____
  - ☐ Brief **SAFETY** and **EMS** on information.
- ☐ CHEMTREC contacted for assistance.
  - ☐ *CHEMTREC Worksheet* completed.
- ☐ Contact manufacturer(s)/shipper(s) for further information.

Company:
Contact Person:
Phone #: ( ) -
Discussed: ☐ Properties ☐ Hazards ☐ Deco ☐ PPE ☐ Handling ☐ Other -
☐ Enroute ETA: Time:

Company:
Contact Person:
Phone #: ( ) -
Discussed: ☐ Properties ☐ Hazards ☐ Deco ☐ PPE ☐ Handling ☐ Other -
☐ Enroute ETA: Time:

Company:
Contact Person:
Phone #: ( ) -
Discussed: ☐ Properties ☐ Hazards ☐ Deco ☐ PPE ☐ Handling ☐ Other -
☐ Enroute ETA: Time:

Prince George's County Fire/EMS Department Hazardous Materials Checklist

Company:
Contact Person:
Phone #: ( ) -
Discussed: ☐ Properties ☐ Hazards ☐ Deco ☐ PPE ☐ Handling ☐ Other -
☐ Enroute ETA: Time:

Company:
Contact Person:
Phone #: ( ) -
Discussed: ☐ Properties ☐ Hazards ☐ Deco ☐ PPE ☐ Handling ☐ Other -
☐ Enroute ETA: Time:

- ☐ Estimate likely harm without intervention.
- ☐ Is Hot Zone adequate? ☐ Yes ☐ No
- ☐ Identify protective clothing required for:
  Hot: ______
  Decon: ______
- ☐ Identify decon requirements.
- ☐ HAZMAT Officers briefed on current status.
- ☐ Assist in development of Mitigation Action Plan.
- ☐ Contact outside agencies, organization, or contractors when requested by **HAZMAT OPS**.

Agency:
Contact Person:
Phone #: ( ) -
☐ Enroute ETA: Time:

Agency:
Contact Person:
Phone #: ( ) -
☐ Enroute ETA: Time:

Agency:
Contact Person:
Phone #: ( ) -
☐ Enroute ETA: Time:

Agency:
Contact Person:
Phone #: ( ) -
☐ Enroute ETA: Time:

Agency:
Contact Person:
Phone #: ( ) -
☐ Enroute ETA: Time:

- ☐ Reevaluate situation with new information collected.

TERMINATION

- ☐ Identify information issues.
- ☐ Attend debriefing of HAZMAT personnel.
- ☐ Provide completed checklist, worksheets, and collected information to **HAZMAT OPS**.

Phone Numbers:

CHEMTREC	(800) 424-9300
National Response Center	(202) 267-2675
MD Department of Environment	(866) 633-4686
Poison Control	(800) 492-2414
Military Munitions	(703) 697-0218
Military Shipments (not munitions)	(800) 851-8061
Pesticide Hotline	(800) 858-7378

Notes:

Figure 9.3 Hazardous materials checklist.

one or more coordinators can move to nearby offices or houses to complete their assignments. Access to comfortable surroundings with telephones, computer or Internet access, office supplies, bathrooms, and other amenities of life makes the job much more endurable and the work product more reliable. Good lighting and an uninterrupted power supply are also essential.

RESOURCE COORDINATION

ORGANIZING RESOURCES

What are resources and how do they differ from information? As previously described, information is knowledge that is based upon data and facts that can be used to support decision making. In contrast, resources are made up of the people, equipment, and supplies required to manage a hazardous materials emergency. As with military operations, managing and moving resources around the incident scene are the basis of strategy and tactics.

Human resources—Include responders, support personnel, technical information specialists, and product or container specialists. Coordinating human resources is important because people provide the thinking power and manual labor required to bring the situation under control. People also represent the greatest financial, legal, political, and technical exposure for the Incident Commander. Chapter Three lists some of the different types of human resources ("players") which may be involved in a hazmat incident.

Equipment resources—Include items that are reusable, such as hand tools, generators, pumps, and monitoring instruments. Hazmat equipment can represent a substantial cost outlay whether it is rented, leased, or owned. Very few organizations are self-sufficient when it comes to the equipment required to resolve a large-scale hazmat incident. Consequently, good coordination is required to assure that the right equipment resources are available when they are needed, that they are tracked throughout the operation, that their use and costs are monitored, and that they are returned to their owner in a timely manner.

Supply resources—Differ from equipment resources in that they are usually considered expendable. In other words, you use them once or twice, then dispose of them. Examples include foam concentrate, decontamination solutions, limited-use protective clothing, product control materials, and most medical supplies. Hazmat supplies require special coordination because they are usually consumed in bulk quantities, may require special decontamination procedures, may take a long time until inventories are replenished, and can be expensive.

In major hazmat incidents the Incident Commander will appoint a Logistics Section Chief to manage the resources required. Within the command structure, the Logistics Section is typically organized into two subgroups that include a Resources Branch and a Support Branch. See Figure 9.5.

As we noted in Chapter 3, a good Logistics Section Chief doesn't wait until the Operations Section Chief needs a resource; they anticipate what will be needed and then arrange for the resource to go to the Staging Area.

Earlier we stated that resources are made up of people, equipment, and supplies. For each resource committed to the incident, there needs to be a support component in place to support it. See Figure 9.6.

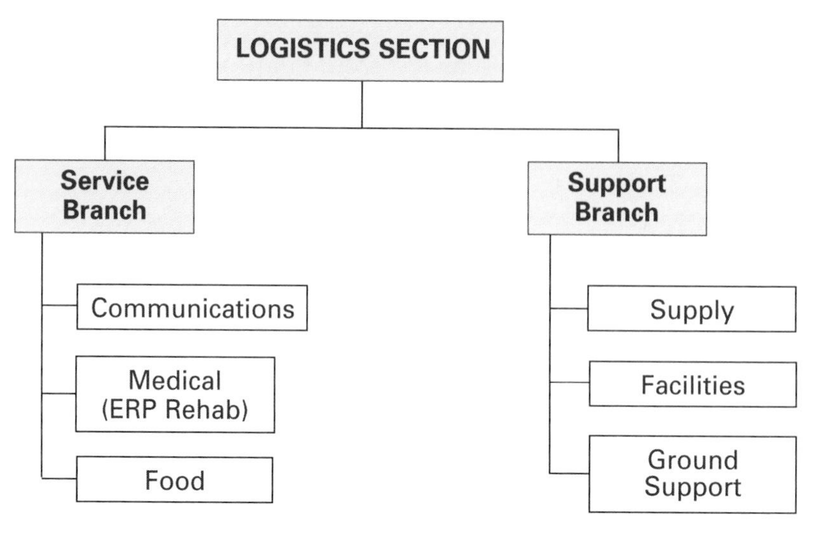

Figure 9.4 Logistics section organizational chart.

RESOURCES AND SUPPORT RELATIONSHIP	
RESOURCE	SUPPORT
People	Food and Water
	Toilet Facilities
	Medical Facilities
	Rehab Facilities
Vehicles	Diesel Fuel and Oil
Generators	Gasoline or Diesel Fuel

Figure 9.5 Each resource must be backed up by a support function.

COORDINATING RESOURCES

Hazardous materials incidents are unique to the emergency response business because of the resources required to mitigate the problem. There are few technical emergencies that involve such a broad spectrum of private and public services. For example, the train derailment previously described may require extensive resources to bring the situation under control. These could include the following:

- Railroad operations and hazmat specialists
- Product and/or container specialists representing a variety of different companies
- Wreck clearing contractors
- Environmental specialists and contractor

worth reviewing some basic problem-solving techniques as they apply to resource coordination.

Most resource coordination problems fall into three categories:

1. Failure to understand or work within the IMS structure
2. Given the type and nature of the incident, failure to anticipate potential problems and "gaps" in information or resources
3. Communications and personality problems between the players

Emergencies don't permit much time for resolving long-standing personality conflicts; they only intensify under stress. If you must bring uncooperative people into the command structure, assign a qualified person as a "chaperone" or liaison to keep them out of trouble. One way to keep uncooperative and unwanted external resources out of your hair is to team them up with someone in your command structure who knows what they are doing.

The IC can resolve many communications and personality problems by using a little psychology and some group leadership techniques at the ICP. Some useful techniques that can be effectively applied in stressful situations include the following:

Listening: Pay attention to others as they communicate. Don't just listen to what people are saying, pay attention to their body posture, gestures, mannerisms, and voice inflections. Identify angry players and their issues and objections. Resolve the problem before the situation gets out of hand and eats up your valuable time.

Clarifying: Clarify what the person's issue or concern is. Identify and sort out individual problems. In most cases (but not always) the person has an issue with some kernel of legitimacy. Big problems are more manageable when they are broken into smaller individual issues. If you can identify what the person's concern is, you can address one issue at a time. Issues can also be handed off to subordinates for resolution.

Summarizing: If large groups of people are involved in a Unified Command setting, the decision-making process can get bogged down or fragmented. If several individuals have a problem, summarizing where you are can be helpful. For example, if the discussion is turning into a debate, the Incident Commander can interrupt and ask each agency representative to briefly state how he or she feels about the issue. The IC can then summarize by identifying some common ground and turn the discussion toward acceptable alternatives. Look for issues to agree on, and then build on those as a platform to work through the unresolved issues.

Empathizing: If a special interest emerges and becomes a problem, try empathizing with the individual to reassure him or her that the concern is valid. For example, protecting sea birds from a massive oil spill created by a burning barge at the dock is a high priority to fish and wildlife officials and environmental activists; however, under the circumstances, life safety is a bigger and more immediate concern. Empathize and show respect for the individual's concern; then get a "buy in" to the fact that your alternatives are limited. A little well-placed empathy goes a long way to building support for your plan for managing the resources available.

Develop a good working relationship with supporting agencies, suppliers, contractors, and consultants before the incident and most of the personality issues will dissolve, or at a minimum they won't get in the way of on-scene operations. The Local Emergency Planning Committee (LEPC) is a good place to start laying the foundation.

SUMMARY

The function of Information Management and Resource Coordination can be viewed from two perspectives. From a strategic perspective, it is a constant that starts with the notification and dispatch of emergency responders and ends with the termination of the incident. From a tactical perspective, it is the transition point from the size-up phase of the hazmat incident to the mitigation and termination phases. Both perspectives share a common bottom line—failure to get the right information to the right people at the right time can jeopardize both the safety of responders and the overall success of the emergency response effort.

As much time, thought, and effort must be put into information management and resource coordination as into any other hazmat function. The way other agencies and the public perceive how the incident was handled generally depends on the way information was managed and people were handled on the scene.

Information and resources must be managed within the framework of the Incident Management System. The checklist system is one of the most effective tools for assuring that information and resources are effectively coordinated both internally and externally.

When resources (people, equipment, and supplies) are committed to an incident, they must be backed up by a support component. The Logistics Section Chief is the "go to" person within the command structure to request and coordinate resources.

REFERENCES AND SUGGESTED READINGS

Chemical Manufacturers Association, SITE EMERGENCY RESPONSE PLANNING GUIDEBOOK, Washington, DC: Chemical Manufacturers Association (1992).

Corey, Gerald, and Marianne Schneider, GROUPS: PROCESS AND PRACTICE (2nd edition), Monterey, CA: Brooks/Cole Publishing Company (1982).

Drucker, Peter F., THE EFFECTIVE EXECUTIVE, New York, NY, Harper and Row (1985).

Engels, Donald W., ALEXANDER THE GREAT AND THE LOGISTICS OF THE MACEDONIAN ARMY, Berkeley and Los Angeles, CA: University Press (1978).

Hersey, Paul and Ken Blanchard, MANAGEMENT OF ORGANIZATIONAL BEHAVIOR: UTILIZING HUMAN RESOURCES (4th Edition), Englewood Cliffs, NJ, Prentice-Hall (1982).

Hildebrand, JoAnne Fish, "Stress Research: Solutions to the Problem, Part-3." FIRE COMMAND (July 1984). Page 23–25.

Pagonis, William, P., Lt. General, and Cruikshank, Jeffrey L., MOVING MOUNTAINS: LESSONS LEARNED IN LEADERSHIP AND LOGISTICS IN THE GULF WAR, Boston, MA: Harvard Business School Press (1992).

Siu, R. G. H., THE CRAFT OF POWER, New York: John Wiley and Sons, Inc. (1979).

Van Horn, Richard L., "Don't Expect Much from Your Computer System." THE WALL STREET JOURNAL ON MANAGEMENT, New York, NY: Dow Jones and Company (1985).

...THE CHIEF WANTS TO KNOW HOW IT'S GOING=
-YOU WANT TO TELL HIM... -OR SHOULD I?

In summary, the operational strategy for the incident is developed based on the IC's evaluation of the current conditions and forecast of future conditions.

This chapter focuses on the various strategic goals and tactical objectives available to the IC to influence outcomes. Topics include basic principles of decision-making, guidelines for determining and implementing strategic goals and tactical objectives, and special tactical problems. We will not use a "how to do" approach that focuses on the specific tasks involved in implementing respective tactical objectives; rather, we focus on the management criteria and guidelines associated with selecting and implementing the appropriate response action.

The authors would like to thank Ludwig Benner, Jr. for permission to reproduce materials from his copyrighted works in the Basic Principles discussion of this chapter. Ludi has contributed greatly to the body of knowledge in hazardous materials emergency response, especially in the areas of hazard and risk assessment, firefighter safety, and accident investigation. His groundbreaking work at the National Transportation Safety Board (NTSB) and at Montgomery College, Rockville, Maryland in the 1970's set the standard for hazard and risk assessment and is still used today in emergency services. No doubt many lives have been saved due to his willingness to share his knowledge with the emergency response community.

BASIC PRINCIPLES: UNDERSTANDING EVENTS

In order to determine which strategy and tactics are best suited to change the outcome for a hazmat incident, responders must be able to understand what has already occurred, what is occurring now, and what will occur in the future. In other words, if emergency responders can visualize what has already happened and understand what is happening now, it may be possible to interrupt the chain of events and favorably change what will be happening in the future.

If you think about every emergency you ever responded to in your career, it could be plotted along a timeline. At some point on that timeline you arrived on the scene, sized up the situation, took some action, and began influencing the outcome in some positive way (e.g., you rescued a victim, extinguished a fire, etc.). See Figure 10.1.

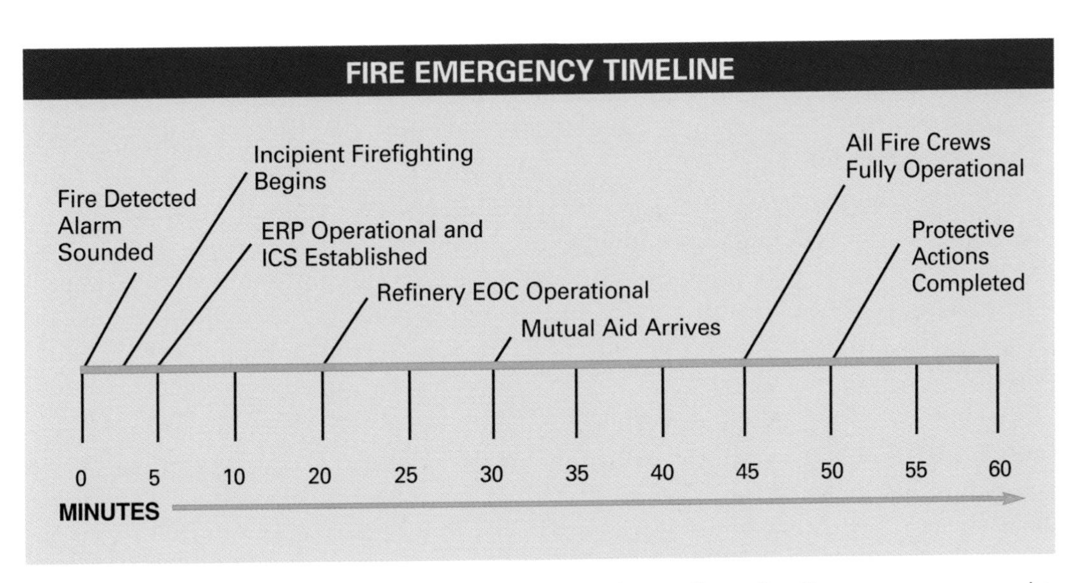

Figure 10.1 Example of a emergency response timeline for a refinery fire. Every emergency can be plotted along a timeline.

The options available to you to change the outcome of an incident has a lot to do with where you are on the timeline when you arrive at the scene. If containers have already breached and produced fatalities and injuries, the options available to influence the outcome in a favorable way are fairly limited. On the other hand, if you arrive on the scene while containers are still being stressed, it may be possible to prevent containers from breaching.

Experience over the last fifty years has been marked by many significant hazmat incidents that produced bad outcomes. Unfortunately, many of these incidents occurred after emergency responders intervened in an incident and tried to change the outcome using offensive tactics. Ironically, by intervening without understanding the risk they were taking, responders actually created a worse outcome by becoming casualties themselves.

A review of NTSB accident investigation reports of 16 major historic hazardous materials incidents occurring between 1968 and 1978 shows that

- There were over 1,400 injuries to both the public and responders.
- Fifty-one (51) emergency responders were killed.
- Over 2,200 homes and commercial buildings destroyed.

Fortunately, there has been a significant reduction in the number of incidents resulting in multiple fatalities and casualties over the last two decades. One of the primary reasons for this reduction is that emergency responders are better trained to understand how hazardous materials behave and what responders can realistically achieve with the resources available.

PRODUCING GOOD OUTCOMES

Producing good outcomes at hazmat emergencies depends a great deal on the ability of emergency responders to visualize the emergency in a chronological sequence of events. The IC should analyze and visualize events at the emergency scene in three phases: (1) what has already taken place; (2) what is taking place now; and (3) what will be taking place in the immediate future. Stated another way, responders must evaluate the incident in terms of past, present, and future events.

PAST EVENTS—During the initial size-up, the IC should develop a mental picture of what has already occurred before the arrival of responders. Examples include whether containers have already breached, which containers have already been stressed, the dispersion patterns created, and the exposures that have already been contacted and harmed.

PRESENT EVENTS—Understanding what has already happened can help the IC focus on what is occurring now and what may be happening in the next 10 to 15 minutes. Getting the "Big Picture" as quickly as possible is an important step in (1) predicting what will be happening in the future, and (2) determining which options are available to influence those events. How quickly the IC develops this "Mental Movie" of what is happening has a lot to do with how quickly command is established and how information is managed to support decision making.

FUTURE EVENTS—If the IC has a good mental picture of what is occurring along a timeline, it may be possible to influence the outcome of the emergency in a variety of different ways. There are five factors that should be evaluated by the IC. These can be remembered by the acronym MOTEL:

Magnitude—Trying to keep the incident as small or limited as possible. For example, rotating a one-ton chlorine container 180° can change a liquid leak to a vapor leak and reduce the size of the vapor cloud.

Occurrence—Preventing a future event from occurring by influencing the current event. For example, activating a water spray system or extinguishing a ground fire to minimize thermal stress upon surrounding containers.

Timing—Trying to change when an event happens and/or how long it lasts. For example, product transfer operations can be initiated to reduce the quantity of product remaining in a fixed facility storage tank while resources are being assembled to implement leak control tactics.

Effects—Trying to reduce the size and/or effects of an event. For example, using vapor dispersion tactics (e.g., water monitors and water spray systems) to reduce the size or magnitude of an anhydrous ammonia release.

Location—Trying to change where the next event occurs. For example, a tank truck or railroad tank car with a minor leak at the dome cover could be moved to a safer, less congested area and handled.

In selecting strategic goals to either stop the current event or prevent future events from occurring, remember these two basic principles:

1. You cannot influence events that have already happened or change the outcomes of those events; and
2. The earlier the events sequence can be interrupted, the greater the probability of a producing a favorable outcome.

STRATEGIC GOAL

STRATEGY DEFINED

Uncertainty is a reality of emergency response. Yet one of the IC's most critical tasks is to minimize uncertainty by using a structured decision-making process to size up the problem and select the safest strategy to make the problem go away. Adopting strategic goals to manage the incident is one of the IC's first priorities before intervening in an incident. Guessing about what the best course of action is when dealing with hazardous materials can teach you some hard lessons—acting on instinct without facts and without predicting both product and container behavior can get people killed. A systematic thought process is important to keeping yourself and your team members safe.

What are strategy and tactics? The terms *strategy* and *tactics* are sometimes used interchangeably, but they actually have very different meanings. A strategy is a plan for managing resources. It becomes the IC's overall goal or game plan to control the incident. Like most goals, a strategy is usually very broad in nature and is selected at the Command Level. Several strategic goals may be pursued simultaneously during an incident. Examples of common strategic goals implemented at hazmat incidents include the following:

- **Rescue**—Finding disoriented, trapped and injured people and removing them to a safer area.

- **Public protective actions**—Protecting-in-place, evacuation, or a combination of both tactics.
- **Spill control (confinement)**—A defensive strategy to keep a released material within a defined or specific area.
- **Leak control (containment)**—An offensive strategy to control a release at its source.
- **Fire control**—Confining, controlling, and sometimes extinguishing the fire.
- **Recovery**—Recovering the hazmat (usually liquids) for recycling or disposal.

In contrast, tactics are the specific objectives the IC uses to achieve strategic goals. Tactics are normally decided at the section or group/division levels in the command structure (see Figure 10.2). For example, tactical objectives to achieve the strategic goal of spill control (confinement) would include absorption, diking, damming, and diversion.

If the IC hopes to have strategic goals understood and implemented, they must be packaged and communicated in simple terms. If strategic goals are unclear, tactical objectives will become equally muddied. Hazmat strategic goals and tactical objectives can be implemented from three distinct operational modes as outlined below. Criteria for evaluating and selecting operational modes include the level of available resources (e.g., personnel and equipment), the level of training and capabilities of emergency responders, and the potential harm created by the hazmat release.

Offensive Mode—Offensive-mode operations commits the IC's resources to aggressive leak, spill, and fire control objectives. An offensive strategic goal/tactical objective is achieved by implementing specific types of offensive operations that are designed to quickly control or mitigate the problem. Although offensive operations can increase the risk to emergency responders, the risk may be justified if rescue operations can be quickly achieved, if the spill can be rapidly confined or contained, or if the fire can be quickly extinguished.

Defensive Mode—Defensive-mode operations commits the IC's resources (people, equipment, and supplies) to less aggressive objectives. A defensive strategic goal/tactical objective is achieved by using specific types of defensive tactics, such as diverting or diking the hazmat. The IC's defensive plan may require "conceding" certain areas to the emergency while directing response efforts toward limiting the overall size or spread of the problem (e.g., concentrating all efforts on building dikes in advance of a spill to prevent contamination of a fresh water supply). As a general rule, defensive operations expose responders to less risk than offensive operations.

Nonintervention Mode—Nonintervention means taking no action other than isolating the area. The basic plan calls for personnel to wait out the sequence of events underway until the incident has run its course and the risk of intervening has been reduced to an acceptable level (e.g., waiting for an LPG container to burn off). This strategy usually produces the best outcome when the IC determines that implementing offensive or defensive strategic goals/tactical objectives will place responders at an unacceptable risk. In other words, the potential costs of action far exceed any benefits (e.g., BLEVE scenario).

In some situations, nonintervention tactics may be implemented until sufficient resources arrive on-scene and an offensive attack can be implemented. Defensive tactics are always preferable over offensive tactics if they can accomplish the same objectives in a timely manner.

Operations will usually begin in the defensive mode. The most important question the IC should ask is, "What happens if I do nothing?" There will also be times when an operation is in a marginal mode. In other words, initial information indicates that it is relatively safe to attempt an offensive tactical objective, yet it is very possible that things may turn for the worse during that process.

STRATEGIES AND TACTICS FOR HAZARDOUS MATERIALS RESPONSE OPERATIONS

ORGANIZATIONAL LEVELS	STRATEGIC GOALS "What are you going to do?"	TACTICAL OBJECTIVES "How are you going to do it?"	OPERATIONAL TASKS "Do it."
DEFINITION	STRATEGY: The overall plan to control the incident and meet incident priorities.	TACTICS: The specific and measurable processes implemented to achieve the strategic goals.	TASKS: The specific activities that accomplish a tactical objective.
KEY ELEMENTS	• Goals • Overall game plan • Broad in nature • Meets incident priorities (life safety, incident stabilization, environmental and property conservation)	• Objective oriented • Specific and measurable • Often builds on procedures	• "Hands-on" work to meet the tactical objectives. • The most important organizational level—where the work is actually performed. • Most problems "go away" as a result of members performing task-level activities.
DECISION MAKERS	• Incident Commander • Section Chiefs	• Operations Section Chief • Hazmat Group Supervisor • Group/Division Supervisors • Sector Officers	• Individual units and individuals
OPTIONS	• Rescue • Public Protective Options • Fire Control • Spill Control • Leak Control • Recovery	• Rescue • Public Protective Options - Evacuation - Protection-in-Place • Fire Control - Exposure Protection - Extinquishment • Leak Control - Neutralization - Overpacking - Patching & Plugging - Pressure Isolation & Reduction - Solidification - Vacuuming	• Spill Control (Confinement) - Absorption - Adsorption - Covering - Diking, Damming and Diversion - Dispersion - Retention - Vapor Dispersion - Vapor Suppression

Figure 10.2

SELECTING STRATEGIC GOALS

Once the IC has run a "mental movie" for each of the options being considered, Command must make a decision to do something or to do nothing—both constitute a decision. Selecting the best strategic goal involves weighing what will be gained against the "costs" of what will be lost—a process often easier said than done.

Determining what will be "gained" involves weighing many different variables, including the following:

- **Potential casualties and fatalities**—Will lives be saved by pursuing aggressive rescue operations? For example, if a civilian is trapped in a confined area contaminated with poison gas, can his or her life be saved by committing personnel to a rescue? Likewise, does the rescue environment present an unreasonable risk to responders?
- **Potential property damage or financial loss**—What will be the financial cost of implementing one option over another? For example, a small chemical spill occurs in an auto assembly facility. Should the entire building be evacuated and production operations be halted ($$$$$)? Of course, if you don't evacuate and the problem quickly worsens, the potential risk to both employees and emergency responders may be unacceptable.
- **Potential environmental damage**—What will be the impact on the environment as a result of your actions? For example, if offensive fire control tactics are implemented at an agricultural chemical warehouse fire, water runoff and pollution will likely result. However, allowing the fire to burn out with no application of water may result in widespread air pollution over a much larger area. This may require additional public protective actions downwind. Which option is the most acceptable?
- **Potential disruption of the community**—Will the community be disrupted to an unacceptable level? For example, a gasoline tank truck has overturned and is burning on a major freeway just before rush hour. Is it an acceptable risk to the community to let the tanker burn for three hours and consume the product, or will extinguishment be required sooner? Which option will actually open the freeway sooner?

DECISION-MAKING TRADE-OFFS

Every decision the IC makes involves making trade-offs between two or more factors that are usually in conflict with one another (e.g., life safety vs. property or environmental damage). Decision-making at a hazmat incident is not a "black and white" process; rather, it is full of lots of gray area.

Conflicting information and values can also add to the uncertainty of the decision-making process. Consider the following:

- **Conflicting or uncertain information**. The more critical the life safety situation, the less likely accurate and reliable information will be available to make an informed decision. Experience shows that the IC is sometimes pressured into making split-second decisions based on incomplete or inadequate information. For example, firefighters have been killed and seriously injured pursuing aggressive rescue operations when told that there is someone trapped inside a building, only to discover later that the information was incorrect.

- The most difficult decisions that responders must make are those related to life safety. Experience clearly shows that the greater the risks to people, the greater the risks responders are willing to take to save them.

  A corollary to this fact is that the worse the potential outcome, the greater the level of uncertainty the IC should be willing to accept in implementing response goals and objectives. Understanding this point and incorporating it in your decision-making process will save lives!

- **Conflicting or competing values**. Every individual has a set of values that they bring to the incident. Regardless of your background, these values influence the decision-making process. Even among similar groups with the same types of values, people have different opinions about the perceived risk involved in carrying out an option as well as the perceived value of what will be gained.

  When lives are at stake, each person can be expected to select a different option as the "best" one in his or her judgment. If you doubt this, conduct an exercise in which you give responders a specific situation and ask them to rate the risks involved in protecting life, property, environment and in disrupting the community. You may be surprised what you learn about how people think and how much risk they are willing to take.

- There is no single best way to evaluate life safety decisions. Pure risk taking (the black area) can lead to self-destruction, while pure safety (the white area) can lead to taking no action at all. Somewhere between these two extremes is the right decision (the gray area).

Figure 10.3 Getting a buy-in from the risk-takers is an important step in getting the IC's game plan implemented.

Getting a buy-in from the risk takers (i.e., the people in the PPE) is an important step in getting the IC's strategic game plan implemented. The success of obtaining agreement depends on

1. The IC's ability to understand the differences in how responders perceive risks

2. The IC's ability to explain the options available to the risk takers
3. Getting the other agencies and organizations involved in the incident to understand the big picture concerning what will be lost versus what will be gained

The more you understand how different organizations and individuals perceive risk and the value of implementing different options before an incident, the quicker you can implement these decisions at the incident. This is why getting to know the people, personalities, and missions of the people you will be working with before the incident is so important.

TACTICAL OBJECTIVES

TACTICS DEFINED

Once the Command Level has committed to a strategy, subordinate groups, divisions and sectors will implement the IC's general game plan by establishing specific tactical objectives. There is no way to make a hazmat emergency go away without tactics!

A well-defined tactic has a stated objective that can be implemented using specific procedures and tasks within a reasonable period of time. The tactical action plan must be easy to understand and be laid out with straightforward objectives. A good standard for evaluating a tactical objective is that it should be able to be communicated without providing too much detail.

An IC who gets involved in making detailed tactical-level decisions like which type of plug to use loses the broad perspective of the incident and develops tunnel vision. This can be a serious problem with new command officers who have worked their way up through the system and just can't let go of the hands-on details (like the fire chief on the nozzle going down the hallway). Our military experience has taught us that generals are good at strategy, captains are good at tactics, and sergeants are good at making problems go away. Things get pretty screwed up when the process works in reverse. Just as a military commander cannot effectively implement tactics from Washington, a sergeant deep in enemy territory cannot get the "big picture" and develop sound strategy.

TACTICAL DECISION-MAKING

Tactical decision-making is about whether to use one tactic over another. Most hazmat incidents cannot be resolved using just one tactic. Effective tactical decision-making requires thinking ahead and planning various tactical options so that the right people, with the right training and the right equipment, are available at the appropriate time.

When deciding which tactic to use, consideration should be given to how long it will take to accomplish the specific objective. Conditions can change rapidly as an incident progresses along a natural timeline. What seemed like a good tactical objective early in the incident may no longer be an option as conditions change. Spills can grow larger and leaks can get worse as responders gather their equipment and prepare for entry. Valuable time can be lost if the entry team has to shift from Plan A to Plan B. The trick is to keep ahead of what the hazmat will be doing when the entry team is ready to go to work.

The length of time required to implement tactical objectives at the task level must be compared to how long the window of opportunity will be open. As the clock ticks, tactical options to deal with the problem will often become more limited. Remember, all hazmat incidents follow a natural timeline. Leaks and spills usually get worse all by themselves before the situation gets better.

Some tactics can be employed to delay events or slow down the clock until entry teams are ready to implement the "final solution." In other words, responders may buy time with less effective, but easy to implement, defensive tactics until the most effective offensive tactic can be implemented.

Examples of tactical options that can be used to buy time include the following:

- **Barriers**—Put a physical barrier between the hazmat, its container, and surrounding exposures. For example, building dikes, retention ponds, or diversion dams well in advance of an oncoming spill can confine the hazmat release to a limited area or slow it down until entry teams are ready to contain the leak.
- **Distance**—Separate people from the hazmat. The further away you are from the problem, the lower the risk. Increasing the size of hazard control zones or moving the problem to an isolated area can further reduce the life safety risk until the entry team is ready to enter the hot zone to resolve the problem.
- **Time**—Reduce the duration of the release, or trade off or rotate persons exposed to the hazmat. For example, if you can reduce the pressure on a leaking pipe or vessel, the magnitude of the problem can be reduced significantly. This may allow responders to buy some time to work out a more complete solution to the problem.
- **Techniques**—These are specific tasks and procedures performed by responders to stop the leak and control the problem (e.g., uprighting a leaking liquefied gas cylinder so that vapors rather than liquid product will be released).

Tactical decision making sometimes involves trial and error. What sounded like a great idea after recon operations may not be very practical after the entry team begins on-site operations. Not everything works in the field the way you planned it. No amount of training and simulation can prepare you for actual field conditions—always have options available. Remember—surprises are nice on your birthday, but not on the emergency scene.

In the next part of this chapter, we will discuss specific strategic goals and the tactics associated with each option.

RESCUE AND PROTECTIVE ACTIONS

RESCUE

Saving lives is our number one mission! Life safety is always the IC's highest priority. One of the first concerns after sizing up the incident is search and rescue. Regardless of the nature of the rescue operation, always remember the "First Law of Hot Zone Operations": All personnel working in the hot zone must

- Be trained to play;
- Be dressed to play;
- Always use the buddy system;

- Always have a back-up capability;
- Have an emergency decon capability, as a minimum; and
- Have Command approval for the entry/rescue operation.

Hazmat rescue problems fall into three general categories:

1. **Searching for and relocating people who will be immediately exposed and harmed by the hazmat as the situation gets worse.**

   This group consists of everyone inside of the Hot Zone who is not wearing protective clothing and equipment rated for the hazards. This can consist of civilians and employees who have left the immediate hazard area on their own and believe that they are now in a safe location. It may also include the curious, the "I'm authorized to be here," or the just plain stupid. This includes people who need to be rescued but just don't know it yet! Normally, all that is needed is a little organization and direction to move these people back and away from the hazard area. This group may also include people who were initially exposed but have not yet shown the signs or symptoms of exposure.

2. **Rescuing victims who have been disoriented or disabled by the hazmat.**

   This group includes individuals or groups of people who have been exposed to the hazmat and are suffering from its harmful effects. Examples include victims who have been burned, poisoned, blinded, and so on. Normally, rescue involves packaging and removing the victim following standard operating procedures. Chapter 11 provides some specific guidelines on how to handle and treat these contaminated patients.

3. **Planning and executing technical rescue.**

   This category includes rescues of one or more victims who have been exposed to the hazmat and require physical extrication. Examples of technical hazmat rescue situations include the following:

   - High-angle rescue situations, including injured or disabled victims found in high areas (e.g., cooling towers and elevated structures in refineries or chemical plants, on top of a large-diameter storage tank, scaffolding, etc.).
   - Victims pinned and trapped inside wreckage or debris (e.g., building collapse, train crew pinned in a train locomotive cab or a tank truck driver inside of an overturned vehicle).
   - Confined space rescue situations (e.g., inside storage tanks, on lowered floating roof tanks, underground vaults, sewers, etc.).

TECHNICAL RESCUE

Technical rescue operations are high-risk operations that often require the response and use of special operations teams and personnel. Consequently, many emergency response organizations have formed specialty teams to provide the expertise required to handle these unique situations.

Technical rescue problems involving hazardous materials have several common elements that make them difficult to plan for and execute. These include the following:

- **Hazardous atmospheres**—Typically involve multiple combinations of flammability, toxicity, and oxygen deficiency or enrichment that change over time. The longer the victim is exposed to these atmospheres, the less likely the chance for survival.
- **Hazardous work areas**—Slippery or uneven walking surfaces, missing or damaged catwalks, ladders, and handrails, as well as sharp and jagged metal edges. Under normal working conditions, these types of physical hazards require special precautions. Add the restricted vision and motion problems created by wearing SCBA and specialized protective clothing, and these damaged work areas become ultra-hazardous.
- **Limited access areas**—Areas with a single access point (only one way out if you are trapped), confined/narrow walkways or ladders, and high angles of egress. If the route to or from the rescue site is blocked or cut off, the rescue team will be trapped.

Historically, the track record of making effective technical rescues has not been very good. Some rescuers have become victims because they either (1) underestimated the hazards and risks, (2) took action without the proper tools or equipment, (3) were not properly trained for the tasks at hand, or (4) failed to understand the emergency response timeline (e.g., if the victim is trapped in an atmosphere above the IDLH, and medical statistics tell you that humans are brain dead in five minutes after breathing has stopped, and it takes you 20 minutes to respond, set up, and start the rescue.....you do the math!). Time simply works against the rescuers and the victims in most technical rescues when hazmat are involved.

Figure 10.4 Confined spaces include many hazards, entering them is a high risk operation.

Remember—**without an oxygen supply to the brain, clinical death occurs in three minutes, and biological death occurs in five minutes. Exposure to hazardous atmospheres accelerates the timeline. Don't turn rescuers into victims.**

CONFINED SPACES RESCUES

Confined spaces present emergency responders with a serious technical rescue challenge. By the nature of their definition, confined spaces are dangerous places. Add in any combination of hazardous materials, and the confined space environment becomes deadly.

According to OSHA, the majority of the fatalities in confined spaces incidents occur from entrapment in (1) toxic atmospheres, (2) asphyxiating atmospheres, and (3) physical hazards inside the confined space.

What is a confined space? According to OSHA (29 CFR 1910.146), a confined space is "any area that has limited or restricted means for entry or exit; is large enough and so configured that an employee can bodily enter and perform assigned work; and is not designed for continuous employee occupancy." Examples of confined spaces and related rescue situations include fixed storage tanks, the interior spaces on mobile tank trucks or rail cars, process vessels, pits, ship and barge compartments, vats, reaction vessels, boilers, ventilation ducts, tunnels, or pipelines. This is not a complete list, but you get the idea—these spaces are confined.

SPECIAL HAZARDS AND RISKS

Confined spaces have a number of hazardous characteristics, including the following:

- **Hazardous atmospheres**—Always approach a confined space rescue problem with the attitude that the atmosphere inside the space is flammable, toxic, and oxygen deficient or enriched. Even if air monitoring instruments indicate that the atmosphere is entirely clean or within acceptable limits, handle the incident as if the atmosphere were contaminated. Never enter a confined space for a rescue operation without using a supplied air respirator (SCBA or airline hose unit). Never remove your facepiece to "revive the victim." If you remove your facepiece in a confined space in a contaminated atmosphere, you can die!
- **Limited egress**—Most confined spaces have only one way in and one way out. Egress is further complicated by restricted access points, such as manways that may be poorly designed and placed at unusual angles. Manways create special problems for entry teams using SCBA and specialized protective clothing. Removing the SCBA bottle and harness from the wearer's back while leaving the facepiece in place is a poor option. This is a very difficult task and is dangerous. Many inspectors and maintenance personnel have been overcome by a hazardous atmosphere when they temporarily removed their facepiece to crawl through a manway. Also, accidentally dropping the SCBA harness while crawling through the hole can yank the facepiece off of the wearer's head. Supplied air respirators are a better option when working in a confined space.
- **Extended travel distances**—Many confined spaces require extended travel distances to enter and access the confined area to conduct search and rescue operations. The longer the travel distance, the greater the rate of air consumption. This ultimately translates to a short-duration stay inside the confined area. Much like a diver who must calculate air consumption time to allow for a safe ascent to the surface, a confined spaces rescuer must use special care not to overextend the stay and risk running out of air. Even 45-minute SCBA units will limit the actual on-

site working time to five to ten minutes. The use of an SAR can make entry into the space easier and extend the search and rescue time, but these units are limited to 300 feet of air hose. This may not be very practical for confined spaces such as pipelines and sewers.

- **Unusual physical hazards**—These can include the hazards of uneven or slippery walking and climbing surfaces, being struck by falling objects, or becoming trapped in inwardly converging areas that slope to a tapered cross section (e.g., a grain elevator chute).
- **Darkness**—Almost every confined space is totally dark—you can't see your hand in front of your face. Search and rescue operations must be carried out using hand lights or portable lights that are brought into the space. The potential for a flammable atmosphere inside most confined areas means that all lighting must be intrinsically safe. In other words, the equipment must be suitable for NEC Class 1, Division 2 locations.
- **Poor communications**—Confined spaces don't like radios. Radio communications are always a problem. Below-grade spaces and steel construction do not promote very good radio reception. Add the protective clothing and SCBA facepiece problem and good radio transmissions become virtually impossible. Like hand lights, radio equipment used in confined spaces must meet NEC Class 1, Division 2 atmosphere requirements.

EVALUATING CONFINED SPACES FOR RESCUE OPERATIONS

The IC should ensure that the following have been evaluated before committing personnel to rescue in a confined space:

1. **Can the confined space be entered safely by emergency responders?**

   Confined spaces may be flammable, toxic, oxygen deficient, or some combination of these.

   Flammable atmospheres—Flammable atmospheres containing 10% or less concentration of flammable vapors are considered within safe limits for conducting rescue operations, but this is not the only factor that requires consideration. Concentrations between 10 and 20% are considered hazardous and should never be entered by rescue teams unless they have the proper PPE and respiratory protection and all electric equipment is rated for Class 1, Division 2 atmospheres. The explosive range of many hydrocarbon vapors range from a 1 to 10% vapor-to-air mixture; however, the explosive range for oxygenated materials like alcohols and glycols is wider. Any mixture of vapor and air between the UEL and LEL will ignite when exposed to an ignition source and should be considered as too dangerous for entry.

   Toxic atmospheres—An atmosphere above the OSHA Permissible Exposure Limit (PEL) or the ACGIH Threshold Limit Value (TLV) does not necessarily prohibit entry into a confined space to perform rescue operations. But realize that the risk to the rescue team increases significantly and there is no margin for error. A rescuer who experiences damaged protective clothing or an air supply problem inside of a confined space with a toxic atmosphere faces almost certain injury or death. Making an entry under these conditions is a judgment call that must be made on a case-by-case basis by the person taking the risk. Risk looks and feels a lot different when you are the person taking the risks. The people taking the risk always have the final say on Go/No Go.

   Oxygen-deficient or enriched atmospheres—An oxygen-deficient confined space has less than 19.5% oxygen. An oxygen-enriched atmosphere is 23.5% or greater in oxygen content. The risk of entering an oxygen-deficient atmosphere is similar to entering a toxic atmosphere. Obviously, the less oxygen content, the greater the risk to the rescue team if there is an air supply problem. Oxygen-enriched atmospheres present rescuers with a significant risk because an increase in oxygen content increases the risk of fire. If a flammable atmosphere is present, the explosive range will become wider.

All potentially hazardous atmospheres in confined spaces should be confirmed by monitoring instruments.

2. **Are you really rescuing someone or recovering a body?**

   When human life is involved this is a serious and tough question to ask and answer. When the human body is deprived of oxygen death occurs in 3 to 5 minutes so the chances for life decline. The longer someone goes without oxygen, the less likely he or she can be revived, even under the best of circumstances. The presence of a toxic atmosphere accelerates death's time-line. If a flammable hydrocarbon atmosphere is present, the PEL and the TLV-TWA will usually be exceeded before 10% of the lower flammable atmosphere is reached. Therefore, the atmosphere will almost always be toxic before reaching the flammable concentration.

   As a general guideline, whenever the victim has been subjected to an oxygen-deficient atmosphere of less than 19.5% or a flammable atmosphere of 10% or greater or a toxic atmosphere above the PEL/TLV for periods longer than 5 to 15 minutes, the IC should consider the possibility that there is no real chance for a successful rescue. The lower the oxygen content and the higher the toxic and flammable atmospheres inside the confined space, the less likely that the victim will survive for periods of exposure exceeding five minutes. As is the case with every medical emergency, the condition of the victim, age, pre-existing health conditions, etc., affect the chance for survival.

   In addition to these basic medical parameters, the IC must consider the amount of time that it will take to set up and safely conduct search and rescue operations. Darkness, limited access, extended travel distances, and difficult working conditions cause delays that work against the victim's chances for survival.

3. **Do you have control of the situation and is there a coordinated incident action plan?**

   Once the frantic pace of a rescue begins, it is very difficult to stop the operation. Emergency responders eagerly do what they do best, save lives. Be sure that you are handling the rescue following standard operating procedures and are not "winging it." As the IC, the objectives of the rescue and your expectations must be communicated to everyone involved in the operation. Make sure your IAP is clear.

To learn more about confined spaces rescue consult the following references:

Air Force Civil Engineer Support Agency (AFCESA) and PowerTrain, Inc. CONFINED SPACE—EMERGENCY RESPONSE SERIES CD-ROM. Tyndall Air Force Base, FL: AFCESA (2002).

Brown, George and Gus Christ, CONFINED SPACE RESCUE. Albany, NY: Delmar–Thomson Publishing (1999).

Emergency Film Group, CONFINED SPACE EMERGENCY. Edgartown, MA: Emergency Film Group (2003). NOTE: This is a three-part series covering: Understanding Confined Spaces. Confined Spaces and the First Responder and Confined Space Technical Rescue.

Rekus, John, COMPLETE CONFINED SPACES HANDBOOK. Washington, DC: National Safety Council, (1994).

Roop, Michael, Thomas Vines, and Richard Wright, CONFINED SPACE AND STRUCTURAL ROPE RESCUE. Saint Louis, MO: Mosby Publishing (1998).

Sargent, Chase. CONFINED SPACE RESCUE. Saddle Brook, NJ: Fire Engineering Books and Videos (2000).

International Association of Fire Fighters, TRAINING FOR HAZARDOUS MATERIALS RESPONSE: CONFINED SPACE OPERATIONS FOR FIRST RESPONDERS. Washington, DC: IAFF (2002).

U.S. Occupational Safety and Health Administration (OSHA), PERMIT REQUIRED CONFINED SPACES FOR GENERAL INDUSTRY STANDARD (29 CFR 1910.146), Washington, D.C.

PUBLIC PROTECTIVE ACTIONS

While we have already discussed public protective actions in detail in Chapter 5, Site Management and Control, some of the basics are worth reviewing here as they relate to implementing response objectives.

Public protective actions (i.e., evacuation or protection-in-place) must be continuously monitored during the course of an incident. Incidents are dynamic events—weather conditions may change, the problem may grow, resources may be used up, and so on. Initial protective actions will often be insufficient as we gather more information, get smarter about the incident, and fully assess the level of hazards and risks. Likewise, the Incident Commander will often be under tremendous political pressure to allow civilians and employees who have been evacuated to return to their homes and work stations.

Responders must maintain strict control of the scene throughout the entire incident. The following activities can help maintain site discipline until the problem has been eliminated and the incident safely terminated:

- **Maintain an Incident Safety Officer throughout the incident.** On campaign operations, rotate this position at regular intervals to maintain alertness. A Safety Officer is especially important during hazardous operations such as leak control and technical rescue.
- **Use formal site safety checklists**. If you have read this text from the beginning, this statement is probably getting a little old. However, checklists don't make mistakes, tired people do. The longer you work at the hazmat scene, the more likely you will overlook something critical, like making sure your suit is zipped up.
- **Enforce isolation perimeter security and the use of hazard control zones**. Make sure that the playing field is clearly known and identified to all of the players. A weak perimeter often leads to loss of site control, which increases the potential for an accident.
- **Establish a crew rotation schedule**. This is especially important during extreme weather conditions where responders sometimes remove their PPE because they are uncomfortable. Don't contribute to the problem by holding personnel in forward positions for an unreasonable time period. Rested people are more alert.

SPILL CONTROL AND CONFINEMENT OPERATIONS

Spill-control strategies and confinement tactics are the actions taken to confine a product release to a limited area. These actions usually occur remote from the spill or leak site and are, therefore, defensive in nature. As a general rule, confinement tactics expose personnel to less risk than containment tactics. If responders can accomplish the same objective using defensive confinement tactics such as diking or remotely closing a valve, then they should be implemented before attempting higher risk, offensive-oriented options.

Confinement operations present several advantages over containment options, including the following:

- Avoid direct personnel exposure.

- Can often be performed without special equipment other than some shovels and dirt.
- Can usually be performed by first responders with minimal supervision.

The decision to use confinement tactics is based on the availability of time, personnel, equipment, and supplies. It also should be made with a review of the potential harmful effects the leaking material will have on personnel downwind of the spill, where most of the spill-control operations normally take place.

For example, a decision to divert a flowing diesel fuel spill from a storm drain to a roadside ditch may be based on the observation that the fuel is flowing too fast and sufficient personnel and equipment are not available to construct a dike. Finally, the fuel will cause substantially less potential damage in the ditch than in the storm system.

Confinement tactics such as diversion can usually begin immediately upon the arrival of first responders trained to the Operations Level. Diking can be started with basic first-responder equipment as more personnel arrive. Retention techniques will then follow as specialty teams and equipment become available.

Don't make the mistake of concentrating all resources on one tactic. It is easy to assign too many responders to the construction of a dike, for example, which may fail and force everyone to move to a safer location to begin again. Recognize that virtually all confinement tactics are "first aid" measures and will eventually fail over time.

CONFINEMENT TACTICS

There are a number of tactical options available to achieve the spill control strategic goal. These include both physical and chemical methods. A summary of the various tactical options is given below.

Absorption is the physical process of absorbing or "picking up" a liquid hazardous material to prevent enlargement of the contaminated area. As the material is picked up, the sorbent will often swell and expand in size. Depending upon the absorbent, it can be used for liquid spills on both land and water.

Operationally, absorbents are effective when dealing with liquids of less than 55 gallons. Larger spills are more difficult to absorb, and often the cost and time exceed the benefits. Materials used as absorbents include clay, sawdust, charcoal, absorbent particulate, socks, pans, pads, and pillows. Absorbent socks and tubes can also be deployed as a circular dike around small spills. When using absorbents, compatibility must be considered (e.g., sawdust used on an oxidizer could start a fire).

Adsorption is the chemical process in which a sorbate (liquid hazardous material) interacts with a solid sorbent surface. Since the sorbent surface is solid, the sorbate adheres to the surface and is not absorbed, as with absorbents. An example is activated carbon. Characteristics of this chemical interaction include the following:

- The sorbent surface is rigid and there is no increase/swelling in the size of the adsorbent.
- The adsorption process is accompanied by the heat of adsorption, whereas absorption is not. As a result, spontaneous ignition may be a possibility with some liquid chemicals.
- Adsorption can only occur when the sorbent has an activated surface, such as activated carbon.

Adsorbents are primarily used for liquid spills on land and should be nonreactive to the spilled material.

Covering is a physical method of confinement. It is typically a temporary measure until more effective control tactics can be implemented. Depending on the product involved, it may be necessary first to consult with a product specialist.

Phil Baker

Figure 10.5 A tarp is placed over an exposed dome cover as a covering tactic.

Examples of covering include

- Placing a plastic cover or tarp over a spill of dust or powder
- Placing a cover or barrier over a radioactive source, normally (alpha or beta) to reduce the amount of radiation being emitted
- Covering a flammable metal or pyrophoric material with the appropriate dry powder agent

Damming is a physical method of confinement by which barriers are constructed to prevent or reduce the quantity of liquid flowing into the environment. Damming consists of constructing a barrier across a waterway to stop/control the product flow and pick up the liquid or solid contaminants.

Phil Baker

Figure 10.6 Damming tactic for diesel fuel spill.

There are two types of dams—overflow and underflow.

- **Overflow dam**. Used to trap sinking heavier-than-water materials behind the dam (specific gravity >1). With the product trapped, uncontaminated water is allowed to flow unobstructed over the top of the dam. Operationally, this is most effective on slow moving and relatively narrow waterways.
- **Underflow dam**. Used to trap floating lighter-than-water materials behind the dam (specific gravity <1). Using PVC piping or hard sleeves, the dam is constructed in a manner that allows uncontaminated water to flow unobstructed under the dam while keeping the contaminant behind the dam. Operationally, this is most effective on slow moving and relatively narrow waterways.

If the pipes are not deep enough on the upstream side of the dam, a whirlpool may be created and pull the hazardous substance through the pipes. This problem can be overcome through the use of a t-siphon on the upstream side. To be effective, several overflow or underflow dams should be placed downstream to catch product that may be missed by the first dam.

Figure 10.7 Diking tactic.

Diking is a physical method of confinement by which barriers are constructed on ground used to control the movement of liquids, sludges, solids, or other materials. Dikes prevent the passage of the hazmat to an area where it will produce more harm.

Dikes are most effective when they can be built quickly. Although any available material will do the job, the best quickly acquired supplies are dirt, tree limbs, boards, roof ladders, pike poles, and salvage covers. Bagged materials such as tree bark, sand, and kitty litter can be found at hardware and garden supply stores when more substantial control is required. However, when really large spills occur, dump-truck-sized deliveries will be required.

Dikes will usually be constructed by first responders using whatever on-scene equipment is available. When considering building a dike, quickly compare your resources to the quantity of the spilled material. Most people overestimate the amount of spill and underestimate the personnel and resources required to complete a dike.

Slow-moving or heavy materials should be confined by use of a circle dike. Faster moving products will require a V-shaped dike located in the best available low-lying area. Always use the land to your advantage.

Dike construction should begin by choosing large, heavier materials for reinforcement, followed by an outer layer of lighter material such as dirt. Operationally, dikes are normally a temporary measure and can begin to leak after a while. Seepage can be minimized by using plastic sheets or tarps at the dike base and within the dike by placing a final layer of dirt along the leading edge between the plastic and the ground. Be aware that plastic sheets may be degraded by certain chemicals.

Factors that can limit dike construction include situations in which:

- The surrounding area is concrete or asphalt with no available soil. Either sacrifice the area for better turf or truck in necessary materials.
- The ground is frozen. Snow may be used in conjunction with materials such as plastic and ladders. Otherwise, truck in necessary materials.
- Essential equipment is unavailable. At least three pointed, long-handled shovels are necessary. When possible, construct dikes upwind in safe areas. Be sure to consider the need for SCBA.

Dilution is a chemical method by which a water-soluble solution, usually a corrosive, is diluted by adding large volumes of water to the spill. There are four important criteria that must be met before dilution is attempted. These include determining in advance that the substance (1) is not water reactive; (2) will not generate a toxic gas upon contact with water; (3) will not form any kind of solid or precipitate; and (4) is totally water soluble.

As a general rule, dilution should only be attempted on liquid and solid substances that are corrosives, and only when all other reasonable methods of mitigation and removal have proven unacceptable. In other words, dilution tactics are a last resort.

Dilution can be effective for small corrosive spills of one quart or less, especially for concentrated corrosives with a pH of 0–2 (acidic) or 12–14 (alkaline). In outdoor situations, local water department or fish and wildlife representatives should be consulted for their approval to use dilution tactics. Federal and most state regulations limit corrosive entries into storm drains and drainage canals to a pH of 6 to 8 as long as no other pollutants are involved that may be harmful to the environment or wildlife.

The major disadvantage to dilution is that it is not well understood by emergency response personnel. It is not a straight linear one-to-one process. It is important to recognize that dilution is actually a logarithmic process (i.e., on a 1 to 10 scale). For example, a 1-gallon spill of an acid with a pH of zero will require 1 million gallons of water just to bring its pH up to 6! That is a lot of water just to dilute 1 gallon of acid. This same rule applies to the full range of corrosives, from a pH of 0 to 14. The following chart provides some guidelines that can help responders determine if dilution is the best tactical option.

Acid Spill of 1 Gallon (pH of 0)

Water to add	
10 gallons	1
100 gallons	2
1,000 gallons	3
10,000 gallons	4
100,000 gallons	5
1,000,000 gallons	6

Alkaline Spill of 1 Gallon (pH of 14)

Water to add	
10 gallons	13
100 gallons	12
1,000 gallons	11
10,000 gallons	10
100,000 gallons	9

A rule of thumb that can be used in the field to determine the volume of water required to bring the pH to the 6–8 range is as follows:

Step 1: Determine the size of the spill to be diluted in gallons (e.g., there are 10 gallons on the ground).

Step 2: Determine the pH of the spilled material using pH paper or a pH meter (e.g., the spill has a pH of 3).

Step 3: Determine the pH that you want to dilute the spill to (e.g., you want to go to a pH of 6 so that the spill can be safely flushed into the storm system).

Step 4: Determine the number of dilution steps between the starting pH and the ending pH. In our example, we started with a pH of 3 and want to end up with a pH of 6. That is three steps.

Step 5: Add three zeros to the beginning gallonage. In our example, we started with 10 gallons, so we add three zeros, which will give us 10,000. This is the number of gallons of water that must be added to the spilled 10 gallons of acid in order to bring the pH up to the desired level of a pH of 6. This rule can be applied to the entire logarithmic scale no matter where you enter it (e.g., you started with a pH of 4 and want to go to 6, that would be two steps, etc.).

Diversion is a physical method of confinement by which barriers are constructed on ground or placed in a waterway to intentionally control the movement of a hazardous material into an area where it will pose less harm to the community and the environment.

A flowing, land-based spill can quickly be diverted by placing a barrier (e.g., dirt) ahead of the spill. As when fighting a fast-moving brush fire, the barrier should be placed well ahead of the actual spill. This may require sacrificing some intermediate territory to the hazmat in order to establish complete control at the final diversion site.

Booms can also be placed across streams and waterways to divert the hazardous substance into an area where it can be absorbed or picked up, such as with vacuum trucks.

In constructing a diversion barrier, consider the angle and speed of the oncoming spill. The greater the speed of the flow, the greater the length and angle of the barrier required to slow and divert the flow. For fast-moving spills, barriers constructed at a 45° perpendicular angle will be ineffective; a barrier angle of 60° or more should be used.

Constructing a diversion barrier requires teamwork. When a team with the right equipment works quickly, a large area can be rapidly controlled. A typical four-person crew can build a 20-yard-by-8-inch diversion wall in about 10 minutes if the proper materials are available.

Dispersion is a chemical method of confinement by which certain chemical and biological agents are used to disperse or break up the material involved in liquid spills on water. The use of dispersants may result in spreading the hazmat over a larger area.

Dispersants are often applied to hydrocarbon spills, resulting in oil-in-water emulsions and diluting the hazmat to acceptable levels. They do not neutralize or make flammable materials become nonflammable. Experience also shows that some dispersants will separate over time. Use of dispersants may require prior approval of the appropriate environmental agencies.

Figure 10.8 A Dakota aircraft drops chemical dispersants, as it sweeps over the stern of the stricken tanker Sea Empress off St. Ann's Head on the Welsh coast,Monday Feb. 19, 1996.

Retention is a physical method of confinement by which a liquid is temporarily contained in an area where it can be absorbed, neutralized, or picked up for proper disposal. Retention tactics are intended to be more permanent and may require resources such as portable basins or bladder bags constructed of chemically resistant materials.

Retention tactics can sometimes be implemented independently and act as a backup to diversion or diking tactics. For example, storm sewer systems can be protected by placing salvage covers or plastic over drains and covering them with dirt. The same procedure can be used for sewer system manways.

When the hazmat is primarily a liquid or slurry, has a specific gravity less than 1.0, and is not water reactive, it may be possible to flood the retention area with water from an engine or hydrant. The hazmat would then float on the water, and any subsequent leakage into the storm system would only be water.

Eric Bachman

Figure 10.9 Retention basin constructed by a responder at a diesel fuel pipeline incident.

Vapor dispersion is a physical method of confinement by which water spray or fans is used to disperse or move vapors away from certain areas or materials. It is particularly effective on water-soluble materials (e.g., anhydrous ammonia), although the subsequent runoff may involve environmental trade-offs. Fans and positive pressure ventilators may also be used if they are rated for the hazardous atmosphere.

When dealing with flammable materials, such as LP gases, the turbulence created by the water spray may reduce the gas concentration and bring the atmosphere into the flammable range.

Figure 10.10 Vapor suppression tactics may involve the use of water fog, foams or chemical vapor suppressants.

Vapor suppression is a physical method of confinement to reduce or eliminate the vapors emanating from a spilled or released material. Operationally, it is an offensive technique used to mitigate the evolution of flammable, corrosive, or toxic vapors and reduce the surface area exposed to the atmosphere. Common examples include the use of firefighting foams and chemical vapor suppressants.

While vapor suppression does not change the nature of a hazardous material, it can greatly reduce the immediate hazard associated with uncontrolled vapors. In addition, it can buy additional time to undertake further measures to control the problem.

LEAK CONTROL AND CONTAINMENT OPERATIONS

Leak control strategies and containment tactics are the actions taken to contain or keep a material within its container. Typically regarded as offensive operations, containment tactics require personnel to enter the hot zone to control the release at its source and should be considered high-risk operations. Examples include uprighting a leaking container, closing and tightening container caps and valves, plugging and patching container shells, and depressurizing vessels by isolating valves or shutting down pumping systems.

Containment tactics are often implemented when defensive options have not produced acceptable results or when citizens or employees are at great risk from potential chemical exposures. These tactics should only be approved after conducting a thorough hazard and risk evaluation. No emergency situation is worth taking unreasonable risks. Rapid withdrawal from the hot zone is always an option; aggressive/offensive does not mean quick and stupid.

Before initiating containment operations, emergency responders should consider the following:

1. What hazardous material(s) are involved?
2. What is its physical form (i.e., solid, liquid or gas)? With the possible exception of dusts, solids are usually the easiest materials to contain. High-vapor-pressure liquids and gases present the most difficult challenge to responders.
3. What are its hazards? Whenever working with or near flammable materials, always have charged hoselines and a reliable water supply.
4. What are the risks to both responders and civilians?
5. What are the training levels and physical abilities of the entry team that will perform the operation?

6. Are special tools and equipment needed for the leak-control operation? Are they available (for example, leak-control kits or nonsparking tools)?
7. Are the responders prepared for emergency care and decontamination if an accident happens?

If these questions cannot be answered positively, leak control operations should be delayed until sufficient information or resources are obtained and the IC feels the operation can be safely conducted.

Although containment operations may pose higher risks, they may be necessary to:

- **Minimize environmental damage**—This is particularly true for hazardous liquids that threaten storm systems or water supplies.
- **Reduce operating response time**—Leaks confined to the area immediately around the container usually limit the spread of the material and minimize the need for evacuation, particularly when faced with a gas or toxic chemical.
- **Reduce clean-up costs**—Contaminants are usually limited to smaller areas or have not entered impacted ground or surface waters.

Situations well suited for aggressive leak control offensive strategies include the following:

1. The hazmat is a vapor or gas and threatens to migrate away from its container.
2. The hazmat is in a solid, powder form and weather conditions threaten to carry it from its original site.
3. Defensive options have been attempted but have not produced the desired results.
4. The situation is getting worse and increasing in risk as time progresses.

Successful offensive operations should be preceded by a thorough reconnaissance. Recon may be as simple as having a trustworthy individual relay his or her observations or as complex as a Recon Team surveying the entire work site with a video camera.

When planning leak control tasks, remember the KISS Principle (Keep it Simple, Stupid)—most leak-control tactics are pretty simple. Consider the following:

- Where possible, reduce the rate of release before containing the leak. For example, relieve pressure on the container.
- If piping is involved, isolate the leak by checking the position of upstream and downstream valves.
- Check the integrity of container openings—tighten caps, bungs, lids, and so on.
- Standing up a leaking liquid container may be sufficient to stop the leak.
- Moving the container to place the hole above the liquid or solid level reduces the hazmat release.
- When dealing with liquefied gases (e.g., chlorine, anhydrous ammonia, LPG), rotate the container to deal with a vapor release rather than a liquid release. If liquid escapes, it will expand the problem and hazard area (e.g., liquid/vapor expansion ratio for chlorine is 460 to 1 and 850 to 1 for ammonia).

When these common-sense techniques aren't effective, try to

- Reduce the container pressure (i.e., limit its magnitude).

- Plug or patch the opening (i.e., control it at the source).
- Use vapor suppression agents such as foam (i.e., limit its vaporization).
- Neutralize using another chemical (after consulting with a product specialist).
- Control the leak and dispose in-place (e.g., controlled burning, flaring).

CONTAINMENT TACTICS

A number of tactical options are available to achieve leak control strategies. These include physical and chemical methods. A summary of the various tactical options is given below.

Neutralization is a chemical method of containment by which a hazmat is neutralized by applying a second material to the original spill that will chemically react with it to form a less harmful substance. The most common example is the application of a base to an acid spill to form a neutral salt.

The major advantage of neutralization is the significant reduction of harmful vapors being given off. In some cases, the hazmat can be rendered harmless and disposed of at much less cost and effort. However, during the initial phases of combining an acid and a base a tremendous amount of energy may be generated, as well as toxic and flammable vapors.

Operationally, many responders recommend that neutralization operations should be limited to spills less than 55 gallons. It is quite easy to use too much neutralizing agent and end up with a large caustic spill instead of the original acid spill.

Before initiating any neutralization techniques, the following conditions should be satisfied: (1) the hazmat has been identified positively; (2) its physical and chemical properties have been researched properly; and (3) the spill has been controlled and confined to prevent runoff after application of the neutralizing agent. Responders should always contact product specialists before initiating neutralization operations for large spills and releases. Sufficient neutralizing agent should be on hand to complete the process once it is begun.

Figure 10.11 Neutralization tactic.

To determine the amount of base necessary for an acid spill, consider the following example: A glass bottle containing 1 gallon of 70% nitric acid falls and breaks open. Responders have an ample supply of neutralizing agent (soda ash) available. How much would be required to bring the pH somewhere close to 7 (neutral)?

1. Determine the weight of 1 gallon of the acid in pounds, using an MSDS or the following formula to determine the information:

Quantity of Acid Spilled	X	Specific Gravity	X	8.33 lbs./gal. (weight of water)	X	Percent Acid	=	1 Gallon Weight in Pounds
Example:								
(1 gallon)	X	(1.5)	X	(8.33)	X	(.70)	=	8.75 lbs.

2. Select the appropriate conversion factor:

Neutralizing Agent	**Sulfuric Acid**	**Nitric Acid**	**Hydrochloric Acid**	**Phosphoric Acid**
Sodium Carbonate (soda ash)	1.082	0.841	1.452	1.622
Calcium Hydroxide (slaked lime)	0.755	0.587	1.014	1.133
Sodium Bicarbonate (baking soda)	1.673	1.302	2.247	2.541

3. Multiply the weight of the spilled acid times the conversion factor for the appropriate neutralizing agent to determine the amount needed to neutralize the acid spill.

   8.75 lbs./gallon x 0.841 = 7.4 lbs. of soda ash would be required

Responders can develop a chart by which they can estimate the amount of neutralizing agent available for specific size spills. Using 70% nitric acid as an example:

1-gallon spill	=	7.4 lbs. of soda ash
10-gallon spill	=	74.0 lbs. of soda ash
100-gallon spill	=	740 lbs. of soda ash
1,000-gallon spill	=	7,400 lbs. of soda ash

When a decision has been made to neutralize a spill, some consideration should be given to the type of neutralizing agent that will be used. Some materials are more environmentally friendly than others. The key concern is biodegradability. The most widely favored neutralizing agents from an environmental perspective are sodium sesquicarbonate (for acid spills) and acetic acid (for alkali spills). Sodium and calcium hydroxide will not produce a biodegradable end product. If the spill is in an environmentally sensitive area, a product specialist should be consulted for advice on which neutralizing agent to use.

Corrosive spills should be neutralized by applying the neutralizing material from the outermost edge inward, thereby protecting responders. Avoid walking through spills, even when wearing proper protective clothing.

Some caustic spills have been neutralized using various types of diluted acids, but this technique should never be attempted without seeking the advice of a product specialist.

Neutralizing agents should be purchased in bulk quantities and stored at key locations. Commercial neutralizing kits are also available for handling small spills; these are normally packaged for smaller laboratory or workshop-type spills of one to five gallons.

Overpacking is a physical method of containment by which a leaking drum, container or cylinder is placed inside a larger undamaged overpack container. Although commonly used for liquid containers, overpacking can also be used for some compressed gas cylinders like chlorine.

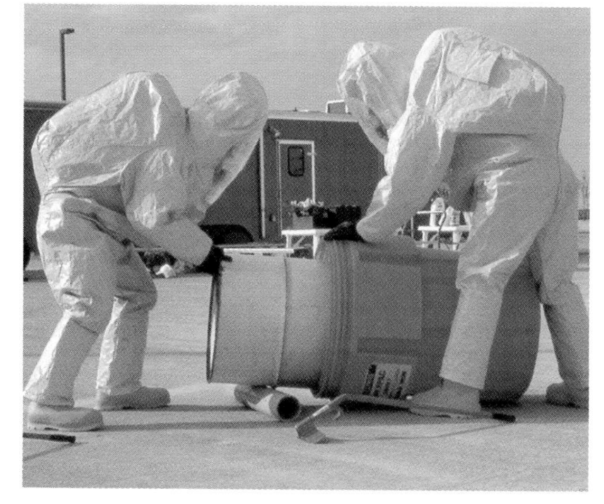

Figure 10.12 Slide in method of overpacking.

Liquid overpacks are constructed of both steel and polyethylene and range from lab packs to large 85-gallon overpack drums. Common sizes include 8, 15, 30, 55, and 85 gallons. When possible, the leak should be temporarily repaired before the container is placed inside of the overpack. Methods of placing a leaking drum into an overpack are as follows:

- *Slip the overpack container over the leaking drum*. However, if the leaking drum is upright, all bungs and lid openings will now be on the bottom after the overpack operation, thereby making pump-off operations more difficult. One tactic to minimize this problem is to first overturn the leaking drum (if safely possible), invert the overpack drum, and then flip both containers upright.
- *The slide-in method can be used if the leaking drum is in a horizontal position*. Place the open end of a horizontal overpack container near the leaking drum, then raise the end of the leaker while your entry partner slides the overpack around the leaking drum. Once the leaking drum is in the overpack, both drums can then be tilted to an upright position. CAUTION: The weight of the leaking container will be a critical issue in the ability of responders to perform this operation.
- *The rolling slide-in or V-roll method* is a variation of the slide-in method that can be used if space permits if the leaking drum is in a horizontal position. Orient

the drums so that they form a wide letter V, with the open end of the overpack drum placed under the rim of the leaking drum. Then push the drums from the apex of the "V" so that the rolling motion causes the leaker to roll into the overpack drum. Finally, tilt both drums to an upright position.

Depending on container size and weight, mechanical equipment (e.g., forklift or hoist) may be required to raise and lower the leaking container into the overpack. A 55-gallon drum of sulfuric acid can weigh over 600 pounds. In addition, the container may have been weakened as a result of the leak. The overpack container must then be labeled in accordance with DOT hazmat regulations if it will be transported from the scene.

Cylinder overpack devices are used by compressed gas companies and industrial response teams for controlling and transporting leaking cylinders. Once a cylinder is overpacked, it can be transported to a facility or other location for handling and disposal. Cylinder overpacks do have several drawbacks, including the fact that they aren't readily available, the mobilization process can be very time consuming, and they are extremely expensive to manufacture or purchase ($20,000+).

Patching and plugging is a physical method of containment that uses chemically compatible patches and plugs to reduce or temporarily stop the flow of materials from small container holes, rips, tears, or gashes. Although commonly used on atmospheric pressure liquid and solid containers, some tactics can also be used on pressurized containers.

Patching—Involves placing a material or device over a breach to keep the hazmat inside of the container. Patches can include both commercial and home-made devices and are used to repair leaks on container shells, piping systems, and valves. Patches must be compatible with the chemicals involved.

Like plugs, patches can also be fabricated on the scene, but you can save a great deal of time by manufacturing a variety of devices before the fact and carrying them on response vehicles. Responders are only limited by their ingenuity. Common examples include toggle bolt compression patches, gasket patches, glued patches, and epoxy putties.

Figure 10.13 Patching tactic.

Container pressure is a critical factor in evaluating the application and use of patching tactics. Leak bandages and leak sealing kits are effective tools for dealing with liquids with a low head pressure or low pressure gases (<100 psi or < 7 bar). Some inflatable air bag patch kits, like the Vetter Bag™, are effective at pressures less than 25 psi (2 bar). The Chlorine A and B Kits have a side patch kit for container shell leaks on 100- and 150-pound and 1-ton chlorine cylinders.

To properly organize a patching operation, consider the following:

- Select a patching device that is at least half a size larger than the breach. Smaller devices using nuts, toggle bolts, T-bolts, and so on, can be drawn inside the container as these closure devices are tightened.
- Ensure that the patch is compatible with the hazmat involved. This is especially true when dealing with corrosives.

- Patching operations should be planned with air supply operating times in mind. Several entries may be required. Overlap entry crews so that one crew is always working, one crew is always ready to step-in (i.e., backup team), and a third is ready to step into the operations flow.
- If the patching operation is complex and time allows, consider having the entry team walk-through the patching operation in the cold zone. Make sure that you have the proper tools and equipment and that all personnel are thoroughly briefed on the entry and patching operation.

When dealing with liquid containers, recognize that plastics and metals behave differently when they are breached and that the methods of repair will vary accordingly.

Plugging involves putting something into a breach or opening to reduce both the size of the hole and the amount of product flow. The plug must be compatible with both the chemical and the container. For example, a small hole in an aluminum MC-306/DOT-406 cargo tank truck can sometimes be plugged by driving a wooden wedge into the opening with a rubber mallet. However, a soft pine plug would not be compatible with a strong acid leak.

Figure 10.14 Plugging tactic.

Figure 10.15 Patching and plugging tactics increase risk because they require hands on work in the hot zone.

Plugs can be fabricated on the scene, but you can save a great deal of time by manufacturing a variety of devices before the fact and carrying them on response vehicles. Plugs can be constructed of various materials, including wood, rubber, and metal. Plugs constructed of soft woods, such as yellow pine or douglas fir, are quite effective for holes whose area is less than 3 square inches. Pneumatic plugs are also available. Preplan potential container leaks and network with other responders to determine what works.

Plugging techniques are usually used in conjunction with synthetic rubber gaskets, lightweight cloth, or special chemical-resistant putty to ensure a good seal by filling the cracks around the plug. Small holes (less than 1/2 inch diameter) not under pressure can be filled with putty or epoxy resin compounds. The longevity of these compounds is limited due to material compatibility, the size of the breach, and the head pressure of the container. These should be viewed as only temporary first aid techniques.

LEAK CONTROL ON LIQUID CARGO TANK TRUCKS

When evaluating leak control options on liquid cargo tank trucks, remember some basic principles:

- Damage to an insulated or double-shell cargo tank truck may only be to the outer shell and may be difficult to assess.
- Small holes or cracks on lined cargo tanks (e.g., MC-312/DOT-412) may get larger when a plug is inserted, as the internal container shell may be weakened over a larger surface area. In addition, inserting a plug into a small crack on some liquid cargo tanks may actually cause the crack to "run" and increase the problem.
- Valves may not always operate properly when in a different orientation from normal operations. This may also be the cause of a valve leak.
- Some liquid products are shipped under pressure of an inert gas (e.g., nitrogen) to prevent the material from reacting with air or moisture. If the container is still under pressure, the rate of release can often be reduced by simply relieving the pressure, which will create a vacuum. However, do not leave the valve open after the pressure is relieved.
- Always make sure that the leak-control materials and equipment are compatible with both the chemical and the container. This is especially critical when dealing with corrosives.
- If you are unsure of anything, always consult product or container specialists.

TYPES OF LEAKS

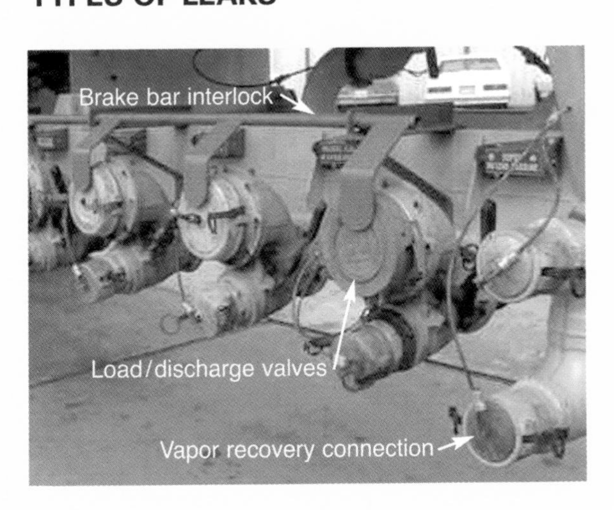

PIPING AND VALVE LEAKS

These are among the most common leaks found on cargo tank trucks, and may be controlled by either closing valves or tightening the valve packing.

If the piping is leaking, try to isolate any valves both upstream and downstream of the leak. While MC-306/DOT-406 and MC-307/DOT-407 are normally under gravity or atmospheric pressure, MC-312/DOT-412 corrosive cargo tank trucks will often be under pressure.

SPLITS, TEARS, PUNCTURES, AND IRREGULARLY SHAPED HOLES

Most breaches on bulk cargo tank trucks can be controlled in the same manner as those on smaller, nonbulk containers. The big differences will be the size of the container, the quantities of product involved, and the potential for higher pressures. In addition, double-shell or insulated cargo tank trucks can have the product flow into the shell interface, further weakening the structural integrity of the container.

If the breach is in the vapor space, containment is often very easy. Providing the materials are compatible with the product, a patch, plug, or bandage can be applied with minimal product contact. If the breach is below the liquid level, significant product release will likely occur during leak control operations. In addition, head pressure and the flow of product will make leak control operations extremely difficult to perform as well as contaminate the entry team.

PRESSURE-RELIEF DEVICES, VENTS, AND RUPTURE DISKS

Pressure relief devices include spring-loaded pressure relief valves, "Christmas Tree" vents, and rupture disks. All are piped into the vapor space and are designed to allow the release of vapors under normal operating conditions. In addition, they are engineered to minimize a liquid product leak if the cargo tank truck is involved in a rollover. Although a pressure relief valve may quickly open and lose up to 1 liter (.26 quarts) of liquid product, it should immediately reseat.

If liquid is being released from a pressure-relief device, it has probably failed. Responders should consult with container specialists to assess the level of risk and control options. Remember that blocking off a pressure relief device may result in further container stress depending on the scenario, especially in a fire event.

Safe Transportation Safety Specialsts, LLC

DOME COVER LEAKS

Dome covers may become loose and begin to leak as a result of a rollover. However, the stability of the container and the pressure of the product flow may increase the risks of leak-control operations.

For MC-306/DOT-406 cargo tank trucks, dome clamps can typically be used to control the leak. For MC-307/DOT-407 and MC-312/DOT-412 cargo tank trucks, the leak can sometimes be controlled by tightening down the wingnuts found around the manway.

Pressure isolation and reduction is a physical or chemical method of containment by which the internal pressure of a closed container is reduced. The tactical objective is to reduce sufficiently the internal pressure in order to either reduce the flow or minimize the potential of a container failure. Pressure reduction tactics are high risk operations that require responders to work in close proximity to the container. Examples include flaring and vent and burn.

Many hazmat containers are designed to store their contents under pressure. Cylinders, process vessels, MC-331 cargo tank trucks, rail cars, and pipelines are examples. It is also possible for nonpressurized containers to become pressurized because of internal chemical reactions, thermal stress, or accidentally diverted pressure.

Pressurized containers are dangerous because:

- They can rupture under stress and travel great distances as fragments or in one piece. This happens quickly and allows no reaction time.
- High pressure kills quickly. You cannot usually determine the operating pressure of a given container without close inspection. High pressure can propel valve caps, breach protective clothing, or sever SCBA air lines. Ultra high pressures (5000 to 15,000 psi) can penetrate the skin and cause an air embolism, which will be followed by death within minutes.
- Cryogenic liquids are stored in pressurized containers at temperatures below –130°F (–90°C). Cryogenics can freeze tissue and damage protective clothing.

Examples of pressure reduction tactics include the following:

Isolating valves—Pressurized containers often leak in and around valves and fittings. Most valves can be closed by turning the valve wheel clockwise, unless it is damaged. If the leak continues after the valve is closed, try tightening the packing nut on the valve.

However, there are exceptions to these procedures. According to the *Handbook of Compressed Gases,* nearly 200 types of compressed gases are commonly shipped annually. More than 12 different cylinder specifications exist with more than 64 different valve outlets used for these 200 gases.

Figure 10.16 Isolating valve tactic.

In-service containers may also have piping leaks. These leaks usually stop when the supply valve is isolated and blocked in. Depending upon the situation, it may be necessary to isolate the valve both upstream and downstream.

Isolating pumps and pressure/energy sources—Some containers are pressurized through a separate compressor or pumping system to move and transfer products. In these scenarios, the magnitude of a leak can be significantly reduced by simply lowering the pressure or completely isolating or shutting down the compressor or pumping system. Product and container specialists familiar with the system must be consulted before any shutdown operations are initiated, as a shutdown may over-pressurize other vessels or create related upstream and downstream problems. Lack of pressure may also produce unstable chemical reactions.

Product specialists and process engineers should also be contacted whenever responders face large, complex process units and related pressure vessels. Shutting down electrical power to pressure systems may ruin chemical batch processing equipment, cause dangerous pressure buildups in other locations, and disable critical safety devices. In some cases, safety systems can kick in and dump hazardous materials to neutralizing scrubbers, flares, or exhaust systems.

Venting is the controlled release of a material to reduce and control the pressure, and decrease the probability of a violent container rupture. The method of venting will depend upon the nature of the hazardous material and process involved. For example, nontoxic materials may be vented directly to air (e.g., steam), while toxic or corrosive materials may be vented to a scrubber or back-up storage vessel. Venting is typically limited to nonflammable gases.

Ray Haring

Figure 10.17 Flaring operations involving an overturned and stressed MC-331 cargo tank truck traansporting propane.

Flaring is the controlled burning of a liquid or gas material to reduce or control pressure inside the tank, dispose of residual vapors, and/or to dispose of a product when transfer operations may be impractical. Flares are designed to burn either liquid or vapor product. Flaring is commonly used in the propane industry to safely burn off product in cylinders, trucks, and rail cars when they have been severely damaged or access to the accident site is impractical for offloading to another vehicle. When dealing with bulk containers, liquid flares are often preferred due to the exceptionally long burn-off times of a vapor flare.

One drawback of flaring is the time required to burn off the product. For example, using a 2-inch-diameter hose of 150 feet in length, it would take approximately 177 hours to flare off 30,000 gallons of propane, 54 hours to burn off 11,500 gallons, and 14 hours to burn off 3000 gallons (assuming the temperature of the propane was 0°F). If the IC decides to go with a flaring option, the time required to accomplish the task must be weighed against other factors such as safety, disruption of transportation systems and businesses, and the safety and speed that other options may present.

Another factor affecting flaring operations involving flammable liquefied gases is auto-refrigeration. Essentially, as the product is burned off the liquid begins to boil inside the container. Boiling requires energy and this energy is obtained through heat from the air or ground surrounding the tank being flared. When the capacity to make up the loss of heat from the surrounding area is exceeded, the temperature of the product drops as energy is drawn from it. The product actually cools itself, boiling slows down, and the product eventually goes into a state of auto-refrigeration. Of course, this decrease in product temperature also causes a decrease in vapor pressure and a subsequent decrease in the size and intensity of the flaring operation.

Hot tapping is used to gain access to bulk liquid or gas tanks, pipeline, or container for the purpose of product removal. It involves the welding of a threaded nozzle to the exterior of a tank or pipeline. A valve is then attached to the threaded nozzle and a hole is drilled through the container shell with a specially designed machine. The drilling machine is equipped with seals that prevent the loss of product during the drilling operation. Hoses are then attached to the valve outlet and the contents are removed.

Hot tapping is commonly used within the petroleum and petrochemical industries. However, it should only be attempted during an emergency by trained hot tapping specialists.

Vent and burn is a process by which shaped explosive charges are placed by explosives demolition specialists on a flammable container, such as a pressurized tank car, to cut a hole (or holes) in it, thereby allowing the contents to flow into a pit constructed adjacent to the breached car where the product can safely burn off. When dealing with pressurized tank cars, two holes are normally required: Hole #1 is at the high end of the tank car and is designed to relieve the internal pressure so that the tank car will not "rocket," while hole #2 is at a low point and is designed to allow the liquid contents to drain into a containment area. This is a highly sophisticated technique that should only be attempted by trained specialists under very specific situations.

AP/Wide World

Figure 10.18 Vent and burn tactics are sometimes used when offloading or moving the container presents a high risk.

Vent and burn is normally an option of last resort and may be used under the following conditions:

- The tank car has been exposed to fire, resulting in elevated internal pressures and possible tank damage.
- Conditions do not allow for the safe transfer, venting, or flaring of the tank car.
- Site conditions prevent rerailing the damaged tank car.
- There has been damage to leaking valves, and fittings cannot be repaired.
- The tank car has been damaged to the extent that it cannot be safely offloaded, rerailed, and moved to an unloading point.

Vent and burn tactics have been used safely and successfully by the railroad industry at a number of major train derailments. In September, 1982, vent and burn was successfully used on eight tank cars—six vinyl chloride monomer, one styrene, and one toluene di-isocyanate—in Livingston, Louisiana.

Solidification is a chemical method of containment whereby a liquid substance is chemically treated so that a solid material results. The primary advantage of this process is that a small spill can be contained relatively quickly and immediately treated.

Solidification is often used for both corrosive and hydrocarbon spills. Commercial formulations are available that can be applied to a liquid acid or caustic spill, neutralize the hazard, and form a neutral salt. Commercially available adsorbents can also be used to solidify nonsoluble oily wastes. The spilled hydrocarbon is adsorbed to granules to form a solid, nonflowing mixture. This resulting mixture is actually safer than the original spilled material and can be easily transported and disposed of at a waste treatment facility.

Bill Hand

Figure 10.19 Vacuuming tactic.

Vacuuming is a physical method of containment by which a hazardous material is placed in a chemically compatible container by simply vacuuming it up. The method of vacuuming will depend upon the hazmats involved. Vacuuming is commonly used for containing releases of certain hydrocarbon liquids, solid particulates, asbestos fibers, and liquid mercury. Vacuuming devices can range from high-efficiency particulate (HEPA) vacuums to large vacuum trucks.

The primary advantage of vacuuming is that there is no increase in the volume of waste materials. In selecting this tactic, care must be taken to ensure that the vacuum and related equipment is compatible with the hazmats involved and that vacuum exhaust vapors are controlled. The use of vacuum trucks is discussed later in this chapter in relation to product transfer methods.

One last thought about containment options is that if two heads are better than one to come up with a solution, then five heads must be even better. Don't blow off ideas on how to solve the problem until you heard from other voices on your hazmat team. What sounds like a dumb idea in hour one of the response might be looking like a really brilliant idea in hour four when nothing else has worked. Like they say in the Army, "If it's stupid and works, then it ain't stupid!"

FIRE-CONTROL OPERATIONS

There is a wide range of hazardous materials that are flammable in nature. While specific tactics to deal with specific materials are beyond the scope of this textbook, in this section we will focus on those three hazmat classes that emergency responders commonly encounter: flammable liquids, flammable gases, and reactive chemicals. These hazmat classes have historically produced the most injuries and fatalities to emergency responders and are encountered in virtually every plant and community. Emphasis will be placed on hazards, risks, and tactical response considerations of each.

The following general factors can be applied uniformly to hazmat fire problems. They should be considered early in the incident as part of the hazard and risk evaluation process.

1. **What hazardous material(s) are involved?** Specifically, are we dealing with flammable liquids, gases, an exotic reactive material, or some combination of flammability and other hazards?

2. **What are its hazards?** The physical and chemical properties of a material significantly influence the selection of tactics and fire extinguishing agents. What works well on one type of fire won't work on another. Critical questions include determining the material's (a) chemical family (i.e., hydrocarbon or polar solvent); (b) water solubility; (c) specific gravity; (d) water reactivity; and (e) control and extinguishing agents. What type of container is involved? This can include storage tanks, pressure vessels, reactors, and pipelines. Responders must also evaluate container features, such as pressure-relief devices, valves and fixed foam systems. Many reactive materials are shipped in specialized containers and may have unusual features.
3. **What are the risks to responders, employees, and the community?** What is the likelihood of the incident growing and involving other containers? Are there any significant environmental impacts?
4. **Are specialized resources required?** What is their availability? This could include personnel, supplies, equipment, extinguishing agents, and/or related appliances.
5. **What will happen if I do nothing?** Remember—this is the baseline for hazmat decision making and should be the element against which all strategies and tactics are compared.

FLAMMABLE LIQUID EMERGENCIES

Flammable liquids are the most common hazard class encountered by emergency responders. The majority of flammable liquid emergencies are relatively small, involve nonbulk containers, and are successfully and safely handled by first responders.

Figure 10.20 Class B firefighting foam operations.

Dealing with large flammable liquids incidents is another story entirely. The fact is that over 70 firefighters have been killed in the line of duty in the last 45 years at incidents involving flammable liquid storage tank fires. No other class of hazardous materials or type of container has killed more firefighters—so pay attention and learn this stuff!

When we take an objective look at the history of fighting large flammable liquids fires it gets ugly. The costs from major losses is very high. For example, according to a 2003 study of losses from 1972 to 2001 conducted by MARSH (a major insurer of petroleum and chemical facilities), indicate that some of the largest losses in the history of the hydrocarbon and chemical industries involved petroleum storage tanks. MARSH data shows that the number of losses at terminals and other points of distribution continued to increase, with recent large losses attributed to the overfilling of storage tanks and natural related incidents. In January 2002 U.S. dollars, these losses totaled $363,000,000. See Scan Sheet 10-C.

SUMMARY OF THE THREE LARGEST PETROLEUM STORAGE TANK LOSSES IN THE UNITED STATES

RTFC

RTFC

NOTE: All loss figures are shown in their original value. The figure inside the parentheses has been adjusted to 2002 dollars.

NOVEMBER 25, 1990
DENVER, COLORADO
$32,000,000 ($40,000,000) LOSS

A fire at a 16-acre tank farm that supplied jet fuel to the Denver International Airport. The fire burned for more than 55 hours, damaging seven storage tanks and consuming more than 1.6 million gallons of jet fuel. The tank farm was comprised of 12 storage tanks.

At approximately 9:20 a.m., the fuel supply company received a "no flow" indication in the pipeline to the tank farm. Shortly after that time, the airport control tower noticed a black column of smoke from the tank farm. An initial fuel leak originating at an operating fuel pump in the valve pit was ignited by the electric motor for the pump resulting in the fire. A cracked fuel supply pipe in the valve pit formed two V shaped streams extending 25 to 30 feet into the air. This provided additional fuel to the pool fire.

As the fire continued to grow, coupling gaskets in the piping deteriorated and more fuel flowed out of the storage tanks, spreading the fire throughout the dike area. The valve controlling the fuel flow to the airport supply line sporadically released fuel into the valve pit. Firefighters were unable to prevent the back flow of fuel from this line since the nearest manual shutoff valve was two miles from the tank farm.

The Denver Airport Fire Department dispatched four aircraft rescue firefighting (ARFF) vehicles and one rapid intervention vehicle to the fire. The second and third alarms provided an additional five engine companies, three truck companies, and one rescue unit from the Denver City Fire Department.

In addition to the foam concentrate on hand at the scene, foam concentrate was received from other local departments as well as the Seattle, Houston, and Chicago Fire Departments. Unknown to the fire department, a pipeline that was reported to have been shut down continued to supply fuel to the fire. After repeated unsuccessful attempts to extinguish the fire by the

Denver Fire Department, Williams Fire and Hazard Control was brought in by the owners to assist the fire department with extinguishing the fire. The pipeline was eventually shut down and the fire was extinguished.

JULY 7, 1983
NEWARK, NEW JERSEY
$35,000,000 ($52,000,000) LOSS

Gasoline was being received by pipeline into a 42,000-barrel internal floating roof tank at a products terminal when an overfill occurred, spilling about 1300 barrels into the tank diked area. A slight wind (1 to 5 mph) carried gasoline vapors about 1000 feet to a drum reconditioning plant, where an incinerator provided the ignition source. The resulting explosion caused $10,000,000 damage to the terminal and up to $25,000,000 in over 2000 claims to rail cars and adjacent properties. Although dikes contained the burning spill to the tank that was overfilled, two adjoining internal floating roof tanks and a smaller tank ignited and were eventually destroyed along with 120,000 barrels of product. Since the burning tanks presented little exposure to other facilities, the decision was made to let the fire burn itself out. This incident resulted in tank overfill prevention requirements in *NFPA-30, The Flammable and Combustible Liquids Code.*

SEPTEMBER 24, 1977
ROMEOVILLE, ILLINOIS
$8,000,000 LOSS

Lightening struck a 190-foot diameter cone roof tank containing diesel fuel. Roof fragments were thrown 240 feet and struck a 100-foot-diameter covered floating roof gasoline tank. An adjacent 180-foot diameter floating roof gasoline tank 80 feet away was also struck by debris. The entire surfaces of the cone and internal floating roof tanks ignited immediately. The rim fire on the floating roof resulted in the roof sinking after about four hours. The two largest tanks were full; the smallest about half full. The two larger tanks and their contents were destroyed. The fire in the internal floating roof tank was extinguished after about two hours.

After burning for approximately 46 hours, the fire was extinguished through both topside and subsurface foam applications. The refinery's five stationary fire pumps supplied up to 10,000 gpm of the estimated 14,000 gpm required during the fire. Thirty-five municipal and industrial fire departments, including a 12,000-gpm fire boat, assisted the refinery fire department. About 22,000 gallons of foam concentrate were consumed during the fire fighting effort.

SOURCE: Marsh, THE 100 LARGEST LOSSES 1972–2001: Large Property Damage Losses in the Hydrocarbon-Chemical Industries, (20th edition).

Incidents involving bulk flammable liquids pose considerable risks to both firefighters and emergency responders. History is filled with case studies where responders have been seriously burned, injured, and killed because of their failure to understand how flammable liquids and their containers behave. In light of this fact, the following material will focus on the tactical problems associated with managing larger flammable liquid fires, such as dealing with bulk liquid storage facilities, transportation containers, and storage tank fires.

HAZARD AND RISK EVALUATION

Flammable liquids are a "good news–bad news" type of event. The bad news is that there will be a lot of fire, tremendous amounts of radiant heat, and exposure problems. The good news is that unlike most other hazmat problems, you can usually see where your problem is, and the fire will often allow enough time to develop a well-thought-out plan of action. Responders will normally have enough time to gather the necessary resources before mounting an aggressive, offensive oriented fire attack. This doesn't mean that defensive tactics and exposure protection won't be required (they will!), but only that the time factors are sometimes a little different from those encountered at structural fires or other types of major incidents where life safety is a big issue and offensive strategies are required.

Keeping track of times and key events at a major flammable liquid fire can be a difficult task. The Incident Commander should acquire the following information during the size-up process:

- Time when the incident started. This may not necessarily be the same time the incident was reported.
- Time at which responders arrived on scene.
- Probability that the fire will be confined to its present size.
- Fuel involved (flammable or combustible liquid), including the quantity, surface area involved, and the depth of the spill.
- Hazards involved, including flash point, reactivity, solubility (e.g., hydrocarbon or polar solvent), and specific gravity.
- Estimated pre-burn time. This will help the IC determine factors such as how "hot" the fuel is, identify and prioritize exposures, consider transfer and pump-off options, determine if a heat wave is developing for crude oils, and so on.
- Layout of the incident, including the following specific points:
  - Type of storage tank(s) involved. Common aboveground liquid storage tanks are cone roof, open floating roof, covered floating roof, and dome roof tanks. See page 231 for additional information.
  - Size of the dike area(s) involved.
  - Valves and piping systems stressed or destroyed by the fire.
  - All surrounding exposures, including tanks, buildings, process units, utilities, and so on. This should include identifying and prioritizing exposures (e.g., flame impingement, radiant heat exposure). Process unit personnel can help with these decisions.

TACTICAL OBJECTIVES

Hazard and risk evaluation is the cornerstone of decision making. Based on the type and nature of risks involved, the Incident Commander will implement the appropriate strategic goals and tactical objectives. Tactical options for flammable liquid emergencies include the following:

Nonintervention—This is a "no win" situation in which responders assume a passive position (i.e., get out the beach chairs, umbrellas, and suntan lotion, and watch the fire burn itself out). This option is sometimes implemented when there are insufficient water supplies, very little product remaining which can be saved, or no exposures in the immediate area.

Defensive tactics—These tactics involve protecting exposures and allowing the fire to burn. In many cases, defensive tactics are a temporary measure until sufficient resources can be assembled to pursue an aggressive, offensive attack.

The primary concerns during defensive operations are direct flame impingement and radiant heat exposures. Flame impingements must be cooled immediately, while radiant heat exposures should be handled as soon as possible. Exposures should be prioritized in the following manner:

- **Primary exposures**—Pressure vessels, closed containers, piping systems, or critical support structures exposed to direct flame impingement. Failure of exposed vessels, tanks, and piping systems is likely unless cooling water is quickly applied.

  If a storage tank is involved, direct flame contact on the tank shell can cause the upper portion of the shell, as well as any associated foam systems, to lose their integrity and fold inward. Streams should cool all surfaces above the liquid level. Remember, cooling water is a valuable resource; don't waste it. See Figure 10.21.
- **Secondary exposures**—Pressure vessels, closed containers, piping systems, or critical support structures exposed to radiant heat. Failure of structural components is possible if cooling water is not applied.

  Be careful of applying water onto open floating roofs—sinking the roof with water lines can be somewhat embarrassing as well as hazardous! Also, remember that radiant heat will pass through structures with clear glass and windows. In addition to applying exterior cooling lines, firefighters should be sent inside to check for any fire extension.
- **Tertiary Exposures**—Noncritical exposures without life safety concerns.

Offensive Tactics—These tactics are implemented when sufficient water and firefighting foam supplies and related resources are available for a continuous, uninterrupted fire attack. Although the primary focus is on fire extinguishment, it may also be necessary to maintain exposure lines.

Time becomes a critical factor for offensive operations. The IC must determine the duration of any fire attack. For example, *NFPA 11—Technical Standard on Low Expansion Foam and Combination Agents*, recommends an application time of 15 minutes for flammable liquid spill fires and 65 minutes for Class I flammable liquid

Figure 10.21 Offensive tactics are being used here to support a rescue of an injured utility worker.

storage tank fires. The IC should document the time foam operations start, the time at which the fire is controlled, and the time at which the fire is extinguished.

A final note about exposure protection. Flammable liquid spill fires confined to a diked area may accumulate a large quantity of water from both fire attack and exposure streams. If these flows and associated runoff are not closely monitored, dikes may overflow and carry the burning flammable liquid outside of the area. In addition, loss of electrical power to the facility may cause the sewer system pumps to lose power and create additional runoff problems. As a general guideline, water streams applied to exposures inside of a diked area (e.g., adjoining storage tank) should be temporarily shut down when they no longer produce steam at the point where water contacts the hot steel surface.

EXTINGUISHING AGENTS

The availability of water and foam concentrate is a critical factor in evaluating the risks involved in a flammable liquid emergency. The IC must determine if:

1. An adequate and uninterrupted water supply is available.
2 There is sufficient capability to deliver the required foam solution to control and extinguish the fire.

If an adequate water and foam supply is not available for both protecting exposures and controlling the fire, the IC should consider implementing defensive or non-intervention tactics until sufficient resources are available.

Class B firefighting foam is the workhorse of flammable liquid firefighting. While other agents, such as dry chemicals, are used to extinguish small fires or deliver the knockout punch for three-dimensional fires (e.g., a flange fire), Class B foam is still the "top gun" for large-scale flammable liquid problems. When dealing with three-dimensional fires, be sure that the fuel source can be shut off when the fire is extinguished. Aqueous film-forming foam (AFFF) may be used to secure the fuel surface area.

SELECTING FOAM CONCENTRATES

Selection of a foam concentrate is an important part of a successful firefighting operation. There are several different types of Class B foam concentrates sold by a variety of reliable manufacturers at different concentrations and for different fire protection applications. The selection of a foam concentrate can be as much a business decision as a technical one.

Firefighting foam concentrates should be selected based on the type of fuel (e.g., hydrocarbon vs. polar solvent), and the type and nature of the hazard to be protected (e.g., spill scenario vs. storage tank). The most common foam concentrates used for flammable liquid storage tank fire protection are as follows:

Fluoroprotein foam—Combination of protein-based foam derived from protein foam concentrates and fluorochemical surfactants. The addition of the fluorochemical surfactants produces a foam that flows easier than regular protein foam. Fluoroprotein foam can also be formulated to be alcohol resistant.

Key characteristics of fluoroprotein foam as it relates to tank firefighting include the following:

- Available in 3% and 6% concentrations
- Are oleophobic (they shed oil) and can be used for subsurface injection

- Compatible with simultaneous application of dry chemical extinguishing agents (e.g., Purple K)
- Form a strong blanket for long-term vapor suppression on unignited spills
- Must be delivered through air aspirating equipment
- Has a shelf life of approximately 10 years

Aqueous Film-Forming Foam (AFFF)—Synthetic foam consisting of fluorochemical and hydrocarbon surfactants combined with high boiling point solvents and water. AFFF film formation is dependent on the difference in surface tension between the fuel and the firefighting foam. The fluorochemical surfactants reduce the surface tension of water to a degree less than the surface tension of the hydrocarbon so that a thin aqueous film can spread across the fuel.

Figure 10.22 AFFF foam application.

Key characteristics of AFFF foam as it relates to tank firefighting include the following:

- Available in 1%, 3%, and 6% concentrations for use with either fresh or salt water
- Very effective on spill fires with a good knock down capability
- Compatible with simultaneous application of dry chemical extinguishing agents (e.g., Purple K)
- Suitable for subsurface injection

Alcohol Resistant AFFF (ARC)—Alcohol-resistant AFFFs are available at 3% hydrocarbon/3% polar solvent (known as 3 x 3 concentrates), although 3% hydrocarbon/6% polar solvent concentrations (known as 3 x 6 concentrates) may also be found. When applied to a polar solvent fuel, they will often create a polymeric membrane rather than a film over the fuel. This membrane separates the water in the foam blanket from the attack of the polar solvent. Then the blanket acts in much the same manner as a regular AFFF.

Key characteristics of alcohol resistant AFFF foam as it relates to tank firefighting include the following:

- Must be applied gently to polar solvents so that the polymeric membrane can form first.
- Should not be plunged into the fuel, but gently sprayed over the top of the fuel.
- Very effective on spill fires with a good knock-down capability.
- May be used for subsurface injection applications.

Film-Forming Fluoroprotein Foam (FFFP)—Based on fluoroprotein foam technology with AFFF capabilities. FFFP combines the quick knock-down capabilities of AFFF along with the heat resistance benefits of fluoroprotein foam.

Key characteristics of FFFP foam as it relates to tank firefighting include the following:

- Available in 3% and 6% concentrations.
- Compatible with simultaneous application of dry chemical extinguishing agents (e.g., Purple K).
- Can be used with either fresh or salt water.

FFFP is also available in an alcohol-resistant formulation. Alcohol-resistant FFFP has all of the properties of regular FFFP, as well as the following characteristics:

- Can be used on hydrocarbons at 3% and polar solvents at 6%. Newer FFFP concentrates can be used on either type of fuel at 3% concentrations.
- Can be used for subsurface injection.
- Can be plunged into the fuel during application.

A few rules should be observed regarding the compatibility of Class B foams:

1. Similar foam concentrates by different manufacturers are not considered to be compatible in storage applications. The exception to this would be Mil-Spec (i.e., military specification) foam concentrates. The Mil Specs are written so that mixing can be done with no adverse effects.
2. Don't mix different kinds of foam concentrates (e.g., AFFF and fluoroprotein) before or during proportioning.
3. On the emergency scene, concentrates of a similar type (e.g., all AFFF's, all fluoroprotein, etc.) but from different manufacturers may be mixed together immediately before application.
4. Finished foams of a similar type but from different manufacturers (e.g., all AFFF's) are considered compatible.

DETERMINING FOAM CONCENTRATE REQUIREMENTS

The availability of water and foam concentrate are critical factors in evaluating the risks involved in a storage tank emergency. The Incident Commander must evaluate the following factors in developing the Incident Action Plan, including

1. Size of the fire (i.e., area involved, spill fire, tank fire, combination tank and dike fire)

2. Type of fuel (i.e., hydrocarbon or polar solvent)
3. Required foam application rate
4. Amount of foam concentrate required on-scene and the ability to resupply it
5. Ability to deliver the required amount of foam/water onto the fuel surface and sustain the required flow rates

If an adequate water and foam supply is not available for both protecting exposures and controlling the fire, the IC should consider implementing defensive or nonintervention tactics until sufficient resources are available. As a general tactical guideline, foam application operations should not be initiated until sufficient foam concentrate is onsite to extinguish 100% of the exposed flammable liquid surface area.

In evaluating specific types of Class B foam concentrate for the protection and/or extinguishment of specific fire scenarios (e.g., spill, tank fire, hydrocarbon vs. polar solvent, etc.), emergency responders should review the technical data package and the minimum foam application rates published by the respective foam manufacturer.

NFPA 11 recommended minimum foam application rates for specific fuels, foams, and applications are listed below.

- 0.10 gpm/ft^2—Fixed system application for hydrocarbon fuels (e.g., cone roof storage tank with foam chambers).
- 0.30 gpm/ft^2—Fixed system application for seal protection on an open-top floating roof tank.
- 0.10 gpm/ft^2—Subsurface application for hydrocarbon fuels in cone roof tanks.
- 0.10 gpm/ft^2 (AFFF, FFFP) to 0.16 gpm/ft^2 (protein, fluoroprotein)—Portable application for hydrocarbon spills (e.g., 1-3/4-inch handlines with foam nozzles).
- 0.16 gpm/ft^2—Portable application for hydrocarbon storage tanks (e.g., portable foam cannons and master stream devices). A foam application rate of 0.18 to 0.20 gpm/ft^2 has been used to successfully extinguish hydrocarbon fires in large diameter tanks using master stream portable nozzles.
- 0.20 gpm/ft^2—Minimum recommended rate for polar solvents. Higher flow rates may be required depending on the fuel involved and the foam concentrate used.

HOW TO CALCULATE FOAM REQUIREMENTS

Refinery Terminal Fire Company

Foam concentrate requirements can be determined by the following process:

1. Determine the type of fuel that is involved— hydrocarbon or polar solvent. This will determine the type of foam concentrate to be used.
2. Determine the surface area involved.
   - Calculate storage tank area = (.785)(Diameter^2) or (.8)(Diameter^2)

     NOTE: The formula (.785)(Diameter^2) is commonly used during the preincident planning process, while the formula (.8)(Diameter^2) is commonly used for field applications.
   - Calculate dike or rectangular area around the tank

     Area = Length x Width
3. Determine the recommended NFPA 11 foam application rate, as noted above.
4. Determine the duration of foam application per NFPA 11.
   - Flammable liquid spill = 15 minutes
   - Storage tank

Flash point 100°–200°F	= 50 minutes
Flash point < 100°F	= 65 minutes
Crude oil	= 65 minutes
Polar solvents	= 65 minutes
Seal application	= 20 minutes

5. Determine the quantity of foam concentrate required. This figure will be determined by the percentage of foam concentrate used.

PROBLEM 1 A 125-ft-diameter open floating roof tank containing gasoline is fully involved in fire. Determine the amount of foam concentrate required to control and extinguish the fire. The fire department is using a 3% x 3% alcohol resistant foam concentrate (ARC).

125 foot diameter open top floating roof tank.

1. *What is burning?*

   Gasoline–hydrocarbon liquid. The 3% x 3% ARC can be used and will be proportioned at a 3% concentration.

2. *Determine the surface area involved.*

   Area = (.8) (Diameter^2)

   Area = (.8) (125 ft)^2

   Area = 12,500 ft^2

3. *Determine the appropriate foam application rate.*

   The fire department is using portable application devices—foam cannons. The foam application rate is 0.16 gpm/ft^2

   Foam application = area x recommended application rate.

   Foam application = (12,500 ft^2) (0.16 gpm/ft^2)

   Foam application = 2,000 gpm

4. *Determine the duration of foam application.*

   Gasoline has a flash point of approximately -45°F. Therefore, the recommended duration is 65 minutes.

   Required foam solution = foam application x duration

   Required foam solution = (2000 gpm) (65 minutes)

Required foam solution = 130,000 gallons of foam solution

4. *Determine the quantity of foam concentrate required.*

   The fire department is using a 3% foam concentrate.
   Required amount foam concentrate = required foam solution x 3%
   Required amount foam concentrate = (130,000 gallons) (0.03)
   Required amount foam concentrate = 3900 gallons of foam concentrate
   Required amount of water = 126,100 gallons

PROBLEM #2 A 150-ft-diameter covered floating roof tank containing gasoline has been overflowed and ignited. Both the tank and the dike area (100 ft x 80 ft) are completely involved in fire. Determine the amount of foam concentrate required to control and extinguish the fire. The fire department is using a 3% x 3% alcohol resistant foam concentrate (ARC).

150 foot diameter covered floating roof tank fully involved with dike fire.

1. *What is burning?*

   Gasoline–hydrocarbon liquid. The 3% x 3% ARC can be used and will be proportioned at a 3% concentration.

2. *Determine the surface area involved.*

Storage Tank	Dike Area
Area = (.8) (Diameter^2)	Area = (Length)(Width)
Area = (.8) (150 ft)^2	Area = (100 ft)(80 ft)
Area = 18,000 ft^2	Area = 8000 ft^2

3. *Determine the appropriate foam application rate.*

   The storage tank is protected with foam chambers designed for full-surface protection and require a foam application rate of 0.10 gpm/ft^2. Portable application devices are required for the dike fire and require a foam application rate of 0.16 gpm/ft^2.

   Storage Tank
   Foam application = area x recommended application rate
   Foam application = (18,000 ft^2) (0.10 gpm/ft^2)
   Foam application = 1800 gpm

   Dike Area
   Foam application = area x recommended application rate
   Foam application = (8000 ft^2) (0.16 gpm/ft^2)
   Foam application = 1280 gpm

4. *Determine the duration of foam application.*

   Gasoline has a flash point of approximately -45°F. Therefore, the recommended duration is 15 minutes for the dike area and 65 minutes for the storage tank.

   Storage Tank
   Required foam solution = foam application x duration
   Required foam solution = (1800 gpm) (65 minutes)
   Required foam solution = 117,000 gallons of foam solution

   Dike Area
   Required foam solution = foam application x duration
   Required foam solution = (1280 gpm) (15 minutes)
   Required foam solution = 19,200 gallons of foam solution

5. *Determine the total quantity of foam concentrate required.*

   The fire department is using a 3% foam concentrate.

   Storage Tank
   Required amount foam concentrate = Required Foam Solution x 3%
   Required amount foam concentrate = (117,000 gallons) (0.03)
   Required amount foam concentrate = 3510 gallons of foam concentrate
   Required amount of water = 113,490 gallons

   Storage Tank
   Required amount foam concentrate = Required Foam Solution x 3%
   Required amount foam concentrate = (19,200 gallons) (0.03)
   Required amount foam concentrate = 576 gallons of foam concentrate
   Required amount of water = 125,524 gallons

   Total amount foam concentrate required = 4086 gallons
   Total amount of water required = 239,014 gallons

PROBLEM #3 A MC-306/DOT-406 cargo tank truck containing 8500 gallons of gasoline has overturned on a four-lane interstate highway. All cells are ruptured upon your arrival and the gasoline tank truck is fully involved in fire. The surface area burning on the street is approximately 150 feet x 300 feet (46 m x 91 m). Determine the amount of foam concentrate required to control and extinguish the fire. The fire department is using a 3% x 3% aqueous film forming foam (AFFF).

Surface fire involving gasoline tank truck.

1. *What is burning?*

   Gasoline—hydrocarbon liquid. The 3% AFFF can be used for extinguishment using 1-3/4-inch handlines with foam nozzles.

2. *Determine the surface area involved.*

   Surface Spill
   Area = (Length) (Width)
   Area = (150 ft) (300 ft)
   Area = 45,000 ft^2

3. *Determine the appropriate foam application rate.*

   The gasoline surface fire can be extinguished using 0.10 gpm/ft^2 AFFF.

   Surface Area
   Foam application = area x recommended application rate
   Foam application = (45,000 ft^2) (0.10 gpm/ft^2)
   Foam application = 4500 gpm

4. *Determine the duration of foam application.*

   Gasoline has a flash point of approximately -45°F. Therefore, the recommended duration is 15 minutes for the contained surface fire.

   Dike Area
   Required Foam Solution = Foam application x Duration
   Required Foam Solution = (4500 gpm) (15 minutes)
   Required Foam Solution = 67,500 gallons of foam solution

5. *Determine the total quantity of foam concentrate required.*

   The fire department is using a 3% AFFF foam concentrate.

   Surface Fire
   Required amount foam concentrate = required foam solution x 3%
   Required amount foam concentrate = (67,500 gallons) (0.03)
   Required amount foam concentrate = 2,025 gallons of foam concentrate
   Required amount of water = 65,475 gallons

NOTE: Scenario 3 provides an example of the problems an MC-306/DOT-406 gasoline tank truck fire will present. Although some steel cargo tanks may still be found, the majority of MC-306/DOT-406 cargo tank trucks are constructed of aluminum. Aluminum shell MC-306/DOT-406 cargo tanks will melt down to the liquid level, and the product will burn off in a controlled manner rather than build up internal pressures that would otherwise lead to a catastrophic breach. As a result, these scenarios usually involve a spill fire of some size or magnitude, and an open pit fire as any remaining product in each compartment burns off.

If the running spill fire can safely be confined using defensive tactics (e.g., diversion and diking), the surface area becomes smaller, the product pooled in the compartments is confined, and less foam concentrate may be required for extinguishment because of the smaller surface area.

Remember back in fifth grade when you told your friends you would never really use math. Guess what? Jimmy Buffet was wrong! Math does not suck. Hazmat responders have to know this stuff and get it right the first time. You have to use the right amount of foam concentrate, at the right application rate, for the right amount of time, or the fire won't go out.

DETERMINING WATER SUPPLY REQUIREMENTS

Flammable liquid storage tank fires can require tremendous quantities of water for a sustained period of time. Before an effective fire attack can be made, it must be determined if the water system is capable of

- Delivering a water flow rate equal to or greater than that required to control the largest potential fire area
- Delivering the required flow rates at pressures that can be used effectively by water application devices such as fixed systems, portable monitors, and handlines

There are several assumptions regarding water supply requirements that are often misunderstood by emergency response personnel. These include:

1. **ASSUMPTION:** The fire water system is known to be adequate for fire attack because all of the past fires at the facility have been successfully controlled.

   REALITY: Most hydrocarbon fires never reach their full potential because they are extinguished in their incipient or small stages. Therefore, they were controlled well before they reached their full-scale potential and never really taxed the water system.

Industrial Emergency Response Corporation

Figure 10.23 Large flammable liquid fires require large fire flows up to 20,000 gpm.

2. **ASSUMPTION:** The water flow rate available in the facility is believed to be adequate because the rated flow of the facility's fixed fire pumps exceeds the maximum foreseeable water flow demand.

   REALITY: Fire pumps are not constant flow and constant pressure devices. Therefore, the sum of their rated flows is not equal to the water flow rate available in the facility. When delivering water, fire pumps must provide sufficient pressure to operate water application devices in the area of the fire and to overcome pressure losses in the piping system. If this pressure is higher than the pumps' rated pressure, they will flow LESS water than the sum of their flows. If the water flow rate demand exceeds the rated flow of the pumps, the pressure available in the fire area will DECREASE, with possible effects on the operation of water application devices.

   As a result, the water flow rate available in each area of a facility varies depending on the area's relative elevation and location with respect to the pumps and the pressure requirements of the water application devices to be used. In addition, experience shows that some water systems have been modified or damaged over time. While they may have been properly designed and installed, years of neglect may have rendered them ineffective.

WHAT IS A BLEVE?

BLEVE is an acronym for Boiling Liquid Expanding Vapor Explosion. A BLEVE is defined as a container failure with a release of energy, often rapidly and violently, which is accompanied by a release of gas to the atmosphere and propulsion of the container or container pieces due to an overpressure rupture.

As a liquefied gas, an LPG tank contains both liquid and vapor. Any external fire creating direct flame impingement on the vapor space will heat the tank shell; any fire on the vapor space will heat the tank shell more rapidly than any fire impingement on the liquid area. As temperatures soar, the tank shell's temperature in the vapor space quickly reaches 752°F (400°C). Above 1,112°F (600°C) steel weakens significantly. By the time the steel reaches 1,800°F (982°C) it has lost 90% of its strength. Thinned and weakened from the fire, the tank will eventually relieve pressure to the outside either through a split in the tank in the form of a jet flame or the container will fail. Not every LPG tank that is exposed to fire fails by BLEVE. Sometimes the tank just splits open at the weld seam and burns off, and sometimes the pressure relief valve functions and burns off the LPG. There is no way to predict how and when any type or size of tank will fail. A functioning pressure-relief valve is not an effective way of determining if a tank will fail, but it is a good indicator that the internal pressure inside of the tank is higher than normal. The pressure relief valve on an ASME tank is set for 250 psi (17.2 bar), while a motor fuel tank relief valve would be set between 312.5 to 375 psi (21.5 to 25.8 bar).

NTSB

Crescent City, IL (1971)

When a tank fails due to a BLEVE, projectiles (i.e., pieces of tank, metal parts, and other debris) travel in all directions (360 degrees around the tank) for great distances. If you are standing (or hiding) between where the tank is headed and where it plans on landing, you are probably not going to survive the experience!

Distance is your friend. Consult the Emergency Response Guide for protective action and evacuation recommendations based on the size of the container. This isn't rocket science—the larger the container involved, the further back you need to be. A nonintervention strategy is often the best option if the tank is on fire when you arrive at the scene. Before a defensive or offensive strategy is employed the IC must be satisfied that there is an adequate water supply to support the fire attack, the risks have been adequately evaluated (gain or loss vs. the risk you will be taking), and both the IC and the responders have been properly trained in handling LPG fires (i.e., specifically in evaluating the hazards and risks for "Go or No-Go" decision making).

Tank exposures to high-velocity jet flame (i.e., pressure-fed fire) will require 500 gallons per minute of water (1892 liters per minute) at the point of impingement. Exposures to radiant heat with no direct flame contact will require 0.1 to 0.25 gallons per minute per square foot (0.4 to 0.9 L per minute) to maintain the integrity of the exposure.

Never extinguish a pressure-fed flammable gas fire unless you can control the fuel supply. Isolate the source of the gas and permit the fire to self-extinguish, thereby consuming any residual gas inside the vessel or piping system. Unignited flammable gases and vapors escaping under pressure will rapidly form an unconfined vapor cloud which will usually be re-ignited by ignition sources in the area. Explosions of unconfined vapor clouds can cause major structural damage and quickly escalate the size of the emergency beyond responder capabilities.

FLAMMABLE GAS EMERGENCIES AT PETROCHEMICAL FACILITIES

Incidents at facilities that manufacture, process, store, or use large quantities of flammable gases can create specialized problems for responders. The IC must evaluate the overall fire and hazmat problem, recognizing that products other than flammable gases may be involved.

Specifically, the Incident Commander should determine the following:

- Were there any abnormal operating conditions immediately before the emergency?
- Were there any equipment problems or changes immediately before the emergency (e.g., maintenance operations, changing over pumps, blinding off lines)?
- Are exposures protected with fixed water spray systems or monitors? Are systems operating?
- Are fixed fire protection and chemical mitigation systems available (e.g., scrubbers and neutralizers)? Have they been activated?
- What is the status of the fire pumps? What is the fire water system pressure?
- Is the process isolated?
- What is the structural stability and potential failure of the unit? Is fireproofing in place? Experience in the petrochemical industry shows that in a major fire:
  - Instrumentation lines can begin to fail within 5 minutes.
  - Pressure vessels and other closed containers can begin to fail within 10 minutes.
  - Structural steel will begin to fail within 15 minutes.
- What is the status of the process unit? Has the process been isolated? Is the process stable (e.g., temperature, pressure, and reactions)?
- What types of safety systems are in place? Have they been activated? These would include emergency shutdown systems, pressure relief devices, flares, scrubbers, and so on.
- What is the status of the utility systems, including electrical, instrument air, steam, fuel gas, and so forth? Isolating utilities without coordinating with facility process personnel may create upstream and downstream secondary and tertiary problems greater than the initial event.

REACTIVE CHEMICALS—FIRES AND REACTIONS

Reactive chemical families include oxidizers, organic peroxides, certain flammable solids, pyrophoric materials, and water reactive substances. Fires involving reactive chemicals have resulted in numerous responder injuries and deaths. Examples include Norwich, Connecticut, Roseburg, Oregon, Kansas City, Missouri, and Newton, Massachusetts.

In Chapter 7 we noted that when a hazmat container breaches, there are only two things jumping out at you—energy and matter. With reactive chemical families, tremendous amounts of energy can be released in an instant. If you are committed to the hot zone when these materials go bad, you probably will not escape without suffering injury.

DOD

Figure 10.27 Reactive chemicals contain lots of energy.

Chemical reactivity can cover many hazards and properties. These may include the following:

- **Stability**—The resistance of a chemical to decomposition or spontaneous change. Unstable materials would include those which can rapidly and/or vigorously decompose, polymerize, condense, or become self-reactive.
- **Incompatibility**—The chemical reactions and the products of the reactions as a result of the incompatibility will vary with the nature of the chemicals involved. Typical hazards may include the release of toxic materials, the release of flammable materials, the generation of heat, and the destruction of materials.

- **Decomposition Products**—May be produced in dangerous quantities as a result of a chemical reaction or thermal decomposition. These would include toxic products of combustion (e.g., carbon monoxide, hydrogen cyanide, hydrogen chloride) as well as off-gases created by a chemical reaction.
- **Polymerization**—When polymerization occurs spontaneously or without controls, it can give off tremendous levels of both heat and pressure. Unplanned polymerization can occur as a result of environmental conditions (excessive heat), the depletion of inhibitors, or the inadvertent introduction of a catalyst and can lead to detonation.

HAZARD AND RISK EVALUATION

The hazard and risk evaluation process for reactive chemicals should focus on the following factors:

- **Hazardous nature of the material involved**. Key factors would include the nature of its reactivity (e.g., water, air, heat, other materials, etc.).
- **Quantity of the material involved**. Although risks are often greater when dealing with bulk quantities, small quantities of highly reactive chemicals can pose significant risks. Remember—there is a very fine dividing line between explosives, oxidizers, and organic peroxides. All are capable of releasing tremendous amounts of energy! See Figure 10.27.
- **Design and construction of the container**. The type of container will vary depending upon whether the chemical is a raw material, an intermediate material being used to form another chemical or product, or the finished product.
- **Fixed or engineered safety systems**. Reactive chemical processes and facilities may have a number of engineered safety systems in place, including explosion suppression systems, explosion venting systems (e.g., blowout panels), holding tanks, flares, and scrubbers. Individual containers may also have pressure relief devices based on the nature of the hazard.
- **Type of stress applied to the hazmat and/or its container**. A number of chemical families will react to the presence of heat and elevated temperatures (not necessarily a fire). Likewise, other chemical families (e.g., oxidizers) can become contaminated with water or dirt, creating heat, and then overpressurize the container.
- **Size and type of area being affected**. The quantity involved will effect the size of the fire or reaction, as well as the likelihood that the problem will be confined to its present size.
- **Identifying and prioritizing exposures**. This includes the proximity of exposures and the rate of release.
- **Level of available resources**. Incidents involving reactive chemicals will typically require the expertise of technical information and product specialists who are inherently familiar with the materials involved.

TACTICAL OBJECTIVES

Responders must carefully weigh the hazards and risks of intervention when selecting the strategic goals and tactical objectives. Unlike flammable liquids and gases that leave responders with a lot of options, the tactical options for reactive materials will often be limited unless the fire or reaction can be handled in its initial stages. See the Garfield Heights, Ohio Case Study on page 482.

Tactical options include the following:

Nonintervention—This is a "no win" situation in which responders cannot positively change or influence the sequence of events. As a result, responders withdraw to a safe distance and allow the incident to run its natural course. A classic example of nonintervention would be a vehicle or structure in a remote area containing oxidizers and organic peroxides that is heavily involved in fire. Given the limited benefits, this should be a "no brainer" type of decision.

Defensive tactics—These tactics involve protecting exposures and allowing the fire to burn or the reaction to run its course. In some instances, responders must play a "wait and see" game. Defensive tactics may include implementing a controlled burn, remotely transferring the product to another container, remotely injecting a stabilizer into the reaction, or disposing of the decomposition products by sending them to a flare or scrubber.

If a building containing reactive chemicals is heavily involved in fire, the only option responders may have is to protect exposures. Do not underestimate the rapid speed at which these fires can move. High temperature accelerants, such as flammable metals and pyrotechnics, have been used in a number of arson fires and terrorists incidents in the United States and Canada. A full-scale test fire in a 30,000 ft^2 (2787 m^2) vacant shopping center in Puyallup, Washington resulted in the flashover of the entire structure within 2 minutes!

Offensive Tactics—These tactics are implemented when sufficient resources are available to control and extinguish the fire. However, unless the fire or chemical reaction is observed in its initial stages, there are often limited offensive tactics that can be implemented to change the sequence of events. In addition, offensive tactics will expose responders to a significantly higher level of risk.

Figure 10.28 Reactive chemicals may be used by criminals and terrorists and require assistance by bomb squad and fire and explosion investigators.

Case Study

MAGNESIUM PLANT FIRE
GARFIELD HEIGHTS, OHIO
DECEMBER 29, 2003

On December 29, 2003 the Garfield Heights Fire Department responded to a magnesium processing plant fire a metal recycling plant. White-hot flames engulfed the complex and explosions shot sparks into the sky. With prior knowledge of the facility through pre-planning, the decision was made by the IC to use a defensive strategy. Evacuation was the first tactical priority, followed by protection of exposures.

Magnesium burns intensely and can explode when it comes into contact with water; consequently the IC decided that no attempt would be made to place water onto the burning metal. Factors that effected decisionmaking included darkness, heavy rain, and the close proximity of a nearby warehouse containing tons of magnesium.

As firefighters concentrated on protecting a 30,000-square-foot building (2787 m^2) across the street, the magnesium began to explode. Unmanned master streams were placed to protect several trailers outside the warehouse that were loaded with magnesium. While master streams were being deployed for defensive operations, one of the trailers exploded, knocking firefighters to the ground, and breaking windows 2,000 feet away.

Hundreds of explosions continued throughout the night. A total of five alarms were sounded bringing 18 departments to provide assistance. Pre-planning, good site command, and a defensive strategy provided a good outcome to a bad situation. There were no fatalities and no serious injuries to firefighters.

SOURCE: This case study is based on an article written by Lt. Tom Lisy, Firehouse.com, January 2004. Photos and article used with permission from the Garfield Heights (Ohio) Fire Department.

Garfield Heights, Ohio Fire Department

TACTICAL CONSIDERATIONS AND LESSONS LEARNED

- Don't wait to ask for help. Unless you work with these chemicals or metals every day—and even if you do—quickly seek out information and assistance from product specialists. The costs of screwing up a reactive chemicals incident will often be measured in lives.
- Some reactive chemical incidents may require that both thermal and chemical protective clothing be used simultaneously. For example, a fire of molten sulfur will also result in high levels of sulfur dioxide in the area. When combined with skin moisture and moisture within the respiratory tract, it will also form a mild acidic solution.
- Water reactive materials can react explosively with no warning when they come into contact with even small quantities of water and moisture. When water is applied, results can include steam explosions, burning metal being thrown in all directions, and the production of flammable hydrogen gas, hydroxide compounds, and related toxic gases. Specialized extinguishing agents (i.e., dry powders) will be required for these situations. When dealing with smaller quantities, the best course of action may be to isolate the material outdoors and allow for a controlled burn.
- When dealing with large quantities of strong oxidizers and organic peroxides, responders should consider treating the incident like an explosives fire. (The only difference between some oxidizers and explosives is the speed of the reaction, which is measured in thousandths of seconds). For example, consider the consequences of the May 4, 1988 ammonium perchlorate plant fire and explosion in Henderson, Nevada. The incident killed two employees, injured 300 civilians, and caused $75 million in property damage. This dramatic explosion was caught on video camera taken miles away on a hill looking over the site. When the plant exploded, the viewer can see a powerful shock wave propagating across the ground, pushing air and dust before it, followed by another large explosion. Remember to use distance to your advantage and start emergency evacuations early in the fire.
- Several major fires have occurred involving pool chemicals, such as calcium hypochlorite and chlorinated isocyanurates. Contamination of these materials can lead to the generation of toxic vapors, heat, and oxygen, which can lead to a fire. If large quantities are involved in fire, manufacturers recommend that large, copious amounts of water be used for fire extinguishment. Of course, this may involve a trade-off between air pollution (decomposition products are toxic and will travel a large distance) versus water pollution (runoff is toxic but may not travel as far and can be controlled).
- Controlled burning may be an appropriate tactical option if extinguishing a fire will result in large uncontained volumes of contaminated runoff, further threaten the safety of both responders and the public, or lead to more extensive clean-up problems. This option is sometimes used at fires involving agricultural chemical facilities or industrial facilities, where runoff may have significant environmental impacts upon both surface water and groundwater supplies.
- Chemical process operations at fixed facilities will often have a series of safety features in place in the event of a chemical reaction or decomposition problem. These may include the injection of certain chemicals to reduce the rate of polymerization

or "kill" a chemical reaction, the use of emergency tanks in which product can be diverted, gas scrubbers to neutralize toxic or corrosive vapors before being released, and flares to burn off flammable vent gases.

PRODUCT RECOVERY AND TRANSFER OPERATIONS

In this section we will discuss the general decision-making criteria that should be made to switch from an emergency response to a recovery operation. We will also discuss basic product recovery and transfer operations and the related safety considerations. Specific product and container recovery and offloading procedures are beyond the scope of this chapter, given the variety of of containers, products, and scenarios that might be encountered. Responders should always consult with product and container specialists to develop a safe and efficient plan for recovering product and the container or transport vehicle.

ROLE OF EMERGENCY RESPONDERS IN RECOVERY OPERATIONS

As we will discuss in more detail in Chapter 12, determining when the emergency response phase of an incident ends and when the recovery and restoration phase begins can sometimes fall into a gray area. As product and container specialists arrive on the scene to assist, the IC needs to make sure that they understand how and where they fit into the incident command system, and their roles and responsibilities for operating safely at the incident scene. The IC is responsible for making sure that the transition from emergency response phase to the post-emergency response phase is formal, safe, and as seamless as possible.

Public safety emergency response teams have a fairly limited scope of responsibility for clean-up recovery and site restoration. But if the hazmat incident is on public property (e.g., highway), police and fire agencies have a responsibility to assure that public safety is safeguarded during these operations.

The extent of emergency responder involvement in making the transition to the restoration, recovery and ultimately post-emergency response operations will vary depending on the type of hazards present, location and extent of the incident, jurisdiction, and so on. Regardless of the situation encountered, the following general activities should be considered during the restoration and recovery phase:

- **PPE requirements**—Personal protective clothing and equipment requirements for restoration should be determined by the IC in consultation with the Incident Safety Officer and other health, safety, or environmental specialists. Requirements should be based upon the results of air monitoring, the potential for re-ignition, the specific tasks to be accomplished, and other related factors. PPE requirements should take into account the overall safety of restoration personnel, including mobility, comfort, and heat stress.
- **Product/container specialists and contractor safety**—Before transferring site control to non-emergency response personnel, the IC should verify that any environmental spill contractors used for clean-up and recovery operations are trained per the requirements of *OSHA 1910.120—Hazardous Waste Operations and Emergency Response.* Many states now require that hazardous waste contractors be licensed and that they carry credentials listing their level of training and certification. There are many excellent clean-up contractors in the field

today; there are also a few bad ones. Don't contribute to an accident by allowing unqualified workers to handle your clean-up. Asking to see credentials at an emergency scene is standard protocol for safety and security reasons. The IC should not hesitate to ask for credentials of clean-up personnel.

Figure 10.29 Product recovery operations require close coordination with the Incident Commander.

- **Sampling**—Residual contamination of the soil, pavement, or surrounding area may require further assessment to determine if they present a threat to public safety. Depending on the nature of the incident, sampling may be used to determine whether the surfaces require decontamination or removal. It may also be necessary to sample soil, surface water, sediments, or groundwater to assess environmental impact.

  Sampling procedures should follow regulatory guidelines, including careful sample technique documentation, chain of custody, and quality assurance/quality control (QA/QC) procedures. If the incident is a crime scene, proper evidence procedures must be followed. If contamination is found that will require specialized clean-up, a written remedial action plan should be developed and coordinated through environmental officials.
- **Disposal concerns**—It may be necessary to dispose of contaminated protective clothing, decontamination solutions, runoff water, or other materials that may be considered as hazardous waste following an emergency. An environmental specialist should be consulted for waste characterization and disposal, as appropriate.

SITE SAFETY AND CONTROL ISSUES

Product removal operations cannot commence until after the incident site is stabilized. Stabilization means that all fires have been extinguished, ignition sources have been controlled, and all spills and leaks have been controlled, as necessary. Specific site safety considerations that should be addressed during this phase of the incident include the following:

- When flammable or combustible liquids are involved, ensure that backup crews with a minimum of two 1-3/4-inch foam handlines and at least one 20- to 30-pound dry chemical fire extinguisher are in place to protect all personnel involved in the offloading and uprighting operation.
- Always have an escape plan with an alternate escape route. Emergency responders have been seriously injured and disabled while attempting to escape during recovery operations that went bad. The emergency escape signal must be clearly understood by everyone. An exit pathway out of the work area should be kept clear at all times for personnel working in the immediate hazard area.

- Continuously monitor the hazard area for flammability, toxicity, and oxygen deficiency, as required by the hazards of the materials involved.
- Ensure that all personnel remain alert. Both public safety and industry response personnel sometimes become sloppy, less attentive, and may attempt shortcuts as the emergency extends over several hours. Frequent rotation and rehab of personnel can usually minimize this problem.

PRODUCT REMOVAL AND TRANSFER CONSIDERATIONS

Specific procedures for product removal and transfer will vary based on the hazmat involved, container design and construction, container stress and actual/potential breach, and the position and location of the container. In addition, shipping documents from the product shipper or manufacturer can also provide guidance and recommendations.

The following are general guidelines and should be used as applicable.

SURVEYING THE CONTAINER

The container should initially be surveyed to determine the safest method of offloading. This is particularly true when dealing with bulk transportation containers, such as cargo tank trucks, railroad tank cars, and intermodal tank containers. It should be noted that offloading may reduces stress on a container but is no guarantee that the container won't fail mechanically (e.g., during lifting). Remember the Waverly, Tennessee disaster described earlier. Factors to evaluate include the following:

- The pitch and position of the container. Containers that are upside down can complicate offloading as access to valves may be blocked. If a cargo tank truck or rail car are involved, the pitch and position of the container—front to back and left to right—are particularly important. It's possible that the container may move as product is pumped off and the product load shifts. Even where the unit appears stable, consideration must be given to bracing. Bracing materials may include timber, jacks, or air bags.
- The position and location of the openings or attachments that will be used for product offloading.
- If the container is a cargo tank truck (e.g., MC-306/DOT-406), the position of the baffle holes.
- The product being offloaded (e.g., flammable vs. corrosive vs. poisonous).
- The level of training, resources, and equipment available for product transfer and container uprighting operations.

BONDING AND GROUNDING CONSIDERATIONS

The foundation and justification for bonding and grounding when flammable liquids and gases are involved is well established in *NFPA-472—Standards for Professional Competence for Responders to Hazardous Materials Incidents.* For more guidance see the NFPA Hazardous Materials Response Handbook (3rd edition). Included in the Handbook is a special supplement entitled "Bonding and Grounding for Emergency Responders."

SCAN 10-F

BONDING AND GROUNDING

What is bonding? Bonding is the process of connecting two or more conductive objects together by means of a conductor; for example, using an approved bonding wire to connect an aircraft being refueled to the fuel truck. Bonding it is done to minimize potential differences between conductive objects, thereby minimizing or eliminating the chance of static sparking.

What is grounding? Grounding is the process of connecting one or more conductive objects to the ground through an earthing electrode (i.e., grounding rod). For example, connecting an aircraft to the ground through an approved grounding wire and connecting the fuel truck to the ground, through a separate grounding wire and grounding rod. Grounding is done to minimize potential differences between objects and the ground. An ohm meter is used to measure the electrical resistance and ensure the electrical continuity of bonding and grounding operations.

What is static electricity? Static electricity is an accumulated electrical charge. In order for static electricity to act as an ignition source, four conditions must be fulfilled:

1. *There must be an effective means of static generation.* This can occur when a flammable or combustible liquid is moved from one place to another through pipes, filtering, or by pouring. Some products like gasoline are good static accumulators and can pick up a static charge as they pass through piping during loading operations. Products that easily accumulate static charges must be loaded at slower flow rates to permit downstream relaxation time for the product to lose its charge.
2. *There must be a means of accumulating the static charge buildup.* Not every product lends itself to accumulating a static charge.
3. *There must be a spark discharge of adequate energy to serve as an ignition source (i.e., incendive spark).* We have all experienced a static discharge at one time or another when we walked across a carpet during the winter and touched a metal object, or exited a car and then touched the car door. Not every static spark carries enough energy to cause ignition, and even if it does, there must be an adequate flammable mixture present (see item 4 below).
4. *The spark must occur in a flammable mixture.* In order for a fire or explosion to occur in the presence of an adequate ignition source (static spark), the fuel-to-air mixture must fall within the flammable range.

By bonding and grounding, you are giving a static charge a pathway in which to travel to earth without creating a spark. The resistance of the grounding field will be affected by weather, type of soil, moisture content of the soil, and the time of year. Agencies should establish an acceptable resistance level for grounding purposes. For example, while the National Electric Code notes that the ground level should be < 25 ohms resistance for residential purposes, a standard that has been adopted by many emergency response agencies. In contrast, a standard of < 10 ohms is used by some petroleum industry organizations.

Bonding and grounding must be established before product removal and transfer operations can begin. Consider the following operational guidelines:

GROUNDING AND BONDING SEQUENCE

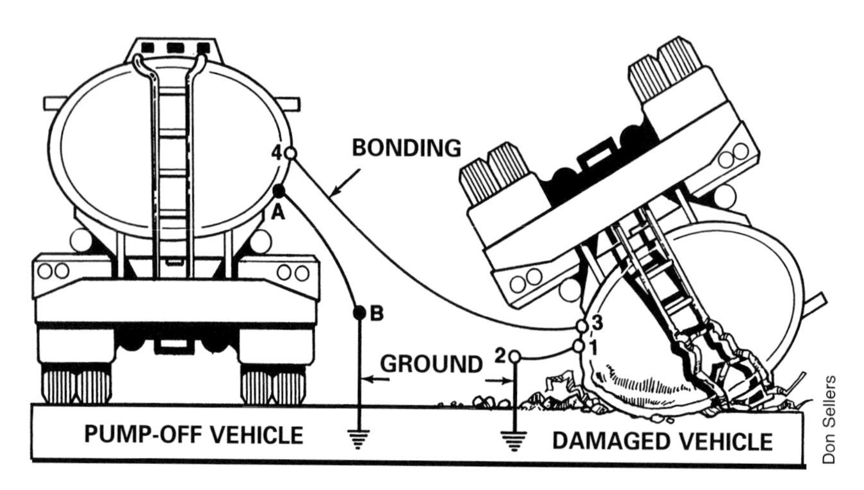

NOTE: This illustration does not depict the proximity or exact spatial layout of the bonding and grounding system.

- Make connections in sequence shown to avoid sparks in potentially flammable areas.
- Connections A and B can be made anytime prior to pump-off.

Figure 10.30

- The pump-off vehicle and all pump-off appliances (e.g., hose couplings, downspouts, and recovery pans and tubes) should be bonded by connecting a bonding cable from the overturned container to the appliance. In all bonding operations involving appliances, the first connection must always start at the damaged unit.
- Bonding cables must be placed on a clean, grease-free, paint-free surface. Ensure that the cables are connected to a part of the vehicle that is connected directly to the vessel holding the product to be transferred. Screw clamps or pressure point clamps (e.g., C-clamps) are preferable to cables with "alligator clips" because they make better connections. Periodically monitor all bonding and grounding cable connections to ensure that they remain in place and connected.
- Rubber hoses with a built-in wire will not necessarily provide bonding protection, as the wire within the hose may become broken or the wire may not be properly tied into the coupling.
- Plastic buckets can pick up static charges and should not be allowed for use as retention basins in an emergency situation. However, there are some conductive plastic liners and containers are commercially available.
- Grounding cables should initially be connected to the damaged container, then moved outward and away from the overturned vehicles. The final connection can be made to a ground rod or a guardrail post, if it's deep enough to carry away the charge. Auger-type T-handle grounding rods can be very effective. See above figure for an illustration of the bonding and grounding sequence.
- If the ground resistance is too high, responders may be able to improve the ground field through several tactics, including
  - Use of water while inserting the grounding rod into the earth. This will lower the resistance to the earth.

- **Grounding**—*API Publication 2219, Safe Operation of Vacuum Trucks in Petroleum Service*, indicates that static electricity does not present an ignition problem with either conductive or nonconductive vacuum truck hoses. However, with nonconductive hoses, any exposed metal, such as a hose flange, can accumulate static electricity and act as an ignition source if the metal touches or comes close to the ground. Since it is often difficult to distinguish between nonconductive and conductive hoses in the field, API recommends that all exposed metal be grounded when any hose is used in other than a closed system with tight connections at both ends of the hose.
- **Venting**—When flammable or toxic liquids are loaded into a vacuum truck, the vacuum pump exhaust should be vented downwind of the truck by attaching a length of hose sufficient to reach an area that is free from hazards and personnel.
- **Personnel safety**—All unnecessary personnel should leave the area during loading. The vacuum truck driver should leave the truck cab and be in proper PPE. Strict control of ignition sources should be maintained within 100 feet of the truck, the discharge of the vacuum pump, or any other vapor source.

UPRIGHTING METHODS

There is no best strategy to handle the offloading or uprighting of hazardous materials containers. Safety should be the number one factor used in making a decision for the best course of action. The decision to offload or upright a rail car, cargo tank truck, or intermodal tank container involving any type of hazmat requires careful consideration and input from a variety of specialists. These should include both product and cargo tank specialists and may also include rigging and heavy equipment specialists. All of these groups should be consulted before a plan of action is implemented.

The safety experience in successfully uprighting most cargo tank trucks and rail cars has been very good when (1) the job is performed within the design limitations of the container and the equipment being used for the operations, and (2) safe operating practices are followed. For example, the MC-331 container is an ASME pressure vessel that can sustain heavy mechanical stress and damage without losing its integrity. See Figure 10.32.

SAFETY CONSIDERATIONS FOR APPROVING LIFTING

Before any operations are initiated, the IC must verify that the personnel performing the lifting and uprighting operation are trained and qualified to perform the expected tasks. History has taught us some ugly lessons that there is a big difference between uprighting an overturned eighteen-wheeler and an MC-312/DOT-412 corrosive tank truck involved in a rollover. While there are many first-rate riggers in the business, there are also plenty of well-meaning tow truck operators and salvage companies who simply do not know what they are doing. The towing industry is doing a very good job addressing the need for standards and has implemented many training and certification programs to qualify their personnel. The bottom line—make sure the people and equipment on your incident scene are qualified for the job.

Factors that responders must consider before approving a lifting or uprighting operation include the following:

1. **Conduct a damage assessment of the container before moving or lifting the tank.**

   Containers such as MC-306/DOT-407 aluminum shell cargo tank trucks are not designed to be lifted and uprighted fully loaded. Container specialists, riggers, and crane operators must have access to damage assessment information. Keep in mind that damage assessment is, at best, a qualitative assessment process that requires input from product and container specialists who have experience in evaluating structural damage. For example, is the tank stable versus nonstable, have go/no go safety factors been evaluated, what is the practical field experience with the type of container involved, and so on.

2. **Qualified and experienced rigging specialists and crane operators must be available to supervise and perform the uprighting or lifting operation.**

   Operating hoisting equipment can be a very complex and dangerous process, especially under emergency conditions. A qualified rigger and crane operator should know a wide range of specialized information, including knowledge of:

   - **What can go wrong and how to deal with situations when they go bad.** Riggers and operators must understand the changing conditions of the job and how these conditions will affect the overall safety of the operation. Examples include changes in wind direction, changes in temperature, formation of ice on the ground or lifting surface, and so on.
   - **Is the load is properly rigged?** A qualified rigger and operator must be able to read a load chart that shows the lifting capacities for various crane operations.
   - **Cables, chains, ropes, and slings.** Riggers and operators must have a thorough knowledge of the rigging used to lift and stabilize the loads and have the basic training to inspect rigging for problems. Rigging is critical since it forms the interface between the hoisting equipment and the load. The main hazards of rigging are failure due to excessive overloading, deterioration or wear, and improper rigging techniques. Load capacity charts are available for materials handling equipment and rigging.
   - **Safe lifting operations.** An operator must have the ability to evaluate the site for safe lifting operations. For example, for a safe lift with a mobile crane, at least seven items of information are needed: (1) Is the vehicle level? (2) Are outriggers properly extended or retracted? (3) Are extended outriggers supported by stable ground? (4)What is the angle of the boom? (5) What is the boom and jib length? (6) What positions will the boom be in during the lift? and (7) How much does the load weigh?

Lancaster, PA Hazmat 29

Figure 10.31 Propane truck rollover.

OPERATIONAL CONSIDERATIONS

The following are general guidelines that emergency responders should consider while overseeing and approving a cargo tank truck uprighting operation.

1. **Offload or Upright?** The decision to either (1) offload the contents and then upright the container, or (2) upright the container while still loaded will be dependent upon a number of variables, including the type of cargo tank truck involved, the nature of container stress and damage, the location of the incident, and resources available to lift the damaged container. Speed, the impact terrain, level of the surface, and weight of the product load will all contribute to the stressing of the container.

   There is only one universal rule—never upright a loaded aluminum-shell MC-306/DOT-406 cargo tank truck. Everything else will be incident specific.

Bill Hand Figure 10.32

2. **Use enough lift.** Experience shows that in most rollovers the cargo tank truck will be lying on its side. Depending upon the type of cargo tank, it may be impossible to remove all of the product load. For example, an MC-331 lying on it side can only have approximately 40 to 50% of the contents removed because the liquid and vapor eduction tubes are at the 3 o'clock/9 o'clock position. While offloading 40% of the load will certainly reduce the weight of the container, it does not change the internal pressure of the contents.

   Make sure you have enough lifting capacity for the situation. An MC-331 transport weighs up to 80,000 pounds when fully loaded. Mobile cranes and heavy wrecker cranes should be used for the uprighting operation, as they typically have boom and lift ratings of 30 tons and higher. Although a 3500-gallon propane bobtail may be uprighted with one wrecker, transports and tractor trailer combinations will require two and sometimes even three wreckers for the uprighting operation.

Ray Haring Figure 10.33

3. **Stabilize the vehicle.** If the container is in an unstable position where it could shift and injure workers or slide further off a roadway, shoulder, bridge, and so on, take time to stabilize the entire rig using cables, cribbing, and the like.

   Tractor and trailer combinations should be lifted as a single unit; separating the tractor from the trailer could contribute to instability. If the tractor becomes separated from the trailer during the rollover, it may be necessary to use a fifth-wheel dolly to support the

Bill Hand Figure 10.34

trailer during the uprighting operation. If the trailer has both a front and rear axle, the axles should be cross-chained together to keep them parallel.

4. **Position all lifting equipment.** If the container is on its side, wrecker personnel can have difficulty in getting lifting straps or cables under the tank. To facilitate this process, "mat jacks" or airbags can be used to raise the tank a few inches so that the lifting straps can be placed under the tank. Mat jacks are usually rated for over 20 tons of lift and can be inflated from the wrecker's air tanks. NOTE: Do not use hydraulic rescue tools to lift the container, as they may cause further damage to the container shell.

Phil Baker

Figure 10.35

Most wreckers use nylon straps that are 12 to 24 inches wide when lifting cargo tank trucks. These allow the weight to be distributed over a larger surface area as the container is uprighted. While cables may be used, they should be attached to the chassis and should not be placed around the tank shell, as they can cause further damage.

When dealing with liquid cargo tank trucks, straps should be placed at the strongest points of the container, such as at the fifth wheel, at tank baffles, and at compartment bulkheads. If straps are positioned between compartment bulkheads, the tank may be "crunched" during the uprighting process.

5. **Upright the damaged container.** Most cargo tank truck uprighting operations require one or two wreckers providing the lift, while another wrecker "holds" the cargo tank to prevent it from tipping over. Responders must be aware of the potential for cables or straps to break and whip around, causing serious injury. One individual should be responsible for controlling the overall uprighting operation and to minimize the number of personnel working in the immediate area.

Keith Hanchett

Figure 10.36

Once the overturned container is uprighted, it should be chocked so that the vehicle does not begin to roll away. Even if the tires are flat, a trailer unit can usually be towed a short distance where it can be safely offloaded. However, if it is badly damaged, it may be eventually placed upon a "low boy" or a flat-bed trailer and transported from the scene.

THINKING BIG AND PREPARING FOR YOUR WORST DAY

LESSONS FROM HISTORY

December 7, 1941 and September 11, 2001: two incidents 60 years apart but with similar hard-learned lessons. As members of public safety agencies, one of the most striking lessons we should learn from both of these historic landmark events is that when America is having her worst day, our citizens still count on their emergency services to be "the best".

Being the best isn't easy and it is going to get even harder to meet that mark in the future if we don't learn from our mistakes and plan and train to meet new threats and challenges. Our definition of what we visualize as our "worst day" must constantly be adjusted as the nature and range of threats continues to change. In short, we need to think big, plan big, and exercise big if we are going to be prepared to meet the expectations of our nations citizens. And yet we must also understand that though we respond to our day-to-day problems, we must also provide a sound foundation upon which our "worst day" response must be built.

The American philosopher and poet, George Santayana (1863–1952) wrote, "Those who cannot remember the past are condemned to repeat it." If you want to know what we must do to prepare for the future, just take a look at the past. History can teach us a great deal about where our level of thinking needs to be in a post 9-11 environment. Let's benchmark our thinking against three events—two modern examples and one from ancient history:

The 1947 Ammonium Nitrate Disaster at Texas City, Texas

The Texas City ammonium nitrate explosion was one of the worst disasters in American history. The explosions of two World War II era Liberty ships loaded with ammonium nitrate on April 16 and 17 killed at least 468 people (25% of the 1947 population of Texas City), injured 3500 people, and destroyed more than $700 million in infrastructure (in today's monetary value). The blast basically destroyed everything within a 2000-foot radius. The explosions totally destroyed the adjacent Monsanto chemical plant ($20 million loss), burned 1.5 million barrels of petroleum products valued at $500 million, and damaged 50 storage tanks and pipelines within area refineries. One third of the towns 1,519 houses were condemned, leaving 2000 people homeless, many of them parentless children.

FOOD FOR THOUGHT: What are the parallels between this disaster, which occurred nearly 60 years ago, and the threat that terrorist may present to our critical infrastructure in the future? More important, what do we need to do to get ready?

SOURCE: THE TEXAS CITY DISASTER, 1947 by Hugh W. Stephens, University of Texas Press, Austin, TX (1997), pp. 1–6. NOTE: Hugh Stevens wrote the definitive history of the famous Texas City Disaster. This is a well-researched, and easy-to-read book with many useful lessons for preparing for large scale hazmat incidents. (ISBN 0-292-77722-1). 1. Santayana's quote comes from the poem, "O World, Thou Choosest Not (1894).

The 1918 Influenza Pandemic

The so-called Spanish Flu epidemic of 1918 rapidly spread worldwide, killing between 20 to 40 million people (nobody knows for sure how many people actually died because of the remote corners of the world that were affected). An estimated 675,000 Americans died of influenza, ten times as many Americans than were killed in World War I! It has been cited as the most devastating epidemic in recorded history. More people died of influenza in a single year than died in the four years of the dreaded Black Death Bubonic Plague from 1347 to 1351.

The virus was incredibly deadly, often killing a victim within hours of contact. The flu was the most deadly for people ages 20 to 40. For example, an estimated 43,000 Americans who were mobilized for World War I died from influenza.

A 1918 quote from third year medical student Isaac Starr from the University of Pennsylvania, is especially chilling:

> As their lungs filled, the patients became short of breath and increasingly cyanotic. After gasping for several hours, they became delirious and incontinent, and many died struggling to clear their airways of a blood-tinged froth that sometimes gushed from their nose and mouth. It was a dreadful business.

When the flu broke out in the United States, many doctors were concerned about starting a panic, and they refused to warn neighboring communities, which helped the epidemic become a pandemic. This is an interesting parallel to how the Chinese government handled the SARS outbreak in 2003.

The flu killed 3,100 people in Washington, D.C., 10,000 in Maryland, and 14,000 in Virginia. In October 1918, about 90 people died from the flu in Washington, D.C. every day. Schools and churches were closed, the military was quarantined in their installations, and it was unlawful for anyone who was sick to appear on a public street.

> FOOD FOR THOUGHT: Epidemiologists warn us that something similar could strike again. Think about the impact an event like this would have on our emergency services and health care system. Are we ready to deal with a terrorist attack using a biological agent like smallpox? Are we really thinking big?

SOURCES: "Autumn 1918: Washington's Season of Death: Dwarfing Modern-Day Epidemics, Spanish Flu Killed Thousands In Mere Weeks," by John F. Kelley, The Washington Post, Southern Maryland News, (February 1, 2004), p-15. Also see "The Flu Pandemic of 1918: Is A Repeat Performance Likely?" by Doug Rekenthaler, Managing Editor, DisasterRelief.org (February 22, 1999), and The Influenza Pandemic of 1918 at Stanford University's Web site www.stanford.edu/group/virus/uda .

PLANNING AND EXERCISES

Planning is to preparedness as exercises are to readiness. Most communities think they are prepared for their worst day because they have planned well for the events, but a critical component to being ready to perform on your worst day is practicing what is in the plan through exercises.

THE ROMANS THOUGHT BIG!

In historian Edward Gibbon's classic book The Decline and Fall of The Roman Empire, the author discusses the training and exercise programs of the famous Roman Legions, who were both respected and feared by the enemies of Rome.

The Romans held the value of military exercises in the highest importance. According to Gibbon, the ancient Romans directly linked the concept of valor to the words skill and practice. In fact, the Latin name for exercise (exercitus) was borrowed in the Roman language (Exercitus ab exercitando), which was synonymous with the word Army. (Army = Exercise, which means Preparedness.)

Roman generals understood that the key to readiness and discipline on the battleground was a strong training and exercise program. For example, new recruits, as well as regular soldiers, were constantly training in basic skills as well as exercising their legion in the Roman order of battle. Training was part of the soldier's daily routine, with drills both in the morning and in the evening. It is interesting to note that even veteran soldiers of "age and knowledge" were never excused from these daily training sessions. Roman officers understood that skills degrade, and the presence of these veterans at drills and exercises sent a strong message about the importance of being ready.

In the winter months, large sheds (modern-day training rooms) were erected wherever the troops were garrisoned, so that training could continue in bad weather. The Romans regularly trained using heavier weapons and equipment (double the weight of what was required in actual combat) to add physical strength, maintain endurance, and build self-confidence.

The Roman military and its citizens held training and exercises of such high importance that it was the policy of the most respected and qualified generals to personally participate in advanced tactical exercises. According to the ancient historian Josephus de Bello Judaico, "The only circumstance that distinguished the difference between a real Roman military field battle from a field exercise was the spilling of blood." These guys didn't just talk the talk, they walked the walk.

The history lesson here is that the safety and honor of the Roman Empire was entrusted to the legions, Roman soldiers and their commanders understood that they had to be "the best" every single day they were on the job. They planned big; they trained big, and for over 1000 years in history THEY WERE BIG! The Roman army was a serious opponent to any adversary because they were always prepared for their worst day. This is a lesson we can learn for the future.

SOURCE: THE DECLINE AND FALL OF THE ROMAN EMPIRE, by Edward Gibbon. Cumberland House, Herfordshire, England (1998), pp. 11-13. NOTE: Author Edward Gibbon published this great historical work between 1776 and 1778, and it is the undisputed masterpiece of English historical writing. The book measures 71 chapters, about 1000 pages.

SUMMARY

Implementing response objectives is the phase in a hazmat emergency when the Incident Commander implements the best available strategic goals and tactical objectives which will produce the most favorable outcome. The operational strategy for an incident is developed based upon the IC's evaluation of the current conditions and forecast of future conditions. The effectiveness of this phase of the incident is directly related to how well the hazards were identified and the risks evaluated.

A strategy is a plan for managing resources. It becomes the IC's overall goal or game plan to control the incident. Primary hazmat strategic goals include rescue, public protective actions, spill control (confinement), leak control (containment), fire control, and transfer and recovery. Tactics are the specific objectives the IC uses to achieve strategic goals. Tactics are normally decided at the section or group/division levels in the command structure. Both strategy and tactics can be implemented by the IC in the offensive, defensive, or nonintervention mode. Usually, the IC uses a combination of tactics to manage the problem.

Saving lives is the IC's number one mission! Life safety should always be the IC's highest priority, but remember that in some cases doing nothing and letting the incident run its course is the smartest and safest strategy. As emergency responders we cannot save everyone, and time often works against the responders and the people you are trying to rescue. The IC must weigh the chance for a successful rescue against the hazards and risks.

Product removal and recovery operations usually begin after the emergency has run its course (e.g., all leaks have been controlled). Product removal and recovery operations should not begin until after the incident site is stabilized and the area has been re-evaluated for hazards and risks. Stabilization means that all fires have been extinguished, ignition sources have been secured, and all product releases have been controlled.

Product removal and transfer operations involve moving the contents from the damaged or overloaded cargo tank(s) into an undamaged and compatible receiving tank(s) such as a tank car, cargo tank truck, intermodal tank, or fixed tank.

Product transfer and removal operations are typically performed by either product/container specialists or environmental contractors working on behalf of the carrier or shipper. However, public safety responders will often continue to be responsible for site safety and will oversee the implementation of all product transfer and removal.

REFERENCES AND SUGGESTED READINGS

Action Video, PLUGGING AND PATCHING DRUMS (videotape), Portland, OR: Action Video (1998).

Air Force Civil Engineer Support Agency (AFCESA) and PowerTrain, Inc, HAZARDOUS MATERIALS INCIDENT COMMANDER EMERGENCY RESPONSE TRAINING CD-ROM, Tyndall Air Force Base, FL: AFCESA (2002).

Air Force Civil Engineer Support Agency (AFCESA) and PowerTrain, Inc, HAZARDOUS MATERIALS TECHNICIAN EMERGENCY RESPONSE TRAINING CD-ROM, Tyndall Air Force Base, FL: AFCESA (1999).

American Petroleum Institute, FIRE PROTECTION CONSIDERATIONS FOR THE DESIGN AND OPERATION OF LIQUEFIED PETROLEUM GAS (LPG) STORAGE FACILITIES, (2nd edition), API Publication 2510-A, Washington, DC: American Petroleum Institute (1996).

American Petroleum Institute, GUIDELINES AND PROCEDURES FOR ENTERING AND CLEANING PETROLEUM STORAGE TANKS, (1st edition), ANSI/API RP 2016, Washington, DC (2001).

American Petroleum Institute, GUIDELINES FOR WORK IN INERT CONFINED SPACES IN THE PETROLEUM INDUSTRY, (2nd edition), API Publication 2217A, Washington, DC (November 1997).

American Petroleum Institute, MANAGEMENT OF ATMOSPHERIC STORAGE TANK FIRES, (4th edition), API Publication 2021, Washington, DC: American Petroleum Institute (2001).

American Petroleum Institute, PREVENTION AND SUPPRESSION OF FIRES IN LARGE ABOVEGROUND ATMOSPHERIC STORAGE TANKS, (1st edition), API Publication 2021A, Washington, DC: American Petroleum Institute (1998).

American Petroleum Institute, GUIDELINES FOR THE SAFE DESCENT ONTO FLOATING ROOF TANKS IN PETROLEUM SERVICE, (2nd edition), API Publication 2026, Washington, DC (1998).

American Petroleum Institute, PROTECTION AGAINST IGNITIONS ARISING OUT OF STATIC, LIGHTNING, AND STRAY CURRENTS, (6th edition), API Publication 2003, Washington, DC: American Petroleum Institute (1998).

American Petroleum Institute, SAFE OPERATION OF VACUUM TRUCKS IN PETROLEUM SERVICE, (2nd edition), API Publication 2219, Washington, DC: American Petroleum Institute (1999).

American Society for Testing and Materials, Fire and Explosion Hazards of Peroxy Compounds, Special Publication No. 394, September 1965.

Andrews, Jr., Robert C., "The Environmental Impact of Firefighting Foam." INDUSTRIAL FIRE SAFETY (November/December, 1992), pages 26–31.

Arnold, David, "Water-Sodium Mix Set off Newton Blast," The Boston Globe, Thursday, October 28, 1993, page 22.

Association of American Railroads—Transportation Test Center (AAR/TTC), "Grounding and Bonding." Student Handout from the Hazardous Materials Response Curriculum, Pueblo, CO: Association of American Railroads (1999).

Benner, Ludwig, Jr., "D.E.C.I.D.E. In Hazardous Materials Emergencies." FIRE JOURNAL (July, 1975), pages 13–18.

Benner, Ludwig, Jr., HAZARDOUS MATERIALS EMERGENCIES (2nd Edition), Oakton, VA: Lufred Industries, Inc. (1978).

Bevelacqua, Armando, HAZARDOUS MATERIALS CHEMISTRY, Albany, NY: Delmar – Thomson Learning (2001).

Bevelacqua, Armando and Richard Stilp, HAZARDOUS MATERIALS FIELD GUIDE, Albany, NY: Delmar—Thomson Learning (1998).

Bevelacqua, Armando and Richard Stilp, TERRORISM HANDBOOK FOR OPERATIONAL RESPONDERS, Albany, NY: Delmar—Thomson Learning (1998).

Bradish, Jay, "The Fatal Explosion," A Special Report of the NFPA Investigations and Applied Research Division on the December 27, 1983 Propane Explosion in Buffalo, New York. Fire Command (March 1984) pages 28–33.

Callan, Michael, STREET SMART HAZMAT RESPONSE, Chester, MD: Red Hat Publishing (2002).

Compressed Gas Association, Inc., HANDBOOK OF COMPRESSED GASES (4th Edition). Boston, MA: Kluwer Academic Publishers (2000).

Crowder, Don, "Emergency Transfers Utilizing Compressors and Pumps." Student Handout from 2003 Propane Emergencies Industry Responders Conference, Washington, DC: Propane Education and Research Council (2003).

Dunn, Vincent, "BLEVE: The Propane Cylinder." FIRE ENGINEERING (August, 1988), pages 63–70.

Emergency Film Group, OIL SPILL RESPONSE (four-videotape series), Edgartown MA: Emergency Film Group (2002).

Emergency Film Group, FOAM (video-tape), Edgartown, MA: Emergency Film Group (2004).

Emergency Film Group, PETROLEUM STORAGE TANKS (video-tape), Edgartown, MA: Emergency Film Group (2003).

Emergency Film Group, THE EIGHT STEP PROCESS: STEP 1—SITE MANAGEMENT AND CONTROL (video-tape), Edgartown, MA: Emergency Film Group (2004).

Emergency Film Group, THE EIGHT STEP PROCESS: STEP 6—IMPLEMENTING RESPONSE OBJECTIVES (video-tape), Edgartown, MA: Emergency Film Group (2004).

Emergency Film Group, INTERMODAL CONTAINERS, Edgartown, MA: Emergency Film Group (1995).

Emergency Film Group, CYLINDERS—CONTAINER EMERGENCIES (videotape), Edgartown, MA: Emergency Film Group (2002).

Emergency Film Group, CONFINED SPACE EMERGENCY, Edgartown, MA: Emergency Film Group (2003).

Emergency Film Group, TERRORISM RESPONSE: Nine-Part Video Series On Weapons of Mass Destruction, Edgartown, MA: Emergency Film Group.

- Detecting Weapons of Mass Destruction (2003)
- Response to Anthrax Threats (2003)
- Terrorism: 1st Response (2002)

- Terrorism: Biological Weapons (1999)
- Terrorism: Chemical Weapons (2000)
- Terrorism: Explosive & Incendiary Weapons (2003)
- Terrorism: Medical Response (2002)
- Terrorism: Radiological Weapons (2004)
- Terrorism: Roll Call Edition (2002)

Epperson, Jimmy C., Jr., "Preparing for Cylinder Emergencies." INDUSTRIAL FIRE SAFETY (November/December, 1992), pages 32–40.

Federal Emergency Management Agency—U.S. Fire Administration, Report on Tire Fires, Emitsburg, MD: FEMA (1999).

Federal Emergency Management Agency—U.S. Fire Administration, Technical Report on High Temperature Accelerant Arson Fires, Emitsburg, MD: FEMA (1991).

Fire, Frank L., THE COMMON SENSE APPROACH TO HAZARDOUS MATERIALS (2nd Edition), Tulsa, OK: Fire Engineering Books and Videos (1986).

Goodland, Larie, "Sorbents Selection" HAZARDOUS MATERIALS MANAGEMENT (February/March, 2002), page 18.

Hanchett, Keith, "Propane Transportation Emergencies." Student Handout from 2003 Propane Emergencies Industry Responders Conference, Washington, DC: Propane Education and Research Council (2003).

Hawley, Chris. HAZARDOUS MATERIALS INCIDENTS. Albany, NY: Delmar —Thomson Learning (2002).

Hawley, Chris, Gregory G. Noll and Michael S. Hildebrand, SPECIAL OPERATIONS FOR TERRORISM AND HAZMAT CRIMES. Chester, MD: Red Hat Publishing, Inc. (2002).

Hawthorne, Edward, PETROLEUM LIQUIDS—FIRE AND EMERGENCY CONTROL. Englewood Cliffs, NJ: Prentice Hall (1987).

Hildebrand, Michael, S., "The American Petroleum Institute Response to the Winchester, Virginia Oil Spill and Tire Dump Fire." A technical paper presented to the API Operating Practices Committee, OPC Paper 9.2-20 (D-120) (1984).

Hildebrand, Michael S. and Gregory G. Noll, PROPANE EMERGENCIES (2nd Edition), Washington, DC: National Propane Gas Association (2002).

Hildebrand, Michael S. and Gregory G. Noll, STORAGE TANK EMERGENCIES, Chester, MD: Red Hat Publishing, Inc. (1997).

Hildebrand, Michael, S., HAZMAT RESPONSE TEAM LEAK AND SPILL CONTROL GUIDE, Stillwater, OK: Fire Protection Publications, Oklahoma State University (1992).

International Association of Fire Chiefs et al., Guidelines for the Prevention and Management of Scrap Tire Fires. Washington, DC: International Association of Fire Chiefs (1993).

International Association of Fire Fighters, TRAINING FOR HAZARDOUS MATERIALS RESPONSE: TECHNICIAN, Washington, DC: IAFF (2002).

International Fire Service Training Association, PRINCIPLES OF FOAM FIRE FIGHTING (2nd Edition), Stillwater, OK: International Fire Service training Association (2003).

Isman, Warren, E. and Gene, P. Carlson, HAZARDOUS MATERIALS, Encino, CA: Glencoe Publishing Company, Inc. (1980).

Kenley, Scott, W. and Meidl, James H., FLAMMABLE HAZARDOUS MATERIALS. Encino, CA: Brady-Prentice Hall Third Edition, (1995).

Lesak, David M. HAZARDOUS MATERIALS STRATEGIES AND TACTICS, Upper Saddle River, NJ: Prentice-Hall Inc. (1999).

Marsh, THE 100 LARGETS LOSSES 1972–2001: Large Propert Damage Losses In The Hydrocarbon-Chemical Industries, (20th edition), API Publication 2510-A, New York, NY (2003).

Meal, Larie, "Static Electricity." FIRE ENGINEERING (May, 1989), pages 61–64.

Meidl, James H., EXPLOSIVE AND TOXIC HAZARDOUS MATERIALS. Encino, CA: Glencoe Publishing Company, Inc. (1978).

Meyer, Eugene, CHEMISTRY OF HAZARDOUS MATERIALS. Upper Saddle River, NJ: Brady Third Edition (1999).

Meyer, Eugene, "Volatile Chemical Drama Was Ended By Secret Convoy." The Washington Post, Thursday, October 25, 1979, page C-6.

Monsanto Chemical Company, Olin Chemicals and PPG Industries, GUIDELINES FOR THE SAFE HANDLING AND STORAGE OF CALCIUM HYPOCHLORITE AND CHLORINATED ISOCYANURATE POOL CHEMICALS, St. Louis, MO: Monsanto Chemical Company et al. (1989).

National Fire Protection Association, FIRE PROTECTION HANDBOOK (19th edition), Quincy, MA: National Fire Protection Association (2003).

National Fire Protection Association, NFPA 30—NATIONAL FLAMMABLE AND COMBUSTIBLE LIQUIDS CODE, Quincy, MA: National Fire Protection Association (2000).

National Fire Protection Association, HAZARDOUS MATERIALS RESPONSE HANDBOOK (4th Edition), Quincy, MA: National Fire Protection Association (2002).

National Fire Protection Association, LIQUEFIED PETROLEUM GASES HANDBOOK (6th edition), Quincy, MA: National Fire Protection Association (2001).

National Fire Protection Association, "Standard for Low Expansion Foam and Combined Agent Systems," NFPA 11, Boston, MA: National Fire Protection Association (2002).

National Fire Protection Association, "Standard on Static Electricity," NFPA 77, Boston, MA: National Fire Protection Association (2000).

National Foam, A FIREFIGHTER'S GUIDE TO FOAM, Exton, PA: National Foam Inc. (1993).

National Institute for Occupational Safety and Health, Control of Unconfined Vapor Clouds by Fire Department Water Spray Handlines, (1987).

Noll, Gregory G., Michael S. Hildebrand and Michael L. Donahue, HAZARDOUS MATERIALS EMERGENCIES INVOLVING INTERMODAL CONTAINERS, Stillwater, OK: Fire Protection Publications—Oklahoma State University (1995).

Noll, Gregory G., Michael S. Hildebrand,, and Michael L. Donahue, GASOLINE TANK TRUCK EMERGENCIES: GUIDELINES AND PROCEDURES, 2nd edition, Stillwater, OK: Fire Protection Publications—Oklahoma State University (1996)

Omans, Leslie P., "Fighting Flammable Liquid Fires—Part 1: Family of Foams." FIRE ENGINEERING (January, 1993), pages 50–61.

Omans, Leslie P., "Fighting Flammable Liquid Fires—Part 2: A Primer." FIRE ENGINEERING (February, 1993), pages 50–58.

Omans, Leslie P., "Fighting Flammable Liquid Fires—Part 3: Family of Foams." FIRE ENGINEERING (March, 1993).

CHAPTER 11

DECONTAMINATION

OBJECTIVES

1. Define the following terms:
   - Safe refuge area
   - Contaminant
   - Contamination
   - Decontamination
   - Decontamination corridor
   - Degradation
   - Disinfection
   - Exposure
   - Sterilization
2. Describe the difference between contamination and exposure and their significance in decontamination operations.
3. Describe the difference between surface contamination and permeation contamination and their significance in decontamination operations.
4. Describe the difference between direct contamination and cross contamination and their significance in site safety operations.
5. Define the following terms and describe their potential harmful effects to the human body relating to contamination:
   - Highly acute toxicity contaminants
   - Moderate to highly chronic toxicity contaminants
   - Embryotoxic contaminants
   - Allergenic contaminants
   - Flammable contaminants
   - Highly reactive or explosive contaminants
   - Water reactive contaminants
   - Etiologic contaminants
   - Radioactive contaminants
6. Describe the phases of the decontamination process and their significance in decon operations:
   - Gross decon
   - Secondary decon
7. Describe the following types of decontamination operations and their application and implementation at a hazardous materials incident:
   - Emergency decon
   - Technical decon
   - Mass decon
8. Identify the advantages and limitations and describe an example where each of the following decontamination methods would be used: [NFPA 472 - 6.3.4(1)]
   - Absorption
   - Adsorption
   - Brushing and scraping
   - Chemical degradation

- Dilution
- Disinfection
- Sterilization
- Evaporation
- Heating and freezing
- Isolation and disposal (i.e., "dry decon")
- Neutralization
- Pressurized air
- Solidification
- Vacuuming
- Washing
- Evaporation

9. Identify three sources of technical information for selecting decontamination procedures and identify how to contact those sources in an emergency. [NFPA 472-6.3.4(2)]
10. Identify considerations associated with the placement, set-up, and operation of the decontamination area. [NFPA 472-6.3.4]
11. List three types of fixed or engineered safety systems that may be used to assist emergency responders in implementing decon operations within special hazmat facilities.
12. Define the term clean-up and its coordination with decontamination operations.
13. Describe four general clean-up concerns when decontaminating equipment.

ABBREVIATIONS AND ACRONYMS

ALARA	As Low As Reasonably Achievable	PAPR	Powered Air Purifying Respirator
APR	Air-Purifying Respirator	PERO	Post-Emergency Response Operations
Decon	Decontamination	PID	Photo Ionization Detectors
HEPA	High-Efficiency Particulate Air	START	Simple Triage and Rapid Treatment Process
mR	Milliroentgens	WMD	Weapons of Mass Destruction
NFPA	National Fire Protection Association		

INTRODUCTION

This chapter will describe the seventh step in the **Eight Step Process©**. Decontamination (Decon). Proper decon is essential to ensure the safety of emergency responders and the public. You cannot conduct safe entry operations if you have no way to perform decontamination. Decon methods and procedures must be considered early in the incident as part of the hazard and risk evaluation process as described in Chapter 7.

When we wrote the first edition of this text in 1988, there were many unknowns about the process of decontamination. As a result, emergency responders tended to use a lot of overkill when approaching decon problems in the field. This overcautious approach usually got the job done safely, but the down side was that hazmat operations took a long time to bring to closure. Fortunately, we have learned a great deal about what works and what doesn't work under field conditions. The anthrax attacks of 2001 have also produced more practical operating experience to improve current decon methods.

For the purposes of this text, we have adopted the decontamination terminology referenced in NFPA 471, 472, and 473. Where standard terms or technical guidance did not exist in these standards, we have adopted terms and methods used in the hazmat response community. A list of these additional reference and training materials can be found at the end of this chapter.

Despite the many improvements to the decon process, the basic principles of decontamination are simple and relatively the same. In fact, the simplicity of decontamination has been compared to changing an infant's diaper: (1) Remove it from others; (2) keep it off yourself; and (3) don't spread it around! In other words, don't get contaminated and you won't have a decontamination problem.

Decon needs to be an adaptive and flexible procedure that respects the hazards and the behavior of the contaminants. The hazards and risks presented by the incident will define the scope, nature, and complexity of decon operations. For example, a minimal hazard, such as petroleum oil on turnout boots, can be decontaminated by simply wiping the oil from the boot and then rinsing it with soap and water. Nothing more elaborate is needed. In contrast, a highly toxic material will require implementing a detailed procedure that includes several intermediate cleaning steps. Exactly how many steps and where they are performed is best determined locally based on response requirements.

BASIC PRINCIPLES OF DECONTAMINATION

The process of decontamination basically involves the physical removal or neutralization of contaminants from personnel and equipment. This procedure is vital to lessen the potential of transferring contaminants beyond the hazard area. Proper decontamination is especially important in those instances where injured personnel must be transported to medical facilities.

At every incident involving hazardous materials, there is a possibility that personnel, their equipment, and members of the general public will become contaminated. The contaminant poses a threat not only to the persons contaminated, but to other personnel who may subsequently have contact with the contaminated individuals or their equipment. The entire process of decontamination should be directed toward confinement of the contaminant within the Hot Zone and removing it within the decon corridor to maintain the safety and health of response personnel and the general public.

The best method of decontamination is to avoid contamination. If you don't get the stuff on you, you don't get hurt. Methods to minimize contamination and to ensure safe work practices should be part of the standard safe operating procedures at any incident involving potentially hazardous substances.

TERMINOLOGY AND DEFINITIONS

To understand the materials in this chapter, let's first review some of the basic terminology pertaining to contamination, decontamination, and the establishment of decon operations. Some of these terms should be familiar to you from your First Responder—Operations level training.

- **Contaminant**—A hazardous material that physically remains on or in people, animals, the environment, or equipment, thereby creating a continuing risk of direct injury or a risk of exposure outside of the Hot Zone.
- **Contamination**—The process of transferring a hazardous material from its source to people, animals, the environment, or equipment, which may act as a carrier.
- **Exposure**—The process by which people, animals, the environment, and equipment are subjected to or come in contact with a hazardous material. People may be exposed to a hazardous material through any route of entry (e.g., inhalation, skin absorption, ingestion, direct contact or injection).
- **Decontamination (a.k.a. decon or contamination reduction)**—The physical and/or chemical process of reducing and preventing the spread of contamination from persons and equipment used at a hazardous materials incident. OSHA 1910.120 defines decontamination as the removal of hazardous substances from employees and their equipment to the extent necessary to preclude foreseeable health effects. Note: Some effects may manifest themselves as one or more disease processes many years after exposure and, by definition, may not be foreseeable at the time of the incident.
- **Degradation**—(a) A chemical action involving the molecular breakdown of a protective clothing material or equipment due to contact with a chemical. (b) The molecular breakdown of the spilled or released material to render it less hazardous.
- **Disinfection**—The process used to destroy the majority of recognized pathogenic microorganisms.
- **Sterilization**—The process of destroying all microorganisms in or on an object.
- **Safe refuge area**—A temporary holding area for contaminated people until a decontamination corridor is set up.
- **Decontamination team**—The Decon Team is managed by the Decon Leader (a.k.a. Decon Officer) and is responsible determining, implementing, and evaluating the decon procedure.
- **Decontamination corridor**—A distinct area within the Warm Zone that functions as a protective buffer and bridge between the Hot Zone and the Cold Zone, where decontamination stations and personnel are located to conduct decontamination procedures. An incident may have multiple decon corridors, depending upon the scope and nature of the incident.

Basic terminology pertaining to the phases and methods of decon includes the following:

- **Gross decontamination**—The initial phase of the decontamination process during which the amount of surface contaminant is significantly reduced.
- **Secondary decontamination**—The second phase of the decontamination process designed to physically or chemically remove surface contaminants to a safe and acceptable level. Depending on the scope and nature of the incident, multiple secondary decon steps may be implemented.

- **Emergency decontamination**—The physical process of immediately reducing contamination of individuals in potentially life-threatening situations with or without the formal establishment of a decontamination corridor.
- **Technical decontamination**—The planned and systematic process of reducing contamination to a level that is As Low As Reasonably Achievable (ALARA). Technical decon operations are normally conducted in support of emergency responder recon and entry operations at a hazardous materials incident, as well for handling contaminated patients at medical facilities.
- **Mass decontamination**—The process of decontaminating large numbers of people in the fastest possible time to reduce surface contamination to a safe level. It is typically a gross decon process utilizing water or soap and water solutions to reduce the level of contamination.

UNDERSTANDING THE BASICS OF CONTAMINATION

Before responders can understand the basics of decontamination, it is first necessary to establish a foundation by learning some important concepts about how contamination occurs. In simple terms, the level of decon will always be based on the level of contamination. Four basic concepts of contamination are as follows:

1. How to prevent contamination
2. Surface versus permeation contamination
3. Direct versus cross contamination
4. Types of contaminants

PREVENTING CONTAMINATION

Emergency responders sometimes get the terms *contamination* and *exposure* confused. They actually have very different meanings. Knowing these differences can help a great deal in understanding the importance of preventing contamination.

Contamination is any form of hazardous material (solid, liquid, or gas) that physically remains on people, animals, or objects. In the emergency response business, contamination generally means any contaminant that is on the outside of PPE or equipment while it is still being worn or after it has been taken off. In simple terms, it means that something is "dirty."

Exposure means that a person has been subjected to a toxic substance or harmful physical agent through any route of entry into the body (e.g., inhalation, ingestion, injection, or by direct contact [skin absorption].) In other words, they have the contaminant ("The Bad Stuff") on the outside or on the inside of their bodies.

A responder who has been "contaminated" when wearing PPE has not necessarily been "exposed." For example, an entry team can be contaminated with pesticide dust on the outside of their PPE without having their respiratory system or skin exposed to the contaminant. In order for exposure to occur, the contaminant must come in direct contact with the person, such as through a breach in either protective clothing or respiratory protection.

Even if the contaminant makes direct contact with the responder inside the PPE, it still does not necessarily mean that the person will be harmed by the contaminant. Remember the basic principles discussed in Chapter 2—harm depends on the dose, the route of exposure, and the hazards and properties of the contaminant.

Figure 11.1 Contamination may result from poor decontamination and improper doffing techniques.

Although exposures can occur through a breach or failure of PPE, the most common cause of contamination comes from poor decontamination and clean-up operations. Decon is a critical safety benchmark, and the reason why so much emphasis is placed on developing and implementing a well-planned, structured, and disciplined approach to decon operations.

If you are good at decon, you will probably be good at preventing exposures. If contact with the contaminant can be controlled, the risk of exposure is reduced, and the need for decon can be minimized. Consider the following basic principles to prevent contamination:

- Stress work practices that minimize contact with hazardous substances. Don't walk through areas of obvious contamination, and stay out of areas that potentially contain hazardous substances. Special care must be taken to avoid slip, trip, and falls into the contaminants. If you don't get the "Bad Stuff" on you, you don't get hurt!
- If contact is made with a contaminant, move contaminated personnel to a Safe Refuge Area within the Hot Zone until they can be decontaminated. Remove the contaminant as soon as possible.
- Keep your respiratory protection on as long as possible during the decon process.
- Use of limited-use/disposable protective clothing or overgarments can significantly "lighten" your decon requirements.
- Use a systematic approach to decon. Don't "wing it."

SURFACE VERSUS PERMEATION CONTAMINATION

Contaminants can present problems in any physical state (i.e., solid, liquid, or gas). There are two general types of contamination—surface and permeation.

Surface contaminants are found on the outer surface of a material but have not been absorbed into the material. Surface contaminants are normally easy to detect and remove to a reasonably achievable and safe level using field decon methods. Examples include dusts, powders, fibers, and so on.

Permeation contaminants are absorbed into a material at the molecular level. Permeated contaminants are often difficult or impossible to detect and remove. If the contaminants are not removed, they may continue to permeate through the material. Permeation through chemical protective clothing could cause an "exposure" inside of the suit. If the material is a tool or piece of equipment, it could lead to the failure of the item (e.g., an airline hose on a supplied air respirator). Remember that permeation can occur with any porous material, not just PPE.

Factors that influence permeation include the following:

- *Contact time*—The longer a contaminant is in contact with an object, the greater the probability and extent of permeation.
- *Concentration*—Molecules of the contaminant will flow from areas of high concentration to areas of low concentration. All things being equal, the greater the concentration of the contaminant, the greater the potential for permeation to occur.
- *Temperature*—Increased temperatures generally increase the rate of permeation. Conversely, lower temperatures will generally slow down the rate of permeation.
- *Physical state*—As a rule, gases, vapors, and low-viscosity liquids tend to permeate more readily than high viscosity liquids or solids.

A single contaminant can present both a surface and permeation threat. This is especially the case when liquids are involved.

Figure 11.2A Surface contamination—Contaminant remains on the surface of the fabric.

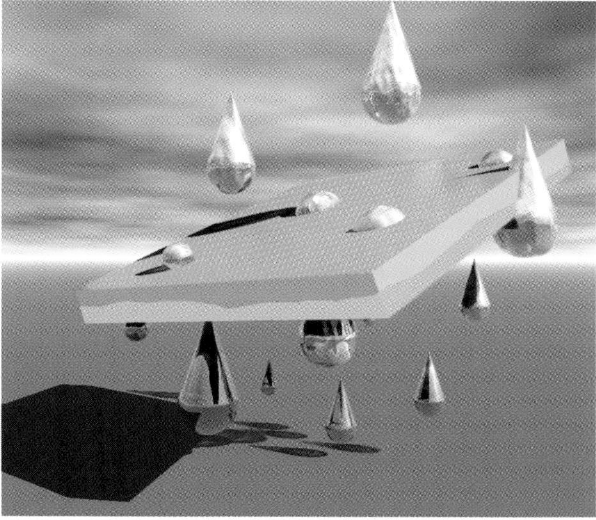

Figure 11.2B Permeation contamination—The contaminant has permeated the fabric.

bleach, sodium hydroxide as a saturated solution (household drain cleaner), sodium carbonate slurry (washing soda), calcium oxide slurry (lime), liquid household detergents, and isopropyl alcohol. Chemical degradation is primarily used to decon structures, vehicles, and equipment and should not be used to decon chemical protective clothing. Degradation chemicals should never be applied directly to the skin!

Technical advice for chemical degradation procedures should be obtained from product specialists to ensure the solution used is not reactive with the contaminant. Potential problems include overly aggressive concentration mixtures (i.e., The Firefighter's Rule: If 0.5% is good, then 5% will be great!) and damage to CPC and equipment from degradation chemicals.

The physical and chemical compatibility of the decon solutions must be determined before they are used. Any decon method that permeates, degrades, damages, or otherwise impairs the safe function of PPE should not be used unless there are plans to ultimately isolate and dispose of the equipment.

NOTE: The use of sodium hypochlorite or bleach solutions can have adverse effects on any firefighting protective clothing or equipment using Kevlar® or Kevlar® blends. Research by protective clothing manufacturers has shown that the level of degradation will be dependent upon the duration of the exposure and temperature and will shorten the life of the garment or material.

- ❑ **Neutralization** is the process used on corrosives to bring the pH of the final solution to somewhere within the range of pH 5 to pH 9. The neutralization process uses an acid substance to neutralize alkalies or an alkali substance to neutralize an acid. Preferably, the less harmful byproduct produced is a neutral or biodegradable salt.

  According to the EPA, the ideal substances to use for neutralization of corrosives in emergencies are citric acid (a powder in 25-pound bags), used for neutralizing alkalies, and sodium sesqucarbonate (a powder in 50-pound bags), used for neutralizing acids. Either one forms neutral salts, and depending on the substance being neutralized, sometimes forms a biodegradable salt. Neutralization is primarily used to decon equipment, vehicles, and structures that are contaminated with a corrosive material.

- ❑ **Solidification** is a process by which a contaminant physically or chemically bonds to another object or is encapsulated by it. This method is primarily used to decon equipment and vehicles. Commercially available solidification products can be used for the clean-up of spills.

  In some situations, large pieces of equipment have been covered with a cement like material so that the contaminant is permanently bonded to the object. The contaminated object can then be buried in a hazardous waste landfill. After the Chernobyl disaster, large contaminated objects were covered with cement and entombed.

- ❑ **Disinfection** is becoming increasingly important due to the threats posed by chemical and biological warfare agents. Disinfection is the process used to inactivate (kill) virtually all recognized pathogenic microorganisms. Proper disinfection results in a reduction in the number of viable organisms to some acceptable level. It does not cause complete destruction of the microorganism you are trying to remove. Consequently, it is important that emergency responders obtain technical advice about disinfection techniques prior to their use. Likewise, some disinfectants work better on certain etiologics than others. Commercial disinfectants usually include detailed information outlining the capabilities and limitations of the product. If you respond to research labs, hospitals, and universities in your area, you should perform a hazard assessment and familiarize yourself with the specific types of biological hazards present and the best disinfectant for the type of hazard you may encounter.

There are two major categories of disinfectants:

- **Chemical disinfectants** are the most practical for field use. The most common types of chemical disinfectants are commercially available, including phenolic compounds, quaternary ammonium compounds, chlorine compounds, iodine, and iodophors.
- **Antiseptic disinfectants** are designed primarily for direct application to the skin. These include alcohol, iodine, hexachlorophene, and quaternary ammonium compounds. Some of these compounds are also classified as disinfectants, but alterations in concentration allow them to be classified as antiseptics.

The terms disinfection and sterilization are sometimes used interchangeably. It is important to recognize that sterilization is not the same as disinfection. A decontamination recommendation from an etiologic specialist to sterilize a piece of equipment must not be misunderstood to mean disinfect it.

- **Sterilization** is the process of destroying all microorganisms in or on an object. The most common method of sterilization is by using steam, concentrated chemical agents, or ultraviolet light radiation. Because of the size of the equipment involved in the sterilization process, it has limited field application and cannot be used to decontaminate personnel, but it does play an important role in decontaminating medical equipment. Contaminated medical equipment is sometimes disinfected at the site, then transported as contaminated equipment to a special facility, where it is then sterilized or discarded. Contaminated emergency response equipment may be sterilized through autoclaving, but the ability for the item to withstand this process has to be confirmed by the manufacturer.

EVALUATING THE EFFECTIVENESS OF DECON OPERATIONS

Decon methods vary in their effectiveness for removing different substances. The effectiveness of any decon method should be assessed at the beginning of the decon operation and periodically throughout the operation. If contaminated materials are not being removed or are permeating through protective clothing, the decon operation must be revised.

Five simple criteria can be used for evaluating decon effectiveness:

1. No personnel are exposed to concentrations above the TLV/TWA.
2. Personnel are not exposed to skin contact with materials presenting a skin hazard.
3. Contamination levels are reduced as personnel move through the decon corridor.
4. Contamination is confined to the hot zone and decon corridor.
5. Contamination is reduced to a level that is as low as reasonably achievable (ALARA).

Methods for assessing the effectiveness of decontamination include the following:

- **Visual observation**—Stains, discolorations, corrosive effects, and so on.
- **Monitoring devices**—Devices such as photoionization detectors (PIDs), detector tubes, radiation detection instruments, and survey meters can show that contamination levels are at least below the device's detection limit. For example, placing CPC in a closed bag after decon will allow any residual organic vapors to accumulate; monitoring devices can then be used later to detect any vapors present.
- **Wipe sampling**—Provides after-the-fact information on the effectiveness of decon. In some cases, wipes can be taken with pH paper or other types of indicator papers. However, most wipe samples must be subsequently analyzed in a laboratory. Wipe sampling can be used to assess decon effectiveness on CPC, equipment, and skin, as well as vehicles and structures.

It should be noted that there is currently no practical way to determine the effectiveness of decontamination in the field for most etiologic hazards.

DECON SITE SELECTION AND MANAGEMENT

The success of decontamination is directly related to how well the Incident Commander and the Decon Officer control on-scene personnel and their operations. Before initiating decontamination, the Hazmat Group Supervisor and the Decon Officer must decide (1) how much and what decon method is required; and (2) to what extent decon will be accomplished in the field.

These decisions should be based on the answers to the following questions:

- Can decon be conducted safely? For example, dilution may be impractical due to cold weather or because it presents an unacceptable risk to emergency personnel.
- Are existing resources immediately available to decon personnel and equipment? If not, where can they be obtained, and how long will it take to get them?
- Can the equipment used be decontaminated? The toxicity of some materials may render certain equipment unsafe for further use. In these cases, disposal may be the only safe alternative.

THE DECONTAMINATION TEAM

At a working hazmat incident, a Decontamination Team should be established to manage and coordinate all decon operations. Decon Team functions include research and development of the decon plan and set-up and operation of an effective decontamination area capable of handling all potential exposures, including entry personnel, contaminated patients, and equipment. If necessary, the Decon Team will also coordinate the establishment of a Safe Refuge Area.

The Decon Team is managed by the Decon Officer, who reports to the HazMat Group Supervisor. The Decon Officer and the Decon Team should be trained to the Hazardous Materials Technician level (or equivalent). The Decon Team should be trained to at least the First Responder Operations, and should be versed in decon techniques and operations.

The Decon Officer performs the following activities:

- Determine the appropriate level of decontamination to be provided.
- Ensure that proper decon procedures are used by the Decon Team, including decon area set-up, decon methods and procedures, staffing, and protective clothing requirements.
- Coordinate decon operations with the Entry Officer and other personnel within the Hazmat Group.
- Coordinate the transfer of decontaminated patients requiring medical treatment and transportation with the Hazmat Medical Unit.
- Ensure that the Decon Area is established before any entry personnel are allowed to enter the Hot Zone, whenever possible.
- Monitor the effectiveness of decon operations.
- Appoint or act as an accountability officer to limit access to personnel entering and operating within the decon area.

The Decon Officer should use a formal checklist to assure that important items are not overlooked.

DECON SITE SELECTION

When a hazmat incident occurs outdoors, the decon site should be accessible from a hard-surfaced road. Water supply, access to safety showers, runoff potential, and proximity to any environmentally sensitive areas, such as streams or ponds, should be considered.

The ideal outdoor decon site is upwind and uphill from the incident and remote from drains, manholes, and waterways, but close enough to the scene to limit the spread of contaminants. Unfortunately, it is often not possible to actually choose such an ideal site. Shifting winds and dispersing vapors can further complicate this selection process. Such real-life problems may force the movement of the decon area once it has been in operation if initial site planning has been hasty.

Houston, TX F.D.

Figure 11.5 The best location for decon is uphill and upwind. The real world often does not provide a perfect location.

If decon will be conducted indoors, consideration should be given to quick access such as hallways, type and slope of floor, floor drains, and ventilation airflows in the area.

Complete decon may be impractical to achieve at a single location, so a combination of onsite and offsite contingencies may be necessary. Any time decon is conducted offsite, the entire operation will become more complicated due to the logistics of moving people offsite. If you intend to use an offsite location for decon, special preparations will need to be made to prevent the spread of contaminants.

DECONTAMINATION CORRIDOR

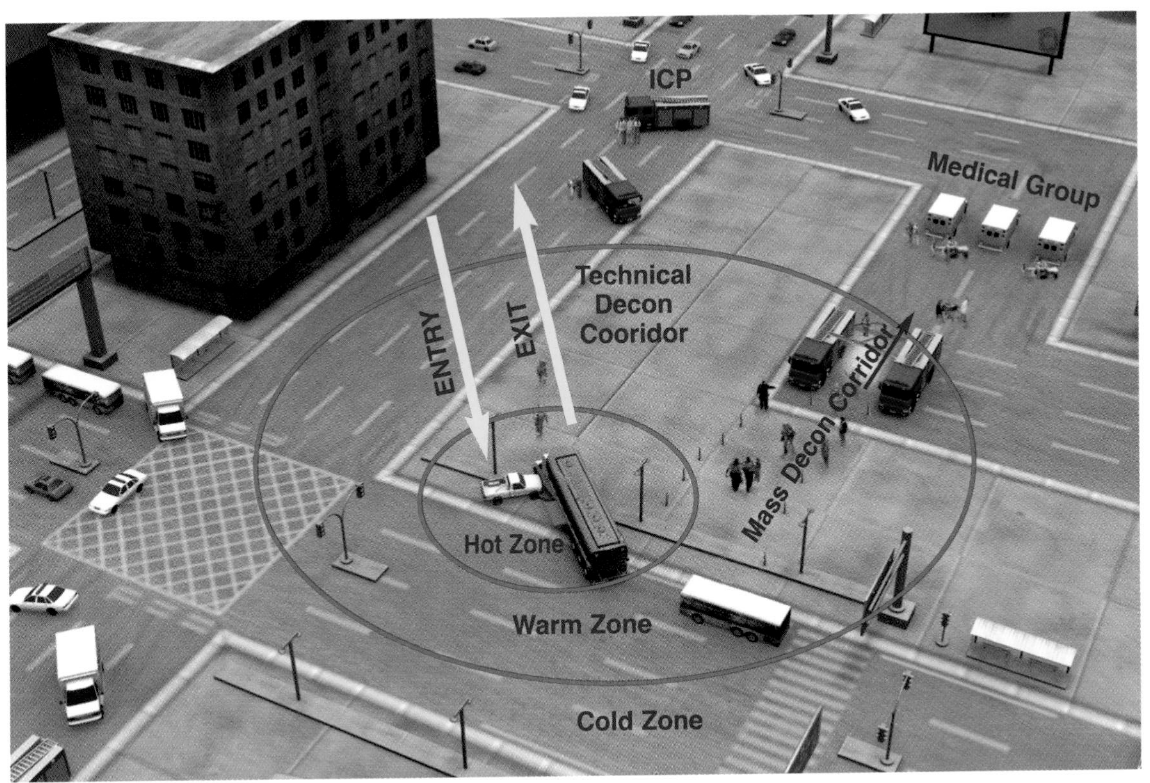

Figure 11.6

Once a decon site has been selected, a decontamination corridor should be marked. The decon corridor is simply a pathway from the Hot Zone into the decon area, with the exit point near the Warm Zone/Cold Zone interface. The decon corridor and the boundaries of the decon area should be clearly identified. Examples include using fence post stakes and colored fire line tape, traffic cones, and so on. Signs may also be used to indicate the entry and exit points to the decon area. Presized tarps (minimum of 20 ft by 40 ft) with various stations marked on them can also help organize the decon area and provide secondary containment for decon runoff (e.g., Imagine a cattle chute going into a corral!).

Some scenarios will require multiple decon corridors. For example, a mass casualty scenario involving hazardous materials or weapons of mass destruction (WMD) could have a decon corridor for both mass decon operations and a technical decon operations. The mass decon corridors would likely be established by first-due fire units. Upon the arrival of a hazmat team, a separate decon corridor for technical decon operations would then be established. Similarly, an incident covering a large area (e.g., World Trade Center, Pentagon) will have multiple decon corridors, simply due to the size and geography of the incident site.

Response scenarios in which the decon area is physically separated from the "problem" by a large distance can also pose additional challenges for responders. These scenarios will often require some form of transportation from the entry control point to the incident scene, such as mules, golf carts, or pick-up trucks. Regardless of the type of vehicle, it must have the capability of carrying multiple personnel, PPE, and any required control equipment. Continuous air monitoring may be required when dealing with flammable substances to ensure that these transport vehicles do not enter an unsafe atmosphere. PPE must also be provided for the vehicle operator, including SCBA; this can be accomplished by placing the SCBA beside the driver on the seat. Placing a plastic tarp in the vehicle bed can minimize the spread of contaminants (hey—nobody said this was going to be easy!).

STAFFING AND PERSONAL PROTECTION OF THE DECON TEAM

When setting up the decon area, consideration must be given to the staffing of the decon operation and the safety of the Decon Team. Decon is a labor-intensive operation. While there are tools and equipment that can expedite the decon process, it still requires time and personnel to set up the decon corridor.

The Decon Officer should be trained to the Hazardous Materials Technician level (or equivalent). The Decon Officer is a supervisor and should *not* be involved in the actual decon process. However, they may still be "dressed out" in the event of an emergency along the decon corridor that requires additional personnel. At a minimum, each staffed decon station should have one decon member assigned. Those stations that require entry personnel to be scrubbed or to remove their PPE should be staffed with two decon members, if possible.

The level of skin and respiratory protection required by Decon Team members will be dependent upon (1) the type of contaminants involved; (2) the level of contamination encountered by entry personnel; and (3) where individuals are working along the decon line.

Tactical and safety considerations will include the following:

- The Decon Team should be dressed in chemical protective clothing and equipment based on the hazards and risks of the contaminants they will be decontaminating. Experience in the field over the last 20 years indicates that the Decon Team can operate safely using chemical protective clothing that is one level down from the PPE being used by the Entry Team. Example: The Entry Team is using Level A, the Decon Team may use Level B. This is just a guideline, there will always be exceptions, but exceptions should be based on a hazard and risk assessment and made in conjunction with the Safety Officer. The risk of heat exhaustion should be factored in to decisionmaking when selecting the right PPE for decon.
- The most common level of protection for the Decon Team is chemical splash protective clothing (i.e., Level B) and self-contained breathing apparatus (SCBA). Most field-based decon operations involve wet decon methods with a splash hazard. In addition, Level B disposable or limited-use garments and SCBA are readily available in most response organizations.
- Extended entry operations will require that the Decon Team be provided with an uninterrupted air supply. The use of airline hose units (see Chapter 8) can provide an extended air supply but can decrease Decon Team mobility.
- Respiratory protection for Decon Team members may be downgraded to PAPRs or APRs providing that air monitoring is conducted and filter cartridges are compatible with the concentration of contaminants present.
- All Decon Team personnel must be decontaminated before leaving the decon area. The extent of this decon process will be determined by the types of contaminants involved and an individual's work station along the decon corridor. The use of disposable or limited-use garments by the Decon Team can simplify the decon process.

In summary, emergency decon is a gross decon operation. It should include the removal of any contaminated clothing as soon as possible and as thorough a washing as possible. Victims requiring follow-up medical treatment or evaluation should still undergo secondary decon either on-scene or at a medical facility.

Figure 11.8 Emergency decon may be required for civilians, responders or the injured before a formal decon corridor is established.

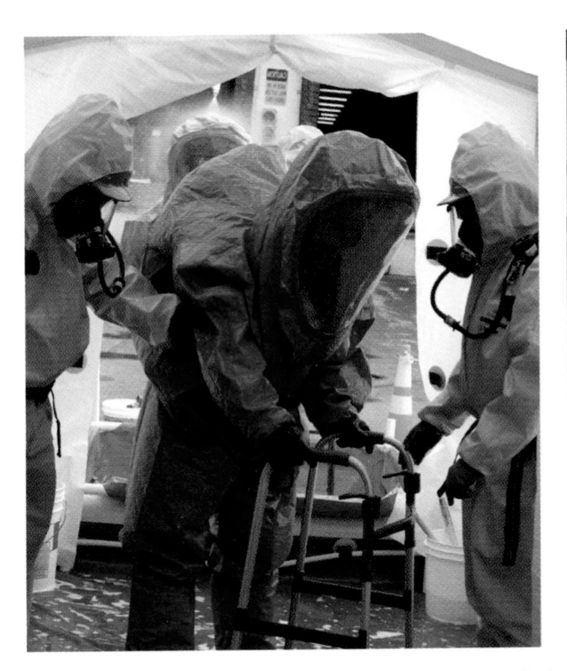

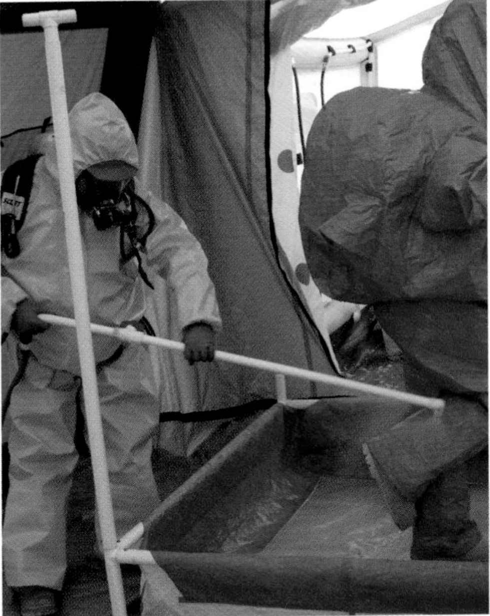

Figure 11.9 For extended operations technical decon areas may be required.

DECONTAMINATION OF INJURED PERSONNEL

Contaminated nonambulatory victims pose unique challenges for emergency responders. At incidents where there are both ambulatory (walking) and nonambulatory (non-walking) contaminated individuals, separate decon corridors should be established to facilitate victim flow.

Decon of injured and/or contaminated individuals should be accomplished in the field before transport to a medical facility—failure to decon these people before transport will only lead to bad outcomes, including cross-contamination to EMS personnel, ambulances, and emergency room facilities.

In some instances, hospitals and medical facilities may encounter "surprise packages" where contaminated victims simply show up at the front door. Approximately 80% of those who sought medical treatment after the 1996 Tokyo subway incident involving Sarin sought treatment outside of the emergency medical system.

Regardless of whether emergency, technical, or mass decon is being provided, there are some basic principles that need to be recognized when dealing with the injured. These include the following:

- Remove all clothing, jewelry, and shoes as soon as possible. If possible, remove clothing from head to foot to limit the risk of inhalation.
- Protect the victim's airway. Blot away any obvious liquids using a soft sponge or washcloth, and/or brush away any obvious solid or dust materials.
- Hazardous materials will tend to enter the body more readily through wounds, the eyes, or mucous membranes than through intact skin. Therefore, these areas should be decontaminated first.
- Begin to rinse the patient around the face and head area; ensure that all fluids flow away from the eyes and respiratory system. Then move to any open wounds, followed by a head-to-toe rinse in a systematic fashion. Pay close attention to skin folds, armpits, genital areas, fingernails, and the feet.
- Water is the universal decon agent. However, if the adherent solids or liquids are not water soluble, a mild liquid detergent can be used to facilitate skin washing. Do not use hot water, as this will cause pores in the skin to open. It is better to use slightly cooler than body temperature water, ideally 30°C (86°F).
- If the eyes are symptomatic, irrigate the eyes continuously.
- Attempt to isolate contaminated areas on the patient if the whole patient is not contaminated. For example, cover uncontaminated areas on the patient with a waterproof material and avoid washing a contaminated area into an open wound. This may require placing goggles, earplugs, or an oxygen mask on the patient.
- Provide initial medical treatment based on the signs and symptoms of the suspected material. Remember the basic ABCs!
- Bag and tag all clothing and possessions. Depending on the scenario, these may need to be disposed of as hazardous waste or they may be treated as evidence if a criminal event is suspected.

TECHNICAL DECON

Technical decontamination is the planned and systematic process of reducing contamination to a level that is As Low As Reasonably Achievable (ALARA). Technical decon operations are normally conducted in support of emergency responder recon and entry operations at a hazardous materials incident and by medical facilities handling contaminated patients. The key variable with technical decon operations is the time and resources required to become operational.

Technical decon is a multistep process in which contaminated individuals are cleansed with the assistance of trained personnel. While the process has changed over the years, it still follows some basic operational concepts. Technical decon is similar to a car wash:

1. There is an entry point and an exit point (i.e., the decon corridor).
2. In between the entry point (i.e., Hot Zone) and the exit point (i.e., Warm Zone/Cold Zone interface), several progressive cleaning steps takes place to remove the dirt (same as the contaminants).
3. Most of the dirt is removed in the initial stage of the car wash (i.e., gross decon), and the car becomes cleaner as it moves through the various cleaning stages (i.e., secondary decon).

Now, replace the words *car wash* with *decon corridor* and you have a technical decon operation. While the car wash facility is permanent, our technical decon process and decon corridor must typically be set up in the field.

Figure 11.10 Just like a car wash, technical decon requires personnel or equipment to go through a planned and systematic process of cleaning. There may be several steps or phases of decon.

THE TECHNICAL DECON PROCESS

There are nine basic steps in technical decon:

- **Tool drop**—Tools that may be reused at the job site are placed here. When the job is completed, they will receive further cleaning. Note: If the incident involves law enforcement agencies, it may also be necessary to incorporate a drop area for evidence and weapons. A weapons safety officer should be established, as appropriate.

Figure 11.11

- **Overglove/overboot drop**—Overgloves and overboots are used to minimize the amount of contamination on PPE. The majority of personal contamination occurs to the hands and feet. Efficient removal of the overgloves and overboots can minimize the potential for secondary contamination.
- **Gross decon**—The initial decon step in which the entry crew is rinsed off. Special attention is given to the hands and feet. A soap and water wash may also be used, followed by a rinse.
- **Secondary decon**—Additional washing and rinsing steps designed to further reduce the level of contamination. Special attention is given to the hands and feet. Based on the level of contamination and the hazards of the contaminant, this phase may not be necessary.
- **PPE removal**—PPE should be removed in a manner that minimizes the potential for the Decon Crew to contact those being decontaminated. Large trash bags make handling and disposal easier. If the materials will require additional postincident handling (e.g., weapons, evidence), clear plastic bags should be used.
- **Respiratory protection removal**—Should always be the last item removed. If individuals being deconned are wearing coveralls, special attention should be given so that the user will continue to wear the facepiece until PPE is removed.
- **Clothing removal**—If necessary, change out of the undergarments. In this case, it will also be necessary to provide an additional change of clothing for personnel.
- **Body wash**—If a breach in a suit occurred, then the whole body, or at least the potentially contaminated area, should be washed. This may be done offsite, but responders should always shower prior to going home.
- **Medical evaluation**—Responders should always be medically evaluated after an entry. Vital signs are taken 10–15 minutes after rest and oral rehydration to

ensure adequate recovery from the stress of entry. Responders who are not recovering appropriately should be further evaluated, hydrated, and possibly transported to an emergency department or other source of definitive medical care for further treatment.

Portable tents and specially designed vehicles or trailers may also be integrated into the decon corridor and the technical decon process. In addition, some hospitals and high-hazard facilities have constructed specialized rooms or areas where technical decon can be completed. There is not necessarily a right way or a wrong way…there are a lot of different options. Use what works for you!

MASS DECONTAMINATION

Mass decontamination is established when large numbers of people (i.e., civilians or responders) need to be decontaminated at the scene of a hazmat emergency. What is "large" is really based on your local resources. From a practical perspective, we suggest starting with a school bus (approximately 20 to 40 students) and working up from there. A major difference between technical and mass decon is the wide range of victims likely to be involved in a mass decon scenario, including ambulatory and nonambulatory, children, and people with disabilities.

Mass decon is a gross decontamination process that relies on the use of water or soap and water solutions to flush the majority of the contaminant from individuals. Mass decon operations are based upon the following basic principles:

1. Removing clothing is a form of decon that can remove the majority of the contaminants. SBCCOM tests with simulated chemical agents show that this can account for the removal of up to 80% of the contaminant. **Note:** If victims are unwilling to remove their clothing, they should be showered while clothed. Responders should not spend any significant length of time trying to convince frightened civilians who do not heed the command to disrobe since this impedes the goal of rapidly decontaminating as many people as possible.
2. Once clothing has been removed, flushing a victim with a low pressure/high-volume shower of water will remove additional amounts of the contaminant. Tests conducted by SBCCOM and the Virginia Department of Emergency Services have shown that high volume, low-pressure streams are most effective. Optimal operating pressures for both automatic and standard fire department nozzles are approximately 30 to 50 psi, with average flows of 150 gpm per nozzle.
3. Water flushing can be delivered in a variety of methods, including the following:
    - Establishment of mass decon corridors using either handlines or master fog streams from fire engines or aerial devices. As more units arrive on scene, they can be added into the process.
    - Use of portable shower set-ups (commercial and available systems).
    - Use of indoor or outdoor swimming pools (victims wade into the water).
    - The use of sprinkler heads in a building with a fire protection sprinkler system.
    - Multiple bathroom showers, such as in a school or health club locker room.

There are many options and methods to provide mass decon. It is our intent to focus on the key concepts, operational issues, and lessons learned that should be incorporated into your mass decon procedure. The key point is to do what works best for YOU!

IT'S ALL ABOUT PEOPLE

The greatest challenge in a mass decon operation will be the management and control of people. Among the lessons learned from research and exercises are the following:

- Timing is everything! In order to create a more favorable outcome, mass decon must be established by the first-arriving responders in a rapid and efficient manner. The new 7,000 GVW specialized decon "assault vehicle" may impress the public and the news media when it arrives on scene, but it probably won't change the outcome if it takes 30 minutes to respond and become operational. Speedy operations are critical to success.
- Law enforcement will play a key role in assisting with mass decon. Depending on the scenario, their training and expertise in controlling and containing large groups of people will be a major factor in a successful operation. More than a few people will want to leave the scene and seek medical treatment on their own, which defeats the goal of rapid decon.
- The ability to establish a safe refuge area and maintain effective site management and control will be critical until the mass decon corridor is established.
- Psychology will play a major factor in a mass decon scenario; a majority of the victims may not be contaminated and are "worried well." For example, approximately 80% of the 5,510 victims who sought medical attention as a result of the 1995 Tokyo subway Sarin incident were not exposed to any significant amount of the chemical agent. These casualties needed to be decontaminated for their own peace of mind as well as for their health care providers. The "worried well" may also complain of symptoms that are similar to that of truly contaminated individuals but which cannot be differentiated in the short time that responders have to decide who warrants emergency decon and treatment.
- When dealing with large numbers of victims, quick air monitoring and detection is crucial. Early identification of the presence of contaminants followed by risk characterization is key to survival.

THE MASS DECON PROCESS

The following tasks and factors should be incorporated into the mass decon procedure:

1. **Protective clothing**—Responders should use either structural firefighting clothing or Level B chemical splash protective clothing with self-contained breathing apparatus. Butyl gloves should be used for hand protection.
2. **Decon prioritization**—Whether you call it triage or decon prioritization, responders must have a system for initially sorting and prioritizing those individuals who are potentially contaminated or injured. In this initial response window, most responders will have a limited amount of personnel and equipment. The general goal of mass decon is to assure the survival of the greatest number of people.

   When people show symptoms and it is evident that they have been contaminated, they will generally willingly submit to some form of decontamination. Don't slow down the overall process to argue with the victims. Studies involving chemical warfare agents show that victims should be washed for at least 3 minutes but no longer than 5 minutes.

People who may have been exposed to the problem but who were not physically contaminated (e.g., most gas and vapor exposures) do not need decon, although they may be "psychologically decontaminated" at a later time. People who have material on them or are severely symptomatic must be decontaminated in a timely manner. Others who were just in the building or in the area in most cases do not need to be decontaminated.

3. **Ambulatory versus nonambulatory victims**—Injured victims should be sorted into ambulatory and nonambulatory groups. Medical triage procedures (e.g., S.T.A.R.T.—Simple Triage and Rapid Treatment Process) can facilitate this process. However, responders should recognize that most triage protocols, including S.T.A.R.T., were designed to be used for victims of acute traumatic injuries. These protocols may not accurately identify and classify ill patients from a hazmat or WMD scenario or predict which patients will clinically deteriorate.
   - **Ambulatory**—Victims are able to understand directions, talk, and walk unassisted. Classified as Priority 3 or minor (walking wounded).
   - **Non-Ambulatory**—Victims who are unconscious, unresponsive, or unable to move unassisted. Classified as Priority 1—immediate (life threatening injury) or Priority 2—delayed (serious non-life-threatening injury). Types of injuries may include:
     - Serious medical symptoms (e.g., shortness of breath, chest tightness)
     - Evidence of liquid on clothing or skin
     - Conventional injuries
     - Casualties reporting exposure to vapor or aerosol
     - Casualties closest to the point of release

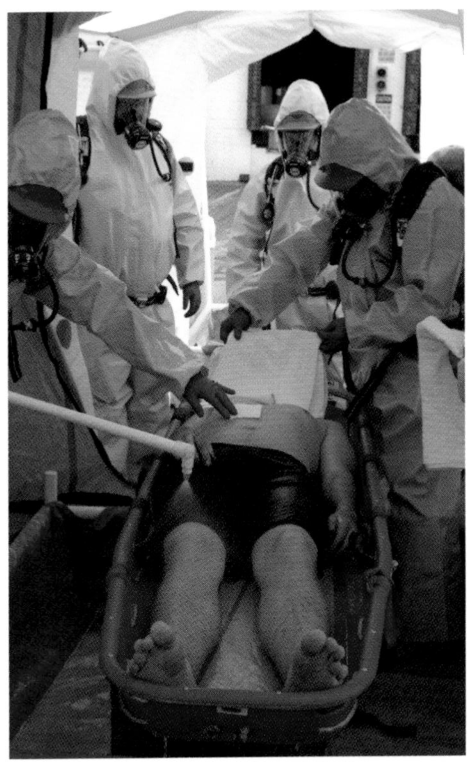

Figure 11.12

4. **Cold weather operations**—Victims can be washed off in cold weather and survive hypothermia if they are then moved to a warm building or vehicle as soon as possible. The rule of thumb is simple: If the victims may die from the contamination, then wash them off, regardless of the theoretical risk of hypothermia. You should then move them to a covered environment with an ambient temperature of at least 70°F (preferably higher) as soon as possible. There is no evidence that cold weather decon will cause permanent injury or harm.

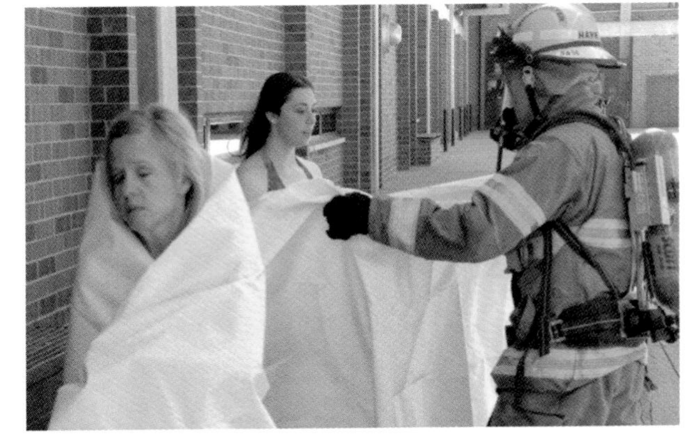

Figure 11.13

Nothing in these preceding statements should imply that these victims will be happy; they will be extremely cold and may be shivering violently. Hypothermia develops when an individual is continuously exposed to a cold stimulus for a period of time, but it is unlikely that responders will cause a significant core temperature drop by performing emergency decon. However, a person who has stopped shiv-

SCAN 11-C

MASS DECONTAMINATION

Direct people to decon.

Remove outter garments.

Go through decon.

Provide cover as soon as possible.

ering or who does not respond to verbal or physical stimuli may be developing hypothermia. Victims who are young, elderly or have chronic health problems will succumb to hypothermia faster than a healthy individual. These people should be moved to a warm area as soon as possible since hypothermia in this group may begin in approximately 10 minutes.

5. **Mass Decon Operations Using Fire Apparatus**—Numerous options using fire apparatus have been developed and tested throughout the fire service. All of these options are typically based upon using the items immediately available to emergency responders—hoselines, water and the fire apparatus itself. SBCCOM's report entitled *Guidelines for Mass Casualty Decontamination During a Terrorist Chemical Agent Incident* is an excellent reference and recommended reading for anyone considering implementation of this mass decon method.

   Mass decon operations can be implemented using a single engine company, multiple engine companies, a truck company, or multiple engine / truck companies in combination with each other. Regardless of the resources available, all involve the establishment of a decon corridor using a combination of overhead and / or side water sprays from fire apparatus. The following information is referenced from the Virginia Department of Emergency Services mass decon training program.

SINGLE ENGINE COMPANY RESPONSE PROCEDURE

- Position the engine to create a herding lane (think cattle) between the apparatus and a building wall or other structure. The lane should be 12 to 16 feet wide. By positioning the apparatus in this manner, victims' modesty can be somewhat protected.
- Use the apparatus PA system to direct the victims to the Safe Refuge Area and the Mass Decontamination Corridor. Instructions must be clear and authoritative (e.g., where to put clothing, what to do in the shower area, where to go after exiting).
- Engine company personnel should attach 2 to 3 nozzles to the opposite side of the pump panel and set them for a wide fog pattern. The operator should engage the pump and maintain a pressure of between 30–50 psi at the panel if using automatic or regular nozzles.
- Upon exiting the decon corridor a cover should be provided to all victims. Black trash bags, disposable gowns, blankets, etc. are acceptable. Tactics to protect personal modesty (e.g., tarps and poles) may be required.
- If banner tape is available, set up lanes to the entrance and from the exit of the decon corridor. At the exit point personnel should be stationed to direct victims to the EMS area to be triaged. If EMS has not yet set up or has not arrived, place victims in a secure area (e.g., building lobbies, schools, large warehouses, etc.). The goal is to get victims out of the weather and into a contained area suitable to deliver medical treatment as soon as possible.

MULTIPLE ENGINE COMPANY RESPONSE PROCEDURE

- If using engine companies with a side-mount pump, the pump panels must face to the street curbside. The engine companies should be positioned approximately 12 to 16 feet apart. Personnel will attach nozzles to all discharges on the side of the pumper facing the decon corridor. If there is only a single discharge, another nozzle should be attached to a section of hose and tied off to either the front bumper or rear of the pumper (e.g., trash line). This corridor will also establish a modesty corridor for the victims. Banner tape and personnel should be positioned to direct victims into the lane (herding).

MULTIPLE ENGINE/LADDER APPARATUS RESPONSE PROCEDURE

- If using engine companies with a side-mount pump, the pump panels must face to the street curbside. The engine companies should be positioned approximately 12 to 16 feet apart. Personnel will attach nozzles to all discharges on the side of the pumper facing the decon corridor. If there is only a single discharge, another nozzle should be attached to a section of hose and tied off to either the front bumper or rear of the pumper (e.g., trash line).
- The truck company should approach from behind the Engine Company so that the operator will be able to work upwind and uphill. The aerial apparatus should position the ladder pipe to form a water shower. The water shower should be directed just ahead of the pumpers to give a final rinse to the victims exiting the primary shower set up by the engine companies.

DECON OPERATIONS INSIDE SPECIAL BUILDINGS

High hazard research and manufacturing facilities are often designed and engineered for specific emergency response scenarios. An increasing number of hospital emergency departments are now being designed to provide immediate decon capability (**Note:** Those seeking additional information on designing decon capability within a health care facility should consult *NFPA 99—Health Care Facilities*, Chapters 10 and 11). Many of these facilities have features that can facilitate the delivery of timely and effective decon operations.

The following factors should be considered when evaluating fixed facilities for decon operations, including:

- **Decon rooms**—The size of the room should be based upon the number of patients the facility expects to handle. The room's floor, walls and ceiling should be coated with an inert material that allows the area to be decontaminated. Electrical fixtures should be rated based upon the expected hazards. If flammable vapors are expected, the room or area should meet *NFPA 70—National Electric Code™* requirements for Class 1, Division 2, Group D classifications.
- **Ventilation systems**—Most facilities engaged in hazardous materials research or manufacturing have specially engineered ventilation systems to reduce employee exposure to hazardous materials. Removal of air contaminants at the source is the most effective method of preventing employee chemical exposure. These systems can be an asset when conducting decon and for determining entry points into a spill area.

  The exhaust system should be filtered to control the materials encountered in the facility (e.g., infectious materials, chemicals, radioactives, etc.). Responders must not shut down the power supply that maintains these special systems.

  In some facilities, fixed monitoring and detection systems (i.e., oxygen, toxicity, flammability, etc.) may also be installed in the ventilation system to provide early warning of a release. These fixed systems may be monitored in a process control room, security center, or by an off-site alarm company.

- **Positive and negative pressure atmospheres**—These systems maintain a negative pressure within the hazard area and a positive air pressure outside of the hazard area. If an incident occurs, the flow of air is from the outside toward the inside, thereby minimizing the spread of contaminants (dust, vapors, and gases) from the involved area.
- **Safety showers**—Safety showers are usually located throughout high risk and medical facilities and they are restricted for emergency use in the event of accidental chemical contamination. If a hazmat incident occurs indoors, consider using these safety showers for gross decon. Safety showers have special "deluge" heads that deliver 30 to 50 gallons per minute, far more than a bathroom shower head. This large flow is essential in the initial stage to sweep away, rather than just to dilute, the strong contaminant. Safety showers are a good gross decon option if they have drains and provisions to confine runoff.
- **Emergency eyewash fountains**—Emergency eyewash fountains or hoses are located throughout most industrial and laboratory facilities. Their location should be identified for responders during the pre-entry briefing.
- **Fixed air supply systems**—Some facilities may provide air outlets for use with airline hose respirators. These may be supplied by a bank of air cylinders, or from a dedicated breathing air compressor.
- **Personal protective clothing and equipment**—Some facilities will maintain an inventory of protective clothing and equipment to be used for emergency response activities

The systems and capabilities outlined above are designed based on the physical, chemical, and toxicological characteristics of the materials being handled or anticipated. Responders should consider the use of these specialized systems as a method of reducing the potential contaminants within the Hot Zone. Activation of fixed ventilation systems may also significantly reduce the contamination level before entry.

Always consult with the facility building engineer and safety personnel before using fixed ventilation options. Activation of a ventilation system not designed and rated for the hazards present under emergency conditions could worsen the situation (e.g., cause an explosion or spread the contaminants to a larger area).

CLEAN-UP OPERATIONS

WHAT IS CLEAN-UP?

The term *clean-up* means different things to different emergency response personnel. For our purposes, clean-up activities consist of any work performed at the emergency scene by emergency responders, which is directed toward removing contamination from protective clothing, tools, dirt, water, and so on. Clean-up may also involve responder decontamination of some debris, damaged containers, and so on.

From a decon perspective, not all work related to restoring the contaminated site to its previous (nonpolluted) state is considered "clean-up." Chapter 10 provides a detailed review of the short- and long-term site restoration and recovery activities associated with hazmat emergency response.

GENERAL CLEAN-UP OPTIONS

Generally speaking, the IC has two clean-up options:

1. Conduct a limited-scale clean-up of key emergency response equipment such as fire apparatus. The objective of this option is to place essential equipment back in service as soon and as safely as possible. Responders may also get involved in the more technical aspects of clean-up by working directly under the supervision of an outside agency or contractor.
2. Conduct clean-up using a qualified and authorized outside contractor. This option is usually exercised when large pieces of heavy equipment have been contaminated and are not part of the emergency response organization's fleet. Examples of such equipment are bulldozers and end loaders used by responders to construct dikes inside the Hot Zone.

Before jumping into clean-up activities, make sure that you have a sufficiently detailed and coordinated plan. Once responders are decontaminated, the rules of the game shift from one of being an "Emergency Responder" to "Hazardous Waste Generator." In addition, there are clear differences between emergency response and postemergency response operations (PERO) under OSHA 1910.120. Make sure that your clean-up activities are conducted within regulatory guidelines.

Agencies that should be consulted in the development of equipment decon plans can include the following:

- **Water/sewage treatment facilities**—Prior arrangements will be necessary before large quantities of waste water can be flushed into storm / sewer systems via drains connected to the street or from facilities such as fire stations, gyms, etc. As a general rule, all waste should be contained until permission is received for disposal.
- **Pollution control (State Environmental Quality, U.S. Environmental Protection Agency, or U.S. Coast Guard)**—The Incident Commander's authority to create a "runoff" situation during an emergency involving life-threatening materials is well established. Creating additional runoff from equipment decontamination is questionable in most cases. Failure to isolate the runoff could result in a citation from regulatory agencies, generate bad publicity, and increase your civil liability.
- **Product specialists**—May be able to provide clean-up recommendations based on their practical knowledge. Some chemical manufacturers are trained and experienced in equipment decon and can be invaluable when their information is correct but very damaging when wrong. Product specialists should be interviewed to establish credibility, as discussed in Chapter 7. They do not have legal authority and cannot assume the responsibility for approving on- or off-site disposal.

EQUIPMENT CLEAN-UP

GENERAL GUIDELINES

Decon of equipment and apparatus can be difficult and very expensive. Liquids can soak into wood and flow into metal cracks and seams or under bolts. Consult product specialists before initiating decon and clean-up operations. A responder with authority should supervise this phase to ensure that proper planning and coordination between all parties takes place.

While decontaminating, avoid direct contact with contaminated equipment. Brooms and sponge mops can be used to apply cleaning agents to equipment. Protective clothing and respiratory protection must be worn unless proven to be unnecessary by technical specialists who have conducted an appropriate analysis of the contaminants.

SMALL AND PORTABLE EQUIPMENT CLEAN-UP

All small to medium-sized equipment, such as monitoring instruments and hand tools, should be decontaminated before leaving the site. The following issues and concerns should be considered:

- **Hand tools**—May be cleaned for reuse or disposed. Cleaning methods include hand cleaning or, more commonly, pressure washing or steam cleaning. You must weigh the cost of the item against the cost of decontamination and the probability that it can be completely cleaned. Wooden and plastic handles on tools should be evaluated to determine if they can be completely decontaminated. The scientific literature referenced at the end of this chapter provides overwhelming evidence that many pesticide-contaminated wooden parts, like shovel handles, cannot be adequately decontaminated. Consult with product specialists for the best advice.
- **Monitoring instruments**—Follow any manufacturer recommendations with respect to decon. If the instrument becomes damaged or disabled during the emergency, most instrument manufacturers will not accept the device for repair unless it has been properly decontaminated.

  If instruments were covered with protective plastic, remove and discard the plastic covering and tape properly.
- **Fire hose**—Should be cleaned following the manufacturer's recommendations. For most materials, detergents will perform adequately. However, strong detergents and cleaning agents may damage the fire hose fibers. The fire hose should be thoroughly rinsed to prevent any fiber weakening. The hose should then be marked and pressure tested before being placed back in service. Severe exposure to some chemicals such as chlorine will result in damage that may require taking the hose out of service (e.g., when the cleaning water turns to hydrochloric acid, $Cl = H_2O = HCL$).

MOTOR VEHICLE AND HEAVY EQUIPMENT CLEAN-UP

If a large number of vehicles need to be decontaminated, consider implementing the following recommendations:

- Establish a decon pad as a primary wash station. The pad may be a concrete slab or a pool liner covered with gravel. Each of these should be bermed or diked with a sump or some form of water recovery system to collect the resulting rinse. Get some engineering help and do it right the first time.
- Completely wash and rinse vehicles several times with an appropriate detergent. Pay particular attention to wheel wells and the chassis. Depending on the nature of the contaminant, it may be necessary to collect all runoff water from the initial gross rinse, particularly if there is contaminated mud and dirt on the underside of the chassis.

- Engines exposed to toxic dusts or vapors should have their air filters replaced. Mechanics sometimes blow dust out of air filters during routine maintenance, exposing themselves unnecessarily to this hazardous dust. Contaminated air filters should be properly disposed.

If vehicles have been exposed to minimal contaminants such as smoke and vapors, they may be decontaminated on site and then driven to an offsite car wash for a second, more thorough washing. Car washes may be suitable if the drainage area is fully contained and all runoff drains into a holding tank. Car washes are not recommended if they drain into the sanitary sewer. Car washes used for decon should be inspected and approved as acceptable in advance.

When a vehicle is exposed to corrosive atmospheres, it should be inspected by a mechanic for possible motor damage. Equipment sprayed with acids should be flushed or washed as soon as possible with a neutralizing agent such as baking soda and then flushed again with rinse water.

POST-INCIDENT DECON CONCERNS

GENERAL CONCERNS

DEBRIEFING

A debriefing should be held for those involved in decontamination and clean-up as soon as practical. (See Chapter 12 for more information on debriefings.) Responders and contractors involved in the operation should be provided with as much information as possible about the delayed health effects of the hazmats.

If necessary, follow-up examinations should be scheduled with medical personnel and exposure records maintained for future reference by the individual's personal physician.

SITE SECURITY AND CUSTODY

In many instances, contaminated materials must remain at the incident scene until they can be removed for offsite cleaning or disposal. In this case, special precautions should be taken:

- Take appropriate security measures. Potential problems that validate the need for security include the potential for vandalism, curious children who may be injured, or folks who simply see an opportunity to add their pile of waste to your pile of waste (this might seem like overkill, but hey, we don't have to make this stuff up! It actually happens and people like you tell us about it). Always ensure that security is provided for hazardous waste and that the proper chain of custody is maintained.
- Make sure appropriate warning signs are posted and labels are attached to containers. Additional lighting may be necessary when the materials remain overnight.
- Make sure containers are properly sealed.

HAZARDOUS WASTE HANDLING AND DISPOSAL

REGULATORY COMPLIANCE

Local, state, and federal regulations require that hazardous wastes be disposed in a specific manner. All personnel involved in the disposal of hazardous waste must be trained in the provisions of the Federal Resource Conservation and Recovery Act (RCRA), any related state or local regulations, and procedures for waste disposal.

HAZARDOUS WASTE CONTAINERS

All containers with materials designated as hazardous waste should be visibly identified with the proper markings. Containers should not be handled or utilized unless the contents are properly identified on the label, per DOT and EPA regulations. Containers used for the accumulation of waste should be labeled so that anyone working in the area will be aware of the contents. Any containers stored outdoors should be labeled in a manner that will withstand the elements.

Only approved chemically compatible containers of sufficient strength should be used for hazardous waste. The containers should be kept covered at all times and arranged so that easy access exists. Care should be taken during all handling to maintain the integrity of the container. Any container stored outdoors must be waterproof.

SUMMARY

Decontamination is the process of making people, equipment, and the environment safe from hazardous materials contaminants. The more you know about how contamination occurs and spreads, the more effective decontamination will be.

The basic concepts of decontamination are relatively simple. If contact with the contaminant can be controlled and minimized, the need for decontamination can be reduced.

The safety and health hazards of the contaminants at any incident will define how complex decon operations will be. The best field decontamination procedures emphasize the need to confine contaminants to a limited area. Establishing a designated decontamination corridor and decontamination area are the first steps in limiting the spread of contaminants.

Regardless of the number of decontamination steps required, decontamination is most effective when it is carried out by a trained Decontamination Team using multiple cleaning stations.

REFERENCES AND SUGGESTED READINGS

Air Force Civil Engineer Support Agency (AFCESA) and PowerTrain, Inc, HAZARDOUS MATERIALS INCIDENT COMMANDER EMERGENCY RESPONSE TRAINING CD-ROM, Tyndall Air Force Base, FL: AFCESA (2002).

Air Force Civil Engineer Support Agency (AFCESA) and PowerTrain, Inc, HAZARDOUS MATERIALS TECHNICIAN EMERGENCY RESPONSE TRAINING CD-ROM, Tyndall Air Force Base, FL: AFCESA (1999).

Black, R. H., "Protecting and Cleaning Hands Contaminated by Synthetic Fallout Under Field Conditions." INDUSTRIAL HYGIENE JOURNAL (April, 1960), pages 162–168.

Bledsoe, Bryan E., D.O., Porter, Robert, S., and Cherry, Richard, ESSENTIALS OF PARAMEDIC CARE, Englewood Cliffs, NJ: Prentice Hall, Inc. (2003).

Borak, Jonathan, M.D., Michael Callan, and William Abbott, HAZARDOUS MATERIALS EXPOSURE: Emergency Response and Patient Care, Englewood Cliffs, NJ: Prentice Hall, Inc. (1991).

Brannigan, Francis, L., "Living with Radiation: Fundamentals No. 1," U.S. Atomic Energy Commission, Washington, DC (Undated).

Callan, Michael, J., "Building a Decon Program." FIRE JOURNAL (November/December, 1993), page 16.

Carroll, Todd, R., "Contamination and Decontamination of Turnout Clothing," Washington, DC: Federal Emergency Management Agency (April 1993).

"Cleaning Pesticide Contaminated Clothing: A Special Report on Safety. PEST CONTROL TECHNOLOGY MAGAZINE (November, 1984) pages 42–44.

Coleman, Ronald, J., "Decontamination: Keeping Personnel Safe." FIRE CHIEF MAGAZINE (June, 1993) pages 43–45.

Dawson, Gaynor, W., and B. W. Mercer, HAZARDOUS WASTE MANAGEMENT, New York, NY: John Wiley and Sons (1986).

Deater, John, William, Medic, USN, "Considerations for Etiological Hazardous Material Exposure." Special report prepared to fulfill the requirements of FS-205, Montgomery College, Rockville, MD (1985).

Docimo, Frank, "Decontamination...Or How Dirty Are We", THE FIREFIGHTER NEWS (June–July 1991).

DOL/HHS Joint Advisory Notice, "Protection Against Occupational Exposure to HBV and HIV," U.S. Department of Labor (DOL) and U.S. Department of Health and Human Services (HHS) Publications (October, 1987).

Easley, J. M., R. E. Laughlin, and K. Schmidt, "Detergents and Water Temperature as Factors in Methyl Parathion Removal from Denim Fabrics." BULLETIN OF ENVIRONMENTAL CONTAMINATION AND TOXICOLOGY (1982), pages 241–244.

Emergency Film Group. DECON TEAM (video tape), Edgartown, MA: Emergency Film Group (2003).

Emergency Film Group, AIDS, HEPATITIS & THE EMERGENCY RESPONDER (video tape), Plymouth, MA: Emergency Film Group (2000).

Emergency Film Group, MEDICAL OPERATIONS AT HAZMAT INCIDENTS (video tape), Plymouth, MA: Emergency Film Group (2000).

Emergency Film Group, TERRORISM: MEDICAL RESPONSE (video tape), Plymouth, MA: Emergency Film Group (2002).

Emergency Film Group, THE EIGHT STEP PROCESS: STEP 7—DECONTAMINATION (video tape), Plymouth, MA: Emergency Film Group (2004).

Federal Emergency Management Agency, Hazardous Materials Workshop for Hospital Staff, Washington, DC, (July, 1992).

Finley, E. L., G. I. Metcalfe, and F. G. McDermott, "Efficacy of Home Laundering in Removal of DDT, Methyl Parathion and Toxaphene Residues From Contaminated Fabrics." BULLETIN OF ENVIRONMENTAL CONTAMINATION AND TOXICOLOGY (1974), pages 268–274.

Finley, E. L., and R. B. Rogillio, "DDT and Methyl Parathion Residues Found in Cotton and Cotton-Polyester Fabrics Worn in Cotton Fields." BULLETIN OF ENVIRONMENTAL CONTAMINATION AND TOXICOLOGY (1962), pages 343–351.

Friedman, William, J., "Decontamination of Synthetic Radioactive Fallout from Intact Human Skin," INDUSTRIAL HYGIENE JOURNAL (February, 1958).

Ganelin, Robert, M.D., Gene Allen Mail, and L. Cueto, Jr., "Hazards of Equipment Contaminated with Parathion." ARCHIVES OF INDUSTRIAL HEALTH (June, 1961) pages 326–328.

Gold, Avram, William A. Burgess, and Edward, V. Clough, "Exposure of Firefighters to Toxic Air Contaminants." AMERICAN INDUSTRIAL HYGIENE ASSOCIATION JOURNAL (July, 1978).

Hawley, Chris, Gregory G. Noll and Michael S. Hildebrand, SPECIAL OPERATIONS FOR TERRORISM AND HAZMAT CRIMES. Chester, MD: Red Hat Publishing, Inc. (2002).

Hildebrand, Michael, S., "Complete Decontamination Procedures for Hazardous Materials: The Nine Step Process, Part-1." FIRE COMMAND (January, 1985), pages 18–21.

Hildebrand, Michael, S., "Complete Decontamination Procedures for Hazardous Materials: The Nine Step Process, Part-2." FIRE COMMAND (February, 1985), pages 38–41.

Hildebrand, Michael, S., HAZMAT RESPONSE TEAM LEAK AND SPILL CONTROL GUIDE, Oklahoma State University, Stillwater, OK (1984).

Hughes, Stephen M., David, W. Berry, and Edward D. Hartin, "What Does A Car Wash and a Baby Have to Do with Hazardous Materials Decon?" An independent technical paper prepared by HazMat-TISI, Columbia, MD, (1992).

Hsu VP, Lukacs and, Handzel, T, Hayslett, J., Harper, S., Hales T, et al., "Opening a Bacillus Anthracis-Containing Envelope, Capitol Hill, Washington, D.C.: The Public Health Response. Emerging Infectious Diseases, Volume 8, No. 10., Center for Disease Control and Prevention, Atlanta, GA. (October 2002).

International Association of Fire Fighters, TRAINING FOR HAZARDOUS MATERIALS RESPONSE: TECHNICIAN, Washington, DC: IAFF (2002).

Isman, Warren, E. John, R. Leahy, Jr. and Roger A. McGary, "Be Prepared for Decontamination at Hazardous Materials Incidents." FIRE ENGINEERING (July, 1982). Pages 12–17.

Kampmier, Craig. "Decon Design," NFPA JOURNAL (March/April, 2000), pages 59–61.

LeMaster, Frank, "Why Protective Clothing Must Be Cleaned." THE VOICE, (August/September 1993), pages 19–20.

Lillie, T. H., R. E. Hampson, Y. A. Nishioka, and M. A. Hamlin, "Effectiveness of Detergent and Detergent Plus Bleach for Decontaminating Pesticide Applicator Clothing." BULLETIN OF ENVIRONMENTAL CONTAMINATION AND TOXICOLOGY (1982), pages 89–94.

Limmer, Daniel, et. al. EMERGENCY CARE (9th edition), Englewood Cliffs, NJ: Prentice Hall, Inc. (2001).

Macintyre, Anthony G., MD, et. al. "Weapons of Mass Destruction Events With Contaminated Casualties—Effective Planning for Health Care Facilities." JOURNAL OF THE AMERCIAN MEDICAL ASSOCIATION (January 12, 2000), pages 242–249.

McGary, Roger, A., "Disinfection of SCBA." THE VOICE (October, 1993), pages 26–28.

Molino, Louis N., Sr. "The Big One: Proven Methods for the Management and Mass Decontamination of a Crowd." HOMELAND FIRST RESPONSE (July/August, 2003), pages 14–21.

National Institute for Occupational Safety and Health,. "Report to Congress on Workers' Home Contamination Study Conducted Under The Workers' Family Protection Act (29 U.S.C. 671a)", Publication No. 95-123, Department of Health and Human Services, Washington, D.C. (September, 1995).

National Fire Protection Association, NFPA 1581—FIRE DEPARTMENT INFECTION CONTROL PROGRAM, Quincy, MA: National Fire Protection Association (2000).

National Fire Protection Association, NFPA 1851—STANDARD ON SELECTION, CARE, AND MAINTENANCE OF FIREFIGHTING STRUCTURAL ENSEMBLES, Quincy, MA: National Fire Protection Association (2001).

National Fire Protection Association, NFPA 471, RECOMMENDED PRACTICE FOR RESPONDING TO HAZARDOUS MATERIALS INCIDENTS, Quincy, MA: National Fire Protection Association (2002).

National Fire Protection Association, NFPA 472, STANDARD FOR PROFESSIONAL COMPETENCE OF RESPONDERS TO HAZARDOUS MATERIALS INCIDENTS, Quincy, MA: National Fire Protection Association (2002).

National Fire Protection Association, NFPA 473, STANDARD FOR COMPETENCIES FOR EMS PERSONNEL RESPONDING TO MATERIALS INCIDENTS, Quincy, MA: National Fire Protection Association (2002).

Olson, Kent R., M.D., POISONING AND DRUG OVERDOSE (San Francisco Bay Area Regional Poison Control Center), Norwalk, CT and San Mateo, CA: Appleton and Lange (1990).

OSHA Instruction CPL 2–2.44B, "Enforcement Procedures for Occupational Exposure to HBV and HIV," U.S. Department of Labor (DOL) and U.S. Department of Health and Human Services (HHS) Publications (February 27, 1990).

Perkins, John, J., PRINCIPLES AND METHODS OF STERILIZATION IN HEALTH SCIENCES (2nd edition), Chicago, IL: Charles Thomas Publishing Co. (1980).

Ronk, Richard, and Mary Kay White, "Hydrogen Sulfide and the Probabilities of Inhalation Through a Tympanic Membrane Defect." JOURNAL OF OCCUPATIONAL MEDICINE (May, 1985), pages 337–340.

Rudner, Glen D. "First Responder Considerations for Decontamination at Mass Casualty Incidents." Student handout material developed for Virginia Department of Emergency Services (January, 2003).

Stutz, Douglas R. and Stanley J. Janusz, HAZARDOUS MATERIALS INJURIES: A Handbook for Pre-Hospital Care (2nd edition), Beltsville, MD: Bradford Communications Corp. (1988).

Teller, Robert, "Developing Proper Decontamination Procedures for Emergency Response." ECON MAGAZINE (March 1993), pages 32–33.

United Kingdom Home Office, THE DECONTAMINATION OF PEOPLE EXPOSED TO CHEMICAL, BIOLOGICAL, RADIOLOGICAL OR NUCLEAR (CBRN) SUBSTANCES OR MATERIAL—STRATEGIC NATIONAL GUIDANCE (1st Edition), London, England: Home Office (February, 2003).

U.S. Army Medical Research Institute of Chemical Defense, MEDICAL MANAGEMENT OF CHEMICAL CASUALTIES (3rd Edition), Aberdeen Proving Ground, MD: USAMIC—Chemical Casualty Care Office (July, 2000).

U.S. Army Medical Research Institute of Infectious Diseases (USAMRID), MEDICAL MANAGEMENT OF BIOLOGICAL CASUALTIES (4th Edition), Fort Detrick, MD: USAMRID (February, 2001).

U.S. Army Soldier and Biological Chemical Command (SBCCOM), GUIDELINES FOR MASS CASULATY DECONTAMINATION DURING A TERRORIST CHEMICAL AGENT ATTACK, Aberdeen Proving Ground, MD (January, 2001).

U.S. Centers for Disease Control (CDC) Publications, A CURRICULUM GUIDE FOR PUBLIC SAFETY EMERGENCY RESPONSE WORKERS: PREVENTION OF HUMAN IMMUNODEFICIENCY VIRUS AND HEPATITIS B VIRUS (February, 1989).

U. S. Centers for Disease Control (CDC) Publications, GUIDELINE FOR HANDWASHING AND HOSPITAL ENVIRONMENTAL CONTROL (1985).

U. S. Centers for Disease Control (CDC) Publications, GUIDELINE FOR PREVENTION OF HIV AND HBV EXPOSURE TO HEALTH CARE AND PUBLIC SAFETY WORKERS (February, 1989).

U. S. Environmental Protection Agency, GUIDE FOR INFECTIOUS WASTE MANAGEMENT (1986).

Walter, Frank G., Raymond Klein and Richard G. Thomas, ADVANCED HAZMAT LIFE SUPPORT (AHLS) COURSE—Provider Manual (3rd Edition), Tucson, AZ: University of Arizona Emergency Medicine Research Center (2003).

CHAPTER 12

TERMINATING THE INCIDENT

Terminating The Incident

OBJECTIVES

1. List the five basic activities that should be completed as part of the termination process.

2. List and describe at least three criteria for terminating the emergency phase of a hazardous materials incident.

3. Identify the steps to be taken to transfer command/control of the incident for post-emergency response operations [NFPA 472-7.6.1].

4. Given a simulated hazardous materials incident, the Hazardous Materials Technician shall [NFPA 472-6.6.1 and 7.6.2]:
   a. Describe the three components of an effective debriefing.
   b. Describe the key topics of an effective debriefing.
   c. Describe when a debriefing should take place.
   d. Describe who should be involved in a debriefing.
   e. Identify the procedures for conducting the incident debriefing.

5. Given a simulated hazardous materials incident, the Hazardous Materials Technician shall [NFPA 472-6.6.2 and 7.6.3]:
   a. Describe the three components of an effective critique
   b. Describe who should be involved in a critique.
   c. Describe why an effective critique is necessary after a hazardous materials incident.
   d. Describe which written documents should be prepared as a result of a critique.

6. Given a simulated hazardous materials incident, the Hazardous Materials Technician shall complete the reporting and documentation requirements consistent with the local Emergency Response Plan and standard operating procedures, and shall meet the following requirements [NFPA 472-6.6.3 and 7.6.4]:
   a. Describe the importance of documentation for a hazardous materials incident, including training records, personnel exposure records, incident reports, and critique reports.
   b. Identify the steps in keeping an activity log and exposure records.
   c. Identify the steps to be taken in compiling incident reports that meet federal, state, local, and organizational requirements.
   d. Identify the requirements for compiling hot zone entry and exit logs.
   e. Identify the requirements for compiling personal protective equipment logs.
   f. Identify the requirements for filing documents and maintaining records.
   g. Identify the procedures required for legal documentation and chain of custody/continuity.

ABBREVIATIONS AND ACRONYMS

IC	Incident Commander
NFIRS	National Fire Incident Reporting System
PIA	Post Incident Analysis
PERO	Post-Emergency Response Operations
PPA	Public Protective Actions
SOP	Standard Operating Procedures

INTRODUCTION

Termination is the final step in the Eight Step Incident Management Process©. It represents the transition between the termination of the emergency phase and the initiation of clean-up, restoration and recovery operations. It also is the phase where responders document incident response operations, including the problem, agencies involved, hazards and risks encountered, safety procedures, site operations, and lessons learned.

Terminating the incident usually consists of five distinct activities:

1. Termination of the emergency phase of the incident
2. Transfer of on-scene command from the Incident Commander of the emergency phase to the individual responsible for managing and coordinating Post-Emergency Response Operations (PERO)
3. Incident Debriefing
4. Post-Incident Analysis
5. Critique

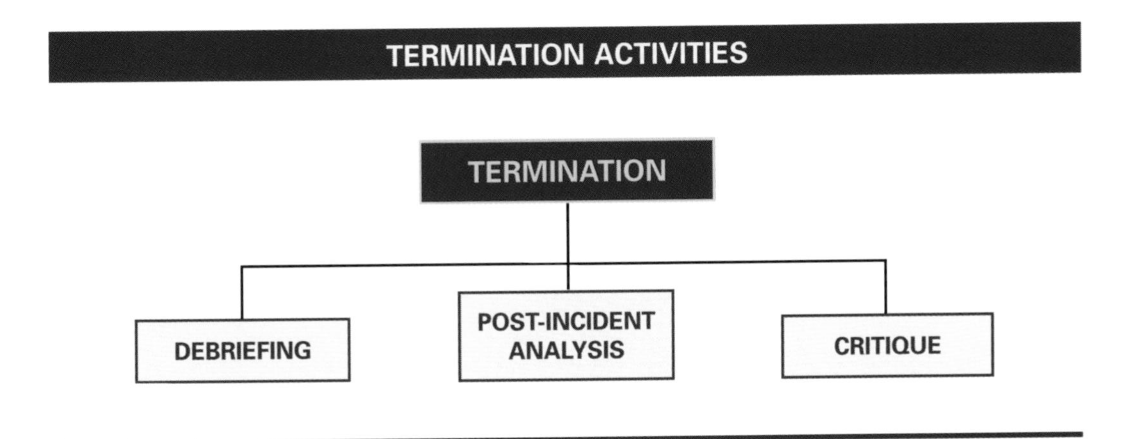

Figure 12.1 Termination activities involve three principal phases.

DECLARING THE INCIDENT TERMINATED

While it may sound a little silly to declare an emergency "terminated," it is an important part of the hazmat response. Unlike fire emergencies, where it is usually obvious that the fire is out (e.g., there is a large smoking hole in the ground where the building used to be), hazardous materials incidents sometimes slowly creep from the emergency phase to the restoration and recovery phase. This type of "mission creep" can lead to an unsafe incident scene. Emergency responders and support personnel sometimes get mixed signals concerning whether there are actually hazards present at this phase of the operation. If spills have been controlled and personnel are standing around waiting for product transfer operations to begin, boredom and complacency sometimes set in. Control of the perimeter becomes relaxed, protective clothing comes off, and the incident scene becomes unsafe or dangerous again.

As the Incident Commander, you should be satisfied with answers to the following questions before the incident is declared terminated:

- Is the incident scene dangerous? (See Chapter 7—Hazard and Risk Assessment.) If the incident scene is still dangerous, this is a strong indicator that you should still be in the emergency response mode and that the emergency response agency should still be in command.
- Is the incident scene unsafe? In some instances, responders may terminate the emergency response phase and transfer command to an environmental agency or contractor (i.e., the PERO Incident Commander) with some hazards remaining providing they have been "stabilized" and are no longer a threat to public safety. However, if the contractor still needs emergency response resources at the incident scene to deal with these hazards, then you should probably still be in the emergency response mode. Retain incident control until the hazards are mitigated to the point that emergency response resources are not needed.
- Is the incident scene safe? Safe means totally safe. If you would feel completely comfortable having your mother sitting on a lawn chair eating lunch where the emergency used to be, then you probably have a safe incident. But guessing is not a good indicator of what is safe. Follow proper standard operating procedures and use instrumentation to evaluate the hazards. The Safety Officer and Hazmat Group Leader should help make this call. Returning to the scene of an incident that you just left is a pretty bad feeling, especially if someone has been injured.

Remember that the requirements of the OSHA Hazwoper Regulation (29 CFR 1910 .120) clearly delineate between "emergency phase" and "post-emergency response operations (PERO)."

TRANSFERRING RESPONSIBILITY OF THE INCIDENT SCENE

When the decision has been made to terminate the emergency response phase and additional work is still required at the scene for restoration and recovery, the Incident Commander should meet with the senior representatives from the agencies or contractors taking over to formally hand off the incident scene. It is unprofessional to pick up and leave an incident scene without briefing the personnel who are responsible for carrying out the next phase of the operation. There may also be liability issues for you and your organization if you do not inform these people of the hazardous materials remaining at the incident scene. Just because emergency response agencies depart the scene doesn't mean that the need for a post-emergency response incident command organization doesn't exist! Although the players may be different, statuatory or regulatory authorities will clearly lay out who are the key player(s).

The IC should make it clear that the emergency response phase is being terminated and then formally transfer command to the PERO Incident Commander. The transfer briefing should cover:

- The nature of the emergency
- Actions taken to stabilize and resolve the emergency
- Names of hazardous materials involved
- Hazards and risks that were mitigated and those that still exist
- Safety procedures
- Relevant documentation and points of contact

Figure 12.2 Contractors should be briefed by the Incident Commander before responders turn over the scene.

If the incident has legal or criminal implications involving potential documentation or evidence, it is critical that chain-of-custody procedures be followed. The IC should ensure that response operations are fully coordinated with law enforcement or investigation agencies involved.

Finally, before leaving the scene, the IC should document the time of departure, names, companies, and contact information for the personnel assuming control of the scene. This information should be placed in the official log. Be sure to leave your contact information with the group that is now taking over responsibility for the event.

INCIDENT DEBRIEFING

The purpose of the incident debriefing in the field is to send accurate information concerning the hazards and risks involved directly to the people who may have been exposed, contaminated, or in some way affected by the response. The debriefing is not a critique of the incident.

An effective debriefing should:

- Inform responders exactly what hazmats they were (potentially) exposed to and their signs and symptoms.
- Identify damaged equipment requiring servicing, replacement, or repair.
- Identify equipment or expended supplies that will require specialized decontamination or disposal.
- Identify unsafe site conditions that will impact the clean-up and recovery phase. Owners and contractors should be formally briefed on these problems before responsibility for the site is transferred.

- Assign information gathering responsibilities for a post-incident analysis and critique.
- Assess the need for a Critical Incident Stress Debriefing.
- Assign a point of contact for incident related issues (concern for delayed symptoms).

Debriefings should begin as soon as the emergency phase of the operation is completed. Ideally, this should be before any responders leave the scene, and it should include the HMRT, emergency response officers, and other key players, such as Information Officers and agency representatives, who the Incident Commander determines have a need to know. On larger incidents, these representatives will return to their personnel and pass on the essential information, including, where applicable, who to contact for more information.

Figure 12.3 Hazmat team debriefing at incident scene.

CONDUCTING DEBRIEFINGS

Debriefings should be conducted in areas that are free from distractions. In poor environmental conditions, such as extremely cold or hot weather or environments with loud ambient noise, the debriefing should be conducted in a nearby building or vehicle.

The debriefing should be conducted by one person acting as the leader. The Incident Commander may not be the best facilitator for the debriefing because special knowledge of the hazards may be required. The IC should at least be present to summarize feelings about the entire incident and to reinforce its positive aspects.

Debriefings longer than 15 minutes are probably too long. The intent is to briefly review the incident and send everyone home, not analyze every action of every player. If more interaction is needed on a specific subject or operation, continue it after the debriefing or schedule it for another day.

Debriefings should cover certain subjects, in the following order:

1. **Health information**

   Describe what you think personnel may have been exposed to and the signs and symptoms of exposure. Some substances may not reveal signs and symptoms of exposure for 24 to 48 hours. When appropriate, cover responsibilities for follow-up evaluations and complete health exposure forms.

2. **Equipment and apparatus exposure review**

   Ensure that equipment and apparatus unfit for service is clearly "red tagged" for repair and plans are made for special cleaning or equipment disposal. Firefighting gear and personal clothing should be laundered upon return to the station. Someone must be delegated the responsibility for assuring that contaminated garments are properly laundered or properly disposed.

3. **A follow-up contact person**

   Ensure that anyone involved after the release of responders from the scene, such as clean-up contractors and investigators, have access to a single information source that can share the needed data. This contact person should also be responsible for collecting and maintaining all incident documents until they are delivered to the appropriate investigator or critique leader.

4. **Problems requiring immediate action**

   Equipment failures, safety, major personnel problems, or potential legal issues should be quickly reviewed on scene. If it is not crucial, save it for the critique.

5. **Thank you**

   Most hazmat incidents are hard work and often test personal endurance and everyone's sense of humor. Reinforcement of the things that went right, a commitment to work on the problems uncovered through the critique process, and a thank you from the boss go a long way. Never end on a sour note.

POST INCIDENT ANALYSIS

The Post Incident Analysis (PIA) is the reconstruction of the incident to establish a clear picture of the events that took place during the emergency. It is conducted to:

- Assure that the incident has been properly documented and reported to the right regulatory agencies.
- Determine the level of financial responsibility (i.e., who pays?).
- Establish a clear picture of the emergency response for further study. Focus on the General Hazardous Materials Behavior Model. (see Chapter 8.)
- Provide a foundation for the development of formal investigations, which are usually conducted to establish the probable cause of the accident for administrative, civil, or criminal proceedings.

There are many agencies and individuals who have a legitimate need for information about significant hazmat incidents. They may include manufacturing, shipping and carrier representatives, insurance companies, government agencies, and even citizens groups. A formal PIA is one method for coordinating the release of factual information to those who have a need to know.

THE POST INCIDENT ANALYSIS PROCESS

The Post Incident Analysis begins with the designation of one person (or office) to collect information about the response. This person is usually appointed during the on-scene debriefing. The PIA Coordinator should have the authority to determine who will have access to information. This method guarantees that sensitive or unverified information (e.g., injured personnel) is not released to the wrong organization or in an untimely manner.

The PIA should focus on six key topics:

- **Command and control**. Was the Incident Management System established and was the emergency response organized according to the existing Emergency Response Plan and/or SOPs? Did information pass from Section personnel to the Incident Commander or through appropriate channels? Were response objectives clearly communicated to field personnel at the task level?

- **Tactical operations**. Were tactical operations completed in a safe and effective manner? What worked? What did not? Were tactical operations conducted in a timely and coordinated fashion? Do revisions need to be made to tactical procedures or worksheets?
- **Resources**. Were resources adequate to conduct the response effort? Are improvements needed to equipment or facilities? Were mutual aid agreements implemented effectively?
- **Support Services**. Were support services adequate and provided in a timely manner? What is needed to increase the provision of support to the necessary level?
- **Plans and Procedures**. Were the Emergency Response Plan and associated Tactical Procedures current? Did they adequately cover notification, assessment, response, recovery, and termination? Were roles and assignments clearly defined? How will plans and procedures be upgraded to reflect the "lessons learned"?
- **Training**. Did this event highlight the need for additional basic or advanced training? Multi-agency training? Were personnel trained adequately for their assignments?

The PIA should attempt to gather factual information concerning the response as soon as possible. The longer the delay in gathering information, the less likely it will be accurate and available. Suggested sources of information include the following:

- Incident reporting forms.
- Activity logs, entry logs and personnel exposure logs.
- Notes and audio recordings from the Incident Command Post.
- Photographs, videos, maps, diagrams, and sketches. If photographs or videotapes are taken, copies should also be obtained for the incident file. If future litigation is a concern, a photo/video log should be made recording the following information:
  - Time, date, location, direction, and weather conditions.
  - Description or identification of subject and relevance of photographs or video.
  - Digital photo disk numbers or the sequential number of photos and film roll number(s). Note that digital photos may not be admissible in some court cases.
  - Camera type and serial number.
  - Name, telephone number, and Social Security number of the photographer.
- Results of air monitoring and sampling, including types of instruments used and calibration information.
- Incident command organizational charts, notes, and completed checklists.
- Business cards or notes from agency, organization, or company representatives.
- Tape recordings from the 911/communications center(s) involved.
- Videotape recordings made by the media. Obtain the unshown, unedited video taken by the media within the first 24 hours after the incident. Only the video used in the broadcast is archived.

- Photographs, film, and videotape taken by responders or bystanders.
- Interviews of witnesses conducted by investigators that may help establish where responders were located at the incident scene.
- Responder interviews.
- Verification of shipping documents or Material Safety Data Sheets (MSDS).
- Owner/operator information.
- Chemical hazard information from checklists, computer printouts, and so on.
- Lists of apparatus, personnel, and equipment on scene.

As soon as practical, construct a brief chronological review of who did what, when, and where during the incident. A simple timeline placing the key players at specific locations at different times is a good start. Cooperation between the PIA Coordinator and other official investigators will save time and combine resources to reconstruct the incident completely.

Once all available data have been assembled and a rough draft report developed, the entire package should be reviewed by key responders to verify that the available facts are arranged properly and actually took place.

Incident Reporting. Each emergency response organization has its own unique requirements for recording and reporting hazmat incidents. These requirements may be self-imposed as administrative and management controls or may be mandatory under federal or state laws.

For private shippers, carriers, and manufacturers, the regulatory reporting requirements for leaks, spills, and other releases of specified chemicals into the environment are significant. These include the following:

- Section 304 of the Superfund Amendments and Reauthorization Act (SARA, Title III)
- Section 103 of the Comprehensive Environmental Response, Compensation and Liability Act (CERCLA)
- 40 CFR Part 110—Discharge of Oil
- 40 CFR Part 112—Oil Pollution Prevention
- 40 CFR Part 302—Reportable Quantities
- Any additional local, state, or regional reporting requirements

Under CERCLA, the responsible party must report to the National Response Center (NRC) any spill or release of a specified hazardous substance in an amount equal to or greater than the reportable quantity (RQ) specified by EPA. In addition, SARA, Title III requires that releases be reported immediately to the National Response Center and the Local Emergency Planning Committee (LEPC).

Many industrial organizations have developed initial incident reporting forms as a way to ensure that key corporate and regulatory reporting requirements are correctly documented. Chapter 1 provides an overview of some of the more important federal laws that have reporting provisions.

Most major public fire departments participate in the National Fire Incident Reporting System. NFIRS is a product of the National Fire Information Council and is sponsored by the U.S. Fire Administration. This system includes a special category for recording hazardous materials incidents.

INITIAL INCIDENT ASSESSMENT AND NOTIFICATION SHEET

ACTIONS

Who have you notified? ______________________

External ______________________

What actions have you taken so far? ______________________

CASUALTIES

How many injured employees? How serious? ______________________

How many injured civilians? How serious? ______________________

RELEASE

Has there been a release? ______________________

What was released? ______________________

Estimated size? ______________________

Worst-case scenario? ______________________

____

Has the release entered drains or waterways? ______________________

Describe the flow ______________________

OTHER CONCERNS

Has there been a fire? ______________________

Confined? ______________________

Potential for further damage ______________________

____

Additional facilities at risk ______________________

Impact on facility operations ______________________

____

Has the incident impacted outside the fence? ______________________

Describe ______________________

News coverage thus far? RADIO / TV / PRINT? ______________________

From where? ______________________

Expected duration of the incident?

Hours ______________________

Days ______________________

Weeks ______________________

Figure 12.4 Initial Incident Assessment and Notification Sheet.

1. Date:	**2.** Time ❑ AM ❑ PM	**3.** Duration of Release:
4. Location:		
5. Identity of released chemical (or its components):		
6. Quantity of material released:		
7. Did any released material leave the company property? ❑ Yes ❑ No If yes, answer a) and b) below. a. How? b. Where?		
8. Medium or media into which release occurred: ❑ Air ❑ Water ❑ Land Description:		
9. Any known or anticipated acute or chronic risks associated with the release and, where appropriate, advice regarding medical attention necessary for exposed individuals?		
10. Any precautions to take as a result of the release, including evacuation:		
11. Name(s) and telephone number(s) of the person(s) to be contacted for information:		
a.	d.	
b.	e.	
c.	f.	
12. Steps being taken to clean up release/spill:		
13. Conclusion: Incident ❑ Does ❑ Does Not* require Section 304 reporting. (*—Accumulate record to Section 313 file)		
14. Notification to Authorities if 304 Release	Contact Name	Time
1) National Response Center (800-424-8802)		
2) County LEPC		
3) EPA		
4) Local Police		
5) County/State Emergency Preparedness		

Figure 12.5 SARA/CERCLA review for evaluating hazardous materials release into the environment.

CRITIQUE

Many injuries and fatalities have been prevented as a result of lessons learned through the critique process. An effective critique program must be supported by top management and is the single most important way for an organization to self-improve over time. OSHA requires that a critique be conducted of every hazardous materials emergency response.

The primary purpose of a critique is to develop recommendations for improving the emergency response system rather than to find fault with the performance of individuals. A good critique promotes:

- Emergency response operations that are system-dependent rather than people-dependent organizations
- A willingness to cooperate through teamwork
- Improvement of safe operating procedures
- Sharing information among emergency response organizations

The critique leader is the crucial player in making the critique session a positive learning experience. A critique leader can be anyone who is comfortable and effective working in front of a group. The critique leader need not necessarily be part of the emergency response team. For example, an organization may select one or two respected and credible individuals to act as neutral parties to critique the larger, more sensitive incidents. Examples of outsiders who may be effective critique leaders include Fire Science or Law Enforcement faculty from a local community college, Regional Fire Training Coordinators, or professionals in your community who are experienced with people skills.

Although every organization has a tendency to develop its own critique style, never use a critique to assign blame (public meetings are the worst time to discipline personnel). Do use it as a valuable learning experience (everyone came to the incident with good intentions).

The critique leader should:

- Control the critique. Introduce the players and procedures, keep the critique moving, and end it on schedule. Critiques lasting longer than 60 to 90 minutes quickly lose their effectiveness, and the quality of the discussion goes down hill.
- Ensure that direct questions receive direct answers.
- Ensure that all participants play by the critique rules.
- Ensure that individual observations are shared with the group.

Publish an agenda for the critique and include the order of presentation so that the participants know when their turn is coming.

When you know going into a critique that a major confrontation between the players is anticipated, set up a meeting before the critique with only the affected parties attending. These types of smaller meetings can be useful to "hash out" the sticky points of a conflict and diffuse the problem. Then during the critique it can be reported that a meeting was held to fix the problem or that a meeting was held to begin working on the problem.

At the end of the critique the leader should sum up some of the positive things learned from the critique and thank everyone for their response to the event and for their involvement in making the next response even more successful.

The following format is a recommended for critiquing large-scale emergency responses:

Participant-level critique. After explaining the rules for the critique, the critique leader calls on each key player to make an individual statement relevant to his or her on-scene activities and what he or she feels are the major issues. Depending on time, more detail may be added. There should be no interruptions during this phase. For obvious time reasons, the leader should limit this phase of the process to two or three minutes per person.

Operations-level critique. After determining a feel for the group, the leader moves on to a structured review of emergency operations. Through a spokesperson, each section/sector presents an activity summary of challenges encountered, unanticipated events, and lessons learned. Each presentation should not exceed five minutes.

Group-level critique. At the end of the operations level critique, the leader moves the meeting into a wider and more open forum. The facilitator encourages discussion, reinforces constructive comments, and records important points.

As the critique draws to a close, the leader should summarize the more important observations and conclusions revealed by the participants. For large groups, the critique leader should have one or two assistants who act as recording secretaries throughout the session. These notes become the beginning point for writing a post-critique report.

Critique reports should be short and to the point. Simply describe what happened in one or two pages and move on to the lessons learned. If recommendations for improvement are appropriate, they should be listed at the end of the report.

When larger incidents are involved or injuries have occurred, formal critique reports should be circulated so that everyone in the response system can share the lessons learned. Other forums that may be appropriate to share lessons learned include trade magazines and technical conferences.

Figure 12.6 Critique being conducted.

It is important that lessons learned that have been identified through the critique process be converted into changes and improvements to the emergency response system. Set aside a quarterly review date to make sure that action items have been addressed. Management must assign someone in the organization to track the implementation or recommendations, otherwise, they tend not to get implemented and the opportunity to improve the system is lost. Failure to change after a bad incident also sends the wrong message to the people in the field (management doesn't care and doesn't want to change).

OPERATIONS SECURITY (OPSEC) ISSUES

In today's environment we need to be concerned about how criminals and terrorists may use the information that we generate in written critiques or After Action Reports. The good news about the critique process is that, if done properly, it improves the emergency response system. The bad news is that in our American democratic society, if we share everything that we learned with everyone in the world, this information can actually teach the bad guys (criminals and terrorists) our vulnerabilities and weaknesses. Think about it. If an adversary understands our weaknesses by reading our critique reports, does this give them a tactical advantage to use this information to hurt us and the community we are sworn to protect? Of course it does! So where do we draw the line on how much information should be released?

When we study the lessons learned from major hazmat and WMD incidents, we have an obligation to share what we know. But we don't need to share everything we know with everyone. Does it make sense to place the complete report of the lessons we learned from a major WMD incident on the Internet so that anyone in the world can access it and download it for evaluation? At the other end of the spectrum is sharing nothing we know with anyone.

A reasonable approach to this problem is to limit information concerning vulnerabilities and weaknesses learned from critiquing the incident to the people who really need to know it (e.g., public safety agencies, investigators, accident review boards). Complete reports can be circulated through secure e-mail, your agency's secure Intranet (restricted to authorized personnel), or distribution of hard copies using controlled and numbered copies.

Sensitive information that can be used by criminals and terrorists to hurt the public or target first responders should be edited from critique reports that are intended for the general public. These edited reports can then be made available to the general public through the department's official Web site.

To learn more on Operations Security see *Special Operations for Terrorism and HazMat Crimes* (Red Hat Publishing, 2002) by Hawley, Noll, and Hildebrand (Chapter 3), or go to the Interagency Operations Security Support Staff Web site at www.ioss.gov for more information on OPSEC.

LIABILITY ISSUES

Many managers express concerns that the critique process can expose weaknesses that can be exploited to build a liability case against emergency responders. There is no question that:

- Civil suits and regulatory citations against government and industry for emergency response operations are a problem.
- Chances of losing the suit or citation increase if you do not meet a standard of care. See Chapter 1 for a discussion of standard of care.
- Fines and award sizes have increased.

Nevertheless, organizations must balance the potential negatives against the benefits that are gained through the critique process. Remember—the reason for doing the critique in the first place is to improve your operations. An organization that does not improve, doesn't meet the standard of care, and performs poorly makes itself a target for lawsuits.

There are five primary reasons for liability problems in emergency response work. They are worth considering as a case for actually building a strong critique program.

1. **Problems with planning**. Plans and procedures are poorly written, out-of-date, and unrealistic. In addition, what is written in the SOP is not followed in the field.
2. **Problems with training**. No training is conducted, training evolutions reinforce unsafe practices, and the training is undocumented.
3. **Problems with identification of hazards**. Hazards were not identified, were not prioritized, or were ignored even though they were known to exist.
4. **Problems with duty to warn**. Warnings concerning safety hazards and design limitations of equipment were not given or were improper.
5. **Problems with negligent operations**. Equipment was not employed properly, plans and procedures were not followed, and equipment was not maintained to an acceptable standard.

A little common sense goes a long way when developing a critique policy. If the critique and follow-up report are properly written in the first place, there will be little useful information that attorneys may use against emergency responders. While the critique report is a critical item, recognize that the legal process of discovery can also reveal organizational shortcomings in other ways. For example, official investigations conducted by OSHA, the fire marshal, or an insurance company can be used to build a case against you or your organization.

Don't let attorneys make management decisions for your organization. Don't let the guy that doesn't have to risk his or her life in the field talk you out of using the critique process to improve the emergency response system. A quality emergency response system is your best liability defense. If you run an operation that meets national standards, you have gone a long way toward reducing your liability exposure.

SUMMARY

Termination is the final step in the Eight Step Incident Management Process©. It is important that every hazmat incident be formally terminated following a formal procedure.

Terminating the incident usually consists of five distinct activities: (1) declaring that the incident is "Terminated" either by radio or in a face-to-face meeting, (2) officially transferring responsibility of the incident scene to another agency or contractor, (3) incident debriefing, (4) post-incident analysis, and (5) critique.

The incident debriefing is done at the incident scene, lasts less than 15 minutes, and focuses on safety and health exposure issues.

The post-incident analysis is conducted after the incident is over and is a focused effort to gather information concerning what actually happened, why it happened, and who the responsible parties are. It also provides a record of resources and events, which may affect the public health, financial resources, and political well-being of a community. Lastly, it provides the data that may be required to comply with local, state, and federal laws and to defend against potential lawsuits.

The critique is usually conducted several days after the incident is over. It is designed to emphasize successful, as well as unsuccessful, operations and to improve the emergency response system. To be successful, management must support the critique process and action items must be tracked to ensure they are implemented.

The critique process can reveal critical information about our weaknesses and vulnerabilities that can be exploited by criminals and terrorists. Some consideration must be given to who has access to critique reports and why they need it.

A strong critique program that is designed to improve the emergency response system reduces potential liability by helping to ensure that the organization meets a standard of care. Critique reports should be written so that they do not become a paper trail for attorneys who may want to use the lessons learned from the incident to support a frivolous lawsuit. The threat of potential lawsuits is a poor excuse for not having a strong critique program.

REFERENCES AND SUGGESTED READINGS

Benner, Ludwig, Jr., and Hildebrand, Michael, S., HAZARDOUS MATERIALS MANAGEMENT SYSTEMS: THE M.A.P.S METHOD, Prentice-Hall, (1981).

Emergency Management Institute, LIABILITY ISSUES IN EMERGENCY MANAGEMENT, Federal Emergency Management Agency, Washington, D.C. (April, 1992).

Energy Research and Development Administration, ACCIDENT/INCIDENT INVESTIGATION MANUAL: The Management and Risk Oversight and Risk Tree, ERDA-76-20, Washington, D.C. (July 1975).

Hawley, Chris, Noll, Gregory, and Hildebrand, Michael, S., SPECIAL OPERATIONS FOR TERRORISM AND HAZMAT CRIMES, Red Hat Publishing, Chester, MD (2002).

Hildebrand, Michael, S., "An Effective Critique Program." FIRE CHIEF, Volume 27, No.4 (April, 1983).

International Fire Service Training Association, PHOTOGRAPHY FOR THE FIRE SERVICE, IFSTA-204, Fire Protection Publications, Oklahoma State University (1976).

Kutner, Kenneth, C., "CISD: Critical To Health and Safety." FIRE ENGINEERING MAGAZINE (April, 1992).

Lorenzo, D. E., "A Manager's Guide to Reducing Human Errors." Washington, D.C., Chemical Manufacturers Association (July, 1980).

Murley, Thomas, E., "Developing a Safety Culture." A technical paper presented at the Nuclear Regulatory Commission, Regulatory Information Conference, Washington, D.C. (April, 1989).

GREAT JOB ON THAT RESCUE LAST NIGHT, KID!
THANKS O.T.! —SEE YOU NEXT SHIFT ... STAY SAFE!
... I THINK THE KID JUST MIGHT MAKE IT...
O.T.'s GOING TO BE O.K...

GLOSSARY OF TERMS

A

Absorbent Material. A material designed to pick up and hold liquid hazardous material to prevent contamination spread. Materials include sawdust, clays, charcoal and polyolefin-type fibers.

Absorption. 1) The process of absorbing or "picking up" a liquid hazardous material to prevent enlargement of the contaminated area. Common physical method for spill control and decontamination. 2) Movement of a toxicant into the circulatory system by oral, dermal, or inhalation exposure.

ACGIH. (See American Conference of Governmental Industrial Hygienists.)

Acids. Compound that forms hydrogen ions in water. These compounds have a pH < 7, and acidic aqueous solutions will turn litmus paper red. Materials with a pH < 2.0 are considered a strong acid.

Activity. The number of radioactive atoms that will decay and emit radiation in 1 second of time. Measured in curies (1 curie = 37 billion disintegrations per second), although it is usually expressed in either millicuries or microcuries. Activity indicates how much radioactivity is present and not how much material is present.

Acute Effects. Results from a single dose or exposure to a material. Signs and symptoms may be immediate or may not be evident for 24 to 72 hours after the exposure.

Acute Emergency Exposure Guidelines (AEGL). Developed by the National Research Council's Committee on Toxicology to provide uniform exposure guidelines for the general public. The Committee's objective is to define AEGLs for the 300+ EHS materials listed in SARA, Title III. AEGLs represent an exposure value specifically developed for a single short-term exposure, such as those encountered in the emergency response community. Three tiers of AEGLs have been developed covering five exposure periods: 10 minutes, 30 minutes,1 hour, 4 hours, and 8 hours.

Acute Exposures. An immediate exposure, such as a single dose that might occur during an emergency response.

Administration/Finance Section. Responsible for all costs and financial actions of the incident. Includes the Time Unit, Procurement Unit, Compensation/Claims Unit, and the Cost Unit.

Adsorption. Process of adhering to a surface. Common physical method of spill control and decontamination.

Aerosols. Liquid droplets, or solid particles dispersed in air, that are of fine enough particle size (0.01 to 100 microns) to remain dispersed for a period of time.

Agency for Toxic Substances and Disease Registry (ATSDR). An organization within the Center for Disease Control, it is the lead federal public health agency for hazmat incidents and operates a 24-hour emergency number for providing advice on health issues involving hazmat releases.

Air Monitoring. To measure, record, and/or detect contaminants in ambient air.

Air Purifying Respirators (APR). Respirators or filtration devices that remove particulate matter, gases or vapors from the atmosphere. These devices range from full-face piece, dual cartridge masks with eye protection, to half-mask, face piece mounted cartridges. They are intended for use only in atmospheres where the chemical hazards and concentrations are known.

ALS. Advanced life support emergency medical personnel, such as paramedics.

Alcohol Resistant AFFF (ARC). Alcohol resistant AFFF's are Class B firefighting foams available at 3% hydrocarbon / 3% polar solvent (known as 3 x 3 concentrates), although 3% hydrocarbon / 6% polar solvent concentrations (known as 3 x 6 concentrates) may also be found. When applied to a polar solvent fuel, they will often create a polymeric membrane rather than a film over the fuel. This membrane separates the water in the foam blanket from the attack of the polar solvent. Then, the blanket acts in much the same manner as a regular AFFF.

Alpha Particles. Type of ionizing radiation. Largest of the common radioactive particles, alpha particles have extremely limited penetrating power. They travel only 3 to 4 inches in air and can be stopped by a sheet of paper or a layer of human skin. Alpha radiation is primarily an internal hazard and the greatest health hazard exists when alpha particles enter the body, such as through inhalation or ingestion.

American Chemistry Council: The parent organization that operates CHEMTREC.

American Conference of Governmental Industrial Hygienists (ACGIH). A professional society of individuals responsible for full-time industrial hygiene programs, who are employed by official governmental units. Its primary function is to encourage the interchange of experience among governmental industrial hygienists, and to collect and make information available of value to them. ACGIH promotes standards and techniques in industrial hygiene, and coordinates governmental activities with community agencies.

American National Standards Institute (ANSI). Serves as a clearinghouse for nationally coordinated voluntary safety, engineering and industrial consensus standards developed by trade associations, industrial firms, technical societies, consumer organizations, and government agencies.

American Petroleum Institute (API). Professional trade association of the United States petroleum industry. Publishes technical standards and information for all areas of the industry, including exploration, production, refining, marketing, transportation, and fire and safety.

Anhydrous. Free from water, dry. For example, anhydrous ammonia and anhydrous hydrogen chloride.

API Uniform Marking System. American Petroleum Institute marking system used at many petroleum storage and marketing facilities to identify hydrocarbon pipelines and transfer points. Classified hydrocarbon fuels and blends into leaded and unleaded gasoline (regular, premium, super). gasoline additives (methyl tertiary butyl ether) and distillates, and fuel oils.

APR. (See Air Purifying Respirator).

Aqueous Film Forming Foam (AFFF). Synthetic Class B firefighting foam consisting of fluorochemical and hydrocarbon surfactants combined with high boiling point solvents and water. AFFF film formation is dependent upon the difference in surface tension between the fuel and the firefighting foam. The fluorochemical surfactants reduce the surface tension of water to a degree less than the surface tension of the hydrocarbon so that a thin aqueous film can spread across the fuel.

Area of Refuge. Area within the hot zone where exposed or contaminated personnel are protected from further contact and/or exposure. This is a "holding area" where personnel are controlled until they can be safely decontaminated or treated.

Aromatic Hydrocarbons. A hydrocarbon containing the benzene "ring" which is formed by six carbon atoms and contains resonant bonds. Examples include benzene (C_6H_6) and toluene (C_7H_8).

Asphyxiation Harm Events. Those events related to oxygen deprivation and/or asphyxiation within the body. Asphyxiants can be classified as simple or chemical.

Association of American Railroads (AAR). Professional trade association which coordinates technical information and research within the United States railroad industry. Publisher of emergency response guidebooks.

Atmosphere-supplying devices. Respiratory protection devices coupled to an air source. The two types are self-contained breathing apparatus (SCBA) and supplied air respirators (SAR).

B

B-End. The end of a railroad car where the handbrake is located. Is typically used as the initial reference point when communicating railroad car damage.

BLS. Basic life support emergency medical personnel, such as emergency medical technicians.

Beta Particles. Type of ionizing radiation. Particle which is the same size as an electron and can penetrate materials much further than large alpha particles. Depending on the source, beta particles can travel several yards in air and penetrate paper and human skin but cannot penetrate internal organs. Depending on their energy, beta particles represent both an internal and external radiation hazard.

Biological Agents and Toxins. Biological threat agents consist of pathogens and toxins. Pathogens are disease-producing organisms and include bacteria (e.g., anthrax, cholera, plague, e coli), and viruses (e.g., small pox, viral hemorrhagic fever). Toxins are produced by a biological source and include ricin, botulinum toxins, and T2 mycotoxins.

Blister Agents. (See Vesicants).

Blood Agents. Chemical agents that consist of a cyanide compound, such as hydrogen cyanide (hydrocyanic acid) and cyanogens chloride. These agents are identical to their civilian counterpart used in industry.

Boiling Liquid Expanding Vapor Explosion (BLEVE). A container failure with a release of energy, often rapidly and violently, which is accompanied by a release of gas to the atmosphere and propulsion of the container or container pieces due to an overpressure rupture.

Boiling Point. The temperature at which a liquid changes its phase to a vapor or gas. The temperature where the vapor pressure of the liquid equals atmospheric pressure. Significant property for evaluating the flammability of a liquid, as flash point and boiling point are directly related. A liquid with a low flash point will also have a low boiling point, which translates into a large amount of vapors being given off.

Bonding. A method of controlling ignition hazards from static electricity. It is the process of connecting two or more conductive objects together by means of a conductor; for example, using an approved bonding wire to connect an aircraft being refueled to the fuel truck. Is done to minimize potential differences between conductive objects, thereby minimizing or eliminating the chance of static sparking.

Boom. A floating physical barrier serving as a continuous obstruction to the spread of a contaminant.

Branch. That organizational level within the Incident Command System having functional/geographic responsibility for major segments of incident operations (e.g., Hazmat Branch). The Branch level is organizationally between Section and Division/Sector/Group.

Breach Event. The event causing a hazmat container to open up or "breach." It occurs when a container is stressed beyond its limits of recovery (ability to hold contents). Different containers breach in different ways - disintegration, runaway cracking, failure of container attachments, container punctures, and container splits or tears.

Breakthrough Time. The elapsed time between initial contact of the hazardous chemical with the outside surface of a barrier, such as protective clothing material, and the time at which the chemical can be detected at the inside surface of the material.

Buddy System. A system of organizing employees into work groups in such a manner that each employee of the work group is designated to be observed by at least one other employee in the work group (per OSHA 1910.120 (a)(3)).

Bulk Packaging. Bulk packaging has an internal volume greater than 119 gallons (450 liters) for liquids, a capacity greater than 882 pounds (400 kg) for solids, or a water capacity greater than 1,000 pounds (453.6 kg) for gases. It can be an integral part of a transport vehicle (e.g., cargo tank truck, railroad tank car, and barges), packaging placed on, or in a transport vehicle (e.g., portable tanks, intermodal portable tanks, ton containers), or fixed or processing containers.

Bung. A threaded plug used to close a barrel or drum bung hole.

C

CAA. (See the Clean Air Act.)

Calibration. The process of adjusting a monitoring instrument so that its readings correspond to actual, known concentrations of a given material. If the readings differ, the monitoring instrument can then be adjusted so that readings are the same as the calibrant gas.

There are four types of calibration:

- Factory Calibration—instrument is returned to a certified factory / facility for testing and adjustment by certified instrument technicians.
- Full Calibration—instrument is shown a calibration gas and the readings are adjusted (automatically or manually) to the certified calibration gas values.
- Field Calibration—instrument is exposed to a known calibration gas and the user verifies that the readings correspond to + 10% of the calibration gas. Field adjustments to the monitor may then be made, as appropriate.
- Bump Test—instrument is exposing to a known calibration gas and the sensors show a response or alarm. If the instrument does not respond appropriately, then it should undergo a field or full calibration.

Canadian Transport Emergency Center (CANUTEC). Operated by Transport Canada, it is a 24 hour, government sponsored hot line for chemical emergencies. (The Canadian version of CHEMTREC.)

Cancer. A process in which cells undergo some change that renders them abnormal. They begin a phase of uncontrolled growth and spread.

Carboy. Glass or plastic bottles used for the transportation of liquids. Range in capacity to over 20 gallons. May be encased in an outer packaging, such as polystyrene boxes, wooden crates, or plywood drums. Often used for the shipment of corrosives.

Carcinogen. A material that can cause cancer in an organism. May also be referred to as "cancer suspect" or "known carcinogens."

Cargo Tanks. Tanks permanently mounted on a tank truck or tank trailer which is used for the transportation of liquefied and compressed gases, liquids and molten materials. Examples include MC-306, DOT-406, MC-307/DOT-407, MC-312/DOT-412, MC-331 and MC-338. May also be any bulk liquid or compressed gas packaging, not permanently attached to a motor vehicle, which because of its size, construction or attachment to the vehicle, can be loaded or unloaded without being removed from the vehicle.

Caustics (Base, Alkaline). Compound that forms hydroxides ions in water. These compounds have a pH > 7, and caustic solutions will turn litmus paper blue. Materials with a pH >12 are considered a strong base. Also known as alkali, alkaline or base.

CAS Number. (See Chemical Abstract Service Number).

Catalyst. Used to control the rate of a chemical reaction by either speeding it up or slowing it down. If used improperly, catalysts can speed up a reaction and cause a container failure due to pressure or heat build-up.

Center for Disease Control (CDC). The federally funded research organization tasked with disease control and research.

CERCLA (See Comprehensive Environmental Response Compensation and Liability Act).

Chemical Abstract Service (CAS) Number. Often used by state and local Right-To-Know regulations for tracking chemicals in the workplace and the community. Sometimes referred to as a chemical's "social security number." Sequentially assigned CAS numbers identify specific chemicals and have no chemical significance.

Chemical Agents. Chemical agents are classified in military terms based upon their effects on the enemy. The intent of using chemical weapons is to incapacitate and to kill. Categories of chemical agents are:

- Nerve agents (neurotoxins)
- Choking Agents (respiratory irritants)
- Blood agents (chemical asphyxiants)
- Vesicants or blister agents (skin irritants)
- Antipersonnel agents (riot control agents)

Chemical Degradation. The process of altering the chemical structure of the contaminant through the use of a second chemical or material. Commonly used degradation agents include calcium hypochlorite bleach, sodium hypochlorite bleach, sodium hydroxide as a saturated solution (household drain cleaner), sodium carbonate slurry (washing soda), calcium oxide slurry (lime), liquid household detergents, and isopropyl alcohol. Chemical degradation is primarily used to decon structures, vehicles, and equipment and should not be used to decon chemical protective clothing.

Chemical Interactions. Reaction caused by mixing two or more chemicals together. Chemical interaction of materials within a container may result in a build-up of heat and pressure, leading to container failure. In other situations, the combined material may be more corrosive than the container was originally designed to withstand and cause the container to breach.

Chemical Protective Clothing (CPC). Single or multi-piece garment constructed of chemical protective clothing materials designed and configured to protect the wearer's torso, head, arms, legs, hands, and feet. Can be constructed as a single or multi-piece garment. The garment may completely enclose the wearer either by itself or in combination with the wearer's respiratory protection, attached or detachable hood, gloves and boots.

Chemical Protective Clothing Material. Any material or combination of materials used in an item of clothing for the purpose of isolating parts of the wearer's body from contact with a hazardous chemical.

Chemical Reactivity. A process involving the bonding, unbonding and rebonding of atoms, that can chemically change substances into other substances. The interaction of materials in a container may result in a build-up of heat and pressure, and may cause container failure. Similarly, the combined materials may be more corrosive than the container was originally designed to withstand and lead to container failure.

Chemical Resistance. The ability to resist chemical attack. The attack is dependent on the method of test and its severity is measured by determining the changes in physical properties. Time, temperature, stress, and reagent may all be factors that affect the chemical resistance of a material.

Chemical Resistant Materials. Materials that are specifically designed to inhibit or resist the passage of chemicals into and through the material by the processes of penetration, permeation or degradation.

Chemical Stress. The result of a chemical reaction of two or more materials. Examples include corrosive materials attacking a metal, the pressure or heat generated by the decomposition or polymerization of a substance, or any variety of corrosive actions.

Chemical Transportation Emergency Center (CHEMTREC™). The Chemical Transportation Center, operated by the American Chemistry Council (ACC), can provide information and technical assistance to emergency responders. (Phone number: 1-800-424-9300).

Chemical Vapor Protective Clothing. Chemical protective clothing ensemble that is designed and configured to protect the wearer against chemical vapors or gases. Vapor chemical protective clothing must meet the requirements of NFPA 1991. This type of protective clothing is a component of EPA Level A chemical protection.

Chlorine Emergency Plan (CHLOREP). Chlorine industry emergency response system operated by the Chlorine Institute and activated through CHEMTREC.

Chlorine Kits. Standardized leak control kits used for the control of leaks in chlorine cylinders (Chlorine A kit), one-ton containers (Chlorine B kit), and tank cars, tank trucks and barges (Chlorine C kit). These kits are commercially available and are built to specifications developed by the Chlorine Institute.

Choking Agents. Chemical agent that can damage the membranes of the lung. Examples include phosgene and chlorine.

Chronic Effects. Result from a single exposure or from repeated doses or exposures over a relatively long period of time.

Chronic Exposures. Low exposures repeated over time.

Clandestine Laboratory. An operation consisting of a sufficient combination of apparatus and chemicals that either have been or could be used in the illegal manufacture/synthesis of controlled substances.

Classes. As used in NFPA 70 —*The National Electric Code,* used to describe the type of flammable materials that produce the hazardous atmosphere. There are three classes:

- Class I Locations—Flammable gases or vapors may be present in quantities sufficient to produce explosive or ignitible mixtures.
- Class II Locations—Concentrations of combustible dusts may be present (e.g., coal or grain dust).
- Class III Locations —Areas concerned with the presence of easily ignitible fibers or flyings (e.g., cotton milling).

Classes. As used in NFPA 1994—Protective Ensemble for Chemical/Biological Terrorism Inci-dents, used to describe the types of protective clothing available for terrorism response. There are three classes:

- Class 1 Ensembles offer the highest level of protection and are intended for use in worst-case circumstances, where the substance creates an immediate threat, is unidentified and of unknown concentrations.
- Class 2 Ensembles offer an intermediate level of protection and are intended for circumstances where the agent or threat may be identified, when the actual release has subsided, or in an area where live victims may be rescued.
- Class 3 Ensembles offer the lowest level of protection and are intended for use long after the initial release has occurred, at relatively large distances from the point of release, or for response activities such as decontamination, patient care, crowd control, traffic control, and clean-up operations.

Clean Air Act (CAA). Federal legislation that resulted in EPA regulations and standards governing airborne emissions, ambient air quality, and risk management programs.

Clean Water Act (CWA). Federal legislation that resulted in EPA and state regulations and standards governing drinking water quality, pollution control, and enforcement. The Oil Pollution Act (OPA) amended the CWA and authorized regulations pertaining to oil spill preparedness, planning, response and clean-up.

Cleanup. Incident scene activities directed toward removing hazardous materials, contamination, debris, damaged containers, tools, dirt, water, and road surfaces in accordance with proper and legal standards, and returning the site to as near a normal state as existed prior to the incident.

Code of Federal Regulations (CFR). A collection of regulations established by federal law. Contact with the agency that issues the regulation is recommended for both details and interpretation.

COFC. (See container-on-flat-car.)

Cold Zone. The hazard control zone of a hazmat incident that contains the incident command post and other support functions as are deemed necessary to control the incident. This zone may also be referred to as the clean zone or the support zone.

Coliwasa (Composite Liquid Waste Sampler). A glass or plastic waste sampling kit commonly used for collecting samples from drums and other containerized wastes.

Colorimetric Tubes. Glass tubes containing a chemically treated substrate that reacts with specific airborne chemicals to produce a distinctive color. The tubes are calibrated to indicate approximate concentrations in air.

Combination Package. Packaging consisting of one or more inner packagings and a non-bulk outer packaging. There are many different types of combination packagings.

Combined Liquid Waste Sampler (Coliwasa). A tool designed to provide stratified sampling of a liquid container.

Combined Sewers. Carries domestic wastewater as well as storm water and industrial wastewater. It is quite common in older cities to have an extensive amount of these systems. Combined sewers may also have regulators or diversion structures that allow overflow directly to rivers or streams during major storm events.

Command. The act of directing, ordering and/or controlling resources by virtue of explicit legal, agency or delegated authority.

Command Staff. The command staff consists of the Public Information Officer, the Safety Officer and the Liaison Officer, who report directly to the Incident Commander.

Community Awareness and Emergency Response (CAER). A program developed by the American Chemistry Council to provide guidance for chemical plant managers to assist them in developing integrated hazardous materials emergency response plans between the plant and the community.

Compatibility. The matching of protective chemical clothing to the hazardous material involved to provide the best protection for the worker.

Compatibility Charts. Permeation and penetration data supplied by manufacturers of chemical protective clothing to indicate chemical resistance and breakthrough time of various garment materials as tested against a battery of chemicals. This test data should be in accordance with ASTM and NFPA standards.

Composite Packaging. Packaging consisting of an inner receptacle, usually made of glass, ceraminc or plastic, and an outer protection (e.g., sheet metal, fiberboard, etc.) so constructed that the receptacle and the outer package form an integral packaging for transport purposes. Once assembled, it remains an integral single unit.

Compound. Chemical combination of two or more elements, either the same elements or different ones, that is electrically neutral. Compounds have a tendency to break down into their component parts, sometimes explosively.

Comprehensive Environmental Response, Compensation and Liability Act (CERCLA). Known as CERCLA or SUPERFUND, it addresses hazardous substance releases into the environment and the cleanup of inactive hazardous waste sites. It also requires those who release hazardous substances, as defined by the Environmental Protection Agency (EPA), above certain levels (known as "reportable quantities") to notify the National Response Center.

Compressed Gas. Any material or mixture having an absolute pressure exceeding 40 psi in the container at 70°F. (21°C), having an absolute pressure exceeding 104 psi at 130°F. (54°C), or any liquid flammable material having a vapor pressure exceeding 40 psi at 100°F. (37.7°C), as determined by testing. Also includes cryogenic liquids with boiling points lower than 130°F. (54°C), at 1 atmosphere.

Computer Aided Management of Emergency Operations (CAMEO). A computer data base storage-retrieval system of preplanning and emergency data for on-scene use at hazardous materials incidents. Developed and maintained by U.S. EPA.

Computerized Telephone Notification System (CT/NS). A computerized autodial telephone system which can be used for notifying a potentially large number of people in a short period of time. CT/NS systems are often used around high hazard facilities to ensure the timely notification of nearby citizens. Systems are capable of making call-backs to unanswered phones, keeping track of both who is notified and the time of notification, and providing prerecorded messages and instructions to residents.

Concentration. The percentage of an acid or base dissolved in water. Concentration is NOT the same as strength.

Confined Space. A space that (1) is large enough and so configured that an employee can bodily enter and perform assigned work; (2) has limited or restricted means for entry or exit (e.g., tanks, vessels, silos, storage bins, hoppers, vaults, and pits are spaces that may have limited means of entry); and (3) is not designed for continuous employee occupancy.

Confined Space (Permit Required). Has one or more of the following characteristics:

1) Contains or has the potential to contain a hazardous atmosphere. A hazardous atmosphere would be created by any of the following, including:
   a) Vapors exceed 10% of the lower explosive limit (LEL).
   b) Airborne combustible dust exceeds its LEL.
   c) Atmospheric oxygen concentrations below 19.5% or above 23.5%.
   d) Atmospheric concentration of any substance for which a dose or PEL is published and which could result in employee exposure in excess of these values.
   e) Any other atmospheric condition which is immediately dangerous to life or health (IDLH).
2) Contains a material that has the potential for engulfing an entrant.
3) Has an internal configuration such that a person could be trapped or asphyxiated by inwardly converging walls or by a floor which slopes downward and tapers to a smaller cross section; or
4) Contains any other recognized serious safety or health hazard.

Confinement. Procedures taken to keep a material in a defined or localized area once released.

Consignee. Person or company to which a material is being shipped.

Consist. A railroad shipping document that list the order of cars in a train.

Contact. Being exposed to an undesirable or unknown substance that may pose a threat to health and safety.

Container. Any vessel or receptacle that holds a material, including storage vessels, pipelines and packaging. Includes both bulk and non-bulk packaging, and fixed containers.

Container-On-Flat-Car (COFC). Intermodal containers that are shipped on a railroad flat cars.

Containment. Actions necessary to keep a material in its container (e.g., stop a release of the material or reduce the amount being released).

Contaminant. A hazardous material that physically remains on or in people, animals, the environment, or equipment, thereby creating a continuing risk of direct injury or a risk of exposure outside of the Hot Zone.

Contamination. The process of transferring a hazardous material from its source to people, animals, the environment, or equipment, which may act as a carrier.

Contingency (Emergency) Planning. A comprehensive and coordinated response to the hazmat problem. This planning process builds upon the hazards analysis and recognizes that no single public or private sector agency is capable of managing the hazmat problem by itself.

Control. The offensive or defensive procedures, techniques, and methods used in the mitigation of a hazardous materials incident, including containment, extinguishment, and confinement.

Controlled Burn. Defensive or non-intervention tactical objective by which a fire is allowed to burn with no effort to extinguish the fire. In some situations, extinguishing a fire will result in large volumes of contaminated runoff or threaten the safety of emergency responders. Consult with the appropriate environmental agencies when using this method.

Corrosive. A material that causes visible destruction of, or irreversible alterations to, living tissue by chemical action at the point of contact.

Corrosivity Harm Events. Those events related to severe chemical burns and/or tissue damage from corrosive exposures.

Covalent Bonding. The force holding together atoms that share electrons.

Crack. Narrow split or break in the container metal which may penetrate through the container metal (may also be caused by fatigue). It is a major mechanism that could cause catastrophic failure.

Crew Resource Management. Originally defined in 1977 by aviation psychologist Dr. John Lauber as "...using all available resources—information, equipment and people—to achieve safe and efficient flight operations." Key components of CRM include command, leadership, and resource management.

Crisis. An unplanned event that can exceed the level of available resources, and has the potential to significantly impact an organization's operability, credibility and reputation, or pose a significant environmental, economic, or legal liability.

Critical Temperature and Pressure. Critical temperature is the minimum temperature at which a gas can be liquefied no matter how much pressure is applied. Critical pressure is the pressure that must be applied to bring a gas to its liquid state. Both terms relate to the process of liquefying gases. A gas cannot be liquefied above its critical temperature. The lower the critical temperature, the less pressure required to bring a gas to its liquid state.

Critique. An element of incident termination that examines the overall effectiveness of the emergency response effort and develops recommendations for improving the organization's emergency response system.

Cross Contamination (a.k.a. secondary contamination). Occurs when a person who is already contaminated makes contact with a person or object that is not contaminated.

Cryogenic Liquids. Gases that have been transformed into extremely cold liquids which are stored at temperatures below -130°F (-90°C). Cryogenic liquid spills will vaporize rapidly when exposed to the higher ambient temperatures outside of the container. Expansion ratios for common cryogenics range from 694 (nitrogen) to 1,445 (neon) to 1.

D

Dam. A physical method of confinement by which barriers are constructed to prevent or reduce the quantity of liquid flowing into the environment.

Damage Assessment. The process of gathering and evaluating container damage as a result of a hazmat incident.

Dangerous Cargo Manifest. A list of the hazardous materials carried as cargo on board a vessel. Includes the location of the hazmat on the vessel.

Dangerous Goods. In international transportation, hazardous materials are commonly referred to as "dangerous goods."

Debriefing. An element of incident termination which focuses on the following factors:

1) Informing responders exactly what hazmats they were (possibly) exposed to, and the signs and symptoms of exposure.
2) Identifying damaged equipment requiring replacement or repair.
3) Identify equipment or supplies requiring specialized decontamination or disposal.
4) Identify unsafe work conditions.
5) Assign information-gathering responsibilities for a post-incident analysis.

Decon. Popular abbreviation referring to the process of decontamination.

Decontamination. The physical and/or chemical process of reducing and preventing the spread of contamination from persons and equipment used at a hazardous materials incident. OSHA 1910.120 defines decontamination as the removal of hazardous substances from employees and their equipment to the extent necessary to preclude foreseeable health effects.

Decontamination Corridor. A distinct area within the "Warm Zone" that functions as a protective buffer and bridge between the "Hot Zone" and the "Cold Zone," where decontamination stations and personnel are located to conduct decontamination procedures. An incident may have multiple decon corridors, depending upon the scope and nature of the incident.

Decontamination Unit Leader. A position within the Hazardous Materials Group which has responsibility for identifying the location of the decontamination corridor, assigning stations, managing all decontamination procedures, and identifying the types of decontamination necessary.

Decontamination Unit. A group of personnel and resources operating within a decontamination corridor.

Defensive Tactics. These are less aggressive spill and fire control tactics where certain areas may be "conceded" to the emergency, with response efforts directed towards limiting the overall size or spread of the problem. Examples include isolating the pipeline by closing remote valves, shutting down pumps, constructing dikes, and exposure protection.

Degradation. (1) A chemical action involving the molecular breakdown of a protective clothing material or equipment due to contact with a chemical. (2) The molecular breakdown of the spilled or released material to render it less hazardous during control operations.

Degree of Solubility. An indication of the solubility and/or miscibility of the material.

Negligible - less than 0.1 percent
Slight - 0.1 to 1.0 percent
Moderate - 1 to 10 percent
Appreciable - greater than 10 percent
Complete - soluble at all proportions.

Dent. Deformation of the tank head or shell. It is caused from impact with a relatively blunt object (e.g., railroad coupler, vehicle). If the dent has a sharp radius, there is the possibility of cracking.

Dermatotoxins. Toxins of the skin which may act as irritants, ulcers, chloracne or cause skin pigmentation disorders (e.g., halogenated hydrocarbons, coal tar compounds).

Detonation. An explosive chemical reaction with a release rate less than 1/100th of a second. This gives responders NO time to react. Examples include military munitions, dynamite, and organic peroxides.

Dike. A defensive confinement procedure consisting of an embankment or ridge on ground used to control the movement of liquids, sludges, solids, or other materials. Barrier which prevents passage of a hazmat to an area where it will produce more harm.

Dilution. Chemical method of confinement by which a water-soluble solution, usually a corrosive, is diluted by adding large volumes of water to the spill. It can increase the total volume of liquid which will have to be disposed of. In decon applications, it is the use of water to flush a hazmat from protective clothing and equipment, and the most common method of decon.

Direct Contact. Direct skin contact with some chemicals, such as corrosives, will immediately damage skin or body tissue upon contact.

Direct Contamination. Occurs when a person comes in direct physical contact with a contaminant, or when a person comes into contact with any object that has the contaminant on it (e.g. contaminated clothing or equipment).

Direct-Reading Instruments. Provide information at the time of sampling. They are used to detect and monitor flammable or explosive atmospheres, oxygen deficiency, certain gases and vapors, and ioninzing radiation.

Disinfection. The process used to inactivate (kill) virtually all recognized pathogenic microorganisms. There are two major categories of disinfectants: chemical and antiseptic.

Dispersants. The use of certain chemical agents to disperse or breakdown liquid hazmat spills. The use of dispersants may result in spreading the hazmat over a larger area. Dispersants are often applied to hydrocarbon spills, resulting in oil-in-water emulsions and diluting the hazmat to acceptable levels. Use of dispersants may require prior approval of the appropriate environmental agencies.

Dispersion. Chemical method of confinement by which certain chemical and biological agents are used to disperse or break up the material involved in liquid spills on water.

Diversion. Physical method of confinement by which barriers are constructed on ground or placed in a waterway to intentionally control the movement of a hazardous material into an area where it will pose less harm to the community and the environment.

Divisions. As used in *NFPA 70—The National Electric Code,* describe the types of location that may generate or release a flammable material. There are two divisions:

> **Division I.** Location where the vapors, dusts or fibers are continuously generated and released. The only element necessary for a hazardous situation is a source of ignition.
>
> **Division II.** Location where the vapors, dusts or fibers are generated and released as a result of an emergency or a failure in the containment system.

Divisions. As used within the Incident Command System, are the organizational level having responsibility for operations within a defined geographic area. A building floor, plant location, or process area may be designated as a division, such as the Division 4 (i.e., 4th floor area) or the Alky Division (i.e., Alkylation Process Unit).

Dome. Circular fixture on the top of a pressurized railroad tank car containing valves, pressure relief valve, and gauging devices.

Dose. (1) The concentration or amount of material to which the body is exposed over a specific time period. The amount of a substance ingested, absorbed and/or inhaled during an exposure period. (2) A quantity of radiation or energy absorbed by the body, usually measured in millirems (mrem).

Dose Rate. The radiation dose delivered per unit of time (e.g., mrem/hour).

Dose-Response Relationship. Basic principle of toxicology. The intensity of a response elicited by a chemical within a biologic mechanism is a function of the administered dose.

Doublegloving. Involves the use of gloves under a work glove. It permits the wearing of the work glove without compromising exposure protection and also provides an additional barrier for hand protection. Doublegloving also reduces the potential for hand contamination when removing protective clothing during decon procedures.

Drums. Cylindrical packagings used for liquids and solids. Constructed of plastic, metal, fiberboard, plywood or other suitable materials. Typical drum capacities range up to 55 gallons.

Ductility. The relative ability of a metal to bend or stretch without cracking.

E

Element. Pure substance that cannot be broken down into simpler substances by chemical means.

Elevated Temperature Materials. Materials which, when offered for transportation in a bulk container, are (1) liquids at or above 212° F. (100° C.); (2) Liquids with a flash point at or above 100° F. (37.8° C.) that are intentionally heated and are transported at or above their flash point; and (3) solids at a temperature at or above 464° F. (240° C.).

Emergency Breathing Apparatus (EBA). Short duration (e.g., 5–10 minutes) respiratory protection devices developed for use by the general public. Typically consist of a small breathing air cylinder and a clear plastic hood assembly which is placed over the head of the wearer to provide a fresh breathing air supply.

Emergency Broadcast System (EBS). The national emergency notification system that uses commercial AM and FM radio stations for emergency broadcasts. The EBS is usually initiated and controlled by Emergency Management agencies.

Emergency Contact. The telephone number for the shipper or shipper's representative that may be accessed 24 hours a day, 7 days a week in the event of an accident.

Emergency Decontamination. The physical process of immediately reducing contamination of individuals in potentially life-threatening situations with or without the formal establishment of a decontamination corridor.

Emergency Medical Services (EMS). Functions as required to provide emergency medical care for ill or injured persons by trained providers.

Emergency Operations Center (EOC). The secured site where government or facility officials exercise centralized direction and control in an emergency. The EOC serves as a resource center and coordination point for additional field assistance. It also provides executive directives to and liaison for government and other external representatives, and considers and mandates protective actions.

Emergency Response. Response to any occurrence which has or could result in a release of a hazardous substance.

Emergency Response Organization. An organization that utilizes personnel trained in emergency response. This would include fire, law enforcement, EMS, and industrial emergency response teams.

Emergency Response Personnel. Personnel assigned to organizations that have the responsibility for responding to different types of emergency situations.

Emergency Response Plan (ERP). A plan that establishes guidelines for handling hazmat incidents as required by regulations such as SARA, Title III and HAZWOPER (29 CFR 1910.120).

Emergency Response Planning Guidelines (ERPG-2). The maximum airborne concentration below which it is believed that nearly all individuals could be exposed for up to one hour without experiencing or developing irreversible or serious health effects or symptoms which could impair an individual's ability to take protective action.

Emergency Response Team (ERT). Crews of specially trained personnel used within industrial facilities for the control and mitigation of emergency situations. May consist of both shift personnel with ERT responsibilities as part of their job assignment (e.g., plant operators), or volunteer members. ERT's may be responsible for both fire, hazmat, medical, and technical rescue emergencies, depending upon the size and operation of the facility.

Emergency Traffic. A priority radio message to be immediately broadcast throughout the emergency scene.

Endothermic. A process or chemical reaction that is accompanied by the absorption of heat.

Engulfing Event. Once the hazmat and/or energy is released, it is free to travel or disperse, engulfing an area. The farther the contents move outward from their source, the greater the level of problems. How quickly they move and how large an area they engulf will depend upon the type of release, the nature of the hazmat, the physical and chemical laws of science, and the environment.

Environmental Protection Agency (EPA). The purpose of the EPA is to protect and enhance our environment today and for future generations to the fullest extent possible under the laws enacted by Congress. The Agency's mission is to control and abate pollution in the areas of water, air, solid waste, pesticides, noise, and radiation. EPA's mandate is to mount an integrated, coordinated attack on environmental pollution in cooperation with state and local governments.

EPA. (See Environmental Protection Agency.)

EPA Levels of Protection. EPA system for classifying levels of chemical protective clothing.

Level A: Chemical vapor protective suit.

Level B: Chemical liquid splash protective suit with SCBA.

Level C: Chemical liquid splash protective suit with air purifying respirator.

Level D: Normal work uniform with appropriate safety equipment such as hard hat, eye protection and safety shoes.

EPA Registration Number. Required for all agricultural chemical products marketed within the United States. It is one of three ways to positively identify an ag chemical. The others are by the product name or chemical ingredient statement. The registration number will appear as a two or three section number.

EPA Waste Stream Number. Indicates the number assigned to a hazardous waste stream by the U.S. EPA to identify that waste stream. NOTE: For all hazardous waste shipments, a Uniform Hazardous Waste Manifest must be prepared in accordance with both DOT and EPA regulations.

ERP. (See Emergency Response Plan).

Etiological Harm Events. Those harm events created by uncontrolled exposures to living micro-organisms. Diseases commonly associated with etiological harm include hepatitis, typhoid, and tuberculosis. It is often difficult to detect when and where the physical exposure to the etiological agent occurred and the route(s) of exposure.

Evacuation. A public protective option which results in the removal of fixed facility personnel and the public from a threatened area to a safer location. It is typically regarded as the controlled relocation of people from an area of known danger or unacceptable risk to a safer area, or one in which the risk is considered to be acceptable.

Evaporation Rate. The rate at which a material will vaporize or change from liquid to vapor, as compared to the rate of vaporization of a specific known material—n-butyl acetate. Is useful in evaluating the health and flammability hazards of a material.

Excepted Packaging. Used to transport material with low levels of radioactivity. Excepted Packaging does not have to pass any performance tests, but must meet specific design requirements spelled out in DOT regulations.

Expansion Ratio. The amount of gas produced by the evaporation of one volume of liquid at a given temperature. Significant property when evaluating liquid and vapor releases of liquefied gases and cryogenic materials. The greater the expansion ratio, the more gas that is produced and the larger the hazard area.

Explosion-Proof Construction. Encases the electrical equipment in a rigidly built container so that (1) it withstands the internal explosion of a flammable mixture, and (2) prevents propagation to the surrounding flammable atmosphere. Used in Class I, Division 1 atmospheres at fixed installations.

Exposure. The process by which people, animals, the environment, and equipment are subjected to or come in contact with a hazardous material. People may be exposed to a hazardous material through any route of entry (e.g., inhalation, skin absorption, ingestion, direct contact or injection).

Exposures. Items which may be impinged upon by a hazmat release. Examples include people (civilians and emergency responders), property (physical and environmental) and systems disruption.

Exothermic. A process or chemical reaction which is accompanied by the evolution of heat.

Extremely Hazardous Substances (EHS). Chemicals determined by the EPA to be extremely hazardous to a community during an emergency spill or release as a result of their toxicities and physical/chemical properties (Source: EPA 40 CFR 355).

F

Failure of Container Attachments. Attachments which open up or break off the container, such as safety relief valves, frangible discs, fusible plugs, discharge valves, or other related appliances.

Film Forming Fluoroprotein Foam (FFFP). Class B firefighting foam based on fluoroprotein foam technology with AFFF capabilities. FFFP combines the quick knockdown capabilities of AFFF along with the heat resistance benefits of fluoroprotein foam.

Fire Entry Suits. Suits which offer complete, effective protection for short duration entry into a total flame environment. Designed to withstand exposures to radiant heat levels up to 2,000°F (1,093 °C). Entry suits consist of a coat, pants, and separate hood assembly. They are constructed of several layers of flame-retardant materials, with the outer layer often aluminized.

Fire Point. Minimum temperature at which a liquid gives off sufficient vapors that will ignite and sustain combustion. It is typically several degrees higher than the flash point. In assessing the risk posed by a flammable liquids release, greater emphasis is placed upon the flash point, since it is a lower temperature and sustained combustion is not necessary for significant injuries or damage to occur.

First Responder. The first trained person(s) to arrive at the scene of a hazardous materials incident. May be from the public or private sector of emergency services.

First Responder, Awareness Level. Individuals who are likely to witness or discover a hazardous substance release who have been trained to initiate an emergency response sequence by notifying the proper authorities of the release. They would take no further action beyond notifying the authorities of the release.

First Responder, Operations Level. Individuals who respond to releases or potential releases of hazardous substances as part of the initial response to the site for the purpose of protecting nearby persons, property, or the environment from the effects of the release. They are trained to respond in a defensive fashion without actually trying to stop the release. Their function is to contain the release from a safe distance, keep it from spreading, and prevent exposures.

Flammable (Explosive) Range. The range of gas or vapor concentration (percentage by volume in air) that will burn or explode if an ignition source is present. Limiting concentrations are commonly called the "lower flammable (explosive) limit" and the "upper flammable (explosive) limit." Below the lower flammable limit, the mixture is too lean to burn; above the upper flammable limit, the mixture is too rich to burn. If the gas or vapor is released into an oxygen enriched atmosphere, the flammable range will expand. Likewise, if the gas or vapor is released into an oxygen deficient atmosphere, the flammable range will contract.

Flaring. Controlled burning of a high vapor pressure liquid or compressed gas in order to reduce or control the pressure and/or dispose of the product.

Flash Point. Minimum temperature at which a liquid gives off enough vapors that will ignite and flash-over but will not continue to burn without the addition of more heat. Significant in determining the temperature at which the vapors from a flammable liquid are readily available and may ignite.

Fluoroprotein Foam. Combination of protein-based foam derived from protein foam concentrates and fluorochemical surfactants. The addition of the fluorochemical surfactants produces a foam that flows easier than regular protein foam. Fluoroprotein foam can also be formulated to be alcohol resistant.

Form. Refers to the physical form of a material - solid, liquid or gas. Significant factor in evaluating both the hazards of a material and tactics for controlling a release. In general, gases and vapor releases cause the greatest problems for emergency responders.

Full Protective Clothing. Protective clothing worn primarily by fire fighters which includes helmet, fire retardant hood, coat, pants, boots, gloves, PASS device, and self-contained breathing apparatus designed for structural fire fighting. It does not provide specialized chemical splash or vapor protection.

Fumes. Airborne dispersion consisting of minute solid particles arising from the heating of a solid material (e.g., lead) to a gas or vapor. This physical change is often accompanied by a chemical reaction, such as oxidation. Odorous gases and vapors should not be referred to as vapors.

G

Gamma Waves. Most dangerous form of ionizing radiation because of the speed at which it moves, its ability to pass through human tissue, and the great distances it can cover. The range of gamma waves depends on the energy of the source material. Gamma radiation penetrates most materials very well, and it is considered a whole body hazard as internal organs can be penetrated and damaged.

Gelation. The process of forming a gel. Gelling agents are used on some hazmat spills to produce a gel that is more easily cleaned up.

Gouge. Reduction in the thickness of the tank shell. It is an indentation in the shell made by a sharp, chisel-like object. A gouge is characterized by the cutting and complete removal of the container or weld material along the track of contact.

Gross Decontamination. The initial phase of the decontamination process during which the amount of surface contaminant is significantly reduced.

Grounding. A method of controlling ignition hazards from static electricity. The process of connecting one or more conductive objects to the ground through an earthing electrode (i.e., grounding rod). For example, connecting an aircraft to the ground through an approved grounding wire and connecting the fuel truck to the ground, through a separate grounding wire and grounding rod. Is done to minimize potential differences between objects and the ground.

Groups. As used in *NFPA 70—The National Electric Code*, are products within a Class. Class I is divided into four groups (Group A-D) on the basis of similar flammability characteristics. Class II is divided into three groups (Groups E-G). There are no groups for Class III materials.

Groups. As used within the Incident Command System, are the organizational level responsible for a specified functional assignment at an incident. Hazmat units may operate as a Hazmat Group. Groups are under the direction of a Supervisor and may move between divisions at an incident.

H

Half-Life. The time it takes for the activity of a radioactive material to decrease to one half of its initial value through radioactive decay. The half-life of known materials can range from a fraction of a second to millions of years.

Halogenated Hydrocarbons. A hydrocarbon with halogen atom (e.g., chlorine, fluorine, bromine, etc.) substituted for a hydrogen atom. They are often more toxic than naturally occurring organic chemicals, and they decompose into smaller, more harmful elements when exposed to high temperatures for a sustained period of time.

Harm Event. Pertains to the harm caused by a hazmat release. Harm events include thermal, mechanical, poisonous, corrosivity, asphyxiation, radiation and etiological.

Hazard. Refers to a danger or peril. In hazmat operations, usually refer to the physical or chemical properties of a material.

Hazard Analysis. Part of the planning process, it is the analysis of hazmats present in a facility or community. Elements include hazards identification, vulnerability analysis, risk analysis, and evaluation of emergency response resources. Hazards analysis methods used as part of Process Safety Management (PSM) include HAZOP Studies, Fault Tree Analysis, and What If Analysis.

Hazard and Risk Evaluation. Evaluation of hazard information and the assessment of the relative risks of a hazmat incident. Evaluation process leads to the development of Incident Action Plan.

Hazard Class. The hazard class designation for the material as found in the Department of Transportation regulations, 49 CFR. There are currently 9 DOT hazard classes which are divided into 22 divisions.

Hazard Communication (HAZCOM). OSHA regulation (29 CFR 1910.1200) which requires hazmat manufacturers to develop MSDS's on specific types of hazardous chemicals, and provide hazmat health information to both employees and emergency responders.

Hazard Control Zones. The designation of areas at a hazardous materials incident based upon safety and the degree of hazard. Many terms are used to describe these hazard control zones; however, for the purposes of this text, these zones are defined as the hot, warm and cold zones.

Hazard Zone. Shipping paper entry, as defined in the ERG, that indicates relative degree of hazard in terms of toxicity (only appears for gases and liquids that are poisonous by inhalation):

- Zone A—LC50 less than or equal to 200 ppm (most toxic).
- Zone B—LC50 greater than 200 ppm and less than or equal to 1,000 ppm.
- Zone C—LC50 greater than 1,000 ppm and less than or equal to 3,000 ppm.
- Zone D—LC50 greater than 3,000 ppm and less than or equal to 5,000 ppm (least toxic).

Hazardous Chemicals. Any chemical that would be a risk to employees if exposed in the workplace (Source: OSHA, 29 CFR 1910).

Hazardous Materials. (1) Any substance or material in any form or quantity that poses an unreasonable risk to safety and health and property when transported in commerce (Source: U.S. Department of Transportation [DOT], 49 Code of Federal Regulations (CFR) 171). (2) Any substance that jumps out of its container when something goes wrong and hurts or harms the things it touches (Source: Ludwig Benner).

Hazardous Materials General Behavior Model (GEBMO). Process for visualizing hazmat behavior. Applies the concept of events analysis which is simply breaking down the overall incident into smaller, more easily understood parts for purposes of analysis.

Hazardous Materials Group. The ICS organizational level responsible for specified hazardous materials-related functional assignments at an incident.

Hazardous Materials Group Safety Officer (i.e., Assistant Safety Officer—Hazmat). Responsible for coordinating safety activities within the Hazardous Materials Group and within the hot and warm zones. This includes having the authority to stop or prevent unsafe actions and procedures during the course of the incident. Reports to the Hazardous Materials Group Supervisor and is subordinate to the Incident Safety Officer.

Hazardous Materials Group Supervisor. Responsible for the management and coordination of all functional responsibilities assigned to the Hazardous Materials Group, including safety, site control, research, entry and decontamination. Depending upon the scope and nature of the incident, will usually report to either the Operations Section Chief or the Incident Commander.

Hazardous Materials Response Team (HMRT). An organized group of employees, designated by the employer, who are expected to perform work to handle and control actual or potential leaks or spills of hazardous substances requiring possible close approach to the substance. A Hazmat Team may be a separate component of a fire brigade or a fire department or other appropriately trained and equipped units from public or private agencies.

Hazardous Materials Specialists. As defined in OSHA 1910.120, individuals who respond and provide support to Hazardous Materials Technicians. While their duties parallel those of the Technician, they require a more detailed or specific knowledge of the various substances they may be called upon to contain. Would also act as a liaison with Federal, state, local and other governmental authorities in regards to site activities.

Hazardous Materials Technicians. Individuals who respond to releases or potential releases of hazardous materials for the purposes of stopping the leak. They generally assume a more aggressive role in that they are able to approach the point of a release in order to plug, patch or otherwise stop the release of a hazardous substance.

Hazardous Substances. Any substance designated under the Clean Water Act and the Comprehensive Environmental Response, Compensation and Liability Act (CERCLA) as posing a threat to waterways and the environment when released (Source: U.S. Environmental Protection Agency [EPA], 40 CFR 302). Note: Hazardous substances as used within OSHA 1910.120 refers to every chemical regulated by EPA as a hazardous substance and by DOT as a hazardous material.

Hazardous Wastes. Discarded materials regulated by the EPA because of public health and safety concerns. Regulatory authority is granted under the Resource Conservation and Recovery Act (RCRA). (Source: EPA, 40 CFR 260–281).

Hazardous Waste Manifest. Shipping form required by the EPA and DOT for all modes of transportation when transporting hazardous wastes for treatment, storage or disposal.

Hazmat. Acronym used for Hazardous Materials.

Hazmat Branch. Responsible for all hazmat operations which occur at a hazmat incident. Functions include safety, site control, information, entry, decontamination, hazmat medical, and hazmat resources.

Hazardous Waste Operations and Emergency Response. Also known as HAZWOPER, this federal regulation was issued under the authority of SARA, Title I. The regulation establishes health, safety and training requirements for both industry and public safety organizations that respond to hazmat or hazardous waste emergencies. This includes firefighters, law enforcement and EMS personnel, hazmat responders, and industrial Emergency Response Team (ERT) members.

HAZCOM. (See Hazard Communication).

Hazmat Branch Director. Officer responsible for the management and coordination of all functional responsibilities assigned to the Hazmat Branch. Must have a high level of technical knowledge, as well as be knowledgeable of both the strategical and tactical aspects of hazmat response. Reports to the Operations Section Chief.

Hazmat Entry Function. Responsible for all entry and back-up operations within the Hot Zone, including reconnaissance, monitoring, sampling and mitigation.

Hazmat Decontamination Function. Responsible for the research and development of the decon plan, set-up and operation of an effective decontamination area capable of handling all potential exposures, including entry personnel, contaminated patients, and equipment.

Hazmat Information Function. Responsible for gathering, compiling, coordinating and disseminating all data and information relative to the incident. This data and information will be used within the Hazmat Branch for assessing hazard and evaluating risks, evaluating public protective options, the selection of PPE, and development of the incident action plan.

Hazmat Medical Function. Responsible for pre- and post-entry medical monitoring and evaluation of all entry personnel, and provides technical medical guidance to the Hazmat Branch, as requested.

Hazmat Resource Function. Responsible for control and tracking of all supplies and equipment used by the Hazmat Branch during the course of an emergency, including documenting the use of all expendable supplies and materials. Coordinates, as necessary, with the Logistics Section Chief.

Hazmat Safety Function. Primarily the responsibility of the Incident Safety Officer and the Hazmat Safety Officer. Responsible for ensuring that safe and accepted practices and procedures are followed throughout the course of the incident. Possess both the authority and responsibility to stop any unsafe actions and correct unsafe practices.

Hazmat Site Control Function. Establish control zones, establish and monitor access routes at the incident site, and ensure that contaminants are not being spread.

HAZWOPER. Acronym used for the OSHA Hazardous Wastes Operations and Emergency Response regulation (29 CFR 1910.120).

Heat Affected Zone. Area in the undisturbed tank metal next to the actual weld material. This area is less ductile than either the weld or the steel plate due to the effect of the heat of the welding process. This zone is most vulnerable to damage as cracks are likely to start here.

Heat Cramps. A cramp in the extremities or abdomen caused by the depletion of water and salt in the body. Usually occurs after physical exertion in an extremely hot environment or under conditions that cause profuse sweating and depletion of body fluids and electrolytes.

Heat Exhaustion. A mild form of shock caused when the circulatory system begins to fail as a result of the body's inadequate effort to give off excessive heat.

Heat Rash. An inflammation of the skin resulting from prolonged exposure to heat and humid air and often aggravated by chafing clothing. Heat rash is uncomfortable and decreases the ability of the body to tolerate heat.

Heat Stroke. A severe and sometimes fatal condition resulting from the failure of the temperature regulating capacity of the body. It is caused by exposure to the sun or high temperatures. Reduction or cessation of sweating is an early symptom. Body temperature of 105°F. or higher, rapid pulse, hot and dry skin, headache, confusion, unconsciousness, and convulsions may occur. Heat stroke is a true medical emergency requiring immediate transport to a medical facility.

Hematotoxins. A toxin of the blood system (e.g., benzene, chlordane, DDT).

Hepatotoxin. A toxin destructive of the liver (e.g., carbon tetrachloride, vinyl chloride monomer).

High Temperature Protective Clothing. Protective clothing designed to protect the wearer against short-term high temperature exposures. Includes both proximity suits and fire entry suits. This type of clothing is usually of limited use in dealing with chemical exposures.

HMRT. (See Hazardous Materials Response Team.)

Hot Tapping. An offensive technique for welding on and cutting holes through liquid and/or compressed gas vessels and piping for the purposes of relieving the internal pressure and/or removing the product.

Hot Zone. An area immediately surrounding a hazardous materials incident, which extends far enough to prevent adverse effects from hazardous materials releases to personnel outside the zone. This zone is also referred to as the "exclusion zone", the "red zone", and the "restricted zone" in other documents. Law enforcement personnel may also refer to this as the inner perimeter.

Housing. Fixture on the top of a non-pressurized railroad tank car designed to provide protection for valves, pressure relief valve, and/or gauging devices.

Hydrocarbons. Compounds primarily made up of hydrogen and carbon. Examples include LPG, gasoline and fuel oils.

Hygroscopic. A substance that has the property of absorbing moisture from the air, such as calcium chloride.

Hypergolic. Two chemical substances that spontaneously ignite upon mixing.

I

IAP. (See Incident Action Plan).

ICS. (See Incident Command System).

ICS Command Staff. Those individuals appointed by and directly reporting to the Incident Commander. These include the Safety Officer, the Liaison Officer, and the Public Information Officer (PIO).

ICS General Staff. Section Chiefs are members of the Incident Commander's general staff, and responsible for the broad response functions of Operations, Planning, Logistics, and Finance/Administration.

Ignition (Autoignition) Temperature. Minimum temperature required to ignite gas or vapor without a spark or flame being present. Significant in evaluating the ease at which a flammable material may ignite.

Immediately Dangerous to Life or Health (IDLH). An atmospheric concentration of any toxic, corrosive or asphyxiant substance that poses an immediate threat to life or would cause irreversible or delayed adverse health effects or would interfere with an individual's ability to escape from a dangerous atmosphere.

Impingement Event. As the hazmat and/or its container engulf an area, they will impinge, or come in contact with exposures. They may also impinge upon other hazmat containers, producing additional problems.

Incident: (1) The release or potential release of a hazardous material from its container into the environment. (2) An occurrence or event, either natural or man-made, which requires action by emergency response personnel to prevent or minimize loss of life or damage to property and/or natural resources

Incident Action Plan (IAP). The strategic goals, tactical objectives and support requirements for the incident. All incidents require an action plan. For simple incidents (Level I) the action plan is not usually in written form. Large or complex incidents (Level II or III) will require that the action plan be documented in writing.

Incident Command System (ICS). An organized system of roles, responsibilities, and standard operating procedures used to manage and direct emergency operations.

Incident Commander (IC). The individual responsible for establishing and managing the overall incident action plan (IAP). This process includes developing an effective organizational structure, developing an incident strategy and tactical action plan, allocating resources, making appropriate assignments, managing information, and continually attempting to achieve the basic command goals. The IC is in charge of the incident site. May also be referred to as the On-Scene Incident Commander as defined in 29 CFR. 1910.120

Incident Command Post (ICP). The "on-scene" location where the Incident Commander develops goals and objectives, communicates with subordinates, and coordinates activities between various agencies and organizations. The ICP is the "field office" for on-scene response operations, and requires access to communications, information, and both technical and administrative support.

Industrial Packaging. Used in certain shipments of Low Specific Activity (LSA) radioactive material and Surface Contaminated Objects, which are typically categorized as radioactive waste. Most low-level radioactive waste, such as contaminated protective clothing and handling materials, is shipped in secured packaging of this type. DOT regulations require that these packagings allow no identifiable release of the material into the environment during normal transportation and handling.

Inert Gas. A nonreactive gas, such as argon, helium, and neon.

Ingestion. The introduction of a chemical into the body through the mouth or inhaled chemicals trapped in saliva and swallowed.

Ingredient Statement. The statement present on all agricultural chemical labels which breaks down the chemical ingredients by their relative percentages or as pounds per gallon of concentrate. "Active" ingredients are the active chemicals within the mixture. They must be listed by chemical name and their common name may also be shown. "Inert" ingredients have no ag chem/pesticide activity and are usually not broken into specific components, only total percentage.

Inhalation. The introduction of a chemical or toxic products of combustion into the body by way of the respiratory system. Inhalation is the most common exposure route and often the most damaging. Toxins may be absorbed into the bloodstream and carried to other internal organs, or they may affect the upper and/or lower respiratory tract. Resulting respiratory injuries include pulmonary edema and respiratory congestion.

Inhibitor. Added to products to control their chemical reaction with other products. If the inhibitor is not added or escapes during an incident, the material will begin to polymerize, possibly resulting in container failure.

Injection. Introduction of a hazardous material directly through the skin and into the bloodstream. Mechanisms of injury include needle stick cuts at medical emergencies and the injection of high pressure gases and liquids into the body similar to the manner in which flu shots are injected with pneumatic guns.

Inorganic Materials. Compounds derived from other than vegetable or animal sources which lack carbon chains, but may contain a carbon atom (e.g., sulfur dioxide—SO_2).

Instability. (See Reactivity).

Instrument Response Time. Also known as "lag time," this is the period of time between when the instrument senses a product and when a monitor reading is produced. Depending upon the instrument, lag times can range from several seconds to minutes. Variables will include if the instrument has a pump and the use of sampling tubing.

Intrinsically Safe Construction. Equipment or wiring is incapable of releasing sufficient electrical energy under both normal and abnormal conditions to cause the ignition of a flammable mixture. Commonly used in portable direct-reading instruments for operations in Class I, Division 2 hazardous locations.

Intermodal Tank Containers. Specific class of portable tanks specifically designed for international intermodal use. Most common types are the IM 101, IM 102 and the DOT Spec. 51 portable tanks.

Ionic Bonding. The electrostatic attraction of oppositely charged particles. Atoms or groups of atoms can for ions or complex ions.

Ionizing Radiation. Characterized by the ability to create charged particles or ions in anything which it strikes. Exposure to low levels of ionizing radiation can produce short-term or long-term cellular changes with potentially harmful effects, such as cancer or leukemia.

Isolating the Scene. The process of preventing persons and equipment from becoming exposed to a actual or potential hazmat release. Includes establishing isolation perimeter and control zones.

Isolation Perimeter. The designated crowd control line surrounding the Hazard Control Zones. The isolation perimeter is always the line between the general public and the Cold Zone. Law enforcement personnel may also refer to this as the outer perimeter.

J

Jacket. Outer metal covering of a railroad tank car that protects the tank's insulation and keeps it in-place.

Joint Information Center (JIC). Single location where Public Information Officers from different agencies work jointly and cooperatively to provide information to the public and other external world groups.

L

Labels. Approximately 4-inch (100 mm) square markings required under DOT regulations and applied to individual hazardous materials packages. They are generally placed or printed near the contents names or are printed on the manufacturing label. When labels cannot be applied directly to the container because of its nonadhesive surface, they are placed on tags or cards attached to the package. The proper label(s) are determined by the product's hazard class.

Lab Pack. An overpack drum or disposal container which contains multiple, smaller chemical containers with compatible chemical characteristics. Absorbent materials are usually placed within the overpack container to minimize potential for breakage and/or leakage.

Leak. The uncontrolled release of a hazardous material which could pose a threat to health, safety, and/or the environment.

Leak Control Compounds. Substances used for the plugging and patching of leaks in non-pressure containers (e.g., putty, wooden plugs, etc.).

Leak Control Devices. Tools and equipment used for the plugging and patching of leaks in non-pressure and some low-pressure containers, pipes, and tanks (e.g., patch kits, Chlorine kits, etc.).

LEPC. (See Local Emergency Planning Committee.)

Lethal Concentration, 50 Percent Kill (LC_{50}). Concentration of a material, expressed as parts per million (PPM) per volume, which kills half of the lab animals in a given length of time. Refers to an inhalation exposure, the LC_{50} may also be expressed as mg/liter or mg/cubic meter. Significant in evaluating the toxicity of a material; the lower the value, the more toxic the substance.

Lethal Concentration Low (LC_{LOW}). The lowest concentration of a substance in air reported to have caused death in humans or animals. The reported concentrations may be entered for periods of exposure that are less than 24 hours (acute) or greater than 24 hours (subacute and chronic).

Lethal Dose, 50 Percent Kill (LD_{50}). The amount of a dose which, when administered to lab animals, kills 50 percent of them. Refers to an oral or dermal exposure and is expressed in terms of mg/kg. Significant in evaluating the toxicity of a material; the lower the value, the more toxic the substance.

Lethal Dose Low (LD_{LOW}). The lowest amount of a substance introduced by any route, other than inhalation, reported to have caused death to animals or humans.

Level I Staging. The initial location for emergency response units at a multiple unit response to a hazmat incident. Initial arriving emergency response units go directly to the incident scene taking standard positions (e.g., upwind, uphill as appropriate), assume command and begin site management operations. The remaining units stage at a safe distance away from the scene, until ordered into action by the Incident Commander.

Level II Staging. Used for large, complex or lengthy hazmat operations. Location where arriving units are initially sent when an incident escalates past the capability of the initial response. It is a tool usually reserved for large, complex, or lengthy hazmat operations. Units assigned to Staging are under the control of a Staging Officer or Staging Area Manager.

Liaison Officer. Serves as a coordination point between the Incident Commander and any assisting or coordinating agencies who have responded to the incident, but who are not part of unified command or are not represented at the Incident Command Post. Member of the Command Staff.

Limited-Use Materials. Protective clothing materials which are used and then discarded. Although they may be reused several times (based upon chemical exposures), they are often disposed of after a single use. Examples include Tyvek™ QC, Tyvek™/Saranex™ 23-P, Hazard-Gard™, I Hazard-Gard™ II, and the Tychem Responder™.

Liquid Chemical Splash Protective Clothing. The garment portion of a chemical protective clothing ensemble that is designed and configured to protect the wearer against chemical liquid splashes but not against chemical vapors or gases. Liquid splash chemical protective clothing must meet the requirements of NFPA 1992. This type of protective clothing is a component of EPA Level B chemical protection.

Local Effect. The health effects of a hazardous materials exposure at the point of contact.

Local Emergency Planning Committee (LEPC): A committee appointed by a state emergency response commission, as required by SARA Title III, to formulate a comprehensive emergency plan for its corresponding local government or mutual aid region.

Lower Detection Limit (LDL). The lowest concentration to which a monitoring instrument will respond. The lower the LDL, the quicker contaminant concentrations can be evaluated.

M

Manifest. A shipping document that lists the commodities being transported on a vessel.

Markings. The required names, instructions, cautions, specifications, or combinations thereof found on containers of hazardous materials and hazardous wastes.

Mass Decontamination. The process of decontaminating large numbers of people in the fastest possible time to reduce surface contamination to a safe level. It is typically a gross decon process utilizing water or soap and water solutions to reduce the level of contamination.

Material Safety Data Sheet (MSDS). A document which contains information regarding the chemical composition, physical and chemical properties, health and safety hazards, emergency response, and waste disposal of the material as required by 29 CFR 1910.1200.

Maximum Safe Storage Temperature (MSST). The maximum storage temperature that an organic peroxide may be maintained, above which a reaction and explosion may occur.

Mechanical Harm Events. Those harm events resulting from direct contact with fragments scattered because of a container failure, explosion or shock wave.

Mechanical Stress. The result of a transfer of energy when one object physically contacts or collides with another. Indicators include punctures, gouges, breaks or tears in the container.

Medical Monitoring. An on-going, systematic evaluation of individuals at risk of suffering adverse effects of exposure to heat, stress or hazardous materials as a result of working at a hazmat emergency.

Medical Surveillance. Comprehensive medical program for tracking the overall health of its participants (e.g., HMRT personnel, public safety responders, etc.). Medical surveillance programs consist of pre-employment screening, periodic medical examinations, emergency treatment provisions, non-emergency treatment, and recordkeeping and review.

Melting Point. The temperature at which a solid changes its phase to a liquid. This temperature is also the freezing point depending on the direction of the change. For mixtures, a melting point range may be given. Significant property in evaluating the hazards of a material, as well as the integrity of a container (e.g., frozen material may cause its container to fail).

Minimum Detectable Permeation Rate (MDPR). The minimum permeation rate that can be detected by the laboratory analytical system being used for the permeation test.

Miscible. Refers to the tendency or ability of two or more liquids to form a uniform blend, or to dissolve in each other. Liquids may be totally miscible, partially miscible, or non-miscible.

Mitigation. Any offensive or defensive action to contain, control, reduce or eliminate the harmful effects of a hazardous materials release.

Mixture. Substance made up of two or more compounds, physically mixed together. A mixture may also contain elements and compounds mixed together.

Monitoring. The act of systematically checking to determine contaminant levels and atmospheric conditions.

Monitoring Instruments. Monitoring and detection instruments used to detect the presence and/or concentration of contaminants within an environment. They include:

Combustible Gas Indicator (CGI): Measures the concentration of a combustible gas or vapor in air.

Oxygen Monitor: Measures the percentage of oxygen in air.

Colorimetric Indicator Tubes: Measures the concentration of specific gases and vapors in air.

Specific Chemical Monitors: Designed to detect a large group of chemicals or a specific chemical. Most common examples include carbon monoxide and hydrogen sulfide.

Flame Ionization Detector (FID): A device used to determine the presence of organic vapors and gases in air. Operates in two modes - survey mode and gas chromatograph.

Gas Chromatograph: An instrument used for identifying and analyzing specific organics compounds.

Photoionization Detector (PID): A device used to determine the total concentration of many organic and some inorganic gases and vapors in air.

Radiation Monitors: An instrument used to measure accumulated radiation exposure. Include both alpha, beta and gamma survey detectors.

Radiation Dosimeter Detector: An instrument which measures the amount of radiation to which a person has been exposed.

Corrosivity (pH) Detector: A meter, paper or strip that indicates the relative acidity or alkalinity of a substance, generally using an international scale of 0 (acid) through 14 (alkali-caustic). (See pH.)

Indicator Papers: Special chemical indicating papers which test for the presence of specific hazards, such as oxidizers, organic peroxides and hydrogen sulfide. Are usually part of a hazmat identification system.

MSDS. (See Material Safety Data Sheet.)

Multi-Use Materials. Based upon the chemical exposure, multi-use materials are designed and fabricated to allow for decontamination and reuse. Generally thicker and more durable than limited-use garments, they are used for chemical splash and vapor protective suits, gloves, aprons, boots, and thermal protective clothing. The most common materials include butyl rubber, Viton, polyvinyl chloride (PVC), neoprene rubber, and Teflon™.

Mutagen. A material that creates a change in gene structure which is potentially capable of being transmitted to the offspring.

N

National Animal Poison Control Center (NAPCC). Operated by the University of Illinois at Urbana-Champaign, this number provides 24-hour consultation in the diagnosis and treatment of suspected or actual animal poisonings or chemical contamination. In addition, it staffs an emergency response team to investigate such incidents in North America and performs laboratory analysis of feeds/animal specimens/environmental materials for toxicants and chemical contaminants.

National Contingency Plan (NCP). Outlines the policies and procedures of the federal agency members of the National Oil and Hazardous Materials Response Team (also known as the National Response Team or the NRT). Provides guidance for emergency responses, remedial actions, enforcement, and funding for federal government response to hazmat incidents.

National Fire Protection Association (NFPA). An international voluntary membership organization to promote improved fire protection and prevention, establish safeguards against loss of life and property by fire, and writes and publishes national voluntary consensus standards (e.g., *NFPA 472—Professional Competence of Responders to Hazardous Materials Incidents).*

National Institute for Occupational Safety and Health (NIOSH). A Federal agency which, among other activities, tests and certifies respiratory protective devices, air sampling detector tubes, and recommends occupational exposure limits for various substances.

National (NIMS) Incident Management System. A standardized systems approach to incident management that consists of five major sub-divisions collectively providing a total systems approach to all-risk incident management.

National Pesticide Information Center (NPIC). Operated by Oregon State University in cooperation with EPA, NPIC provides information on pesticide-related health/toxicity questions, properties, and minor clean-up to physicians, veterinarians, responders, and the general public. Hours of operation are 6:30 am to 4:30 pm (Pacific Time), 7-days a week.

National Response Center (NRC). Communications center operated by the U.S. Coast Guard in Washington, DC. It provides information on suggested technical emergency actions, and is the federal spill notification point. The NRC must be notified within 24 hours of any spill of a reportable quantity of a hazardous substance by the spiller. Can be contacted at (800) 424-8802.

National Response Team (NRT). The National Oil and Hazardous Materials Response Team consists of fourteen federal government agencies which carry out the provisions of the National Contingency Plan at the federal level. The NRT is chaired by EPA, while the vice-chairperson represents the U.S. Coast Guard.

National Transportation Safety Board (NTSB). Independent federal agency charged with responsibility for investigating serious accidents and emergencies involving the various modes of transportation (e.g., highway, pipeline, air), as well as hazardous materials. Issues investigation reports and non-binding recommendations for action.

NCP. (See National Contingency Plan).

Nephrotoxins. Toxins which attack the kidneys (e.g., mercury, halogenated hydrocarbons).

Neurotoxins. Toxins which attack the central nervous system (e.g., organophosphate pesticides).

Neutralization. A chemical method of containment by which a hazmat is neutralized by applying a second material to the original spill which will chemically react with it to form a less harmful substance. The most common example is the application of a base to an acid spill to form a neutral salt. May also be used for the decontamination of equipment, vehicles and structures that are contaminated with a corrosive material.

Neutron particles. A form of high-speed particle radiation that consists of a "neutron" emitted at a high speed from the nucleus of a radioactive atom. There are few natural emitters of neutron radiation; the natural background of neutron radiation comes from cosmic rays from outer space interacting with gas molecules in the atmosphere. Neutrons are considered a whole body hazard.

NFPA. (See National Fire Protection Association).

NIMS. National Incident Management System.

Nonbulk Packaging. Any packaging having a capacity meeting one of the following criteria:

- Liquid—internal volume of 119 gallons (450 L.) or less;
- Solid—capacity of 882 lb. (400 kg) or less; and
- Compressed Gas—water capacity of 1,001 lb. (454 kg) or less.

Nonbulk packaging may consist of single packaging (e.g., drum, carboy, cylinder) or combination packaging—one or more inner packages inside of an outer packaging (e.g., glass bottles inside a fiberboard box, infectious disease sample containers).

Nonintervention Tactics. Essentially "no action." It is useful at certain fire emergencies where the potential costs of action far exceed any benefits (e.g., BLEVE scenario).

Nonionizing Radiation. Waves of energy, such as radiant heat, radio waves and visible light. The amount of energy in these waves is small as compared to ionizing radiation. Examples include infrared waves, microwaves and lasers.

Non-Persistence. Refers to the length of time a chemical agent remains as a liquid. A chemical agent is said to be "non-persistent" if it evaporates within 24 hours.

Normal Physical State. The physical state or form (solid, liquid, gas) of a material at normal temperatures [68°F (20°C) to 77°F (25°C)]. Determining the physical state of a material can allow responders to assess potential harm.

Normalized Breakthrough Time. A calculation, using actual permeation results, to determine the time at which the permeation rate reaches 0.1 $\mu g/cm^2/min$. Normalized breakthrough times are useful for comparing the performance of several different protective clothing materials. Note that in Europe, breakthrough times are normalized at 1.0 $\mu g/cm^2/min$., a full order of magnitude less sensitive.

Not Otherwise Specified (N.O.S.). A shipping paper notation which indicates that the material meets the DOT definition for a hazardous material, but is not listed by a generic name within the DOT Regulations. The technical name of the material must be entered in parenthesis with the basic description. For example, Flammable Liquid, n.o.s. (contains methanol).

NRT. (See National Response Team).

O

Occupational Safety and Health Administration (OSHA). Component of the United States Department of Labor; an agency with safety and health regulatory and enforcement authorities for most United States industries, businesses and states.

Odor Threshold (TLV_{ODOR}). The lowest concentration of a material's vapor in air that is detectable by odor. If the TLVodor is below the TLV/TWA, odor may provide a warning as to the presence of a material.

Offensive Tactics. Aggressive leak, spill and fire control tactics designed to quickly control or mitigate the problem. Although increasing risks to emergency responders, offensive tactics may be justified if rescue operations can be quickly achieved, if the spill can be rapidly confined or contained, or the fire quickly extinguished.

Oil Pollution Act (OPA). Amended the Federal Water Pollution Act, OPA's scope covers both facilities and carriers of oil and related liquid products, including deepwater marine terminals, marine vessels, pipelines and railcars. Requirements include the development of emergency response plans, training and exercises, and verification of spill resources and contractor capabilities. The law also requires the establishment of Area Committees and the development of Area Contingency Plans (ACPs) to address oil and hazardous substance spill response in coastal zone areas.

On-Scene Coordinator (OSC). The federal official pre-designated by EPA or the USCG to coordinate and direct federal responses and removals under the National Contingency Plan.

On-Scene Incident Commander. (See Incident Commander).

OPA. (See Oil Pollution Act).

Operations Section. Responsible for all tactical operations at the incident. The Hazmat Branch falls within the Operations Section.

Organic Materials. Materials that contain carbon. Organic materials are derived from materials that are living or were once living, such as plants or decayed products. Most organic materials are flammable.

Organic Peroxide. Strong oxidizers, often chemically unstable, containing the -o-o- chemical structure. May react explosively temperature and pressure changes, as well as contamination.

OSC. (See On-Scene Coordinator).

Other Regulated Materials D (ORM D). A material, such as a consumer commodity, which presents a limited hazard during transportation due to its form, quantity or packaging.

Overflow Dam. Spill control tactic used to trap sinking heavier-than-water materials behind the dam (specific gravity >1). With the product trapped, uncontaminated water is allowed to flow unobstructed over the top of the dam. Operationally, this is most effective on slow moving and relatively narrow waterways.

Overgarments. Protective clothing ensembles that are worn over chemical vapor protective clothing to provide either additional flash protection or low temperature protection.

Overgloving. The wearing of a second glove over the work glove for additional chemical and/or abrasion protection during entry operations.

Overpack. 1) A packaging used to contain one or more packages for convenience of handling and/or protection of the packages; 2) a term used to describe the placement of damaged or leaking packages in an overpack or recovery drum; 3) the outer packaging for radioactive materials.

Overpacking. Use of a specially constructed drum to overpack damaged or leaking containers of hazardous materials for shipment. Overpack containers should be compatible with the hazards of the materials involved.

Oxidation Ability. The ability of a material to (1) either give up its oxygen molecule to stimulate the oxidation of organic materials (e.g., chlorate, permanganate and nitrate compounds), or (2) receives electrons being transferred from the substance undergoing oxidation (e.g., chlorine and fluorine). Result of either activity is the release of energy.

Oxidizer. A chemical, other than a blasting agent or an explosive, that initiates or promotes combustion in other materials. This action may either cause the material to ignite or release oxygen or other gases which causes the ignition of other surrounding materials.

Oxygen Deficient Atmosphere. An atmosphere that contains an oxygen content less than 19.5 % by volume at sea level.

P

Packaging. Any container that holds a material (hazardous and non-hazardous). Packaging for hazardous materials includes non-bulk and bulk packaging.

Packing Group. Classification of hazardous materials based on the degree of danger represented by the material. There are three groups: Packing Group I indicates great danger, Packing Group II indicates medium danger, and Packing Group III indicates minor danger.

PAPR. (See Powered Air Purifying Respirator).

Patching (Plugging). A physical method of containment which uses chemically compatible patches and plugs to reduce or temporarily stop the flow of materials from small container holes, rips, tears, or gashes. Although commonly used on atmospheric pressure liquid and solid containers, some tactics can also be used on pressurized containers.

PCB Contaminated. Any equipment, including transformers, that contains 50 to 500 ppm of PCB's.

Penetration. The flow or movement of a hazardous chemical through closures, seams, porous materials, and pinholes or other imperfections in the material. While liquids are most common, solid materials (e.g., asbestos) can also penetrate through protective clothing materials.

Permeation. The process by which a hazardous chemical moves through a given material on the molecular level. Permeation differs from penetration in that permeation occurs through the clothing material itself rather than through the openings in the clothing material.

Permeation Rate. The rate at which a chemical passes through a given chemical protective clothing material. Expressed as micrograms per square centimeter per minute (μgm/cm2/min). For reference purposes, .9 μgm/cm2/min. is equal to approximately 1 drop/hour.

Permissible Exposure Limit (PEL). The maximum time-weighted concentration at which 95% of exposed, healthy adults suffer no adverse effects over a 40-hour work week and are comparable to ACGIH's TLV/TWA. PEL's are used by OSHA and are based on an eight-hour, time-weighted average concentration.

Persistence. Refers to the length of time a chemical agent remains as a liquid. A chemical agent is said to be "persistent" if it remains as a liquid for longer than 24 hours and non-persistent if it evaporates within that time. Among the most persistent chemical agents are VX, tabun, mustard and lewisite.

Personal Protective Equipment (PPE). Equipment provided to shield or isolate a person from the chemical, physical, and thermal hazards that may be encountered at a hazardous materials incident. Adequate personal protective equipment should protect the respiratory system, skin, eyes, face, hands, feet, head, body, and hearing. Personal protective equipment includes: personal protective clothing, self-contained positive pressure breathing apparatus, and air purifying respirators.

pH (Power of Hydrogen). Acidic or basic corrosives are measured to one another by their ability to dissociate in solution. Those that form the greatest number of hydrogen ions are the strongest acids, while those that form the hydroxide ion are the strongest bases. The measurement of the hydrogen ion concentration in solution is called the pH (power of hydrogen) of the compound in solution. The pH scale ranges from O to 14, with strong acids having low pH values and strong bases or alkaline materials having high pH values. A neutral substance would have a value of 7.

Physical State. The physical state or form (solid, liquid, gas) of the material at normal ambient temperatures (68° F. to 77° F.).

Placards. Approximately 10.75 inch (273 mm) square markings required under DOT regulations and applied to both ends and each side of freight containers, cargo tanks, and portable tank containers. Factors such as the individual package labels, the size of individual packages, and the total quantity of the product will determine the correct placard to be used.

Planning Section. Responsible for the collection, evaluation, dissemination and use of information about the development of the incident and the status of resources. Includes the Situation Status, Resource Status, Documentation, and Demobilization Units as well as Technical Specialists.

Plume. A vapor, liquid, dust or gaseous cloud formation which has shape and buoyancy.

Pneumatic Hopper Trailer. Covered hopper trailers that are pneumatically unloaded and used for transporting solids. Have a capacity up to 1,500 cubic feet.

Polymerization. A reaction during which a monomer is induced to polymerize by the addition of a catalyst or other unintentional influences, such as excessive heat, friction, contamination, etc. If the reaction is not controlled, it is possible to have an excessive amount of energy released.

Portable Bin. Portable tanks used to transport bulk solids. Are approximately 4 feet square and 6 feet high, with weights up to 7,700 pounds. Normally loaded through the top and unloaded from the side or bottom.

Portable Tank. Any packaging (except a cylinder having 1,000 lbs. or less water capacity) over 110 gallons capacity and designed primarily to be loaded into, on or temporarily attached to a transport vehicle or ship, and equipped with skids, mountings or accessories to facilitate handling of the tank by mechanical means.

Post-Emergency Response Operations. That portion of an emergency response performed after the immediate threat of a release has been stabilized or eliminated, and the clean-up of the site has begun.

Post-Incident Analysis. An element of incident termination that includes completion of the required incident reporting forms, determining the level of financial responsibility, and assembling documentation for conducting a critique.

Powered-Air Purification Respirators (PAPR). Air-purifying respirators that use a blower to force the ambient air through air-purifying elements to a full-face mask. As a result, there is a slight positive pressure in the facepiece that results in an increased protection factor. Where an APR has a protection factor of 50:1, a PAPR will have a protection factor of 1,000:1.

Pressure Isolation and Reduction. A physical or chemical method of containment by which the internal pressure of a closed container is reduced. The tactical objective is to sufficiently reduce the internal pressure in order to either reduce the flow or minimize the potential of a container failure.

Private Sector Specialist Employee A. Those persons who are specially trained to handle incidents involving chemicals and / or containers for chemicals used in their organization's area of specialization. Consistent with the organization's response plan and standard operating procedures, the Specialist Employee A shall have the ability to analyze an incident involving chemicals within the organization's area of specialization, plan a response to that incident, implement the planned response within the capabilities of the resources available, and evaluate the progress of the planned response.

Private Sector Specialist Employee B. Those persons who in the course of their regular job duties, work with or are trained in the hazards of specific chemicals and / or containers for chemicals used in their individual area of specialization. Because of their education, training or work experience, the Specialist Employee B may be called upon to gather and record information, provide technical advice, and provide technical assistance (including work within the hot zone) at an incident involving chemicals consistent with their organization's emergency response plan and standard operating procedures and the local emergency response plan.

Private Sector Specialist Employee C. The Specialist C should be able to provide information on a specific chemical or container and have the organizational contacts needed to acquire additional technical assistance. This individual need not have the skills or training necessary to conduct control operations. This individual is generally found at the command post providing the IC or their designee with technical assistance.

Process Safety Management (PSM). The application of management principles, methods and practices to prevent and control releases of hazardous chemicals or energy. Focus of both *OSHA 1910.119 - Process Safety Management of Highly Hazardous Chemicals, Explosives and Blasting Agents and EPA Part 68—Risk Management Programs for Chemical Accidental Release Prevention.*

Product Name. Brand or trade name printed on the front panel of a hazmat container. If the product name includes the term "technical," as in Parathion Technical, it generally indicates a highly concentrated pesticide with 70% to 99% active ingredients.

Proper Shipping Name. The DOT designated name for a commodity or material. Will appear on shipping papers and on some containers. May also be referred to as shipping name.

Protection In-Place. Directing fixed facility personnel and the general public to go inside of a building or a structure and remain indoors until the danger from a hazardous materials release has passed. It may also be referred to as in-place protection, sheltering-in-place, sheltering, and taking refuge.

Protective Clothing. Equipment designed to protect the wearer from heat and/or hazardous materials contacting the skin or eyes. Protective clothing is divided into four types:

- Structural fire fighting protective clothing
- Liquid splash chemical protective clothing
- Vapor chemical protective clothing
- High temperature protective clothing

Proximity Suits—designed for exposures of short duration and close proximity to flame and radiant heat, such as in aircraft rescue firefighting (ARFF) operations. The outer shell is a highly reflective, aluminized fabric over an inner shell of a flame-retardant fabric such as Kevlar™ or Kevlar™/PBI™ blends. These ensembles are not designed to offer any substantial chemical protection.

PSM. (See Process Safety Management).

Public Information Officer. Point of contact for the media, or other organizations seeking information directly from the incident or event. Member of the Command Staff.

Public Protective Actions. The strategy used by the Incident Commander to protect the general population from the hazardous material by implementing a strategy of either (1) Protecting-in-Place, (2) Evacuation, or (3) a combination of Protection-In-Place and Evacuation. This strategy is usually implemented after the IC has established an isolation perimeter and defined the Hazard Control Zones for emergency responders.

Purging. Totally enclosed electrical equipment is protected with an inert gas under a slight positive pressure from a reliable source. The inert gas provides positive pressure within the enclosure and minimizes the development of a flammable atmosphere. Used in Class I, Division 1 atmospheres at fixed installations.

Pyrophoric Materials. Materials that ignite spontaneously in air without an ignition source.

R

Radiation Harm Events. Those harm events related to the emission of radioactive energy. There are two types of radiation—ionizing and nonionizing.

Radioactivity. The ability of a material to emit any form of radioactive energy.

Rail Burn. Deformation in the shell of a railroad tank car. It is actually a long dent with a gouge at the bottom of the inward dent. A rail burn can be oriented circumferentially or longitudinally in relation to the tank shell. The longitudinal rail burns are the more serious because they have a tendency to cross a weld. A rail burn is generally caused by the tank car passing over a stationary object, such as a wheel flange or rail.

RCRA (See Resource Conservation and Recovery Act).

Reactivity / Instability. The ability of a material to undergo a chemical reaction with the release of energy. It could be initiated by mixing or reacting with other materials, application of heat, physical shock, etc.

Recommended Exposure Levels (REL). The maximum time-weighted concentration at which 95% of exposed, healthy adults suffer no adverse effects over a 40-hour work week and are comparable to ACGIH's TLV/TWA. REL's are used by NIOSH and are based upon a 10-hour, time-weighted average concentration.

Regional Response Team (RRT). Established within each federal region, the RRT follows the policy and program direction established by the NRT to ensure planning and coordination of both emergency preparedness and response activities. Members include EPA, USCG, state government, local government, and Indian tribal governments.

Rehabilitation (Rehab). Process of providing for EMS support, treatment and monitoring, food and fluid replenishment, mental rest and relief from extreme environmental conditions associated with a hazmat incident. May function as either a sector or group within the Incident Management System.

Release Event. Once a container is breached, the hazmat is free to escape (be released) in the form of energy, matter, or a combination of both. Types of release include detonation, violent rupture, rapid relief, and spills or leaks.

Reportable Quantity (RQ). Indicates the material is a hazardous substance by the EPA. The letters "RQ" (reportable quantity) must be shown either before or after the basic shipping description entries. This designation indicates that any leakage of the substance above its RQ value must be reported to the proper agencies (e.g., National Response Center). Regardless of which agencies are involved, the legal responsibility for notification still remains with the spiller.

Reporting Marks and Number. The set of initials and a number stenciled on both sides and both ends of railroad cars. These markings can be used to obtain information on the contents of the car from either the railroad or the shipper.

Residue. The material remaining in a package after its contents have been emptied and before the packaging is refilled, or cleaned or purged of vapor to remove any potential hazard.

Resource Conservation and Recovery Act (RCRA). Law which establishes the regulatory framework for the proper management and disposal of all hazardous wastes, including treatment, storage and disposal facilities. It also establishes installation, leak prevention and notification requirements for underground storage tanks.

Respiratory Protection. Equipment designed to protect the wearer from the inhalation of contaminants. Respiratory protection includes positive-pressure self-contained breathing apparatus (SCBA), positive-pressure airline respirators (SAR's), powered air purifying respirators (PAPR's) and air purifying respirators (APR's).

Respiratory Toxins. Toxins which attack the respiratory system (e.g., asbestos, hydrogen sulfide).

Response. That portion of incident management in which personnel are involved in controlling (offensively or defensively) a hazmat incident. The activities in the response portion of a hazmat incident include analyzing the incident, planning the response, implementing the planned response, and evaluating progress.

Responsible Party (RP). A legally recognized entity (e.g., person, corporation, business or partnership, etc.) that has a legally recognized status of financial accountability and liability for actions necessary to abate and mitigate adverse environmental and human health and safety impacts resulting from a non-permitted release or discharge of a hazardous material. the person or agency found legally accountable for the clean-up of an incident.

Retention. A physical method of confinement by which a liquid is temporarily contained in an area where it can be absorbed, neutralized, or picked up for proper disposal. Retention tactics are intended to be more permanent and may require resources such as portable basins or bladder bags constructed of chemically resistant materials.

Reusable Garments. Chemical protective clothing garments designed and fabricated to allow for decontamination and re-use. Generally thicker and more durable than limited-use garments, they are used for liquid chemical splash and vapor protective suits, gloves, aprons, boots, and thermal protective clothing. Reusable garment materials are usually made from chlorinated polyethylene (CPE), vinyl (plasticized polyvinyl chloride—PVC), fluorinated polymers (Teflon®), and rubber-like fabrics, such as butyl rubber, neoprene rubber, and Viton™.

Riot Control Agents. Usually solid materials that are dispersed in a liquid spray and cause pain or burning on exposed mucous membranes and skin. Common examples include Mace™ (CN) and pepper spray (i.e., capsaicin).

Risks. The probability of suffering a harm or loss. Risks are variable and change with every incident.

Risk Analysis. A process to analyze the probability that harm may occur to life, property, and the environment and to note the risks to be taken to identify the incident objectives.

Risk Management Programs. Required under EPA's proposed 40 CFR Part 68, risk management programs consist of three elements: (1) hazard assessment of the facility; (2) prevention program; and (3) emergency response considerations.

RMP. (See Risk Management Program.)

Roentgen. A measure of the charge produced in air created by ioninzing radiation, usually in reference to gamma radiation.

Roentgen Equivalent Man (REM). The unit of dose equivalent; takes into account the effectiveness of different types of radiation.

Runaway Cracking. Cracking occurring in closed containers under pressure, such as liquid drums or pressure vessels. A small crack in a closed container suddenly develops into a rapidly growing crack which circles the container. As a result, the container will generally break into two or more pieces.

S

Safety Officer. Responsible for the safety of all personnel, including monitoring and assessing safety hazards, unsafe situations, and developing measures for ensuring personnel safety. The Incident Safety Officer (ISO) has the authority to terminate any unsafe actions or operations, and is a required function based upon the requirements of OSHA 1910.120 (q).

Safe Refuge Area. A temporary holding area within the hot zone for contaminated people until a decontamination corridor is set up.

Sampling. The process of collecting a representative amount of a gas, liquid or solid for evidence or analytical purposes.

Sampling Kit. Kits assembled for the purpose of providing adequate tools and equipment for taking samples and documenting unknowns to create a "chain of evidence."

Sanitary Sewer. A "closed" sewer system which carries wastewater from individual homes, together with minor quantities of storm water, surface water, and ground water that are not admitted intentionally. May also collect wastewater from industrial and commercial businesses. The collection and pumping system will transport the wastewater to a treatment plant, where the wastewater is processed.

SAR. (See Supplied Air Respirator).

SARA. (See Superfund Amendments & Reauthorization Act.)

Saturated Hydrocarbons. A hydrocarbon possessing only single covalent bonds. All of the carbon atoms are saturated with hydrogen. Examples include methane (CH_4), propane (C_3H_8) and butane (C_4H_{10}).

SCBA. (See Self Contained Breathing Apparatus.)

Scene. The location impacted or potentially impacted by a hazard.

Score. Reduction in the thickness of the container shell. It is an indentation in the shell made by a relatively blunt object. A score is characterized by the reduction of the container or weld material so that the metal is pushed aside along the track of contact with the blunt object.

Secondary Contamination. The process by which a contaminant is carried out of the hot zone and contaminates people, animals, the environment or equipment outside of the hot zone.

Secondary Decontamination. The second phase of the decontamination process designed to physically or chemically remove surface contaminants to a safe and acceptable level. Depending upon the scope and nature of the incident, multiple secondary decon steps may be implemented.

Section. That organization level within the Incident Command System having functional responsibility for primary segments of incident operations, such as Operations, Planning, Logistics and Administration/ Finance. The Section level is organizationally between Branch and the Incident Commander.

Sector. As used within the Incident Command System, are the organizational level having responsibility for operations within a defined geographic area OR with a specific functional assignment. Examples of functional sectors within the Operations Section would be the Safety, Rescue, or Medical Sectors. A location may also be designated as a geographic sector, such as the North Sector or Marine Dock Sector.

Self-Accelerating Decomposition Temperature (SADT). The temperature at which an organic peroxide or synthetic compound will react to heat, light, or other chemicals, and release oxygen, energy and fuel in the form of an explosion or rapid oxidation. When this temperature is reached by some portion of the mass of an organic peroxide, irreversible decomposition will begin.

Self Contained Breathing Apparatus (SCBA). A positive pressure, self-contained breathing apparatus

(SCBA) or combination SCBA/supplied air breathing apparatus certified by the National Institute for Occupational Safety and Health (NIOSH) and the Mine Safety and Health Administration (MSHA), or the appropriate approval agency for use in atmospheres that are immediately dangerous to life or health (IDLH).

Sensitizer. A chemical that causes a substantial proportion of exposed people or animals to develop an allergic reaction in normal tissue after repeated exposure to the chemical. Skin sensitization is the most common form, while respiratory sensitization to a few chemicals is also known to occur.

SERC. (See State Emergency Response Commission.)

Shipper. A person, company or agency offering material for transportation.

Shipping Documents/Papers. Generic term used to refer to documents that must accompany all shipments of goods for transportation. These include Hazardous Waste Manifest, Bill of Lading, and Consists, etc. Shipping documents should provide the following:

- Proper shipping name.
- Hazard classification.
- Four digit identification number(s), as required.
- Number of packages or containers.
- Type of packages.
- Total quantity by weight, volume, and/or packaging.

Shipping Name. The proper shipping name or other common name for the material; also any synonyms for the material.

Single Trip Container (STC). Container that may not be refilled or reshipped with a DOT regulated material except under certain conditions.

Site Management and Control. The management and control of the physical site of a hazmat incident. Includes initially establishing command, approach and positioning, staging, establishing initial perimeter and hazard control zones, and public protective actions.

Size-Up. The rapid, yet deliberate consideration of all critical scene factors.

Skilled Support Personnel. Personnel who are skilled in the operation of certain equipment, such as cranes and hoisting equipment, and who are needed temporarily to perform immediate emergency support work that cannot reasonably be performed in a timely fashion by emergency response personnel.

Skin Absorption. The introduction of a chemical or agent into the body through the skin. Skin absorption can occur with no sensation to the skin itself. Do not rely on pain or irritation as a warning sign of absorption. Skin absorption is enhanced by abrasions, cuts, heat and moisture. The rate of skin absorption can vary depending upon the body part that's exposed.

Slurry. Pourable mixture of a solid and a liquid.

Sludge. Solid, semi-solid or liquid waste generated from a municipal, commercial, or industrial waste treatment plant or air pollution control facility, exclusive of treated effluent from a waste water treatment plant.

Solidification. Process by which a contaminant physically or chemically bonds to another object or is encapsulated by it. May be used as a chemical method of confinement or decontamination.

Solubility. The ability of a solid, liquid, gas, or vapor to dissolve in water or other specified medium. The ability of one material to blend uniformly with another, as in a solid in liquid, liquid in liquid, gas in liquid, or gas in gas. Significant property in evaluating the selection of control and extinguishing agents, including the use of water and firefighting foams.

Solution. Mixture in which all of the ingredients are completely dissolved. Solutions are composed of a solvent (water or another liquid) and a dissolved substance (known as the solute).

SOP. Standard Operating Procedure.

Specialist Employee. Employees who, in the course of their regular job duties, work with and are trained in the hazards of specific hazardous substances, and who will be called upon to provide technical advice or assistance to the Incident Commander at a hazmat incident.

Specific Gravity. The weight of the material as compared with the weight of an equal volume of water. If the specific gravity is less than one, the material is lighter than water and will float. If the specific gravity is greater than one, the material is heavier than water and will sink. Most insoluble hydrocarbons are lighter than water and will float on the surface. Significant property for determining spill control and clean-up procedures for water-borne releases.

Specification Marking. Found in various locations on railroad tank cars, intermodal portable tanks, and cargo tank trucks, it indicates the standards to which the container was built.

Spill. The release of a liquid, powder, or solid hazardous material in a manner that poses a threat to air, water, ground, and to the environment.

Stabilization. The point in an incident at which the adverse behavior of the hazardous materials is controlled.

Staging. The management of committed and uncommitted emergency response resources (personnel and apparatus) to provide orderly deployment. The safe area established for temporary location of available resources closer to the incident site to reduce response time. See Level I Staging and Level II Staging.

Staging Area. The designated location where emergency response equipment and personnel are assigned on an immediately available basis until they are needed.

Standard of Care. The minimum accepted level of hazmat service to be provided as may be set forth by law, current regulations, consensus standards, local protocols and practice, and what has been accepted in the past (precedent).

Standard Transportation Commodity Code (STCC). A number which will be found on all shipping documents accompanying rail shipments of hazmats. A seven-digit number assigned to a specific material or group of materials and used in determination of rates. For a hazardous material, the STCC number will begin with the digits "49." Hazardous wastes may also be found with the first two digits being "48." It will also be found when intermodal containers are changed from rail to highway movement.

State Emergency Response Commission. Formed under SARA, Title III, the SERC is responsible for developing and maintaining the statewide hazmat emergency response plan. This includes ensuring that planning and training are taking place throughout the state, as well as providing assistance to local governments and LEPCs, as appropriate.

Statement of Practical Treatment. Located near the signal word on the front panel of an agricultural chemical or poison label, it is also referred to as the "First Aid Statement" or "Note to Physician." It may have precautionary information as well as emergency procedures. Antidote and treatment information may also be added.

Static Electricity. An accumulated electrical charge. In order for static electricity to act as an ignition source, four conditions must be fulfilled:

- There must be an effective means of static generation.
- There must be a means of accumulating the static charge build-up.
- There must be a spark discharge of adequate energy to serve as an ignition source (i.e., incendive spark).
- The spark must occur in a flammable mixture.

Sterilization. The process of destroying all microorganisms in or on an object. Common methods include steam, concentrated chemical agents, or ultraviolet light radiation.

Storm Sewer. An "open" system that collects storm water, surface water and ground water from throughout an area, but excludes domestic wastewater and industrial wastes. A storm sewer may dump runoff directly into a retention area which is normally dry or into a stream, river or waterway without treatment.

Strategic Goals. The overall plan that will be used to control an incident. Strategic goals are broad in nature and are achieved by the completion of tactical objectives. Examples include rescue, spill control, leak control, and recovery.

Street Burn. Deformation in the shell of a highway cargo tank. It is actually a long dent that is inherently flat. A street burn is generally caused by a container overturning and sliding some distance along a cement or asphalt road.

Strength. The degree to which a corrosive ionizes in water. Those that form the greatest number of hydrogen ions are the strongest acids (e.g., $pH < 2$), while those that form the hydroxide ion are the strongest bases ($pH > 12$).

Stress Event. An applied force or system of forces that tend to either strain or deform a container (external action) or trigger a change in the condition of the contents (internal action). Types of stress include thermal, mechanical and chemical.

Structural Fire Fighting Protective Clothing. Protective clothing normally worn by firefighters during structural fire fighting operations. It includes a helmet, coat, pants, boots, gloves, PASS device, and a hood to cover parts of the head not protected by the helmet. Structural fire fighting clothing provides limited protection from heat, but may not provide adequate protection from harmful liquids, gases, vapors or dusts encountered during hazmat incidents. May also be referred to as turnout or bunker clothing.

Sublimation. The ability of a substance to change from the solid to the vapor phase without passing through the liquid phase. An increase in temperature can increase the rate of sublimation. Significant in evaluating the flammability or toxicity of any released materials which sublime. The opposite of sublimation is deposition (changes from vapor to solid).

Subsidiary Hazard Class. Indicates a hazard of a material other than the primary hazard assigned.

Superfund Amendments & Reauthorization Act (SARA). Created for the purpose of establishing Federal statutes for right-to-know standards, emergency response to hazardous materials incidents, reauthorized the Federal Superfund program, and mandated states to implement equivalent regulations/requirements.

Supplied Air Respirator (SAR). Positive pressure respirator that is supplied by either an airline hose or breathing air cylinders connected to the respirator by a short airline (or pigtail). When used in IDLH atmospheres, require a secondary source of air supply.

Synergistic Effect. The combined effect of two or more chemicals which is greater than the sum of the effect of each agent alone.

System Detection Limit (SDL). The minimum amount of chemical breakthrough that can be detected by the laboratory analytical system being used for the permeation test. Lower SDL's result in lower (or earlier) breakthrough times.

Systemic. Pertaining to the internal organs and structures of the human body.

Systemic Effect. The health effects of a hazardous materials exposure when the material enters the bloodstream and attacks target organs and internal areas of the human body.

T

Tactical Objectives. The specific operations that must be accomplished to achieve strategic goals. Tactical objectives must be both specific and measurable.

Tasks. The specific activities that accomplish a tactical objective.

Technical Decontamination. The planned and systematic process of reducing contamination to a level that is As Low As Reasonably Achievable (ALARA). Technical decon operations are normally conducted in support of emergency responder recon and entry operations at a hazardous materials incident, as well for handling contaminated patients at medical facilities.

Technical Information Centers. Private and public sector hazardous materials emergency "hotlines" that (1) provide immediate chemical hazard information; (2) access secondary forms of expertise for additional action and information; and (3) act as a clearinghouse for spill notifications. Include both public (e.g., CHEMTREC) and subscription-based systems.

Technical Information Specialists. Individuals who provide specific expertise to the Incident Commander or the HMRT either in person, by telephone, or through other electronic means. They may represent the shipper, manufacturer or be otherwise familiar with the hazmats or problems involved.

Technical Name. Identifies the recognized chemical name currently used in scientific and technical handbooks, journals and texts.

Teratogen. A material that affects the offspring when the embryo or fetus is exposed to that material.

Termination. That portion of incident management where personnel are involved in documenting safety procedures, site operations, hazards faced, and lessons learned from the incident. Termination is divided into three phases: debriefing, post-incident analysis, and critique.

Thermal Harm Events. Those harm events related to exposure to temperature extremes.

Thermal Stress. Hazmat container stress generally indicated by temperature extremes, both hot and cold. Examples include fire, sparks, friction or electricity, and ambient temperature changes. Extreme or intense cold, such as those found with cryogenic materials, may also act as a stressor. Clues of thermal stress include the operation of safety relief devices or the bulging of containers.

Threshold. The point where a physiological or toxicological effect begins to be produced by the smallest degree of stimulation.

Threshold Limit Value/Ceiling (TLV/C). The maximum concentration that should not be exceeded, even instantaneously. The lower the value, the more toxic the substance.

Threshold Limit Value/Short Term Exposure Limit (TLV/STEL). The 15-minute, time-weighted average exposure which should not be exceeded at any time, nor repeated more than four times daily with a 60-

minute rest period required between each STEL exposure. The lower the value, the more toxic the substance.

Threshold Limit Value/Skin (Skin). Indicates a possible and significant contribution to overall exposure to a material by absorption through the skin, mucous membranes, and eyes by direct or airborne contact.

Threshold Limit Value/Time Weighted Average (TLV/TWA). The airborne concentration of a material to which an average, healthy person may be exposed repeatedly for 8 hours each day, 40 hours per week, without suffering adverse effects. The young, old, ill and naturally susceptible will have lower tolerances and will need to take additional precautions. TLVs are based upon current available information and are adjusted on an annual basis by organizations such as the American Conference of Governmental Industrial Hygienists (ACGIH). As TLVs are time weighted averages over an eight-hour exposure, they are difficult to correlate to emergency response operations. The lower the value, the more toxic the substance.

Threshold Planning Quantity (TPQ). The quantity designated for each extremely hazardous substance (EHS) that triggers a required notification from a facility to the State Emergency Response Commission (SERC) and the Local Emergency Planning Committee (LEPC) that the facility is subject to reporting under SARA Title III.

TOFC. (See trailer-on-flat-car.)

Toxic Products of Combustion. The toxic byproducts of the combustion process. Depending upon the materials burning, higher levels of personal protective clothing and equipment may be required.

Toxicity. The ability of a substance to cause injury to a biologic tissue. Refers to the ability of a chemical to harm the body once contact has occurred.

Toxicity Harm Events. Those harm events related to exposure to toxins. Examples include neurotoxins, nephrotoxins and hepatotoxins.

Toxicity Signal Words. The signal word found on product labels of poisons and agricultural chemicals which indicates the relative degree of acute toxicity. Located in the center of the front label panel, it is one of the most important label markings. The three toxicity signal words and categories are DANGER (high), WARNING (medium), and CAUTION (low).

Toxicology. The study of chemical or physical agents that produce adverse responses in the biologic systems with which they interact.

Trailer-on-Flat-Car (TOFC). Truck trailers which are shipped on a railroad flat cars.

Transfer. The process of physically moving a liquid, gas or some forms of solids either manually, by pump or by pressure transfer, from a leaking or damaged container. The transfer pump, hoses, fittings and container must be compatible with the hazardous materials involved. When transferring flammable liquids, proper bonding and grounding concerns must be addressed.

Transportation Index (TI). The number found on radioactive labels which indicates the maximum radiation level (measured in milli-roentgens/hour - mR/hr) at 1 meter from an undamaged package. For example, a TI of 3 would indicate that the radiation intensity that can be measured is no more than 3 mR/hr. at 1 meter from the labeled package.

Type-A Packaging. Packaging used to transport small quantities of radioactive material with higher concentrations of radioactivity than those shipped in Industrial Packaging. Designed to ensure that the package retains its containment integrity and shielding under normal transport conditions. However, they are NOT designed to withstand the forces of an accident.

Type-B Packaging. Packaging used to transport radioactive material with the highest levels of radioactivity, including potentially life-endangering amounts that could pose a significant risk if released during an accident. Must meet all of the Type A requirements, as well as a series of tests which simulate severe or "worst case" accident conditions. Accident conditions are simulated by performance testing and engineering analysis.

U

Unified Command. The process of determining overall incident strategies and tactical objectives by having all agencies, organizations or individuals who have jurisdictional responsibility, and in some cases those who have functional responsibility at the incident, participate in the decision-making process.

Unified Commanders (UC). Command level representatives from each of the primary responding agencies who present their agency's interests as a member of a unified command team. Depending upon the scenario and incident timeline, they may be the "lead" Incident Commander or play a supporting role within the command function. The unified commanders manage their own agency's actions and make sure all efforts are coordinated through the unified command process.

Underflow Dam. Spill control tactic used to trap floating lighter-than-water materials behind the dam (specific gravity <1). Using PVC piping or hard sleeves, the dam is constructed in a manner that allows uncontaminated water to flow unobstructed under the dam while keeping the contaminant behind the dam. Operationally, this is most effective on slow moving and relatively narrow waterways.

UN/NA Identification Number. The four-digit identification number assigned to a hazardous material by the Department of Transportation; on shipping documents may be found with the prefix "UN" (United Nations) or "NA" (North American). The ID numbers are not unique and more than one material may have the same ID number.

Unsaturated Hydrocarbons. A hydrocarbon with at least one multiple bond between two carbon atoms somewhere in the molecule. Generally, unsaturated hydrocarbons are more active chemically than saturated hydrocarbons, and are considered more hazardous. May also be referred to as the alkenes and alkynes. Examples include ethylene (C_2H_4), butadiene (C_4H_6), and acetylene (C_2H_2).

V

Vacuuming. A physical method of confinement by which a hazardous material is placed in a chemically compatible container by simply vacuuming it up. The method of vacuuming will depend upon the hazmats involved. Vacuuming is also a physical method of decontamination.

Vapor. An air dispersion of molecules in a substance that is normally a liquid or solid at standard temperature and pressure.

Vapor Density. The weight of a pure vapor or gas compared with the weight of an equal volume of dry air at the same temperature and pressure. The molecular weight of air is 29. If the vapor density of a gas is less than one, the material is lighter than air and may rise. If the vapor density is greater than one, the material is heavier than air and will collect in low or enclosed areas. Significant property for evaluating exposures and where hazmat gas and vapor will travel.

Vapor Dispersion. A physical method of confinement by which water spray or fans is used to disperse or move vapors away from certain areas or materials. It is particularly effective on water-soluble materials (e.g., anhydrous ammonia), although the subsequent runoff may involve environmental tradeoffs.

Vapor Pressure. The pressure exerted by the vapor within the container against the sides of a container. This pressure is temperature dependent; as the temperature increases, so does the vapor pressure. Consider the following three points:

1) The vapor pressure of a substance at 100°F. is always higher than the vapor pressure at 68°F.
2) Vapor pressures reported in millimeters of mercury (mm Hg) are usually very low pressures. 760 mm Hg. is equivalent to 14.7 psi. or 1 atmosphere. Materials with vapor pressures greater than 760 mm Hg. are usually found as gases.
3) The lower the boiling point of a liquid, the greater vapor pressure at a given temperature.

Vapor Suppression. A physical method of confinement to reduce or eliminate the vapors emanating from a spilled or released material. Operationally, it is an offensive technique used to mitigate the evolution of flammable, corrosive, or toxic vapors and reduce the surface area exposed to the atmosphere. Common examples include the use of firefighting foams and chemical vapor suppressants.

Vent and Burn. The use of shaped explosive charges to vent the high pressure at the the top of a pressurized container and then, with additional explosive charges, release and burn the remaining liquid in the container in a controlled fashion. This is a highly sophisticated technique that is only used under very controlled conditions.

Venting. The controlled release of a liquid or compressed gas to reduce the pressure and diminish the probability of an explosion. The method of venting will depend upon the nature of the hazmat.

Vesicants (Blister Agents). Chemical agents that pose both a liquid and vapor threat to all exposed skin and mucous membranes. These are exceptionally strong irritants capable of causing extreme pain and large blisters upon contact. Examples include mustard, lewisite, and phosgene oxime.

Violent Rupture. Associated with chemical reactions having a release rate of less than one second (i.e., deflagration). There is no time to react in this scenario. This behavior is commonly associated with runaway cracking and over-pressure of closed containers.

Viscosity. Measurement of the thickness of a liquid and its ability to flow. High viscosity liquids, such as heavy oils, must first be heated to increase their fluidity. Low viscosity liquids spread more easily and increase the size of the hazard area.

Volatility. The ease with which a liquid or solid can pass into the vapor state. The higher a material's volatility, the greater its rate of evaporation. Significant property in that volatile materials will readily disperse and increase the hazard area.

W

Warm Zone. The area where personnel and equipment decontamination and hot zone support takes place. It includes control points for the access corridor and thus assists in reducing the spread of contamination. This is also referred to as the "decontamination", "contamination reduction", "yellow zone", "support zone", or "limited access zone" in other documents.

Water Reactivity. Ability of a material to react with water and release a flammable gas or present a health hazard.

Waybill. A railroad shipping document describing the materials being transported. Indicates the shipper consignee, routing and weights. Used by the railroad for internal records and control, especially when the shipment is in transit.

Wheel Burn. Reduction in the thickness of a railroad tank shell. It is similar to a score, but is caused by prolonged contact with a turning railcar wheel.

INDEX

A

B

C

D

E

F

G

H

I

K

L

M

N

O

P

Q

R

S

T

U

V

W

Y